America's Scientific Treasures

America's Scientific Treasures

A Travel Companion

Second Edition

STEPHEN M. COHEN AND BRENDA H. COHEN

OXFORD
UNIVERSITY PRESS

Oxford University Press is a department of the University of Oxford. It furthers
the University's objective of excellence in research, scholarship, and education
by publishing worldwide. Oxford is a registered trade mark of Oxford University
Press in the UK and certain other countries.

Published in the United States of America by Oxford University Press
198 Madison Avenue, New York, NY 10016, United States of America.

Library of Congress Cataloging-in-Publication Data
Names: Cohen, Brenda H., 1940– author. | Cohen, Stephen M., 1963– author. |
Cohen, Paul S., 1938–2004 America's scientific treasures.
Title: America's scientific treasures : a travel companion / Stephen Cohen, Brenda Cohen.
Description: Second edition. | New York, NY : Oxford University Press, [2021] |
Revised edition of: America's scientific treasures : a travel companion /
Paul S. Cohen and Brenda H. Cohen. c1998.
Identifiers: LCCN 2020022512 (print) | LCCN 2020022513 (ebook) |
ISBN 9780197545508 (paperback) | ISBN 9780197545522 (epub)
Subjects: LCSH: Science museums—United States—Guidebooks. |
Historic sites—United States—Guidebooks. |
National parks and reserves—United States—Guidebooks. | United States—Guidebooks.
Classification: LCC Q105.U5 A64 2020 (print) | LCC Q105.U5 (ebook) | DDC 507.4/73—dc23
LC record available at https://lccn.loc.gov/2020022512
LC ebook record available at https://lccn.loc.gov/2020022513

1 3 5 7 9 8 6 4 2

Printed by Marquis, Canada

Contents

Preface to the Second Edition   xvii

1. New England   1
   Connecticut   1
     Bristol   1
       American Clock and Watch Museum   1
     Mystic   3
       Mystic Aquarium   3
       Mystic Seaport: The Museum of America and the Sea   5
     New Haven   8
       Peabody Museum of Natural History   8
   Maine   10
     Augusta   10
       Maine State Museum   10
     Bangor   12
       Cole Land Transportation Museum   12
     Bath   15
       Maine Maritime Museum   15
     Brunswick   17
       Peary-MacMillan Arctic Museum   17
     Freeport   18
       The Desert of Maine   18
     Kennebunkport   19
       Seashore Trolley Museum   19
   Massachusetts   22
     Boston   22
       Museum of Science   22
       New England Aquarium   24
     Brookline   26
       Frederick Law Olmsted National Historic Site   26
     Cambridge and Boston   28
       Harvard Museums of Science and Culture, Warren Anatomical
         Museum, and Arnold Arboretum   28
         The Harvard Museum of Natural History   28
         Peabody Museum of Archaeology & Ethnology   29
         Harvard Semitic Museum   29
         Collection of Historical Scientific Instruments   30
         Arnold Arboretum and Warren Anatomical Museum   30
     Lowell   31
       Lowell National Historical Park and Lowell Heritage State Park   31
     Nantucket   33
       Maria Mitchell Association   33
       Whaling Museum   35

New Bedford   38
    New Bedford Whaling Museum   38
Plymouth   40
    Plimoth Plantation   40
Saugus   42
    Saugus Iron Works National Historic Site   42
Springfield   44
    Springfield Armory National Historic Site   44
Sturbridge   46
    Old Sturbridge Village   46
New Hampshire   48
  Canterbury   48
    Canterbury Shaker Village   48
  North Salem   50
    America's Stonehenge   50
Rhode Island   52
  Pawtucket   52
    Old Slater Mill National Historic Landmark 52
Vermont   54
  Rutland   54
    New England Maple Museum   54
  St. Johnsbury   56
    The Fairbanks Museum and Planetarium   56
  Windsor   58
    American Precision Museum   58
  Woodstock   60
    Billings Farm and Museum and Marsh-Billings-Rockefeller National Historical Park   60

2. Mid-Atlantic   64
Delaware   64
  Dover   64
    Johnson Victrola Museum   64
  Wilmington   65
    Delaware Museum of Natural History   65
    Hagley Museum and Library   67
Maryland   70
  Annapolis Junction   70
    National Cryptologic Museum   70
  Baltimore   72
    Baltimore Museum of Industry   72
    B&O Railroad Museum   74
    National Aquarium   76
  Linthicum   77
    National Electronics Museum   77
  Silver Spring   79
    National Museum of Health and Medicine   79
  St. Michaels   81
    Chesapeake Bay Maritime Museum   81

New Jersey   83
    Ewing   83
        The Sarnoff Collection   83
    Morristown   85
        Historic Speedwell   85
    Ogdensburg and Franklin   87
        The Sterling Hill Mining Museum and the Franklin Mineral Museum 87
            The Sterling Hill Mining Museum   87
            Franklin Mineral Museum   89
    Roebling   89
        Roebling Museum   89
    Wall   91
        InfoAge Science History Learning Center   91
    West Orange   93
        Thomas Edison National Historical Park   93
New York   95
    Albany   95
        New York State Museum   95
    Corning   96
        Corning Museum of Glass   96
    Hammondsport   99
        Glenn H. Curtiss Museum   99
    Mount Lebanon   100
        Shaker Museum and Library   100
    New York   101
        American Museum of Natural History   101
        Brooklyn Botanic Garden   104
        Intrepid Sea, Air & Space Museum   106
        Metropolitan Museum of Art   108
        The New York Botanical Garden   110
        Staten Island Museum   112
        Wildlife Conservation Society   113
            Bronx Zoo   113
            Central Park Zoo   114
            Prospect Park Zoo   115
            New York Aquarium   115
            Queens Zoo   116
    Oyster Bay   116
        Planting Fields Arboretum State Historic Park   116
    Rhinebeck   118
        Old Rhinebeck Aerodrome   118
    Rochester   119
        George Eastman Museum   119
    Syracuse   121
        Erie Canal Museum   121
Pennsylvania   122
    Columbia   122
        National Watch & Clock Museum   122

Cornwall   124
   Cornwall Iron Furnace   124
Doylestown   126
   The Mercer Mile: Mercer Museum, Moravian Pottery and Tile Works,
     and Fonthill   126
Elverson   128
   Hopewell Furnace National Historic Site   128
Northumberland   130
   Joseph Priestley House   130
Philadelphia   133
   Bartram's Garden   133
   College of Physicians of Philadelphia and the Mütter Museum   135
   The Franklin Institute   136
   Pennsylvania Hospital   139
   Philadelphia Zoo   140
   Science History Institute   142
   Wagner Free Institute of Science   144
Pittsburgh   146
   Carnegie Museums of Pittsburgh   146
     Carnegie Museum of Natural History   146
     Carnegie Science Center   147
   Pittsburgh Zoo & PPG Aquarium   148
Scranton   149
   Pennsylvania Anthracite Heritage Museum and Scranton Iron Furnaces,
     and the Lackawanna Coal Mine   149
   Steamtown National Historic Site   151
Titusville   153
   Drake Well Museum and Park   153
York   155
   Harley-Davidson, Inc.   155
Washington, DC   157
   Bureau of Engraving and Printing   157
   Smithsonian Institution   159
   Smithsonian's National Zoo and Conservation Biology Institute   161

3. Southeast   164
  Alabama   164
    Birmingham   164
      Sloss Furnaces National Historic Landmark   164
    Tuskegee   165
      Tuskegee Institute National Historic Site   165
  Florida   167
    Fort Myers   167
      Edison and Ford Winter Estates   167
    Homestead   169
      Everglades National Park   169
    Key Largo   171
      John Pennekamp Coral Reef State Park   171
    Miami   173
      Phillip and Patricia Frost Museum of Science   173
      Miami Seaquarium   175

Tampa   177
    Busch Gardens Tampa   177
Titusville   178
    Kennedy Space Center Visitor Complex, Merritt Island National
        Wildlife Refuge, and Canaveral National Seashore   178
        Kennedy Space Center Visitor Complex   179
        Merritt Island National Wildlife Refuge and Canaveral National
            Seashore   180
Georgia   181
  Atlanta   181
    Atlanta Botanical Garden   181
    Fernbank Science Center and Fernbank Museum of Natural History   183
        Fernbank Museum of Natural History   184
        Fernbank Science Center   186
  Brunswick   186
    Hofwyl-Broadfield Plantation State Historic Site   186
  Jefferson   188
    Crawford W. Long Museum   188
North Carolina   189
  Asheboro   189
    North Carolina Zoological Park   189
  Charlotte   192
    Discovery Place   192
        Discovery Place Science   192
        Discovery Place Nature   193
  Kill Devil Hills   194
    Wright Brothers National Memorial   194
  Raleigh   196
    North Carolina Museum of Natural Sciences   196
South Carolina   198
  Charleston   198
    Middleton Place   198
  Georgetown   200
    The Rice Museum   200
Virginia   202
  Alexandria   202
    Stabler-Leadbeater Apothecary Museum   202
  Chantilly   204
    Steven F. Udvar-Hazy Center   204
  Charlottesville   206
    Monticello   206
  Fort Eustis   208
    U.S. Army Transportation Museum   208
  Hampton   210
    Virginia Air and Space Center   210
  Luray   211
    Shenandoah National Park   211
  Newport News   213
    Mariners' Museum and Park   213
  Virginia Beach   215
    Virginia Aquarium & Marine Science Center   215

Williamsburg   216
   Colonial Williamsburg   216
   Jamestown Settlement and Historic Jamestowne   218
     Jamestown Settlement   219
     Historic Jamestowne   220
West Virginia   221
  Milton   221
   Blenko Glass Company   221

4. Mississippi Valley   223
  Arkansas   223
   Hot Springs   223
    Hot Springs National Park   223
  Kentucky   225
   Lexington   225
    Kentucky Horse Park and American Saddlebred Museum   225
     Kentucky Horse Park   225
     American Saddlebred Museum   227
    Monroe Moosnick Medical and Science Museum and Hopemont   228
     Monroe Moosnick Medical and Science Museum   228
     Hopemont   229
   Louisville   230
    Louisville Zoological Gardens   230
    Museum of the American Printing House for the Blind   231
    WaterWorks Museum   232
   Nicholasville   234
    Harry C. Miller Lock Collection   234
  Louisiana   236
   New Orleans   236
    New Orleans Pharmacy Museum   236
    The Audubon Nature Institute   238
     Audubon Aquarium of the Americas   238
     Audubon Zoo   239
     Audubon Louisiana Nature Center   240
     Audubon Butterfly Garden and Insectarium   240
  Mississippi   241
   Jackson   241
    Mississippi Agriculture and Forestry Museum and the
     National Agriculture Aviation Museum   241
  Tennessee   243
   Memphis   243
    Mississippi River Museum   243
   Oak Ridge   244
    American Museum of Science and Energy and Oak Ridge
     Site–Manhattan Project National Historical Park   244

5. Midwest   248
  Illinois   248
   Batavia   248
    Fermi National Accelerator Laboratory   248

Brookfield 251
    Brookfield Zoo 251
Chicago 253
    Adler Planetarium 253
    Field Museum of Natural History 254
    International Museum of Surgical Science 256
    John G. Shedd Aquarium 258
    Museum of Science and Industry 260

Indiana 261
Indianapolis 261
    Indiana Medical History Museum 261
Muncie 264
    National Model Aviation Museum 264

Iowa 266
Waterloo 266
    John Deere Company 266

Michigan 268
Bloomfield Hills 268
    Cranbrook Institute of Science 268
Dearborn 270
    The Henry Ford 270
        Henry Ford Museum of American Innovation 270
        Greenfield Village 271
        Ford Rouge Factory Tour 272
Negaunee 273
    Michigan Iron Industry Museum 273

Minnesota 275
Grand Rapids 275
    Forest History Center 275
Soudan, Chisholm, and Hibbing 276
    Lake Vermilion–Soudan Underground Mine State Park, Minnesota
        Discovery Center, and the Hull-Rust Mahoning Mine 276
        Lake Vermilion–Soudan Underground Mine State Park 276
        Minnesota Discovery Center 278
        Hull-Rust Mahoning Mine 279
St. Paul 280
    Science Center of Minnesota and Mississippi River Visitor Center 280
        Science Center of Minnesota 280
        Mississippi River Visitor Center 281

Ohio 281
Cleveland 281
    Dittrick Medical History Center 281
Dayton 283
    Carillon Historical Park 283
New Bremen 286
    The Bicycle Museum of America 286
Toledo 288
    National Museum of the Great Lakes 288
Wright-Patterson Air Force Base 290
    National Museum of the United States Air Force 290

Wisconsin 292
  Appleton 292
    Paper Discovery Center 292
  Ashwaubenon 294
    National Railroad Museum 294
  Madison 295
    University of Wisconsin–Madison Geology Museum 295
  Manitowoc 297
    Wisconsin Maritime Museum 297
  Milwaukee 299
    Discovery World 299
    Milwaukee Public Museum 301
  Prairie du Chien 302
    Fort Crawford Museum 302

6. Great Plains 305
Kansas 305
  Bonner Springs 305
    National Agricultural Center and Hall of Fame 305
  Hutchinson 307
    Cosmosphere International Science Education Center & Space Museum 307
  Lawrence 309
    KU Natural History Museum 309
  Overland Park 311
    Museum at Prairiefire 311
  Wichita 313
    Sedgwick County Zoo 313
Missouri 314
  Diamond 314
    George Washington Carver National Monument 314
  St. Joseph 317
    Glore Psychiatric Museum 317
  St. Louis 319
    Missouri Botanical Garden 319
    Saint Louis Science Center 321
    Saint Louis Zoo 323
Nebraska 325
  Hastings 325
    Hastings Museum 325
  Lincoln 326
    University of Nebraska State Museum 326
North Dakota 328
  Medora 328
    Theodore Roosevelt National Park 328
Oklahoma 330
  Bartlesville 330
    Woolaroc Museum & Wildlife Preserve 330
  Oklahoma City 332
    Science Museum Oklahoma 332
South Dakota 334
  Custer 334
    Jewel Cave National Monument 334

Hot Springs	336
Mammoth Site	336
Lead	338
Black Hills Mining Museum and Sanford Lab Homestake Visitor Center	338
Black Hills Mining Museum	338
Sanford Lab Homestake Visitor Center	340
7. Southwest	342
Arizona	342
Flagstaff	342
Lowell Observatory	342
Sunset Crater Volcano National Monument and Wupatki National Monument	344
Sunset Crater Volcano National Monument	345
Wupatki National Monument	345
Grand Canyon	346
Petrified Forest National Park	346
Grand Canyon, Arizona; Springdale, Utah; and Bryce, Utah	348
Grand Canyon, Zion, and Bryce Canyon National Parks	348
Grand Canyon National Park	348
Zion National Park	350
Bryce Canyon National Park	351
Oracle	352
University of Arizona Biosphere 2	352
Phoenix	354
Desert Botanical Garden	354
Musical Instrument Museum	356
Phoenix Zoo	357
Sahuarita	359
Titan Missile Museum	359
Superior	361
Boyce Thompson Arboretum	361
Tucson	363
Arizona-Sonora Desert Museum	363
Kitt Peak National Observatory	364
Pima Air & Space Museum	366
Nevada	368
Hoover Dam	368
Hoover Dam	368
Reno	370
University of Nevada, Reno, and Rancho San Rafael Regional Park	370
W. M. Keck Earth Science and Mineral Engineering Museum	370
Fleischmann Planetarium and Science Center	372
Wilbur D. May Botanical Garden and Arboretum	373
New Mexico	374
Albuquerque	374
The National Museum of Nuclear Science & History	374
New Mexico Museum of Natural Science and History	377
Los Alamos	379
Bradbury Science Museum	379

Texas 381
Beaumont 381
Edison Museum 381
Spindletop Gladys City Boomtown and Texas Energy Museum 382
Spindletop Gladys City Boomtown 383
Texas Energy Museum 384
Fort Sam Houston 385
U.S. Army Medical Department Museum 385
Fort Worth 387
Fort Worth Zoo 387
Galveston 389
Moody Gardens 389
Houston 390
Houston Museum of Natural Science 390
Space Center Houston 391
The Printing Museum 393
San Antonio 394
Natural Bridge Caverns 394
San Antonio Botanical Garden 396
San Antonio Zoo 397
Utah 399
Bryce 399
Bryce Canyon National Park 399
Salt Lake City 399
Natural History Museum of Utah 399
Springdale 401
Zion National Park 401

8. Rocky Mountains 402
Colorado 402
Boulder 402
National Center for Atmospheric Research Mesa Lab Visitor Center 402
Colorado Springs 403
Cheyenne Mountain Zoo 403
Colorado Springs Pioneers Museum 405
Denver 407
Denver Botanic Gardens 407
Denver Museum of Nature & Science 409
Denver Zoo 410
Denver and Georgetown 412
History Colorado Center and Georgetown Loop Railroad® & Mining Park 412
History Colorado Center 412
Georgetown Loop Railroad® & Mining Park 413
Estes Park 415
Rocky Mountain National Park 415
Golden and Idaho Springs, Colorado 417
Mines Museum of Earth Science and Edgar Mine 417
Mines Museum of Earth Science 417
Edgar Mine 418
Idaho 419
Pocatello 419
Idaho Museum of Natural History 419

Montana	421
Bozeman	421
Museum of the Rockies	421
Gardiner	422
Yellowstone National Park	422
Wyoming	422
Yellowstone National Park	422
Yellowstone National Park	422
9. Pacific	426
Alaska	426
Anchorage	426
Anchorage Museum	426
Fairbanks	428
University of Alaska Museum of the North	428
Haines	429
Hammer Museum	429
California	430
Arcadia	430
Los Angeles County Arboretum and Botanic Garden	430
Escondido	432
San Diego Zoo Safari Park	432
La Jolla	434
Birch Aquarium at Scripps	434
Los Angeles	435
Natural History Museum of Los Angeles County and La Brea Tar Pits	435
Natural History Museum of Los Angeles County	435
La Brea Tar Pits	437
Mariposa	438
California State Mining and Mineral Museum	438
Mineral	440
Lassen Volcanic National Park	440
Monterey	442
Monterey Bay Aquarium	442
Mount Hamilton	443
Lick Observatory	443
Mountain View	446
Computer History Museum	446
Oakland	448
Oakland Museum of California	448
Pasadena	449
Mount Wilson Observatory	449
Sacramento	451
California State Railroad Museum	451
San Diego	453
San Diego Air & Space Museum	453
San Diego Zoo	455
San Francisco	457
Cable Car Museum	457
California Academy of Sciences	458
Exploratorium	460
San Francisco Botanical Garden	462

San Francisco Maritime National Historical Park   464
San Francisco Zoo & Gardens   466
San Marino   468
The Huntington Library, Art Collections, and Botanical Gardens   468
Santa Rosa   471
Luther Burbank Home & Gardens   471
Tulelake   473
Lava Beds National Monument   473
Yosemite National Park   474
Yosemite National Park   474
Hawai'i   476
Hawaii National Park   476
Hawai'i Volcanoes National Park   476
Hilo   478
'Imiloa Astronomy Center of Hawai'i   478
Pacific Tsunami Museum   479
Honolulu   481
Bernice Pauahi Bishop Museum   481
Oregon   483
Astoria   483
Fort Clatsop & Visitor Center   483
Cave Junction   485
Oregon Caves National Monument & Preserve   485
Florence   486
Sea Lion Caves   486
Portland   488
Oregon Zoo   488
Winston   490
Wildlife Safari   490
Washington   492
Ashford   492
Mount Rainier National Park   492
Castle Rock   494
Mount St. Helens National Volcanic Monument   494
Grand Coulee   496
Grand Coulee Dam   496
Mukilteo   498
Future of Flight Aviation Center & Boeing Tour   498
Seattle   500
Washington Park Arboretum   500
Woodland Park Zoo   502
Seattle and Everett   503
Museum of Flight   503
Vantage   505
Gingko Petrified Forest   505

Registered Trademarks   507
Index   509

Preface to the Second Edition

Over 20 years have passed since the First Edition was published in 1998; in that time, changes both technological and human have occurred. Regarding human changes, we regret the death of one of the coauthors, Dr. Paul S. Cohen, just as preparations began in 2004 for the Second Edition. Hence this book has a new coauthor, Dr. Stephen M. Cohen. The basic premises and criteria for inclusion in our book, however, have not altered.

The GPS has rendered the art of map-reading nearly obsolete, and many sites have added instructions to their facilities. Thus there was no longer a need for us to include detailed directions. In addition, the Internet now provides most of us with a ready and quick fountain of site information, sometimes accurate, and occasionally not. This information includes details about most of the sites we have chosen to include in the Second Edition, such as temporary exhibits, changeable daily schedules of lectures, and guided tours. In fact, many websites even provide driving directions. At the request of our readers, we have included information about how to go to sites by public transportation.

In the Second Edition, we continue our focus on providing an ancillary background to each site, including some of the unusual, unique, or important aspects of its collection. As before, at least one of the authors has visited practically every site we discuss. We now include each site's own website and unusual directions such as latitude and longitude to correct for occasional errors in GPS directions or when a site may be reached by water as well as land. Because this book showcases science and technology for a worldwide readership, we now indicate metric units as well as English units for all measurements. This edition now lists treasures in *all 50 states*.

As to other changes since the First Edition, we include the expected ones, such as new museums of interest happily appearing, others regrettably closing, and those sites whose focus or presentation has changed pleasantly for the better (i.e., we now decide to include them in the Second Edition), or sadly for the worse (they have shifted their subject away from science or the adult visitor). New sites worthy of our review often appear, and we wish we could include the newest ones that open after this manuscript is completed, and—in fact—every site devoted to science. We do clarify that although the title of the book mentions "scientific" treasures, this book has always included technological and engineering sites of interest.

Among the changes in sites themselves, we have seen a general tendency for sites of scientific or technological focus to "dumb down" their exhibits in order to supposedly increase their family-friendly traffic. While we applaud all attempts to bring science and technology to as many people as possible, in many cases these efforts have resulted in less information and a lower intellectual level presented to the public. Therefore, in the Second Edition, we have eliminated some sites that were listed in the First Edition. We hope that those sites veer away from this current trend.

Another problem we note is a significant number of sites that have unreachable staff members for discussions as to details about their collection or exhibits. In certain cases, we were unable to make contact (despite repeated attempts) with staff members, and therefore could neither confirm the accuracy of our descriptions, nor ask about future plans for those sites. In other instances we found staff members who initially agreed to help us—but such help never appeared. Such a phenomenon is new to us, and we hope that readers do not encounter this difficulty.

For dates we occasionally use the terms B.C.E. (Before the Common Era) and C.E. (Common Era), for inclusiveness to all people, as well as these being the preferred designations in the academic world. For our non-American readers, we use Système International conventions of spaces between digits to indicate thousands and millions, thus avoiding commas, which have a different meaning depending on geography.

Many museums, zoos, and other sites belong to consortiums. These groups share ideas, exhibits, and membership lists. Joining a facility may prove to be a valuable activity for the both the organization and you. Members often receive special privileges at a facility, as well as reduced fees or free admission and gift-shop discounts at other sites that belong to the consortium. Such groups include the Association of Science-Technology Centers (ASTC, www.astc.org) and the Association of Zoos & Aquariums (AZA, www.aza.org). Ask your local science museum, zoo, or other sites about these memberships.

Sites frequently change their days and hours of operation, telephone numbers, exhibits, food offerings, and other variables. Many sites do not update their websites as frequently as we, the visitors, would like. We apologize in advance for any inconvenience caused by any changes, and urge you to always call ahead and confirm the facts so that your plans will go smoothly. Many sites may be reached with public transportation; we suggest that visitors confirm the best route to the site, because bus and train routes often change. When citing handicapped accessibility, we usually accepted the self-definition of the site. In a few cases, however, we note that a site has limited accessibility even though the site may not recognize the limitations.

We gratefully acknowledge those treasures that have donated a photograph to illustrate their section of the book, and those that have helped with our queries. We also thank Jeremy Lewis at Oxford University Press for allowing our unique presentation to continue to assist travelers from all over the United States and the world with finding scientific treasures throughout the United States.

From the Preface to the First Edition

Why write a book on travel to places with scientific content? The reasons are many. We love to travel, and to travel with a purpose; targeted travel, we feel, offers many rewards. We also love museums of all types, and enjoy science exhibits. When we searched for a book to help us on one of our travel ventures, we found none targeted for science travel, so we decided that a book on this subject was in order. What better way can we indulge ourselves in our own passion, than writing this book?

Travel in the United States is growing rapidly for both American and foreign visitors, and both groups want information about the areas they are about to see. Many people are science-lovers; gardeners, birdwatchers, museum buffs, aquarium devotees, zoo aficionados, and just ordinary folks seeking something different to see. The love of science may come from formal training, from hobbies, or out of curiosity. Many individuals travel, whether for business or pleasure, and seek unusual and worthwhile escapes. When away from their home town, they often do not know where to find the great science sites. In fact, many people do not know some of the treasures in their own hometowns!

When we set out to describe America's scientific museums, zoos, aquariums, and other sites, we naively assumed that the relatively short history of the United States would provide us with a relatively short list of sites. What we found was a treasure trove of magnificent science exhibits, natural or manmade, far too numerous for one volume to hold. The diversity of subjects, the depth of content, and the imagination of exhibit creation is superb. An all-inclusive publication would end up being an encyclopedia too large to be useful as a take-along book, and too expansive to be readable.

The task of selecting those scientific sites that we deemed to be treasures proved difficult. One of our criteria was that every site included in the book must be one that we personally enjoyed; we did not want to rely only on others for their evaluations. In the end, our tastes dictated the sites that we included.

While it was possible to fill this book with discovery centers or zoos alone, we felt this volume should cover a wider scope. We decided to provide a variety of sites on a wide assortment of subjects that would meet a broad spectrum of reader interests. Science is not a narrow subject; it includes a broad expanse of material.

We also decided that, to be considered a scientific treasure, a site must meet one or more of the following minimum qualities:

- The content or completeness of the collection must be special. To be suitable for inclusion, the site must have one or more of the following: a single collection of national stature; a comprehensive collection; items worthy of note; a one-of-a-kind specimen; or be representative of an area, industry, or museum type.
- The site must provide an educational component. The visitor should come away with a greater understanding of the material exhibited, and the exhibit should provide instruction about the science material that is presented.
- The presentation of the exhibits must be beyond the ordinary. This quality might be fine teaching, original displays, or completeness of subject coverage.

Our intended audience is the adult or teenage science-lover, age 15 and up. Science sites are often viewed as a place intended to entertain the younger visitor. While it is true that most of the sites we selected have something to offer children, we reviewed all sites with adults in mind.

While putting this volume together, we spoke with museum personnel or met with site staff, read materials prepared for the staff and press, and most importantly, we visited each site personally. No site paid a fee to receive space in this book. To assure our readers the most accurate and up-to-date information, we provided each site with an opportunity to review an early draft of our work for accuracy and completeness.

We realize that to produce a timely book, we could not see all the possible places that might merit inclusion. Some states had so many potentially worthy sites but, with great regret, we simply had to omit some of them. We apologize for omitting any sites that are indeed treasures.

1

New England

Connecticut

Bristol

American Clock and Watch Museum

Science, technology, and art come together at this museum devoted to the history of American horology and America's role in clock-making, particularly in Connecticut. The museum is housed in a Federal-style building, constructed in 1801, with a sundial garden, located in the historic "Federal Hill" district of Bristol.

The eight galleries display over 1500 clocks and watches, about 100 of which are kept running at this facility. This museum exhibits one of the most extensive collections displayed in the United States, with over 5500 pieces. (See National Watch & Clock Museum, Columbia, Pennsylvania.)

The earliest watch dates from 1595 and the earliest clock from 1680, up through the present; however, the majority of the collection ranges from 1800 to 1940. Such pieces on exhibit include grandfather clocks, ornately decorated shelf clocks, cuckoo clocks, sundials, and more.

The museum explores the three main periods of American clock-making: the Colonial, or pre-1800 period, when clocks were handcrafted objects; the period from 1800 to 1860, when machinery and factories were introduced and Connecticut became a prominent clock-making region; and 1860 to World War II, when mass manufacturing developed. In the mid-19th century, central Connecticut was known worldwide as a clock- and watch-manufacturing region. Some 250 companies in the state produced either parts or whole timepieces. Today very few manufacturers in Connecticut produce any parts or timepieces.

In the mid-1950s, Bristol created a museum to inform the public and preserve the history of the clock- and watch-making industry, using the Federal-style home of Miles Lewis, built in 1801. The Ebenezer Barnes Memorial wing was added, using original paneling and beams from other historic local properties. Another wing, the Ingraham Memorial Wing, was added in the 1980s.

The Ingraham Wing exhibit, Connecticut Clock Making and the Industrial Revolution, introduces clock-making, from handmade at the beginning of the 19th century to a mostly machine-made product by century's end. Production costs were significantly cut, and the product became available to a much larger market. Time-keeping equipment for the world went from Europe to America, particularly to Connecticut. Eli Terry revolutionized clock-making by founding the first mass-producing American factory, making clocks and their components. In 1810 Terry sold his shop in Terryville to Silas Hoadley and Seth Thomas, who also gained major recognition in the clock business. Seth Thomas gave his name to the town of Thomaston, as well as the company he founded. A two-story tower clock in the exhibit is large enough that visitors can have a good, close-up look at the inner workings of a clock.

Historic timepieces are shown primarily in the Ebenezer Barnes Wing. The oldest clock in the collections is an English lantern clock built by Henry Webster of Aughton, Lancashire, England, ca. 1680–1700. The clock, with an eight-day time movement and a strike movement, is powered by a single weight suspended by a chain. Its dial is made of cast brass. A "30-hour" tall clock designed by Benjamin Franklin in 1780 is also exhibited. Thomas Harland, the father of Connecticut clock-making, has a tall clock displayed. After emigrating from London to Connecticut in 1773, his fame as a craftsman drew many apprentices. One apprentice, Daniel Burnap, has a tall clock on view, as does Burnap's apprentice, Eli Terry. On view is the patent for the shelf clock that Terry developed in 1815, a timepiece about two feet (0.6 m) tall, the first clock affordable by the general public.

Other interesting time-keeping equipment includes grandfather clocks made in Bristol around 1800, to the shelf clocks so popular in the 1930s. A 19th-century masterpiece displayed is an astronomical clock, known as the Raingo Orrery clock, made in 1825 in Paris, France, by Raingo Frères. This orrery provides the time of day, the phase of the moon, the day of the lunar month, and the relative position of the moon, sun, and Earth. Fewer than 12 of these are known to exist, and only one other Raingo clock has a music box. A Japanese clock with a copper dial, probably from late 17th to the early 18th century, known as a Japanese pillar clock, is on display. The Braille clock on view was invented by John W. Hamilton of Rochelle Park, New York, to help his young blind son.

A collection of regulator clocks reminds visitors of the important role the railroads played in timekeeping. Until 1883, each town kept its own time based on the sun's position in the sky. Coherent railroad timetables were impossible because of the confusing array of times across the nation. International time zones were established to bring some uniformity, but clocks still varied from one to another because they stopped for several seconds while they were being wound. The regulator clock solved that problem—it had a pendulum that kept running while the clock was being wound—and railroad time became the standard.

The Miles Lewis wing exhibits a clock shop of the 1890s containing original timepieces and sales-cases with prices of the day marked on the items. There is also a reassembled 1825 clock shop similar to the one used by Edwards and Willard of Ashby, Massachusetts. They were in business from 1793 to about 1830 making handcrafted clocks in the traditional 18th-century manner, using simple tools. Unfinished clock components on display, such as hands, faces, and gears, come from a recently found collection. Clockmakers in shops of this type needed to be mechanics, carpenters, founders, metalworkers, and engravers. Dutch and English craftsmen who came to America brought their clock-making skills with them and built such superior timepieces that many are still in use.

Not all clock inventions were positive. For example, one alarm clock on view had a match that popped up to strike against an abrasive stone and light a small lamp. Sometimes, however, the match threw sparks that set fire to the house.

The museum has created an authentic American sundial garden landscaped with seasonal herbs and flowers.

A large research center is maintained by the museum, including books, photographs, and other literature. The materials are organized in two sections: a basic reference area where anyone may review the existing literature, and a vault containing rare books and photographs, available by contacting the museum staff.

Watch for temporary exhibits.

Address: American Clock and Watch Museum, 100 Maple Street, Bristol, CT 06010.
Phone: (860) 583-6070.
Days and hours: Daily 10 a.m.–5 p.m. Closed Thanksgiving, 12/24–12/25, 1/1. Hours may vary seasonally.
Fee: Yes.
Tours: Self-guided, or guided on demand; introductory video.
Handicapped accessible: Mostly. One gallery (lower level) is only accessible by stairs.
Public transportation: Take a local bus #502 from New Britain, CT, then walk about 0.6 mile (1 km) to site; taxis are also available to the site.
Food served on premises: None on site; restaurants are in the area.
Website: www.clockandwatchmuseum.org/

MYSTIC

Mystic Aquarium

This museum of aquatic life, the Mystic Aquarium, opened in 1973, and over the years has grown to be one of the country's leading museums and research centers that study sea life. Displayed here are both traditional exhibits and unique exhibits. The research and education center, open to visitors, is involved with marine-life rehabilitation. The aquarium is devoted to researching marine life and helping the public to gain a deeper appreciation for the dignity and uniqueness of aquatic animals.

The facility has a range of exhibits representing a wide variety of places around the world. Displays include penguins, sea lions (with daily shows in the Foxwoods Marine

Mystic Aquarium's Main Gallery. Photo Ryan Donnell.
Courtesy of Mystic Aquarium.

Theater), beluga whales, more than 30 species of frogs, and piranha. Some rescued injured animals are also on view. The exhibits range from the very large to the very small.

A wonderful exhibit, the one-acre (0.4-hectare) outdoor habitat, Arctic Coast, is home to beluga whales. It is the largest outdoor whale exhibit in the United States and the only Beluga whale on exhibit in New England. The exhibit area is surrounded with jagged rocks, glacial streams, and northern evergreens in keeping with the natural habit of these whales. There are underground caves with bubble-shaped viewing windows at various levels. Several rocky overlooks provide an above-ground vantage point. The name "beluga" originates in the Russian word *belii*, which means white, the characteristic color of these whales. Belugas are playful, social creatures that can spit upon and soak their aquarium handlers. They can grow up to 16 feet (4.8 m) long. The aquarium was one of the first to try an insemination program with beluga whales to help promote the survival of this species. Currently the aquarium researches the behavioral ecology and health of these whales.

The Mystic Aquarium has designed an exhibit with three pools, re-creating the California coast and the Pribilof Islands, off the coast of Alaska in the Bering Sea. This exhibit is home to Northern fur seals and Steller sea lions. There are only three facilities in the United States that, as of this writing, display these animals. Steller sea lions are threatened animals, found in the waters from Alaska to California. They are the largest of the eared seals. (See Sea Lion Caves, Florence, Oregon.)

A colony of South African penguins (also called black-footed penguins) lives in the "Roger Tory Peterson Penguin Exhibit," named for the famed American naturalist and ornithologist. The presentation allows the public to observe above and below the water-line. There are underwater viewing windows for visitors to see these birds swim and socialize. Black-footed penguins are listed as an endangered species under the Endangered Species Act. The aquarium is actively involved in the African Penguin Species Survival Plan®.

The studio of Roger Tory Peterson (1908–1996), who wrote the well-known *Field Guide to Birds*, has been replicated here. At this exhibit we have a view of Peterson's world, including his desk, drawing table and implements, and briefcase.

Jellies: The Ocean in Motion currently displays a number varieties of jellies (or jellyfish), such as moon jelly, comb jelly, mangrove upside-down Jelly, South American nettle, blue bubbler, Mediterranean (fried egg) jelly, and crystal jelly. Each of these creatures has its own unique characteristics. For example, the comb jelly lacks nematocysts that sting, and the upside-down jelly lives mainly as a bottom-dweller as an adult and glows in the dark.

White-spotted sharks are a nearly threatened species as listed by the International Union for Conservation of Nature's Red List. They inhabit the waters of the Indo-Pacific west oceans, mostly in tropical waters and coral reefs, as well as Shark Encounters at Mystic Aquarium. This exhibit is shallow and allows visitors to touch these fish. Their food source is small fish and marine invertebrates; predators include larger fish and marine animals, as well as humans. In the wild these sharks can live up to 25 years, although they do live longer in zoos and aquariums. The sharks have muscular pectoral fins that allow them to "crawl" around reefs.

Larger sharks, sand tiger sharks and nurse sharks, are found in the deeper waters of Shark Lagoon. These sharks grow to as much as six feet (1.8 m) long. There are over 30 at the aquarium.

Stingray Lagoon not only houses Atlantic, Southern, and cow-nose rays; it also is home for a green sea turtle named Charlotte who is part of the aquarium's rehabilitation program. The cow-nose rays are part of the shark family, as indicated by cartilage and a small dorsal fin similar to sharks. These rays swim near the top of the tank, whereas other species of rays prefer the bottom.

The beautiful Coral Reef exhibit, with a 30 000 gallon (110 000 L) tank, provides above- and below-water viewing. Displays in this area present clownfish, anemones, loggerheads, and more.

Using sliding color filters, coral reefs can be illuminated, and the aquarium presents an interesting Fluorescent Corals exhibit. Research is in process to explore the dwindling populations of corals as well as medical treatments of illnesses like Alzheimer's disease and cancer using the fluorescent proteins discovered in coral.

Within the Hidden Amazon are poisonous frogs, electric eels, piranhas, iguanas, and lots more. Learn of the importance of conserving this area and its relevance to the health of planet Earth.

Moray eels and barracuda are also on display.

The Mystic Aquarium's reptile-encounter program, Scales and Tales, showcases bearded dragons, pythons, and American alligators, and offers visitors a chance to touch these creatures.

Address: Mystic Aquarium, 55 Coogan Boulevard, Mystic, CT 60355.
Phone: (860) 572-5955
Days and hours: Mar. 1–31: 9 a.m.–5 p.m.; Apr. 1–Labor Day: 9 a.m.–5 p.m.; Sep.–Nov.: 9 a.m.–4 p.m.; Dec. 1–Feb. 28: 10 a.m.–4 p.m.; closed Thanksgiving and 12/25. Hours do vary seasonally, so check in advance.
Fee: Yes. Additional fees may apply for some encounters, rides, and shows.
Tours: Self-guided; some exhibits have docents available.
Public transportation: Peter Pan Bus Lines offers daily, round-trip service to Mystic from Providence, Boston, and New York City. Amtrak offers train service to Mystic and New London along the Washington-Boston corridor.
Handicapped accessible: Yes.
Food served on premises: Café.
Special note: Be sure to check the times of shows and feeding times.
Website: www.mysticaquarium.org

Mystic Seaport: The Museum of America and the Sea

Parts of two towns on opposite banks of the Mystic River—Groton on the west bank and Stonington on the east—make up the community of Mystic Seaport, which, since the 17th century, has built some of the world's finest ships. It is a small seaport, covering only six square miles (16 km^2), but it has produced a larger tonnage of quality ships than any other US port of equal size.

Mystic Seaport Museum was the vision of three local men, Charles K. Stillman, Edward E. Bradley, and Carl C. Cutler, who formed the Maritime Historical Association, Inc., in 1929 and created a living museum of the sea to preserve the skills, values, and traditions of shipbuilding and seafaring in the 19th century. Today the museum's purpose has expanded to include educating the public about US maritime history and its effect on the economic, social, and cultural life of the country.

The living-history museum consists of three parts: a re-created village and harbor dotted with historic ships, a working preservation shipyard, and several exhibit galleries. Mystic Seaport holds about 1000 artifacts, not all of which are on display at any one time.

This 19th-century coastal New England village transports visitors to a time when ships were a major part of life. The original town, known as Greenmanville, was an industrial village that produced ships. Today this town is a complex of indoor and outdoor exhibits that takes up 17 acres (6.9 hectares) and contains about 60 exhibit buildings, four large vessels, and more than 500 small craft. Buildings that make up the village are original period structures brought from various coastal locations of New England. Maritime-related trades are represented, such as shipsmiths, coopers, woodcarvers, riggers, and a ropewalk, as well as trades and shops typical of the time, including a drugstore, general store, and much more.

In 1824, Bourne Spooner, founder of the Plymouth Cordage Company, built the Ropewalk in Plymouth, Massachusetts. This facility continued operations until 1947 when more modern methods were put into use. Part of the original thousand-foot-long (300-m) rope-making machine with three rope-making grounds is a 250-foot (76-m) segment of the factory that was moved to Mystic Seaport. Sailing ships need rope at least 100 fathoms (600 feet; 183 m) long. This exhibit explains the story of rope and the sea. All rope manufacture was performed in a straight line, hence the long building.

Early rope was made of either American or Russian hemp. By the 1830s abaca (manila) was being imported from the Philippines and rapidly became the preferred fiber by virtue of its better durability, flexibility, and easier care.

The Nautical Instruments Shop was an important business for those using ships to navigate the sea. This equipment had to be in good shape and precise; charts and tables needed to be accurate. Tools on display include quadrants, sextants, telescopes, compasses, and marine chronometers. A knowledgeable interpreter is there to demonstrate and explain the equipment. The Drugstore and Doctor's Office were also important to a seafaring community. The Drugstore also sold spices, dyes, kerosene, tobacco, and some groceries and household items.

Several houses are filled with authentic period furniture and boast Victorian-era gardens. The grounds are staffed with costumed docents and expert craftspeople practicing their skills.

For sailors, knowledge of celestial navigation—using stars, planets, and other heavenly bodies of the season—was a must. Thus, in 1960, the Treworgy Planetarium was designed and built by Armand Spitz and was equipped with a Spitz A3p Star Projector. Shows are projected onto the 30-foot (9.1-m) diameter dome; the lobby has a permanent exhibit on 19th-century navigation that includes an orrery (a mechanical scale-model of the solar system). The display shows how sailors navigated the waters without modern conveniences.

At the port, four permanently docked ships are designated as National Historic Landmark vessels. The interiors of the four ships are accessible to the public.

The best-known of these ships is the *Charles W. Morgan*, built in 1841 in New Bedford, Massachusetts, the last surviving wooden whaling ship, which made 37 voyages during its lifetime. The ship was retired in 1921 after sailing from New Bedford to the Arctic and the South Seas, and was used in two silent films: *Down to the Sea in Ships* (1922) and *Java Head* (1930). The meticulous preservation work at the museum's shipyard has restored this proud vessel.

L.A. Dunton, a fishing schooner built in 1921, was one of the last engineless schooners and was used to catch haddock and halibut. Several other schooners of this type were built primarily for racing. In 1923 a 100 hp (75 kW) Fairbanks, Morse, and Co. C-O engine was installed.

The *Sabino* is one of the oldest wooden, coal-fired steamboats still operating. It was built in 1908 to transport passengers across the Damariscotta River. Today visitors at Mystic Seaport can cruise on the steamboat.

The fourth of these National Historic Landmark ships is the *Emma C. Berry—Noank Smack*. One of the oldest surviving commercial vessels in America—built in 1866—the vessel had a truncated pyramidal construction so that water could flow through and contained holes at the bottom of the hull-planking for the water to leave. Thus the ship could bring live fish to market.

Both the square-rigged *Joseph Conrad* and the wooden schooner *Brillant* are used for educational purposes. The *Conrad* is a training ship and an exhibit. The *Brillant* serves as an off-shore classroom for sailing educational programs. Among other boats on display is the *Gerda III*, a 1926 lighthouse tender, which was used to ferry Jews away from Nazi-occupied Denmark, and was donated by Denmark to the Museum of Jewish Heritage in New York City. Mystic Seaport is the official berth for this boat.

At the Mystic Seaport Preservation Shipyard, watch the bygone art of wooden shipbuilding from high above in the visitors' gallery. Ships are brought to this carpenters' shop for repair. See the 85-foot (26-m) spar lathe and a rigging loft. Here the watercraft collection is restored and preserved. A series of exhibits and demonstrations explain the process.

Other work areas include a paint shop, a metalworking shop, a documentation shop, a lumber shed, and a saw mill. Historic methods are always used for historic vessels.

A major national collection of exhibits and maritime galleries exploring the 19th-century world of the sea is also located at the Seaport. The Maritime Gallery displays the work of leading maritime international artists, and The Museum of America and the Sea houses a variety of international artifacts. See such exhibits as a Mystic River scale model, Stories of America and the Sea, Figureheads, and a Whale Boat exhibit.

The Collections Research Center (CRC), across the street from the Seaport, holds various sea charts, diaries, maritime art, artifacts, tools, buildings, photographs, ship registers, audio histories and interviews, and maritime-related films. The CRC, which has limited hours, recommends calling in advance of a visit.

Address: Mystic Seaport: The Museum of America and the Sea, 75 Greenmanville Avenue, Mystic, CT 06355–0990.
Phone: (860) 572-5315 or (888) 973-2767. For the CRC, call (860) 572-5367.
Days and hours: May–Jun. and Aug.–Oct.: 9 a.m.–5 p.m.; Jul.–Aug.: 9 a.m.–8 p.m.; Nov.–Apr.: 9 a.m.–4 p.m. Closed 12/25.
Tours: Self-guided.
Public transportation: Peter Pan Bus Lines runs buses to Mystic. Amtrak trains come to Mystic and New London along the Northeast Corridor. The Mystic train station is about a 10-minute walk to the Seaport, or take a taxi from the station to the Seaport.
Directions by boat: Contact Mystic Seaport for information.
Handicapped accessible: Limited. Wheelchairs are available free of charge. Some historic buildings and the ships are not handicapped accessible.
Food served on premises: Café and bakeshop, restaurants, tavern.

Fee: Yes.

Special note: Check schedule for special fees, special events, programs, and demonstrations. There are extra charges for planetarium and cruises. Lodging is available on site.

Website: www.mysticseaport.org

NEW HAVEN

Peabody Museum of Natural History

The Peabody Museum of Natural History, founded in 1866, is one of the oldest museums of natural history in the nation, with extensive and important collections. Today it holds more than 11 million specimens and objects in anthropology, botany, zoology, paleontology, entomology, ornithology, biology, and historical scientific instruments. Many of these materials may be seen in the displays on the three floors of public exhibit areas used by Yale University's departments of anthropology, biology, geology, and geophysics. This institution is part of Yale University and is used as a major component of research.

Merchant and philanthropist George Peabody donated the funds for this institution dedicated to education, research, publication, and exhibits of natural history, archaeology, and ethnology. The 1926 neo-Gothic building that is now the museum's home clearly fulfills the requirements laid out by Peabody.

Othniel Charles Marsh (1831–1899) was an anthropologist and the first professor of paleontology in North America, as well as the nephew of George Peabody. Marsh convinced Peabody to contribute the funds for the museum. In 1867, O. C. Marsh became the first director of Yale's Peabody Museum of Natural History. On display are many of Marsh's famous finds.

Several of the collections that the Peabody holds are of major international importance. Of these collections, perhaps the most significant is the vertebrate paleontology collection. It is among the largest, most extensive, and most historically significant fossil groups of artifacts in the United States. Most of these artifacts were amassed by O. C. Marsh. On display are many of the most famous of his finds.

The two-story Great Hall of Dinosaurs displays samples from the Marsh finds in the United States, some of which are the first skeletons collected in North America. Dominating the hall is the giant dinosaur of the genus Apatosaurus, which Marsh found in 1877 and named "Brontosaurus." It is 110 feet (34 m) long. The name's Greek roots translate to "thunder lizard," reflecting the animal's giant size and the thundering sound it probably made as it walked about, shaking the land. Marsh found the skeleton in a fossil bed, where it had lain buried for millions of years. He was able to retrieve 80% of the skeleton, making it one of the most complete samples of its kind. The *Apatosaurus*, found in 1879, is actually an immature "Brontosaurus" and of the same genus, but it was reported publicly before "Brontosaurus" so that the name of the genus thus became *Apatosaurus*. Hence the Peabody claims to be the only museum in the world that can accurately use the term "Brontosaurus."

Also displayed here is an immature dinosaur of the genus *Camarasaurus*, one of the few young mounted specimens in the world. On view is the largest turtle skeleton known, the *Archelon ischyros*, which lived 100 million years ago. Among the more recent exhibits is a fleshed-out model of *Deinonychus*, or "terrible claw," a dinosaur found in 1964. Surrounding and overlooking the displays is the Pulitzer Prize–winning mural "The Age of Reptiles," by Rudolph F. Zallinger, painted over the course of over four years (1943–1947), depicting these dinosaurs in their natural habitat. The mural begins on the right in the Devonian Period (362 million years ago), and continues leftward to the Cretaceous Period (65 million years ago), incorporating the knowledge available in the 1940s. Large trees aesthetically divide the sequential geologic periods.

The matching 60-foot (18-m) "The Age of Mammals" mural overlooks the Hall of Mammalian Evolution. Here there are artifacts from several world-important collections, including the vertebrate paleontological collections, and some of the most extensive, largest, and historically important fossil collections in the United States. These fossils include the Otisville Mastodon (*Mammut americanus*) and *Brontotherium*, a large mammal related to horses.

Be sure to see a huge and interesting model of a giant squid that hangs in the main hall.

The ornithology collection is one of the world's largest and most taxonomically inclusive. Associated with this collection is the William Robertson Coe Ornithology Library, which is one of the best libraries of its type in the United States. Some 722 mounted specimens occupy the Birds of Connecticut Hall. These specimens represent 300 of the 382 species that frequent Connecticut on a regular basis. Three of these species have become extinct in the last 100 years: Labrador duck, heath hen, and passenger pigeon. The exhibits represent the shore, the bog, and the forest biomes.

Other exhibits include the Hall of Minerals, Earth, and Space, which incorporates the extensive Museum collections and the latest scientific research. The Earth and the Solar System is another handsome display. Here we learn of the formation of the solar system, volcanic eruptions, and more. Also see the meteorites.

The Invisible Art: The Yale Peabody Museum displays 11 dioramas, considered masterpieces, offering remarkable views into the ecosystems of the natural world. J. Perry Wilson and Francis Lee Jaques painted the backgrounds, and Ralph C. Morrill created the foreground. The extraordinary depictions of these scenes include both Southern New England and North America. The dioramas, created in the 1940s, were restored in the early 1990s.

Daily Life in Ancient Egypt explores the richness, complexity, and worldly knowledge of this ancient civilization from the Predynastic Period of ca. 4500–3100 B.C.E. through the Greek and Roman Periods of 332 B.C.E.–640 C.E. The highlight of this exhibit is the mummy from the Late Period of the 7th to the 4th centuries B.C.E. Here we learn about current Yale research into Egyptian civilization.

Other interesting exhibits include the Hall of Human Cultures, which displays artifacts collected from the Plains Indians and from natives of New Guinea and Polynesia. In the Mexico to Peru exhibit are objects from the museum's meso-American and South American collections.

Address: Peabody Museum, Yale University, 170 Whitney Avenue, New Haven, CT 06511.
Phone: (203) 432-5050.
Days and hours: Mon.–Sat. 10 a.m.–5 p.m.; Sun. and holidays, noon–5 p.m. Some halls may be closed during the academic year weekday mornings; we recommend visiting after 1 p.m. Closed 1/1, Easter Sunday, Thanksgiving, 12/24, and 12/25.
Fee: Yes (Free Thu. 2 p.m.–5 p.m. between Sept. and June).
Tours: Self-guided, audio tours available; 45 min. docent tours available Sat. and Sun. 12:30 and 1:30.

Public transportation: The museum is on the New Haven "J" route bus line and convenient to Amtrak, MTA Metro-North Railroad, and Shore Line East passenger rail service in New Haven.
Handicapped accessible: Yes.
Food served on premises: Not on site, but nearby in New Haven.
Special note: Check schedule for special events, exhibits, and programs.
Website: peabody.yale.edu/

Maine

AUGUSTA

Maine State Museum

When people think of Maine, the state's lush forests and rocky seacoast typically come to mind. Here New England, Maritime Canada, and the sea come together. It is a land that has been occupied by humans for some 11 000 years from the Paleo-Indian hunters following the woolly mammoths to the vacationers of today. But Maine also has an important history that reaches beyond its geography, as this museum demonstrates. The Maine State Museum's goal is to instill and maintain a respect and appreciation for Maine's past. It showcases not only Maine's natural environment, but also its manufacturing heritage and social history in exhibits containing archaeological specimens, objects made in Maine, and tools used by Maine workers and in the state's industries.

This Land Called Maine displays natural-habitat dioramas reflecting different environments in the state: Woodland, Mountains, Inland Waters, Marshland, and Rocky Coast and Saltwater Marsh. These life-size scenes contain plants and animals of the state and reflect seasonal changes. Mountains in Autumn includes a stream with live trout. One of the displays is of Maine's gems and gemstones, which includes a tourmaline in a cleavelandite matrix from Oxford County, Maine, discovered in 1820. Tourmaline, a silicate crystal, is the state's official mineral. The Mount Mica mine, reopened in 2003, has uncovered many pockets of varied colors of tourmaline, and the museum has a number of recently discovered fine specimens on display.

The rich natural resources of the state have defined its economic and social history, making an impact on its technical development. Maine Bounty traces the 400-year tradition of shipbuilding. The exhibit includes a reconstructed 40-ft (12 m) section of the hull and some of the rigging of the *St. Mary*, one of the last square-rigged down-easters. This type of sailboat, built in 1890 in Phippsburg, is unique to the New England area. Shipbuilding tools used to construct the ships and sew the sails complete the display. Maine's important fishing industry is explored, along with the tools to do the many related jobs, including seining, trawling, line fishing, and lobster trapping. On display are examples of saltwater fishing gear, hand-line cod fishing, and sardine-canning equipment.

The lumbering exhibit features the Lion, a steam locomotive constructed in 1846. This piece of equipment transported lumber in Washington County. The Lombard, a log-hauler made in Waterville, is a gasoline-powered engine used for 50 years to transport its cargo from the sawmills to the wharves of the Allagash region. Lumbering dates back to the 1750s in Maine. Some of the wooden products made in Maine are discussed, as is the manufacturing of clapboards.

Quarrying of granite, once a major industry in Maine, is explained in a display showing the tools for cutting the stone, a derrick used to lift the stone from the quarry, and a wagon used to move the stone from the quarry site. Another exhibit illustrates ice harvesting. Before the advent of refrigeration, ice harvesting was important in Maine. Hand and horse-drawn equipment are enhanced by photo murals showing the ice-harvesting process and the storage of ice in large icehouses.

Farming Maine's rugged landscape has never been easy, but it was especially difficult before the advent of modern machinery. Displayed are early equipment typically used to coax agricultural products from the land, including the famous Maine potato.

An extensive archaeological exhibit looks at 12 000 years of Maine's pre-history and history, beginning with the last Ice Age and ending in the 19th century. One of the highlights is a full-size diorama showing the meat cache in northwest Maine of Paleo-Native Americans who lived about 11 000 years ago. Another display is an 1870s seagoing birch canoe, equipped with a canvas sail.

Made in Maine is an extensive exhibit that covers 19th-century manufacturing and the many goods produced in Maine for use around the world. Highlights include a water-powered woodworking mill and a two-story textile factory. Maine's abundance of natural harbors, navigable rivers and estuaries, and waterpower for machinery laid the foundation for its prosperity. Fourteen re-created work settings contain tools, machinery, and other equipment—many of them operating. The products of Maine's industries are shown in context. Work environments include homes, shops, mills, and factories.

Dioramas of home scenes, factories, shops, and mills include everything from hand-sawn lumber to toothpicks, locomotive engines, tombstones, and glass cylinder pumps. The most valuable finished products made in Maine in 1810 were cotton and woolen cloth, which were produced in cottage industries. This trade, along with shipbuilding and other key industries, is carefully illustrated.

As steam and waterpower became more available, manufacturing was increasingly mechanized, as demonstrated by a mill for carding and spinning wool, built around 1850. Other implements in this collection include an 1850s water-powered woodworking mill with a band saw, a shingle saw, and barrel-stave saw.

A depiction of a gun shop owned around 1816 by John Hall—who patented the first breech-loading firearm—illustrates the transition from handcraft to mechanization. Hall produced guns from a stockpile of component parts rather than building whole guns from scratch. He both made guns and sold them. A fishing-rod shop from early in the 20th century that specialized in bamboo rods and a home from 1920 in which yarn was spun have both been reconstructed here.

By the 1840s many Maine homes were engaged in sewing ready-to-wear clothing. The state's easy access to the markets of Boston and other Atlantic coast ports meant an ever-expanding industrial base. Maine's ocean-port towns, such as Biddeford and Lewistown, underwent big changes as cotton factories, shoe manufacturers, iron furnaces, and other flourishing industries built mills along the riverbanks.

At Home in Maine is a large permanent exhibit that explores the often forgotten aspect of technology, the home. Areas of work such as housekeeping and indoor plumbing and their effects on the Maine way of life are looked at through artifacts and home movies in this display. Other exhibits include a changing display of flags from 1675 to 1842, which examines Maine's political and geographic development. Many types of glass were made and or used in Maine and thus an exhibit looks at this glass. At Home in Maine explores how people lived over the years and the equipment they used.

Address: Maine State Museum, 230 State Street, Augusta, ME 04330.
Phone: (207) 287-2301.
Days and hours: Tue.–Fri. 9 a.m.–5 p.m.; Sat. 10 a.m.–4 p.m. Closed on state holidays.
Fee: Yes.
Tours: Self-guided; guided tours available by advance reservation.
Food served on premises: None in the museum; cafeterias are available in nearby government buildings. The front desk maintains a list of local restaurants.
Public transportation: Concord Bus Company runs service to Augusta. At the bus terminal check for local buses or take a taxi.
Handicapped accessible: Yes.
Website: mainestatemuseum.org/

BANGOR

Cole Land Transportation Museum

Maine is a land that has special people with their own way of life and state of mind. The Cole Land Transportation Museum provides a window into their lives through the use of equipment and development of the land. Among its highlights are probably the largest collection of snowplows under one roof in the United States, plus the largest collection of vehicles from a single trucking company in one museum.

The Cole Land Transportation Museum began with Galen Cole, who was part of the family that owned Coles Express, a company founded in Enfield, Maine, which then expanded throughout the state. As a child Galen rode with Coles Express drivers and with his father in the REO Royal Coupe. It was then that he developed a love affair with land transportation and began to dream of preserving old vehicles to keep them from disappearing.

When, in 1992, Coles Express was sold after 75 years in existence under the ownership of one family, some of the money realized went to create and maintain the Cole Transportation Museum. This institution had several purposes. Here some of the quickly disappearing land-transportation equipment would be collected, preserved, and displayed so that future generations could see this equipment and learn about how they were made and used. Later generations would also learn how pioneers cleared the land, built highways and rail lines to develop Maine inland from the seacoast, and how these people kept the railroads operating in all weather. Younger generations could be inspired to work toward their dreams and use their abilities to the utmost. With a collection of US military memorabilia, the museum could also help inspire in young people a love of history. This institution may well be the only land transportation museum where most of the equipment displayed originates within its geographic area.

When the Maine Central Rail Road Station at Enfield was declared surplus, Galen Cole was able to obtain it and place it in his museum. The station serves to put some of the vehicles—trains—and other equipment in context. Exhibits here include many items from the early 20th century: a safe, lanterns, telegraph, ticketing equipment, ledgers,

Bangor and Aroostook BL2 diesel locomotive 557.
Courtesy of Cole Land Transportation Museum.

order board and controls, message wands, and more. The Saponac Lake Post Office has been incorporated into the exhibit. On the walls are displays of maps, pictures, and additional railroading artifacts. In front of the building is a restored Railway Express baggage complete with period luggage and dock carts. Other settings for artifacts include a blacksmith shop and a covered bridge on Museum-operated land.

With the transition from steam to diesel-electric-powered locomotives came locomotive #557, one of 59 made by the Electro-Motive Division of General Motors. This piece of equipment weighed less than the steam-powered units and could operate over otherwise restricted trestles. A separate electric motor rotated each pair of wheels. All the electric motors got their power from a 16-cylinder engine generator, a 1500 hp (1100 W) diesel engine. In winter the engine needed to be kept running 24 hours a day to keep the coolant from freezing. The "Railfan" passenger train was pulled by this engine for almost 20 years from North Maine Junction to Caribou, Maine, and back. On the rear of the train was a caboose, No. 660, built by Western Maryland Railroad in 1940. This car was distinguished for its comforts, which included some important technical innovations for a train, such as a flush toilet, electric lights, oil stove, three bunks, water cooler, and icebox. Here trainmen slept while they had overnight trips, and business was conducted with railroad customers. Trainmen used the cupola to watch for signs of problems as the train went around curves.

Among other noteworthy vehicles preserved is a selection of fire trucks, including a 1907–1908 Steam Pumper manufactured by the Manchester (New Hampshire) Locomotive Works, which could pump 1100 gallons (4200 L) per minute. Originally three horses pulled the truck, but in 1924 the horses were replaced by a tractor. The Steam Pumper and the tractor are displayed together. Another interesting piece of equipment is the 1931 American LaFrance quadruple-ladder fire truck, removed from active duty in 1988. This truck includes ladders built by the Bangor Extension Ladder Company, which patented "tormentor poles," which were installed on the fire truck. With these poles, fewer firefighters were necessary to lean the 300-pound (140 kg) ladder against a building.

In Maine, farming was and still is an important industry. Thus tractor and farm equipment comprise a large portion of the collection. Among the displayed collection is an 1896 treadmill, nicknamed "Baby Junior," which was used to grind grain, with a Shetland pony-sized animal to provide power. As the pony walked on the treadmill, it turned a pulley by means of a belt. The restored 1920s-vintage potato planter, manufactured by Aspinwall & Watson, was horse-drawn. The device could plant three to four acres one day, an important piece of equipment to Maine's large potato industry.

Of the many wagons and carts represented at the Museum, perhaps the oldest item on view is the late 1600s Brazilian ox-cart, from the area of Rio Grande do Norte. It is one of the few pieces not made in Maine. The cart is entirely made of wood, except for the iron rims (possibly a later addition) and an iron ring. Connections between the parts are via sisal rope and wooden pegs. The solid-plank wheels do not freely rotate on the axles, so the cart skids when turning a corner. This is probably an early step in the evolution of the wheel and axle. The cart was used for carrying sisal leaves in rope baskets (as shown in the display) to a place for juice extraction and fiber production.

Dr. Henry A. Mansfield, a dentist in Jonesport, Maine, serviced many of his patients several times a year by transporting his dental chair and pedal-powered drill to these clients on a spring-seat wagon (on display at the Museum) to outlying areas. This type of wagon differs from a buckboard, which has a flexible board.

A collection of baby and doll carriages shows the evolution from wooden wheels and hubs to wire wheels on iron axles. Another type of manual-powered vehicle is the bicycle. The Museum owns one of the oldest bicycles in New England, a homemade "boneshaker" dating from the 1860s. This example has no brakes, and the pedals are attached to the front wheel. The Museum's collection includes other 20th-century models such as an aluminum Silver King from about 1937, and a "Western Flyer" chain-driven tricycle.

Automobiles and trucks are also on view. The tiny "King Midget," built by Midget Motors in Ohio in the 1950s and 1960s, advertised as "a 500-pound car for $500," is represented. These machines have automatic transmissions, steel bodies and aluminum doors, heating systems, 12-volt electrical systems, a 9.2 hp (6.9 kW) one-cylinder engine, and four-wheel hydraulic brakes. A 1912 REO light delivery truck with hard rubber wheels, the very one in which Galen Cole rode as a child, is on display. This machine has a one-cylinder engine, curb-side crank for ignition, and chain-drive two-speed transmission.

Some early highway equipment in the collection includes a horse-driven grader built by the Western Wheeled Scraper Company in about 1900. It has seven controls, including a center wheel to move the grader blade horizontally; left and right wheels to raise and lower the sides of the blade; and foot pedals that act as brakes. Even several early gasoline pumps can be seen, such as a Mobil glass-cylinder (circa 1900), a Tydol hand pump (also circa 1900), and an Atlantic electric pump from about 1920.

Because Galen Cole served in the army during World War II, there is a display of military vehicles and a collection of military uniforms and effects, as well as insignia, weaponry, medals, and a piece of the Berlin Wall.

Address: Cole Land Transportation Museum, 405 Perry Road, Bangor, ME 04401.
Phone: (207) 990-3600.
Days and hours: May 1–Nov. 11: 9 a.m.–5 p.m.
Fee: Yes.
Tours: Docents are available on-site to guide visitors.
Public transportation: Concord Bus Lines serves Bangor, then take a taxi to the site. Other bus companies also service this area.
Handicapped accessible: Yes. Wheelchairs are available on request.
Food served on premises: None.
Website: www.colemuseum.org/

BATH

Maine Maritime Museum

This maritime museum, founded in 1962, and which stands on 10 acres (4 hectares) of galleries and exhibits along 25 miles (40 km) of the scenic Kennebec River in Bath, chronicles Maine's long history of seafaring, fishing, and shipbuilding. The large complex consists of a Maritime History Building (1989), five historic buildings, and two launching ways of the Percy & Small Shipyard constructed between 1897 and 1920, all set on open space and marshland.

In 1607, more than a decade before the Pilgrims landed at Plymouth Rock, a trading vessel was built along the banks of the Kennebec River. Ships like these sailed around the world carrying lumber and oil. They returned to Europe with silk, tea, and trinkets of all sorts. Local residents fished the area's waters for lobster, haddock, and cod.

Galleries in the Maritime History Building have displays of art and artifacts selected from a collection of over 18 000 items that tell the story of shipbuilding in Maine. These exhibits include three-dimensional boat models as well as half-hull and full-rigged pieces, scrimshaw—painstakingly decorated and carved whalebone and walrus tusks—and other art to help tell the story of the sea and those who sailed it. Exhibits also explore the men and their families who built ships and risked their lives sailing them, show specialized shipbuilding tools and implements of other maritime trades, and highlight the recreational use of Maine waterways. Historical materials from the Bath Iron Works, still in operation a short distance from the museum, are on display. The museum building also has changing exhibits, a museum store, meeting and eating facilities, as well as 130 watercraft in its collection.

A major component of the museum complex is the Percy and Small Shipyard, the only surviving US shipyard that built large wooden sailing vessels. Captain Samuel Percy and Frank Small constructed 41 four-, five-, and six-masted wooden schooners in this yard between 1894 and 1920. The yard built seven of the only 12 six-mast ships ever constructed in the eastern United States. Five of the Shipyard's original buildings have been preserved: the sawmill, the paint shop, the treenail shops, the caulking shed, and the mold loft.

Each structure contains exhibits about building wooden ships and displays historic tools and machines. For example, the mold loft (built in 1917 during the First World War's boom in shipbuilding) was the first step in constructing a ship, scaling the shape up from a mold to the actual size of the ship's hull. The caulker's shed (dating from 1899) stored bales of oakum (tarred fiber), cotton yarn, and pitch stored in kegs. The treenail shop (the oldest structure, from 1897) was the area for shaping and storing wooden dowels (usually white oak or black locust woods). The mill shop (1899) houses tools for shaping large items, including a Daniels planer—for smoothing irregular sections of timbers— and bevel-cutting jigsaw. The Daniels planer was patented in 1834 by Thomas Daniels of Worcester, Massachusetts. It had a reciprocating carriage to which the lumber was attached. Cutters were placed on the ends of a horizontal bar hung from a rotating vertical spindle above the lumber. This removed wood from the boards, flattening the upper face of the lumber. Manufactured until about 1900, and cumbersome because the machinery needed to be twice as long as the stock, few of these Daniels planers survive today.

The second story was used for finished woodwork such as cabins and deckhouses. Demonstrations illustrate various shipbuilding tasks. Between April and October there are guided tours of the shipyard. During the rest of the year visitors may do self-guided tours. The shipyard also offers apprenticeship and yearlong courses in traditional wooden boat-building. While visitors watch the work in progress, they can ask staff members questions about classic boat design and construction.

Lobstering and the Maine Coast is an exhibit that covers Maine's most famous maritime industry. Visitors learn how lobsters are trapped, canned, and shipped. The exhibit includes a lobsterman's workshop, dioramas of lobstering in a Maine harbor, and various lobster boats, ranging from rowboats to sailing craft. A historic film, written and narrated by writer E. B. White, describes lobstering in the 1950s.

On the museum grounds are historic vessels, some of which can be visited by the public, such as the *Sherman Zwicker* (in season), a 142-foot (43.3 m) Grand Banks fishing diesel-powered schooner first launched in Nova Scotia. With a cruising speed of 9.5 knots (19 km/h), this ship plied the North Atlantic from 1942 to 1968. Its engine house and crew quarters are open for inspection, and it is the largest wooden ship within Maine's waterways. In season, visitors can learn about the Kennebec River via a 50-minute narrated tour on an excursion boat.

In the Victorian-era Donnell House, adjacent to the Percy & Small Shipyard, the shipbuilder and his family lived, and we can view family life of this period. Displays include family period pieces and other similar pieces. It is open Monday through Saturday, 11 a.m. to 4 p.m., from Memorial Day to Columbus Day.

Other exhibits include 19th-century sailing and trade, clipper ships, and a variety of watercraft used in Maine. There is a children's play area, a research library and archives related to nautical studies, and a building with amenities for visiting boaters moored at the site.

Address: Maine Maritime Museum, 243 Washington Street, Bath, ME 04530.

Phone: (207) 443-1316.

Hours: Daily 9:30 a.m.–5 p.m.; closed Thanksgiving, 12/25, and 1/1. During winter months the Percy & Small Shipyard may be closed because of weather or ground conditions.

Fee: Yes, ticket allows two visits within one week. Special Tours and cruise require additional fees.

Public transportation: Concord Bus Company from Portland and other cities.

By water: 10 miles (16 km) up the Kennebec River from the Gulf of Maine (43° 53.871′ N, 69° 48.889′ W), the mooring field, floating-dock space, and a 75-foot (23 m) pier for vessels 60 to 300 feet (18–91 m) in length. Docking fee includes two tickets to the Museum.

Food served on premises: Café.

Website: www.bathmaine.com

BRUNSWICK

Peary-MacMillan Arctic Museum

This small museum in Hubbard Hall on the Bowdoin College campus is a tribute to Admirals Robert E. Peary and Donald B. MacMillan, famous Arctic explorers who both graduated from Bowdoin. The college has a long-standing interest in Arctic exploration dating from 1869, when professor of chemistry and natural science Paul A. Chadbourne and 20 students from Bowdoin and Williams Colleges repeated the voyage of Norsemen who explored the coasts of Labrador and Greenland.

General Thomas Hubbard, a Bowdoin graduate of the class of 1857, gave his name to the building in which the museum resides, along with large financial contributions to Peary's Arctic ventures. One focus of the museum is on Peary, who in 1908 was the first person to reach the North Pole. MacMillan accompanied him on his last Arctic expedition in 1908–1909 and continued to work in the Arctic until 1954. Other important goals of the museum include presenting Arctic exploration and the indigenous people of the Arctic.

The museum material examines Arctic exploration, cultures of the indigenous people, and biology. Exhibits in the larger gallery change approximately every two years, while the smaller galleries offer new displays more frequently, in order to display as much of the collection as possible.

The first gallery of the museum is the permanent exhibit Robert E. Peary and His Northern World, which focuses on the many people who made important contributions to his expeditions, including his family, his chief assistant Matthew Henson, and the Inuit. Objects on display include the odometer, barograph, thermometers, sidereal watch, and survey instruments used in the expeditions. Peary designed his own camp stoves to burn only six ounces of fuel and heat ice into boiling water in just nine minutes, then extinguish themselves. Examples of these special camp stoves are also shown. An admiralty-style model of the SS *Roosevelt*, a relatively small but maneuverable ship that incorporated a steam-driven propeller as an icebreaker, is also showcased.

The second gallery studies contemporary Inuit art, drawing from the museum's growing collection of art dating from the second half of the twentieth century to the present. These exhibits change annually. The third, and largest, gallery looks at a wide variety of topics, from the history of Arctic exploration, to Arctic clothing and climate change, to Inuit views of the natural world, on a temporary, rotating basis.

In the foyer of Hubbard hall, the museum presents small photographic exhibits, along with the 14-foot (4.3-m) Hubbard Sledge, a mixture of Western and Inuit technology, one of the five sledges used on the April 1909 trek, fabricated of steel runners and an oak body lashed with rawhide, and able to transport 500 pounds (230 kg) of supplies. There is a depiction of the GISP2 ice core from Greenland Ice-Sheet Project. There is a diorama depicting birds (eiders, terns, puffins, guillemots, and more) on the northern shore of the St. Lawrence River. Above the galleries, a number of taxidermied mammals (caribou, polar bear, seals, walrus, muskox) of the Arctic regions are displayed.

Address: Peary-MacMillan Arctic Museum, Hubbard Hall, Bowdoin College, 9 South Campus Drive, Brunswick, ME 04011.
Phone: (207) 725-3416.

Days and hours: Tue.–Sat. 10 a.m.–5 p.m., Sun. 2 p.m.–5 p.m. Closed Mon. and legal holidays.
Fee: Free, donation suggested.
Tours: Self-guided; guided tours available by advance reservation.
Public transportation: Amtrak Downeaster train to Brunswick, then walk about half a mile (1 km) to the Museum.
Handicapped accessible: Yes.
Food served on premises: None. Restaurant on the college campus, others mostly beyond the northern end of campus.
Website: academic.bowdoin.edu/arctic-museum/

FREEPORT

The Desert of Maine

In an area rich with vegetation, it seems strange to find this patch of barren, sandy landscape covered with sand dunes. The terrain, known as the Desert of Maine, was created by glaciers during the last Ice Age. To learn about this unique ecosystem, a Guided Glacier Desert Tour that crosses the desert every half-hour is available to visitors. A walk along the nature trails provides an opportunity to enjoy the beauty of the desert and surrounding forest as well as time to investigate the typical Maine forest that surrounds the desert. Many varieties of lichens and mosses work to rebuild the ancient forest floor.

Information in the museum and in the site's literature explains that glaciers pushed southwest, creating striations in the region's bedrock. As the glaciers scraped across the land, their massive weight pulverized the rock underneath and carried this newly created sand-like substance and minerals along with it. The glacial silt or sandy deposit in the Desert of Maine was left behind as the glaciers melted some 11 000 years ago at the end of the Pleistocene Period, the last Ice Age. Note that this site is not considered a true desert (which receives less than 25 cm of rain annually) because the area receives an abundance of rain.

The sand is composed of many types of minerals that were ground up together as the glacier slid over bedrock: quartz, biotite and muscovite mica, feldspar, garnet, hornblende, granite, granite pegmatite, gneiss, and basalt, among others. Rocks found 25 to 30 miles (40–48 km) northeast of the site match these mineral types. Gradually, soil and thick vegetation covered the huge sea of sand, giving a lush appearance to the land.

Among the European settlers who migrated to this area were William Tuttle and his family. In 1797 they took possession of 400 acres (120 hectares) and created a farm on what appeared to be fertile land. Not understanding the fragile ecosystem or how to care for the veneer of soil on the sand below, the Tuttles cleared parts of their acreage and turned it into pasture for animal grazing. As the animals grazed, the plant life became closely cropped and could no longer hold the soil in place. Eventually portions of the sod were pulled out by the roots. The thin, unprotected layer of soil that was left was removed by wind and rain over the years. By 1897 a small portion of the underlying sand had become exposed. The Tuttles could no longer farm and were forced to leave. Now, over a century later, we can see more than 50 acres (20 hectares) of sand, as well as dunes caused by the wind. As time passes, the sand and dunes move.

Most of the Tuttle farm structures are now gone. A barn built in 1783, before the Tuttles took title to the land, is the last remnant of the farm. It has been restored and now houses a museum displaying farm implements, sands from around the world, a geological explanation of the sand on this land, and a mineral collection.

The site has been a tourist mecca for about 80 years. Every half-hour a tram takes visitors across the "Desert of Maine" and provides a good explanation about desert formations. The angle of the sand on the dune is determined by wind direction; the steeper slope faces the wind. The dunes migrate toward the southeast. The finest particles blow the farthest and leave behind pebbles or "desert pavement." Interestingly, the rate of new plant growth at the edge of the desert seems to just match the sand movement, so the system appears to be relatively stable.

Surrounding the sandy area is a forest filled with indigenous flora and fauna. Marked walking trails go through this area so that visitors may enjoy the Maine scenery. There are walking tours of this area.

There is a Butterfly Room on the property. The types of butterflies change with season and commercial availability.

Address: Desert of Maine, 95 Desert Road, Freeport, ME 04032.
Phone: (207) 865-6962.
Days and hours: May–Oct.: daily 9 a.m.–5 or 5:30 p.m.
Public transportation: Amtrak Downeaster train to Freeport, ME (limited service only), then a 10-minute taxi ride to the Desert of Maine.
Tours: Tram tours run every half-hour from 9:30 a.m. until approximately 4:30 p.m.
Fee: Yes.
Special note: Camping available next to the Desert. Professional sand-designers work on-site.
Food served on premises: Picnic area and small convenience shop.
Handicapped accessible: Yes.
Website: www.desertofmaine.com/

KENNEBUNKPORT

Seashore Trolley Museum

The importance of trolley transportation should not be underestimated. Though steam railroads carried people over long distances, they were more costly. Trolley transportation, however, made travel available to the public for short distances and at a cost affordable to the general population. This was the first time in human history that people could commute to work, shop, and participate in culture and other recreation at a location not very far from where they lived. Thus society was changed forever.

The first electric trolley began operating in 1887. Only 52 years later, in 1939, these street cars were rapidly disappearing at such a rate that the Seashore Trolley Museum opened its doors as the Sea Shore Electric Railway Museum in order to preserve these vehicles and related artifacts.

The site is located where the Mousam River Railroad, a small electric line only two miles (3 km) long, began operation in 1893. This line hauled coal and other necessary

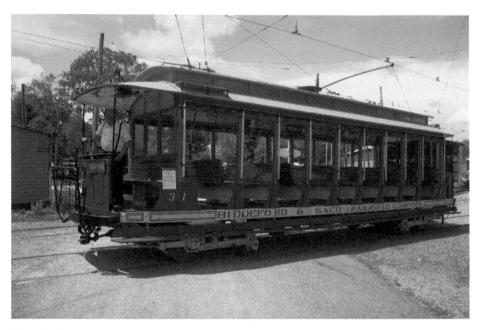

Biddeford open car 31.
Courtesy of Seashore Trolley Museum.

raw materials to power mills and to bring the finished products back to ship to market by railroad.

Because Ohio's Lake Shore Electric Railway, the major interurban railway of the period, closed in 1938, this Museum chose the name Sea Shore Electric Railway Museum to commemorate the demise of this institution.

Today the museum, in three barns on about 300 acres (120 hectares), is possibly the largest electric railway museum in the world. The site stores over 250 vehicles, most of which are electric trolleys, including 30 on display. While the majority of the vehicles and other artifacts come from the United States, there are many from Canada and other places around the world. Most of the exhibits span 1890 through 1930, the most active era of electric trolley transportation. Those pieces that are in good operating condition are used on the Museum's Heritage Railway, which enables the public to experience trolley car travel.

Notable is a three-mile (5-km) demonstration ride on authentic, restored streetcars, on the original roadbed (1904–1927) of the Atlantic Shore Railway, which ran from Biddeford to Kennebunkport, Kittery, and Sanford. The Town House Shop, a restoration and repair shop, allows visitors to view refurbishment and maintenance of streetcars in the collection.

Horse-drawn vehicles began providing local service in 1832 in New York City. These horsecars were a modification of stagecoaches into omnibuses, with a more rectangular body, so that they could carry at least twice the number of passengers. One example here is the closed version No. 34 built in New Utrecht, New York, for the New Bedford [Massachusetts] & Fairhaven Street Railway in 1880. Originally a "somewhat closed" omnibus, this horsecar was electrified in 1894 and became a post-office car in 1902. The operator's platform was enclosed in 1934, as required by labor laws. An ocean surge during a hurricane damaged the electric motors in 1938. In 1983 one motor was rebuilt, allowing the car to operate. Note the archaic roofline on No. 34.

Cable cars and electric trolleys became popular after a devastating cattle disease, the "Great Epizootic" (an outbreak of equine influenza), destroyed herds of streetcar horses and mules throughout the United States, Canada, and Cuba in 1872. Other effects of this epidemic included lack of commercial and passenger transportation, reduced delivery of coal for steam locomotives, decimation of the US Cavalry, and lack of fire-fighting equipment. San Francisco had the additional impediment of steep hills which horses could not handle. Thus, in 1873 the first demonstration of a rope railway occurred in San Francisco. (See San Francisco Cable Car Museum.) Other American cities adopted cable cars, including Denver, Chicago, and Seattle. The small No. 5 cable car was acquired from Dunedin, New Zealand, from the Mornington Line, after it was last used in 1957.

The first successful electric trolley is usually considered to have been built by Frank Sprague in Richmond, Virginia, in 1887–1888. His innovation was the "wheelbarrow suspension." In this scheme the motor pinion gear and axle bull gear always were synchronized because one end of the motor's casing was mounted on the wheel-axle sleeve bearings, while the other end of the casing was hung from the car's chassis. Bearings on the axle-side allowed the motor to move while maintaining perfect alignment. Bouncing of the trolley had no effect on the meshing of the gears. Sprague's scheme became the standard electric-railway system.

An early horsecar converted to electric trolley is the Museum's No. 724, constructed in 1884 for a Boston railway, the oldest known surviving equipment from the Boston transit system. For most of its life the trolley was used as a rail-grinding car, to smooth out rough areas on active rails and remove rust from stored rails.

The popularity of electric trolleys was almost immediate. In Boston, for example, the Boston Elevated Railway experimented with larger bodies for cars to accommodate growing ridership. In 1890 the standard length for larger cars was determined to be 25 feet (7.6 m). No. 396 was built for the Boston system 10 years later in St. Louis as an open car, but was enclosed sometime between 1901 and 1905. The chassis is mounted on a pair of four-wheel swivel trucks (the "Boston Special"). Eventually No. 396, the sole survivor of this style, became part of the Museum's collection in 1954, and was used in the film *The Cardinal* after refurbishment in 1963.

Open trolley cars, or "breezers," include the Museum's first acquisition, No. 31, built in 1900 by Brill for the Biddeford & Saco line, to hold 60 passengers. The chassis rides on two patented "maximum traction" trucks. Each truck included two large motor-driven wheels on an axle, plus two smaller pony wheels which face inward rather than outside. More weight was distributed on the powered axle for the maximum traction while avoiding spinning the wheels. A roofless breezer is No. 2, built in 1906 in Montréal, with the unusual feature of theater seating, where the seats are progressively higher toward the rear of the trolley. Controls are only at one end of the car, so that a U-turn is required to return in the opposite direction, rather than switching direction at the rear of the car.

Convertible cars (closed windows in winter; open sides in summer) are also on display. From Baltimore's United Railways and Electric Company hails No. 5748, built in 1917. Acquired without the trucks, special 27G gauge (5'4" [1.63 m] wide) wheels were later found for this trolley. This type of trolley car was converted to one-driver usage in the 1930s. Some were converted to single-end driving (like the roofless car described earlier). Such single-ended cars were repainted from red to yellow as a color code for easy identification by waiting passengers.

A specialized type of electric trolley used either overhead or in-ground wires for electric current. Such a method was employed in Manhattan, London, Paris, as well as Washington, DC. The Museum's No. 197 was run between the Capitol and White House in the early 20th century, where overhead wiring was banned from the Capitol environs. Therefore the Washington transit system included an underground plow collection scheme. A collector enters a slot similar to that for cable cars, to contact the wires, with special controllers without ground return circuitry.

Other interesting trolley cars include several British double-deckers, a center-entrance streetcar from postwar Berlin, and No. 1700 from Sydney, Australia. This car, used from 1926 to 1960, has eight separate compartments with entrances for each. Trackless trolleys, begun following political pressure to remove deteriorating rails from city streets, are also represented at the Museum, such as examples from Seattle, Halifax (Nova Scotia), Philadelphia, and Cleveland.

The Seashore Trolley Museum has a satellite facility in downtown Lowell, called the Seashore Trolley Museum of Lowell, Massachusetts. The National Street Car Museum at Lowell is an exhibit of this satellite campus, and is next to the Lowell National Historic Park. (See Lowell National Historical Park and Lowell Heritage State Park, Lowell, Massachusetts.) The Lowell National Historical Park runs trolley service in season around Lowell free of charge.

Address: Seashore Trolley Museum, 195 Log Cabin Road, Kennebunkport, ME 04046.
Phone: (207) 967-2900.
Days and Hours: Daily 10 a.m.–5 p.m., Memorial Day to Columbus Day; Sat. and Sun. only in May and October.
Fee: Yes.
Public transportation: Amtrak trains stop in Wells and Saco, about 10 miles (16 km) away. Local taxis travel between the train stations and the Museum.
Food served on premises: Snacks, picnic facilities.
Handicapped accessibility: Visitor center is accessible; the restoration center is not. Marginal accessibility in the exhibit barns. Contact the Museum in advance for assistance.
Website: www.trolleymuseum.org

Massachusetts

BOSTON

Museum of Science

The contemporary granite building that houses the Museum of Science stands atop the Charles River Dam, between Boston and Cambridge, in an area called Science Park. It is the world's first museum to explore all sciences in one facility, and one of the few that still provides exhibits in so wide a range of sciences. Although it was founded in 1830 as the Boston Society of Natural History, its scope was broadened after World War II, resulting in the museum we know today.

In addition to more than 500 exhibits and a wide range of hands-on, interactive displays, the museum offers live demonstrations and educational programs, all designed to make people comfortable with today's scientific and technological world. The museum emphasizes observation, theory-testing, and creation of models.

Outside the main entrance is the extensive Rock Garden, which features rocks and minerals from around the world. The large samples are labeled and originate from all over the world. See petrified wood from Arizona, Egyptian granite, a boulder from the Grand Canyon, limestone from the Rock of Gibraltar, and much more.

It is not surprising that an institution that began as a natural history museum should still have a great collection of displays in the natural sciences. Here are some highlights of these exhibits:

New England Habitats explores the sights, noises, and smells of this unique region. All of these habitats are within a 150-mile (240-km) radius of Boston, yet their climate, vegetation, and wildlife vary significantly. One of the ideas studied here is life in the soil during summer and winter; that is, life of micro-organisms, what happens to seeds, and how animals live beneath the soil's surface. The exhibit includes hoof, antler, and fur displays to touch.

Arcadia National Park is in New England. The Museum presents a virtual tour of this park that includes a specimen of every bird found in New England. Insights into bird behaviors are provided. We learn of the natural predators and listen to bird language. Several interactive displays help the learning process.

The Human Body Connection provides information about human anatomy and physiology, variations of humans within the species, and evolution of the species over time. How humans are connected to other species is also explored.

The exhibit How Your Life Began starts at the beginning of human development, with a single cell. Displays include male and female reproductive systems, menstrual cycle, and growth of the human embryo, complete with heart rhythms at various stages of fetal development. Follow a baby emerging from the birth canal. Learn about the Apgar score for newborns' health, and see short videos on different types of birth.

Displays in Human Evolution examine the theory and understanding of why the body has evolved to its present form. Touchable fossil casts of human skulls and ancient footsteps of human ancestors, and taste-tests with phenylthiocarbamide (PTC, an organic compound that some people perceive as bitter) are among the interactive exhibits. Genetic makeup determines the ability to taste PTC.

In 1964 a 65-million-year-old fossil, a triceratops, was found in the Dakotas. It is only one of four nearly complete examples of these fossils publicly displayed in the world. The triceratops adds to the wonderful collection of dinosaurs at the Museum. Dinosaurs: Modeling the Mesozoic explores our ever growing knowledge of these ancient creatures. One of the prizes of this exhibit is the full-size model of *Tyrannosaurus rex* standing in the center of the display. A variety of fossils such as bones, footprints, dinosaur dung, and more helps visitors learn about these ancient creatures and their relationship with other animals that lived and are still living on Earth.

Do not forget to see the Butterfly Garden and the Bee exhibits.

Many other disciplines of science are covered in the museum. In the Thomson Theater of Electricity, observe the effects of high-voltage electricity and artificial lightning. The 2.5-million-volt Van de Graaff lightning-bolt generator is used to give live demonstrations. This lightning-bolt generator, the world's largest, was built in 1933 for the Massachusetts Institute of Technology.

The Computer Museum of Boston (closed in 1999) merged several of its exhibits with the Museum of Science, and the rest of its artifacts are now housed at the Computer

History Museum in Mountain View, California. (See Computer History Museum, Mountain View, California.) A result of this merger is a computer exhibit, Cahners Computerplace. Here visitors can build their own computer model, talk with virtual human guides, design and program animations and games, build electronic inventions, and much more.

Ecology is an important topic that the Science Museum explores by using turbines to transform wind into green energy in the exhibit Catching the Wind. The Museum has its own wind laboratory with nine turbines mounted on the Museum roof. The displays examine the factors to be considered when using turbines, including what it means when we choose one energy source over another. Wind Power is closely tied with the Sun Power exhibit.

The Museum of Science is now an accredited member of the Association of Zoos and Aquariums and is home to over 100 animals, rescued and rehabilitated from various dangerous situations.

Other exhibits, too numerous to describe in detail, cover principles of physics, animal behavior, and astronomy. See a large collection of shells, a wave tank, or visit the Special Effects Stage to experience walking on the moon or flying over Boston. Museum demonstrators present live animals and discuss their behavior on the Live Animal Stage. Exhibits change and are moved. Temporary exhibits are always opening while others close. Check for temporary exhibits on display and any changes in the permanent exhibits.

The museum has the most advanced planetarium in New England with a Zeiss Starmaster projector (the only one on the East Coast), installed in 2011, which illuminates the stars and planets with high-power LEDs (including simulating the flickering of stars), and accommodates 209 people. The Gilliland observatory, with an 11-inch (28-cm) reflector telescope, is open Friday evenings (weather permitting) to the public. On the site is the only OMIMAX Theater in New England.

Address: Museum of Science, 1 Science Park, Boston, MA 02114–1099.
Phone: (617) 723-2500.
Days and hours: Daily 9 a.m.–5 p.m., Fri. to 9 p.m. Summer hours: Sat.–Thu. 9 a.m.–7 p.m. Closed Thanksgiving and 12/25.
Fee: Yes.
Tours: Self-guided. Docent and volunteers available.
Public transportation: Accessible by subway; Green line, Science Park Station (not handicapped-accessible; check with the MBTA for alternate routes)
Handicapped accessible: Yes.
Special note: Additional fees for planetarium, OMNIMAX, and some exhibits.
Food served on premises: Café.
Website: www.mos.org/

New England Aquarium

With the opening of this facility in 1969, an entirely new idea in aquarium design came to the historic Central Wharf in Boston. Traditionally the role of aquariums was to display aquatic life as curiosities. This facility changed that role when it introduced habitat and large-scale exhibits with broad collections of exotic and familiar water life.

The centerpiece exhibit of this aquarium is the Giant Ocean Tank. This cylindrical environment, four stories high and holding approximately 200 000 gallons (750 000 L) of salt water heated to 74°F (23°C) re-creates a Caribbean coral reef. Fiberglass and resin models of 35 major species of coral and sponges found in the Caribbean simulate caves, coral forests, and habitats for hundreds of reef fish, sharks, turtles, moray eels, and other tropical animals. Follow a spiral ramp that encircles the tank to view sea creatures and their habitat at various depths through some 52 windows at a variety of angles.

As visitors walk along a ramp surrounding the Giant Ocean Tank, they can also view the Penguin Tray populated with penguins, which surrounds the base of the Giant Ocean Tank. The penguin population consists of 80 birds from three species of warm-water birds living in the 150 000 gallon (570 000 L) tank of the Penguin Tray. These rockhopper, little blue, and African penguins (also known as black-footed penguins) are unusual because they breed in warmer temperatures than is usual for most other penguins. Several artificial islands within the exhibit are home for the penguins.

Three levels of exhibit space surround the Atrium, each with a different focus. The Temperate Gallery displays such interesting fish as the Goliath grouper, usually found around coral reefs in waters up to 165 feet (50 m) in shallow tropical waters. Also on view are some ancient fish, sea dragons, several coastal habitats, and a variety of schooling fish.

New England river systems and South American water systems are compared in the Freshwater Gallery. Exhibits here include piranhas, anacondas, electric eels, and Atlantic salmon.

The Gulf of Maine Gallery explores the waters from Cape Cod to Boston Harbor with six habitats. This cold-water world is filled with lobsters, cod, goosefish, and giant sea stars.

A look into the six tanks of the Amazon Rainforest provides a view into one of the world's most diverse habitats. On display are several hundred live plants and fiberglass replicas of huge rainforest trees, vines, and termite mounds. The tanks are populated with piranhas, electric eels, stingrays, poison darts, cardinal tetra fish, and more.

Northern Waters of the World Gallery explores the relationships between the Pacific Northwest marine life and that of New England marine life. Such creatures as shore-birds, lobsters, goosefish, Giant Pacific Octopus, and much more can be viewed here. To see colorful tropical fish like cuttlefish, venomous fish, lionfish, scorpion fish, and living corals, enter the Tropical Gallery.

At the tide pool, known as Edge of the Sea, visitors are encouraged to handle aquatic life. Sea life in the pool includes sea stars, horseshoe crabs, sea urchins, snails, and hermit crabs. The New England Aquarium was the first of its kind to have a touch tank, and it set a standard for other aquariums.

This facility also delights visitors with additional exhibits that include such interesting sea creatures as sea dragons and moon jellies.

Exhibits begin outside of the Aquarium with some harbor seals on the harborside terrace. These sleek mammals and their antics may be viewed without charge. Next to the aquarium building is the floating Marine Mammal Pavilion, where trainers demonstrate the behavior of dolphins and sea lions. The aquarium offers whale-watching trips from April through October. On the four-and-a-half-hour expeditions, learn about the ecology of Boston Harbor and Massachusetts Bay, make field observations, and get hands-on experience in natural history and boat navigation. Because this saltwater region is rich in plankton and fish life, it is an excellent feeding ground for humpback whales, finback whales, right whales, minke whales, and white-sided dolphins. On the trip, herring gulls, greater shearwaters, and other seabirds that follow whales are often visible.

The Aquarium Medical Center, opened in 1997, is a working animal hospital for sick or stranded animals from New England's shorelines and the Aquarium itself. The Medical Center is the first such exhibit open to the public. Features include: diagnostic testing and imaging, an examination and treatment room, a critical-care unit, plus interactive stations and video monitors for the public to learn more about caring for sick animals.

The West Wing was constructed in 1998. This steel-and-glass structure includes the outdoor exhibits, changing exhibits, ticketing, gift shop, café, and lobby.

Address: New England Aquarium, 1 Central Wharf, Boston, MA 02110.
Phone: (617) 973-5200.
Days and hours: Mon.–Fri. 9 a.m.–5 p.m.; Sat.–Sun. 9 a.m.–6 p.m. Closed Thanksgiving and 12/25.
Fee: Yes.
Tours: Self-guided; guided.
Daily events: Animal feeding, demonstrations, lectures, and more.
Mass transportation: Subway; nearest stop is MBTA Blue Line Aquarium Station, a short walk from the Orange and Green Lines' Haymarket Station. Within walking distance of the North End, Government Center, and Financial District. Also available is taxi service.
Handicapped accessible: Yes; wheelchairs for rent.
Food served on premises: Café and picnic area.
Special note: Whale-watching cruises offered for an extra fee at limited times from April through October; reservations required. IMAX* theater has an extra fee.
Website: www.neaq.org

BROOKLINE

Frederick Law Olmsted National Historic Site

Landscape architecture—the integration of architecture, engineering, and gardening—was defined in the United States by Frederick Law Olmsted (1822–1903). Olmsted, who lived and worked on this property in Brookline, just outside Boston, starting in the late 1870s, left a significant imprint on the look of America. He is considered the father of American landscape architecture and urban planning.

During the 1850s, when he traveled to England and Europe, Olmsted was deeply impressed by the public areas, parks, and estate grounds that he saw. He became convinced of the need to preserve open green spaces in the growing US cities, which would maintain the public's health and sanity and encourage social interaction between the classes. Ultimately, he thought, open green spaces would help safeguard American democracy.

The US National Park Service, established in 1916, is based in part on his principles. For a time Olmsted managed an estate, ranch, and mine in Mariposa, California, not far from Yosemite Valley. Here he realized the importance of safeguarding natural wilderness areas, and soon became a major force behind the creation of state and national parks.

Olmsted is most famous for creating urban parks all over the country. To build them he drained swamps; blasted rock; excavated land to make hills and valleys; installed water systems, roads, and bridges; and planted trees and shrubs. Central Park in New York City, which he planned jointly with French landscape architect Calvert Vaux, is his first and best-known design. He also designed the grounds for hospitals, colleges, government buildings, and other institutions.

Olmsted, who believed strongly in the need for suburbs, created a new concept for these communities: They should be in easy reach of urban centers, allow for the inhabitants' individuality, and have many gardens. Rather than follow a gridiron pattern, the layout should preserve the land's natural features. Communities designed according to his principles extend from Illinois to Georgia. (See Washington Park Arboretum, Seattle, Washington; Smithsonian's National Zoo and Conservation Biology Institute, Washington, DC; Planting Fields Arboretum, Oyster Bay, New York; and Wildlife Conservation Society, New York, New York.)

By the time Olmsted moved to this property, which he bought in 1883 and named Fairsted after his ancestral home in England, most of his work was in the Boston area. The history of this small estate dates to 1722, when it was settled by Dudley Boylston. A farmhouse built on the land in 1810 became Olmsted's home and office.

After Olmsted retired in 1895, the firm continued under his stepson, John Charles, and his son, Frederick Law, Jr. Several additions to Fairsted between 1889 and 1925 made it a rambling complex. It remained the firm's principal office until 1980, when the US National Park Service acquired it.

The 36-room house and office complex sits on less than two acres (0.8 hectare) of grounds landscaped by Olmsted over a century ago. The facilities have been restored to look as they did in the late 1920s, when the firm was at its peak. A major renovation to the estate was performed to address safety issues, between 2005 and the fall of 2007.

A short introductory film about the site's history and Olmsted's work begins the tour. The visitor then walks through the offices, dark-wood-paneled drafting rooms, photography developing room, and other areas, which contain original furniture and equipment. There is a Wagenhorst copier (Youngstown, OH) for copying blueprints using sunlight. The blueprint would be put on specially treated paper and placed on a large glass table that was rolled onto the porch, where sunlight developed the chemicals on the page. Also on view are some of the 140 000 documents (now completely catalogued) related to the 5000 landscape projects the firm completed: plans, drawings, photographs, plant listings, financial records, lithographs, designs, and correspondence. They provide a rare look at the development of a professional office.

The estate has almost 200 varieties of plants, arranged according to Olmsted's landscaping principles. Represented here are the meadow, wild garden, and woodland, with transitional elements, just as he used in the many city parks for which he is so famous. The focal point of Fairsted's meadow is the elm, a signature tree he often planted to indicate that a site is an Olmsted creation.

Address: Frederick Law Olmsted National Historic Site, 99 Warren Street, Brookline, MA 02445.

Phone: (617) 566-1689.

Days and hours: Tours given 10 a.m.–3 p.m. Fri., Sat. Other times by appointment.

Fee: Free.

Tours: Guided.

Public transportation: Accessible (with a short walk) by bus or subway. Contact the MTA.

Handicapped accessible: Very limited.

Food served on premises: None on the premises.

Website: www.nps.gov/frla/

CAMBRIDGE AND BOSTON

Harvard Museums of Science and Culture, Warren Anatomical Museum, and Arnold Arboretum

Harvard, the oldest university in the United States and one of the most prestigious, operates several museums that are opened to the public. These collections of art, archaeology, technology, natural history, mineralogy, and other materials are intended to be more than thematic groupings. Their purpose is to present a wider view of the human race by showcasing scientific and artistic achievement. For our purpose we focus on the science museums at Harvard.

The Harvard Museum of Natural History

In 1998 three research museums consolidated their public galleries to form the Harvard Museum of Natural History. These museums consist of the Museum of Comparative Zoology, the Harvard University Herbaria, and the Mineralogical and Geological Museum. The mission of the newly formed facility is to educate the public in its comprehension of the natural world, its value, and our place in the world in which we live. The natural history museum is housed in one complex that is a key repository of science-related information.

The Museum of Comparative Zoology was the first great university museum in the nation. It was established in 1859 by Louis Agassiz, the Swiss-born scientist who believed in learning by observing nature rather than relying on written authority, as was traditional. This museum, then popularly called the Agassiz Museum, began with his personal collection of artifacts, and now oversees more than 21 million specimens.

Harvard Museum of Natural History. © President & Fellows, Harvard College.
Courtesy of Harvard Museum of Natural History.

Although Agassiz never accepted Charles Darwin's theory of the origin of the species, the museum went on to become one of this century's leading centers of research on evolution. Today's researchers also study ecology and sociology, both of which are based on evolutionary principles and theory. The public portion of the museum explains the evolution of animals, their habits, their similarities, and their differences. Exhibits range from early fossils to today's live examples. Among the museum's rarer specimens, some of which are usually on display, are the largest fossil turtle ever found; the extinct marine reptile, Kronosaurus; George Washington's pheasants; ancient reptile eggs; and the Harvard mastodon, found by a farmer in his field in Hackettstown, New Jersey, in 1844. A group of Boston-area citizens donated money to bring the mastodon's bones to Harvard College.

The Harvard University Herbaria, founded in 1858 as the "Museum of Vegetable Products," contains a unique exhibit (in the Harvard Museum of Natural History) known as the garden in glass, or formally as the Ware Collection of Blaschka Glass Models of Plants. Four thousand lifelike models of more than 840 plant species were made out of glass between 1887 and 1936 to illustrate botanical principles to Harvard students. Before that time, teaching models were fashioned of crude papier-mâché or wax. These beautiful, scientifically accurate pieces of flame-worked glass were crafted by Leopold Blaschka and his son Rudolf in Hosterwitz, a town near Dresden, Germany. Some were made with colored glass and others were painted with a thin wash of colored ground glass or metallic oxides that were fused to the model. The Herbaria also has a rare collection of Precambrian plant fossils gathered from across the world. They are evidence of the earliest life on earth yet discovered.

The extensive collections in the Mineralogical and Geological Museums were begun modestly when Professor Benjamin Waterhouse, using "half a bushel of rocks," started teaching earth science in 1788. Today the museum's minerals, rocks, ores, and meteorites represent virtually all types and are world-renowned. Harvard Museum of Natural History Earth and Planetary Sciences exhibits popular with the public include the faceted gems, mineral crystals, carved rocks and gems, and the 1642-pound (744.8-kg) amethyst geode. Also popular is the meteorite collection. Exhibits feature specimens up to 4.3 billion years old, and a simulation globe focusing on climate change through the history of the Earth.

Peabody Museum of Archaeology & Ethnology
Attached to the Harvard Museum of Natural History is the Peabody Museum of Archaeology & Ethnology. This facility was founded in 1866 and is the oldest museum in the Western Hemisphere that focuses singularly on archaeology and ethnology. It displays almost 3000 artifacts from prehistoric and historic cultures worldwide. These objects are only a fraction of the more than 6 million objects the museum holds. The Hall of North American Indians and the Pre-Columbian Gallery are particular highlights. Next door is the world's largest anthropology library.

Harvard Semitic Museum
Classified with science museums is the Harvard Semitic Museum. This archaeology museum, covering three floors of galleries, holds more than 40 000 artifacts of various types from museum-sponsored excavations in Israel, Jordan, Iraq, Egypt, Cyprus, and Tunisia. Teaching, research, and publishing are the purposes of this museum.

Collection of Historical Scientific Instruments
The Collection of Historical Scientific Instruments (CHSI) at Harvard University is housed in the university's Science Center. These instruments, from all over the world, have been gathered by the university since 1672 and now contain more than 20 000 items from the 15th century to the present. Many of the early artifacts stem from a restocking of teaching equipment after a fire in 1764, and were purchased with the aid of Benjamin Franklin, who was living in London at the time. The "official" date of the collection, however, is around 1947, begun under the direction of David Pingree Wheatland (1898–1993), a Harvard alumnus and staff member of the Physics Department. Since 1987 this collection has been under the auspices of the Department of the History of Science to be used for the purposes of teaching and research. Instruments include disciplines such as astronomy, navigation, horology, surveying, geology, calculating, physics, biology, medicine, electricity, and more, presented in glass cases shaped like prisms, an association with the refraction of science into multiple fields. Many are one-of-a-kind or belonged to a major scientist, such as Galileo and Volta. For example, the collection has a geometrical and military compass manufactured by Galileo's instrument maker, Marc'Antonio Mazzoleni, about 1604. On display is an astronomical regulator tall clock, ordered by Franklin for Harvard, built in 1765, and used by John Winthrop to observe the 1769 transit of Venus. An unusual piece of equipment designed in 1890 by physiologist Henry Bowditch illustrates the "waterfall illusion," when hikers, after staring at waterfalls, look away toward nearby cliffs and perceive the rocks rising upward. The fabric used was from a professor's bathing suit. Don't miss the large section of the Mark I electromechanical computer, the first programmable machine, finished in 1944, and now displayed in the Science Center's central atrium. The CHSI mounts changing exhibits on a regular basis.

Arnold Arboretum and Warren Anatomical Museum
Harvard University also has several off-campus science sites that are opened to the public. Woody plants from temperate climates, displayed naturalistically in a landscape designed by Frederick Law Olmsted (see Frederick Law Olmsted National Historic Site, Brookline, Massachusetts), are the focus of the Arnold Arboretum, covering 281 acres (114 hectares). The Fisher Museum at the Harvard Forest has, among its treasures, 23 dioramas on the history of conservation and management of central New England forests. The Warren Anatomical Museum, located within the Harvard Medical School's Countway Library of Medicine, is the Center for the History of Medicine. It has one of the world's leading collections on the history of healthcare and medicine, which include many medical subspecialties.

Addresses: Harvard Museum of Natural History, 26 Oxford Street, Cambridge, MA 02138. Peabody Museum of Archaeology & Ethnology, 11 Divinity Avenue, Cambridge, MA 02138. Harvard Semitic Museum, 6 Divinity Avenue, Cambridge, MA 02138. Warren Anatomical Museum, Countway Library of Medicine, 10 Shattuck St., Boston, MA 02115. Arnold Arboretum, 125 Arborway, Boston, MA 02130–3500.
Phone: (617) 495-8149. For information about the scientific instruments call (617) 495–2779.
Days and hours: Harvard Museum of Natural History and the Peabody: Daily 9 a.m.–5 p.m. Closed 1/1, Thanksgiving, 12/24, and 12/25. Collection of Scientific Instruments and Harvard Semitic Museum: Mon.–Fri. and Sat. 11 a.m.–4 p.m.; Sun. 1 p.m.–4 p.m., closed

on various holidays. Arnold Arboretum: Daily, sunrise to sunset; Visitor Center is closed Wednesdays. Warren Anatomical Museum: Mon.–Fri. 9 a.m.–5 p.m., closed on certain Harvard holidays.

Fee: Yes. (No fee for Harvard Semitic Museum, Collection of Scientific Instruments, or Warren Anatomical Museum.)

Tours: Self-guided. Guided tours by prior arrangement (for a fee).

Public transportation: The Red line of the MBTA stops at Harvard Square at the edge of the Harvard campus, then a very short walk to the Museums on campus.

Handicapped accessible: Yes.

Food served on premises: The Science Center and Northwest Buildings have cafeterias; the town of Cambridge has many local restaurants; food trucks are on the Science Center Plaza.

Websites: www.hmnh.harvard.edu/ (Harvard Museum of Natural History); www.peabody. harvard.edu/ (Peabody Museum of Archaeology & Ethnology); semiticmuseum.fas.harvard.edu/ (Harvard Semitic Museum); chsi.harvard.edu/ (Collection of Historical Scientific Instruments); arboretum.harvard.edu/ (Arnold Arboretum); www.countway.harvard.edu/chom/warren-anatomical-museum (Warren Anatomical Museum).

LOWELL

Lowell National Historical Park and Lowell Heritage State Park

Francis Cabot Lowell, a Boston industrialist, saw a great opportunity to develop a textile industry in Massachusetts after examining the English industry in 1813. After looking for an appropriate site for a factory, he settled on a location at the confluence of the Concord and Merrimack rivers. The excellent waterway and 32-foot drop at the area's powerful Pawtucket Falls were major enticements. The town of East Chelmsford eventually came to be called Lowell, Massachusetts. Construction of the textile mills and power canals began in 1822.

As the mills brought prosperity to this small rural town, Lowell grew into the nation's first great industrial city. To preserve what US National Park Service literature refers to as "historically and culturally the most significant planned industrial city in the United States," a good part of the city-center has become a national park and living-history museum. In this park the Industrial Revolution in America is commemorated.

Lowell's industry was the first to combine labor, capital, and technological innovations in manufacturing to mass-produce cotton cloth on a large scale. While Slater Mill in Pawtucket, Rhode Island, began earlier (in 1793), Slater Mill only produced the thread; the actual weaving was done by local villagers in their own homes. (See Old Slater Mill National Historic Landmark, Pawtucket, Rhode Island.)

A good starting point for a tour is the National Park Visitor Center, located in what was once the Lowell Manufacturing Company mill complex. For a good introduction to the park, there is a multi-image video and several additional introductory exhibits. Register here for regularly scheduled tours, find out other information about the park, and other services in the area helpful for your visit.

To get around town, visitors can walk, ride a trolley, or take a canal boat. The Middlesex Canal, a 27-mile (44-km) barge canal, connected the Merrimack River with the port of Boston and the Charles River. It was built between 1795 and 1803. The Pawtucket Canal was designed for transportation around the Pawtucket Falls. The Merrimack became the feeder canal for a 5.6-mile (9-km) long system of power canals based around the falls. Lowell's vast waterway system was also the primary feeder and power canal. The Pawtucket Waterfall was harnessed by building 5.6 miles (9 km) of a canal system that channeled a continuous surge of kinetic energy, which at its peak in 1888 ran the machinery in 175 mills that produced 4.7 million yards (4.3 million m) of cloth a week. Of the five million square feet (460 000 m^2) of historic mills, almost 80% have been rehabilitated. Here visitors have an in-depth look at the 19th-century textile industry and how the waterpower of the Merrimack River helped to make this industry happen.

Today the canal is used for recreation and hydropower. Parts of the canal on view are the guard locks with the Great Gate, the sluice gatehouse, and the Pawtucket gatehouse, the canal system's largest gatehouse.

Available to visit at the Suffolk Mill are some of the housing in which workers lived, and a restored 19th-century textile mill, Spindle City. The Suffolk Mills Turbine Exhibit shows how waterpower is transferred through a turbine to a power loom that produces cotton cloth. Learn of the changes in waterpower technology, of the work Lowell engineers did, and how important this work was and the impact it still has. For example, in the late 1840s, James Francis, an engineer in Lowell, invented a high-efficiency turbine that replaced the existing waterwheels. The turbine's inlet is shaped like a spiral (the "volute"), with guides that direct the water to the turbine wheel (the "runner") at a tangent. He also worked out new methods of calculation to improve turbine design and engineering.

Kirk Boott, for whom the Boott Mills are named, played a large role in the establishment and operations during the early years of the mill. It is here, in this museum, that we learn of the history of the mill, its operations, and the equipment used. The mill was closed in 1955 after more than 120 years of operation.

To reach Boott Cotton Mill, cross the same bridge that has spanned the Eastern Canal for more than 100 years. This complex explores the use of water power, steam power, and electric power, all of which were in use during the life of the mill. The original part of this mill complex was built in 1836. An addition built in 1873, Boott Mill #6, is a museum devoted to industrial history with exhibits that feature authentic mill equipment.

The Weave Room is an exhibit that recreates a 1920s weave room at the height of its productivity. The exhibit covers one entire floor with a collection of 88 industrial-grade Draper looms (built in Hopedale, Massachusetts, in 1913 and 1920) running at high speed. A Crompton & Knowles shuttle-changing loom for weaving plaid patterns is also on display. Thus visitors can see, feel, and hear what was actually happening during the weaving process. This area is one of the largest industrial exhibits in the nation. Tour guides and videos help visitors to understand the manufacturing process in the room.

Other exhibits at the Boott Mill Museum include interactive exhibits and video programs about the Industrial Revolution, labor, and how Lowell came to be a great industrial city, how it declined, and how it has begun to again be a prosperous city. Visitors learn of the impact of industrialization and the interrelated roles of technology, labor, economics, government, and society. A multimedia presentation, The Great Debate, explains how Alexander Hamilton and Thomas Jefferson saw the nation's industrial future.

To explore the human aspects of the Industrial Revolution in Lowell, plan a visit to the Patrick J. Morgan Cultural Center. Learn of the mill girls and immigrants who worked in the mills.

Although Lowell declined in the 1920s as the textile industry moved south, it now has a diversified commercial base and is prospering again. The Work in the 21st Century exhibit explores the role of electronics and other high-tech industries, and looks ahead to the effect they will have on the region.

The National Street Car Museum at Lowell is a satellite facility of the Seashore Trolley Museum of Lowell and is next to the Lowell National Historic Park.

Addresses: Lowell National Historic Park (LNHP), 167 Kirk Street, Lowell, MA 01852; Lowell Heritage State Park (LHSP), 160 Pawtucket Boulevard, Lowell, MA 01852.
Phones: LNHP, (978) 970-5000; LHSP, (978) 369-6312.
Days and hours: Daily 9 a.m.–5 p.m. Closed 1/1, Thanksgiving.
Fee: Yes.
Tours: Self-guided. Guided tours by reservation (extra fee; sign up the same day as your visit).
Public transportation: Commuter rail from Boston (contact the MTA), then shuttle bus.
Handicapped accessible: Yes.
Food served on premises: None. Restaurants are in Lowell.
Websites: LNHP, www.nps.gov/lowe/; LHSP, www.mass.gov/locations/lowell-heritage-state-park

NANTUCKET

Maria Mitchell Association

The Maria Mitchell Association of Nantucket has two goals: to promote the legacy of Maria Mitchell, and to educate the public about the beauty and history of Nantucket. With these aims, the Association operates two observatories, a natural-science museum, an aquarium, and a research center, and preserves Maria Mitchell's birthplace.

Maria Mitchell (1818–1889) was a scientist, an astronomer, and the first professional American female astronomer. This daughter of Nantucket was born into a seafaring culture were men spent prolonged time at sea hunting whales. Thus the women managed the home and were very involved with Nantucket government, which gave the women a great deal of independence and equality. A distant relative of Benjamin Franklin, Maria was brought up in a Quaker household where the education of women was considered important. Although she finished her formal schooling at age 16, she became the first professor of astronomy at Vassar College (male or female) in 1865, and then director of Vassar College observatory. She taught there until she retired in 1888.

Maria's father, William Mitchell, was a schoolteacher and later ran his own school. As a child she helped her father with surveying and navigation, hence her interest in astronomy developed. When she was 12, she calculated the position of the family home by observing a solar eclipse. At 14 she performed vital navigational computations for whaling journeys of the sailors.

Her girlhood home, the Historic Mitchell House, is set up in the style of the time period in which it was built (1790), and the time in which she lived there (1818–1836). The structure displays many of her personal items. On exhibit are beer mugs, opera glasses, and more. The Dollard telescope, a 2.75-inch (70-mm) brass refractor from the 1880s, used by Maria when she discovered a comet, called "Miss Mitchell's Comet," or C/ 1847 T1—which made her famous—is one of these items displayed. For this discovery, King Frederick VII of Denmark presented Mitchell with a gold medal in 1848, and she

Mitchell House.
Courtesy of Maria Mitchell Association.

became the first woman elected to the American Academy of Arts and Sciences. Also exhibited is her mid-19th-century 5-inch (130-mm) Clark telescope, as well as some other scientific equipment. At the house is Maria's 1869 membership certificate to the American Philosophical Society and William's membership certificate to the American Academy of Arts and Sciences, dated 1842.

In 1836, as the cashier, William Mitchell managed the Pacific National Bank located on 61 Main Street. With employment came an apartment on the second floor of the Bank. The Mitchell family lived here from 1836 to 1861. William set up an observatory on the roof of the bank. It is here that Maria discovered the comet on October 1, 1847. This bank is still in existence and remains a working bank, a short distance from Maria's childhood home. (The bank is not part of the Maria Mitchell Association.) The Research Center in the Mitchell House stores Maria Mitchell's papers and other family-related items. It is not opened to the public, but only for research purposes.

The Vestal Street Observatory is mainly geared toward providing public lectures, public programs, and special events. On the property is a scale model of the Solar System, information on how to operate a sundial, and (weather permitting) you can observe a sundial. This facility does hold a 17-inch (430-mm) PlaneWave Dall-Kirkham telescope, installed in 2008.

The Loines Observatory has two domes: one completed in 1968 and the other finished in 1998. Each dome holds a telescope. The earlier instrument is the 7.5-inch (190-mm) Alvan Clark refractor, built in the mid-1800s. Clark was descendant of a Nantucket whaling family; his company, Alvan Clark & Sons, built many telescopes still in use today. Mitchell used the Clark telescope at the Vassar observatory and brought it to Nantucket when she retired. The other instrument is the large 24-inch (610-mm) Ritchey–Chrétien reflector for research. The Loines is opened for public tours, programs, lectures, and special events.

The Association also operates an aquarium and natural-history museum. Both of these facilities are small and focus on children. Each museum contains some interesting collections, and research is an important component of both facilities.

The Natural History Museum has a collection of over 900 bird specimens representing most species on Nantucket. Some are displayed in seasonally changing exhibits. Research continues on such creatures as owls, ospreys, and American burying beetles, as well as snakes, spiders, and more. Biodiversity is an important concern to the museum.

The aquarium displays are of local species and change yearly, based on what is collected during the summer. Research is done on local life, particularly Nantucket Bay scallops and horseshoe crabs. Biodiversity is important to this facility as well.

Address: Loines Observatory, 59 Milk St.; Vestal Street Observatory, 3 Vestal St.; Historic Mitchell House, 1 Vestal St.; Aquarium and Museum Shop, 28 Washington St.; Natural Science Museum & Museum Shop at Hinchman House, Corner of Milk and Vestal Sts., Nantucket, MA 02554.
Phone: (508) 228-9198.
Days and hours: All facilities are open from June 10 to Labor Day. Loines Observatory, call for calendar; Vestal Street Observatory, Tours Mon.–Fri. 2 p.m., Sat. 1 p.m.; Historic Mitchell House, Mon.–Sat. 10 a.m.–4 p.m.
Fee: Yes.
Tours: 2 p.m., Mon.–Fri., from mid-June until Labor Day. Some properties are only available by guided tour.
Public transportation: Shuttle (Nantucket Regional Transit Authority), and taxi service from airport to downtown. Note that local streets are cobblestone.
Handicapped accessible: Limited access in some cases; contact individual facilities for details.
Food served on premises: None; there are many eating options nearby in town.
Website: www.mariamitchell.org/

Whaling Museum

Nantucket, an island 30 miles (48 km) south of Cape Cod, received its name from Wampanoag, an Eastern Algonquian language of southern New England. Nantucket translates as "in the midst of waters" or "faraway island."

This island was settled by the English in about 1659. In 1835 Obed Macy wrote that pre-1675 island records indicate a type of whale called "scrag" entered the harbor; settlers pursued and killed the whale. And so the whaling industry began. From the description written in 1725, the scrag whale may have been a gray whale, well-known on the West Coast of America.

The 18th century saw the organizing of the whale industry in Nantucket. The first sperm whale was killed in 1712, and by 1715 there were six sloops engaged in whaling. At the end of the 18th century, whaling was a major business. Between 1750 and 1850, whaling in Nantucket became a global business.

Sperm whales seemed to favor this section of the Atlantic Ocean and so their population was high. Spermaceti, a secretion mostly of wax esters and triglycerides used to produce a brightly burning oil, is produced in the head of the many species of whales; however, the sperm whale produces a much greater amount. Because the oil made from the spermaceti was said to burn brighter and last longer than any other whale oil in the world, these whales became a major catch and made Nantucket an important whaling center. The availability of whale oil also fueled the growth of oil refineries and candle factories in the town.

Nantucket streets were paved with cobblestone in 1838 and this is how the streets remain. In spring the mud of New England became a problem for the growing population to get around. Heavy wheeled vehicles that transported the whale-oil industry's supplies and products would sink into the muddy roads, so the cobblestones assisted the movement of goods.

The industry here began to decline, however, because of several factors: the harbor began silting in, thus preventing access for many large ships; railroads on the mainland provided a cheaper and more efficient way to transport whale oil than ships; and New Bedford began to gain importance as a whaling port. (See New Bedford Whaling Museum, New Bedford, Massachusetts.) The "Great Fire" of 1846 accelerated Nantucket's downward slide.

The "Great Fire" of 1846 burned through Nantucket Town. It was fueled by both the whale oil and the wood of the docks and warehouses. The town was devastated. People were homeless, poverty was extensive, and many people left the island.

Shortly after the fire the Mitchell family built a candle factory (out of brick) even though the whaling industry in Nantucket was on the decline. This structure eventually became home for the Nantucket Whaling Museum.

Interior of Nantucket Whaling Museum.
Courtesy of the Nantucket Historical Association.

Two businessmen, William Hadwen and Nathaniel Barney, purchased the candle works and operated it into the 1860s. The last whaling ship left Nantucket in 1869. The building was used for a variety of businesses until 1929, when it was converted to the Nantucket Historical Association's Whaling Museum.

Exhibits present the history and technology of Nantucket, the diverse people here, business, island characters, and most of all, candle-making. This major exhibit provides an overview of the production of candles. The centerpiece is the two-story original beam-press, which squeezed oil from sacks of spermaceti. It is the only beam-press in the world that is complete and still in its original site. The original foundation of the tryworks is also on display. The tryworks (a furnace-like structure) were a major feature on the whaling ships. Detailed records and illustrations of how this machinery worked are included in the museum's exhibit, whose design is "in the tradition of the Mercer Museum in Doylestown, Pennsylvania," as quoted—as of this writing—from the Whaling Museum's website. (See The Mercer Mile in Doylestown, Pennsylvania.)

The rooftop observation deck overlooks the Nantucket harbor. A main feature of the museum is a 46-foot (14-m) skeleton of a baby bull sperm whale, complete with teeth. It is positioned as though diving through water. This sea animal beached itself on a Siasconset beach, Nantucket, in 1998.

Standing majestically by the elegant staircase is the second-order Fresnel lens made in 1849, put in place in 1850, originally 150 feet (46 m) above sea level, in the Sankaty Head lighthouse, at Siasconset Nantucket. August-Jean Fresnel, a French physicist and engineer, has been given credit for developing a multipart lens for use in lighthouses. It used a central lens, called a bull's eye, surrounded by concentric rings of glass projecting beyond one another. This type of lens was much thinner and lighter by making it in separate sections, mounted in a single frame. The design reduced the size and weight while making available a large aperture to collect more of the source's emitted light by using reflectors. The Fresnel lens allowed five times more light to be transmitted to sea than the conventional light made with concentric lenses. This particular lens was manufactured by Henri Lepaute of Paris and cost $10 000. It sat on a rotating clockwork-driven platform. The mechanism required manual winding several times a day until an electric motor was installed in the 1930s. The lighthouse was the first in the United States to be equipped with the Fresnel light as original equipment. The lamp consumed nearly 400 gallons (1500 liters) of whale oil a year with one wick, until replaced with a kerosene lamp in 1885. The light could be seen as far as 20 miles (32 km) away. It was used for a hundred years until it was replaced in 1950 by a modern rotating beacon.

Another interesting exhibit hall is the Scrimshaw Gallery. On display are a variety of engraved images on whale teeth, bone, and walrus tusks. Scrimshaw was a folk art practiced by the men aboard whaling ships during the 19th century. The collection in this museum is one of the most important of its type in the world. Other collections include the lightship-basket collection, decorative arts, paintings, samplers, portraits of seafaring men and women, and items of Nantucket history and prehistory. The 1881 Nantucket Town clock's mechanism is also on display.

The Research Library holds more than 5000 volumes, 50 000 photographs, and a variety of archival documents that include ships' logs, account books, family papers, and scrapbooks.

Address: Whaling Museum, 13 Broad Street, Nantucket, MA 02554; Research Library, 7 Fair Street, Nantucket, MA 02554.

Phone: (508) 228-1894.

Days and hours: Mid-Feb. to early Apr.: Sat. and Sun. 11 a.m.–3 p.m.; mid-Apr. to late May: daily 11 a.m.–4 p.m.; late May through Oct.: daily 10 a.m.–5 p.m.; Nov. 1–Nov. 23: Sat. and Sun. 11 a.m.–4 p.m.; hours vary in Dec. Closed Thanksgiving and the day after Thanksgiving, 12/25, 1/1, Jan. through mid-Feb.

Fee: Yes.

Tours: Self-guided and guided. Check with site for times. We recommend tours to enhance your visit.

Public transportation: Shuttle (Nantucket Regional Transit Authority), and taxi service from airport to downtown. Note that some local streets are cobblestoned.

Handicapped accessible: Yes.

Food served on premises: None; there are many eating options nearby in town.

Website: www.nha.org/sites/

New Bedford

New Bedford Whaling Museum

For 200 years, from the 18th to the 20th centuries, New Bedford was America's whaling capital. Herman Melville contributed to this city's fame when he published *Moby-Dick* in 1851, which described the local whaling industry and New Bedford at its economic peak. The Whaling Museum is devoted to the history of whaling during the age of sail, as well as modern whale ecology and conservation. It is across the street from Seamen's Bethel, where Melville worshipped and which is mentioned in *Moby-Dick*. The church and museum are in the core of a restored historic district on Johnny Cake Hill, just two blocks from the largest fishing fleet on the East Coast.

The museum, the largest facility in America concerning the history of whaling in America and the rest of the world, was founded in 1903 by the Old Dartmouth Historical Society. It is made up of several buildings along a block, containing exhibit galleries, a library, and more. Many of the materials on display are from Old Dartmouth, the original name for the region settled in 1660 that now comprises the towns of Dartmouth, Fairhaven, Westport, Acushnet, and New Bedford. Visitors enter through the Wood Building, which contains scrimshaw, paintings, ship models, harpoons, navigation equipment, and other exhibits.

With over three-quarters of a million items, it claims to have the world's most comprehensive collection of art, artifacts, original documents of whaling history, technology, industry, books and ethnographic materials relating to whaling, maps and charts, as well as clothing and textiles, records of business, financial records, and a large, diverse social history collection.

One of the important exhibits in this museum are five completely articulated whale skeletons. One of these skeletons is a rare fetal skeleton of the endangered North Atlantic right whale. The William A. Watkins Collection of Marine Mammal Sound Recordings can be heard in the museum.

The Jonathan Bourne Building was constructed in 1916 to house the half-scale model of the bark *Lagoda*. Here the public can board this whaling ship, the largest and most

The New Bedford Whaling Museum occupies an entire city block overlooking the waterfront in the historic heart of New Bedford, Massachusetts.
Courtesy of the New Bedford Whaling Museum.

impressive of the models on display, and see the tools and gear that whalers used. The room also holds two full-sized whaleboats, many harpoons, and exhibits concerning whaling in the Azores and Cape Verde.

The Bourne Building often displays sections one of the longest paintings of this type in the world, *Panorama of a Whaling Voyage 'Round the World*, at 1275 feet (389 m) long. It was created in 1848 to document the life of the intrepid sailors of this period.

The art collection, displayed throughout the museum, has works related to whaling and local history, as well as items manufactured locally or owned by local residents. This extensive collection includes works by major artists who lived or worked in local area. Much of this art was funded by the wealth the whaling industry produced in the 19th century.

At the Mariners' Home there is an exhibit on *Moby-Dick* as portrayed in film, the history of the Mariners' Home, domestic life during the age of sail, and New Bedford as a working port.

Address: New Bedford Whaling Museum, 18 Johnny Cake Hill, New Bedford, MA 02740.
Phone: (508) 997-0046.
Days and hours: Jan.–Mar.: Tue.–Sat. 9 a.m.–4 p.m., Sun. 11 a.m.–4 p.m., select galleries open until 8 p.m. every 2nd Thu. of the month; Apr.–Dec.: daily 9 a.m.–5 p.m., select galleries open until 8 p.m. every 2nd Thu. of the month. Open holiday Mondays; closed Thanksgiving, Christmas Day, and New Year's Day.
Fee: Yes.
Tours: Self-guided. Guided tours by arrangement in advance; audio tours.
Handicapped accessible: Yes.

Food served on premises: None. Restaurants are in town.
Public transportation: The city is accessible by bus, train, air, and boat, as well as a ferry from Martha's Vineyard. A bus runs in town.
Website: www.whalingmuseum.org

PLYMOUTH

Plimoth Plantation

Plimoth Plantation is a living-history museum set in the 17th century, when America's most famous immigrants landed near Plymouth Rock. This complex consists of three parts: a re-created English village set in 1627, a re-created Native American village located in this area during the 17th century, and the *Mayflower II*, the reconstructed ship that brought the pilgrims to America in 1620. Although today the word "plantation" refers to an estate where crops are raised, often by resident workers, in the 17th century it meant a colony or new settlement. The site was founded in 1947 by stockbroker Henry Hornblower II, an amateur archaeologist. The life of America's forefathers and the Native American culture they discovered in the New World are explained at Plimoth Plantation. The complex overlooks the Eel River, which flows into Plymouth Bay three miles south of downtown Plymouth. How and why these English people came is examined at the *Mayflower II* that is located at the waterfront in Plymouth.

An important stop upon venturing into the plantation complex is the Visitor Center. Two films are available; a 14-minute orientation film about the plantation and a two-and-a-half-hour film about the ocean crossing of the *Mayflower II*. Both of these films were made primarily on location at the plantation. Exhibits explore the life and work of the Europeans who came to this world and the native people who already lived here. The *Mayflower II* has an exhibit called "Provisioning a Ship." Watch craftspeople at the Craft Center using 17th-century techniques to make baskets, furniture, pottery, and more. These crafts are available for purchase in the shops.

The 1627 English village is made up of authentically reproduced buildings constructed in the 1950s. The time period is almost seven years after the settlers first arrived. Because each interpreter has taken on the mantle of a particular person who lived in the original village, the visitor hears 17 regional dialects of 17th-century England. Visitors may watch and ask questions as the inhabitants do seasonal and daily chores such as baking, cooking, building houses, or making handcrafted objects. House repair must constantly take place, and those walking about the village will see the typical "wattle-and-daub" construction, which is a grid of wooden strips called wattle daubed with mud, animal dung, clay, sand, and straw or a mixture thereof. English 17th-century buildings of this sort used wooden framing, with the wattle and daub as filler. The surgeon might be found in his house, which also served as his office, or in his medicinal-herb garden (typically plants such as horehound, savory, sage, oregano, calendula, and hyssop); people might be tending their kitchen gardens. During that era, vegetables such as onions and lettuce, and even melons were considered herbs. Many varieties of flowers popular at the time are grown at Plimoth Plantation. The livestock has been back-bred to resemble the 17th-century animals brought from Europe. The Nye Barn conserves and back-breeds livestock from around the world, such as the Wiltshire sheep, Milking Red Devon cow, and small-sized Arapawa Island goat. Tourists may visit to see the work that is being done in this area.

Also on site is Wampanoag Homesite, which is a re-creation of a Native American village located in this area in the 17th century. The buildings are also authentic reproductions inhabited by docents, dressed in typical native clothing of the period, who do all the tasks necessary to support life in this village. At the home site visitors learn about the Wampanoag people native to southeastern New England, who lived here before and during the time of the original plantation. The village contains replicas of Wampanoag *wetuash* (dwellings), gardens, and artifacts such as the home of Hobbamock, a Pokanoket Wampanoag man who, along with his family, lived near Plimoth Plantation. He and others of his village helped the Pilgrims understand the land and the people who already inhabited it. Native-American interpreters demonstrate native crafts, the making of dug-out canoes, cooking, tending gardens, making baskets, pottery, wooden bowls, stone tools, and more. The history, culture, and technology of these people from a modern perspective are discussed.

For example, the frames of *wetuash* were constructed from swamp-based saplings, for the young flexible trees are able to be bent into rounded forms. The bark for covering homes was also collected in the spring, for tree sap is running, causing the bark to be moist, flexible, and easily peeled from a tree. The sheets of bark were placed on the wooden frame in two-inch-thick (5 cm) shingles, except for the smokeholes. Bark also was used to tie the frame together. As the bark dried it shrank, tightening and securing the structure. Cattails were harvested and sun-dried. The cattails were woven into mats, acting as rain-gutters for the houses.

Modern-day craftspeople exhibit their expertise, using tools, materials, and techniques of the past, teaching the cultural realities of the objects and lifestyles of the past. These include textiles and hide-clothing and bags, hand-coiled clay cooking pots, porcupine-quill decorations, weapons, and tools. Some of these items are available for purchase in the Visitor Center. The craftspeople are very happy to answer questions and discuss production techniques. In 2002 a reproduction of a 17th-century English wood-fired pottery kiln was built, which relies on small glazed or unglazed "draw rings" placed inside to measure temperature. The rings are "drawn out" via rods when the fire is hot. Examination of the rings serves as a gauge of temperature in a pre-thermometer era.

The *Mayflower II* is a full-size reproduction of the 17th-century merchant vessel that carried 102 passengers and their food, goods, and livestock to the New World on a 66-day trip. Though the exact items the people brought are generally undocumented, a list compiled in 1630 by Francis Higginson called "A Catalogue of such needful things as every planter doth or ought to provide to go to New-England ... for one man," may be a good guide of what was on board. Higginson mentioned "Victuals for a whole year, (to be used after arrival in the colony), apparel, arms, tools, household implements, spices, books, nets, hooks and lines, cheese, bacon, kine and goats & etc."

As the visitor walks through the ship's living and working quarters, interpreters tell about life at sea and conditions in England in 1621. Passengers explain what they expect life to be like in this new land, and sailors speak of their return to England. Exhibits serve as an introduction to the ship and its passengers, and detail the centuries-old techniques used to construct this beautifully reproduced vessel, including English oak timbers, hand-forged nails, hand-sewn canvas sails, hemp ropes, and 17th-century Stockholm tar (a major Swedish export of the time), and how a 21st-century counterpart is constructed. The 17th-century reconstruction with four masts (mainmast, foremast, mizzen, sprit), and six sails was built in Brixton, England, and sailed across the Atlantic in 1957. The ship has solid oak timbers, tarred hemp rigging, and is 106 feet (32 m) long by 25 feet (7.6 m) wide. Its displacement is 236 tons (240 tonnes).

Address (GPS): Plimoth Plantation, 26 River Street, Plymouth, MA 02362; Mayflower II, State Pier (across from 74 Water Street), Plymouth, MA 02360.
Phone: (508) 746-1622.
Days and hours: Daily 9 a.m.–5 p.m.
Fee: Yes.
Tours: Self-guided.
Public transportation: Commuter train or bus from Boston (contact the MTA), then transfer to local bus or taxi to sites.
Handicapped accessible: Yes in Visitor Center and Crafts Center. Limited in historic sites.
Food served on premises: Café (some Native American foods) and restaurant at the Visitor Center.
Website: www.plimoth.org

SAUGUS

Saugus Iron Works National Historic Site

Until the late 1630s there was a great migration from England to the Massachusetts Bay Colony, but this population movement then slowed considerably. Fewer ships came in, making iron products scarce and costly. In the 1640s the colony started using its own abundant raw materials—water, iron ore, wood, and calcium carbonate—to launch its first successful ironworks, which operated on this site from 1646 to 1668. Saugus Iron Works became the first integrated ironworks in North America.

This living history museum sits on nine acres of land nestled along the along the banks of the Saugus River just 8.5 miles (14 km) from Boston. The facility examines the beginnings of the iron industry in the colonies and how iron was made. A key figure in the founding of the iron works was the son of the colony's governor, John Winthrop. John Winthrop, Jr. (the Younger), was a broadly educated man who knew alchemy and metallurgy. In the 1640s he brought the iron industry to Massachusetts by recruiting ironworkers from England who knew the latest technology. He also stimulated investors' interest in his fledgling operation by describing the colony's ample raw materials. The ever-growing shortage of wood in England was pushing up the cost of iron, making Saugus Iron Works a very attractive investment. The workers passed their skills on to the next generation. Some of these offspring relocated and introduced iron-making technology to other colonies. Saugus Iron Works was instrumental in changing North America from a Stone-Age culture to an Iron-Age culture.

Saugus used local bog ore, limonite (a sedimentary mix of iron oxide and hydroxide), as the chief source of iron. In the nearby town of Nahant the ore was found, and by burning the rock with cordwood fires, ore was obtained. The iron was then shipped up the Saugus River to the Iron works. Traditionally, iron was produced in a direct "bloomery" process. The iron never became liquid, but went from iron ore to wrought iron in one direct step. At Saugus, where pig iron (mottled iron) was produced, a more modern process was used; higher heat in the furnace (>2600°F or 1400°C) produced a liquid that was cast into long bars. A forge then converted the cast iron into malleable wrought-iron bars. Although the process was more expensive, the yield was much higher.

Saugus instituted another innovation, a slitting mill, which flattened and slit some of the wrought iron into stock for nails. Developed in the previous century, only a dozen such slitting mills are thought to have existed in the world in 1650. A slitting mill was powered by water wheels. Flat bars of iron, approximately three inches by half an inch, were the initial material. Water-powered shears cut off the end of a bar, and heated this piece of wrought iron in a furnace. Flat rollers then compressed and formed the piece into a thick plate. A second set of rollers ("cutters") sliced the iron into rods.

The blast furnace, forge, rolling and slitting mill, iron house, blacksmith shop, warehouse, and dock are reconstructions on the original sites.

This manufacturing complex demonstrates the engineering and design techniques as well as iron-making technology and operations. Visitors learn about local and overseas trade, in addition to life and work in the Massachusetts Bay colony.

The Iron Works House is the only original structure that has survived. It was a mansion house built in the late 1680s, about a decade after the iron works ceased production. The House was part of what was probably America's first company town, Hammersmith. It was not only the finest house in the community but also its social and business center. Many of the nails made at the ironworks were used to build the house, which has been restored to its original condition—a fine, colonial home. Five rooms are open to the public, three of which contain authentic and reproduction furnishings. The other two rooms have exhibits about the house and its restoration, 17th-century architecture, building hardware, tools, and artifacts excavated at the site. Included in the display is an example of a "notched bar," an analytical tool to measure the quality of the iron. In this museum visitors learn about the complex's major industrial contributions and see a 12-minute video, "Iron Works on the Saugus," that provides a good introduction to the site.

Joseph Jenckes, featured in one of the exhibits, worked at the Saugus Iron Works. In 1646 he received the first industrial patent issued in the colonies, which gave him sole rights for seven years to make saws, scythes, axes, and edge tools by waterpower and a "new invented" sawmill, probably a not a mill for sawing wood, but rather a mill for making saws themselves.

Outside visitors can walk along a half-mile (0.8-km) nature trail that crosses the property's restored tidal marsh and woodlands, where over 200 species of plants, 74 bird species, 11 species of mammals, and several species each of reptiles, amphibians, and fish (including rainbow smelt, yellow and white perch, alewives, and sticklebacks) dwell.

Address: Saugus Iron Works National Historic Site, 244 Central Street, Saugus, MA 01906.
Phone: (978) 740-1650.
Days and hours: May–Oct.: daily 9 a.m.–5 p.m.; closed Nov.–Apr.
Fee: Free. Special events, demonstrations and certain additional programs may require a fee.
Tours: Self-guided. Ranger-led guided tours lasting about 90 minutes are offered Apr.–Oct. A downloadable audio-tour podcast is available.
Demonstrations: Interpreters in period dress demonstrate and explain the various steps in making iron. Check the times of presentation.
Public transportation: From Boston take the MTA orange line to the Malden Subway Station, then catch the local bus #430 from Malden daily (except Sundays).
Handicapped accessible: Mostly.
Website: www.nps.gov/sair/

SPRINGFIELD

Springfield Armory National Historic Site

Sitting on a hill overlooking the Connecticut River is a complex of buildings on approximately 55 acres (22 hectares) that comprise the Springfield Armory National Historic Site. It holds what is probably the world's largest collection of US and foreign military small arms, plus several early industrial exhibits. Here the development of interchangeable precision manufacturing methods is traced.

The collections are housed in original armory structures, most of which are brick buildings of pre–Civil War construction. General George Washington and his Chief of Artillery, Colonel Henry Knox, selected this location for the first US arsenal in 1777 because there were preexisting transportation routes converging and it was far enough upstream to be safe from enemy attack. Supplies, a good source of water power for manufacturing, and skilled people needed for the work were also available locally. For more than 190 years the armory was a storehouse for muskets. It continued to expand as it also became one of the major places in the world for the design, development, and production of military small arms.

In addition to being a storage facility for muskets, cannon, and other weapons during the Arsenal's early years, paper cartridges were also manufactured here. The Arsenal continued to expand. Storehouses, shops, and barracks were constructed on site. However, no arms were produced until after the Revolution, and arms for future use were housed here.

By 1794, a federal decision was made to produce muskets so as not to be dependent on foreign arms, and it became the country's first armory that manufactured US military shoulder arms. President Washington selected the Springfield Armory as one of two Federal Armories (the other was at Harper's Ferry, in what was then Virginia, now West Virginia) to produce the equipment, and a new building was constructed. The complex quickly became a center for invention and development. During the mid-19th century the industrial countries of Europe came to study and buy the newest and most innovative machinery, much of which was designed by the Springfield Armory. They also came to examine the "American System of Manufacture," the mass production of arms (also called the "Armory System"). The first weapon made at the arsenal was the model 1795 musket. Popular during the Revolution, it was modeled after the Charleville French musket. In 1819 Thomas Blanchard designed a lathe that could turn out identical irregular shapes and thus make identical gun stocks. This lathe is still displayed here.

The 1840s saw the introduction of a percussion ignition system for better reliability of long arms. After the Civil War, Master Armorer Erskin Allin invented the "Allin Conversion." This process incorporated an advanced design of breech-loading into obsolete muzzle-loaders, thus extending their service. One of the most accurate rifles ever made was the model 1903 Rifle used by American troops in World War I. In 1919 John C. Garand began developing the M-1 semi-automatic rifle here. The M-14 was the last small arms developed at this site.

In 1830, Col. George Bomford proposed a new type of gun storage system, which could store as much as 60 000 weapons arranged in such a way that there is easy access to any of the weapons. In 1833, money for 36 double racks was appropriated. The Main Arsenal was completed in 1850 and new storage racks were put in place on three floors of the structure. Nearly 300 000 weapons, or 1100 in each rack, were stored. When Henry

Blanchard Lathe.
Courtesy of Springfield Armory NHS and James Langone.

Wadsworth Longfellow and his wife visited the armory, his wife remarked that these racks, filled with rifles, looked like an organ, inspiring Longfellow to write "The Arsenal at Springfield." Only one rack remains, partially filled with 645 Springfield Model 1861 muskets that are in excellent condition and can be seen by the public.

The army first opened a museum on site in 1866 and available for public visitation five years later, making it one of the oldest public museums in the country. In 1968 the Defense Department closed this facility as an economic measure. The armory evolved from a facility for craftsmen to build muskets one at a time to a center that used mass-production techniques and ultimately into a research and development center.

In 1978 this complex became a National Park Service site because of the historical and national significance of antique arms in its collections. The armory holds 74% of all antique arms in the National Park Service collections and has the largest collection of US military arms in the world. Because of the vast holdings of the Armory, the entire collection cannot be on view at one time. Thus approximately 20% of the artifacts are on exhibit at any one time and items rotate.

Today this facility is a technical research center of arms and associated archives containing 8000 objects and 500 000 associated records. Within these walls, the public can examine the major role that the US government played in the development of interchangeable precision manufacturing procedures during the Industrial Revolution.

In addition to its significance as a major armory site, it is interesting to note that the Springfield Armory was also the site of an early dinosaur discovery. In the mid-1800s a blasting operation on the grounds conducted by William Smith to build a dam uncovered the fossil bones of *Anchisaurus polyzelus*, dating from the early Jurassic period, about 200 million years ago. The actual date of the discovery is not known, but records indicate the find was before 1856. This specimen was curated into the Vertebrate Paleontology Collection of the Pratt Museum of Natural History at Amherst College (catalog number is ACM 41109).

Address: Springfield Armory National Historic Site, One Armory Square, Suite 2, Springfield, MA 01105.
Phone: (413) 734-8551.
Days and hours: Jun.–Oct.: daily 9 a.m.–5 p.m.; Nov.–May: Wed.–Sun. 9 a.m.–5 p.m. Closed 1/1, Thanksgiving, and 12/25.
Fee: Free.
Tours: Guided. Ranger tours available occasionally. Check in advance for special programs.
Public transportation: Bus service and rail service is available to Springfield, MA. Check MTA and Amtrak. Approximately 5-minute ride from downtown by taxi.
Handicapped accessible: Yes.
Food served on premises: None on site. Springfield has a number of restaurants.
Website: www.nps.gov/spar/

STURBRIDGE

Old Sturbridge Village

Old Sturbridge Village is one of the largest living-history museums in the United States. It was first opened to the public in 1946 with the goal of exploring life in a New England village during the period from 1790 to 1840. This remains its mission. The village is on 200 acres (81 hectares) of rolling meadows and woodland along the Quinebaug River in central Massachusetts. This land was the early 19th-century farm owned by David Wright. A sawmill, gristmill, and a millpond on the original property still exist on the site. The millpond dates to 1795.

Today 59 historic buildings, three authentic water-powered mills, and two covered bridges sit on this land. These structures are either restored or recreated structures that were moved to the site from other places in New England. For those who do not wish to walk, a stagecoach takes visitors around the village. When there is snow on the ground a horse-drawn sleigh is in service. Heirloom gardens and fields, some with back-bred plants, present a picture of early agriculture. Farm animals roam the area, as they did in the early 19th century. Antiques in many areas of the complex help interpret history. Visitors can participate in making crafts. Costumed interpreters re-enact and explain the daily life, work, and celebrations of early 19th-century New England. Check at the visitor center for days and times of interpretations.

The layout of the Living History Museum is divided into three parts: the Center Village, the Countryside, and the Mill Neighborhood. As is typical of New England towns, the Center Village is built around a green space, or common. Off the common is a mill neighborhood, a rural crossroads, and open farm country.

The buildings of the Center Village contain household objects, furniture, and tools from all parts of New England. These structures include a Friends Meeting House for the Quakers who lived in the area; a Center Meetinghouse for town meetings, lectures, and political events; law offices; a Parsonage that was the home of the Congregational minister; the Asa Knight Store that originally was located in Vermont; the Thompson Bank from Thompson, Connecticut; a printing office, shoe shop, several houses, and more.

The Towne House in Center Village is an early Federal-style home built in 1796 for Salem Towne, Sr., and kept in the family. Elements of the house were based on the *Practical Builder*, a book written by William Pain a few years earlier. Both Towne and his son were prosperous citizens involved in surveying, business, community leadership, and "scientific farming" using the latest techniques. Surrounding the house is one of two farms in Old Sturbridge. The farm has orchards, a barn, outbuildings, and livestock, including merino-type sheep, prized for their fine fleece. The cider mill seasonally presses apples into cider, a favorite New England beverage.

The Countryside has the second farm of the village, the Freeman Farm, a rural enterprise, originally belonged to Pliny Freeman, who lived in Sturbridge and built this residence in 1802. Here conservative farming methods, more typical of the traditional farmer, are in use.

At both farms, crops and animals are back-bred so they look similar to those typically found during this period. Staff members use historic agricultural methods and tools to cultivate the land.

Other facilities in the Countryside include a working blacksmith shop, the Bixby House (the blacksmith's home), a working cooper shop, a working pottery shop, the district school, and a covered bridge.

A walk in the Mill Neighborhood brings visitors in contact with early industrialization, commercial structures, and a mill pond. The site was chosen for its waterpower, important to all New England mills of the period. The gristmill demonstrates 19th-century waterpower technology: the water enters a breast wheel near the center, which throws the wheel off balance and makes it turn and power a 3000-pound (1400-kg) millstone. In the carding mill visitors can see new technology of the early Industrial Revolution as rare rotary carding machines dating from the 1820s process wool by combing, smoothing, and straightening the fibers. With these machines, operations that once took an entire day were completed in a few minutes. The operating sawmill, reconstructed according to an 1830 New Hampshire design patented by a New England mechanic, uses the up-and-down cutting technique of the day, a reaction-type waterwheel.

In the 1830s, shoemaking was a rising new industry in New England. The shoe shop represents a bridge between traditional techniques and the large, centralized factories that were to follow. There are also a blacksmith shop and a cooper shop.

Old Sturbridge Village has interpreters playing the role of craftsmen, known as "mechanics" in the early 19th century, plying their trades. At the printing office, built in the

1780s in Worcester, visitors learn how type was set by hand, how proofs were corrected, and how materials were printed on an iron-frame press. The kiln, woodshed, clay mill, and pot shop, where traditional redware pottery is produced, represent an important New England industry that began to decline in the early 1800s. American pottery faced serious challenges from mass-produced English ceramics and the increasing use of tinware. Production of tin items (actually tin-plate, that is, iron coated with tin) became a fast-growing craft. In the tin shop you can watch demonstrations of machinery that cuts and shapes shiny, rust-resistant, durable products.

Among household tasks carried out are food preparation, canning, preserving, and cloth weaving. The advent of the Industrial Revolution meant that women who wove goods at home started to weave for a living, and eventually worked in the textile mills. By the 1820s, flax production had almost completely stopped in New England because converting flax to linen was very labor-intensive. Wool-processing continued in the home because it was aided by carding mills, the use of which is demonstrated by interpreters at the site.

One section of the village has been set aside for growing plants. The extensive Herb Garden contains medicinal and culinary herbs, vegetables, and flowers back-bred to be similar to those from the post-Revolutionary period as well as herbs for dying cloth and making traditional crafts. In another part of the village are demonstrations of crafts and trades such as textile production, broom making, wool coloring, baking, and glassmaking. Several buildings in the Village have exhibits of a variety of collections, including firearms, uses of herbs, lighting devices and their fuel, early American time-pieces, and three categories of glass: blown glass, molded glass, and pressed glass.

Address: Old Sturbridge Village, 1 Old Sturbridge Village Road, Sturbridge, MA 01566.
Phone: (508) 347-3362.
Days and hours: May–Oct.: daily 9:30 a.m.–5 p.m.; Nov.–Apr.: Wed.–Sun. 9:30 a.m.–4 p.m., closed Mondays. Closed 1/1 and 12/25.
Fee: Yes.
Tours: Self-guided.
Public transportation: Bus service from Boston; hire a taxi service.
Handicapped accessible: Yes.
Food served on premises: Café.
Website: www.osv.org

New Hampshire

CANTERBURY

Canterbury Shaker Village

This once-thriving community of the United Society of Believers in Christ's Second Appearing, known as shaking Quakers or simply Shakers, was established in England in 1776. The Canterbury community began in the 1780s as the fifth of 19 such communities of Shakers in this country. These Shaker communities reached from Maine to Kentucky. The Canterbury Shaker Community was designated a National Historic Landmark in

1993. It is one of the most intact and authentic surviving Shaker sites, a nonprofit museum of decorative, mechanical, and fine arts, and living-history museum founded in 1969. The site examines a simpler way of life according to the Shaker philosophy. The name Shaker comes from the frenetic dances done at religious meetings. (See Shaker Museum, Mount Lebanon, New York.)

At its peak in the 1860s the village had more than 100 buildings, 250 inhabitants, and 4000 acres (1600 hectares) of land. The community supported itself by farming, selling seeds and herbs, manufacturing medicines, and making crafts. The last resident died in 1992 at age 96. Today there are 25 beautiful original structures, most built in the late 18th century and the first half of the 19th century, and four reconstructed period buildings situated on 694 acres (281 hectares) of fields, wood lots, and three mill ponds. Also on the property are an early 19th-century waterpower system and ruins of mills and water-control structures.

Although Shakers lived an unadorned life, they did not retreat from modern technology or shun convenience. In fact, they tried to stay on the cutting edge of efficiency, developing all manner of conveniences to do this. They believed that labor was sacred and that each object should be constructed to last a thousand years. If its maker were to die tomorrow, someone else would have the benefit of good equipment. In their effort to re-create heaven on earth, they developed many industrial and agricultural improvements.

The Meeting House, built in 1792, has a gambrel roof, windows and doors insulated with birch bark, and walls insulated with moss. The interior contains exhibits of products invented or perfected by the Shakers, such as medicinal lozenges, toothache pellets, and woodworking tools, boxes for seed packets, clothespins, flat brooms, packaged medicinal herbs, and sugar-coated pills. The earliest medicinal products were "simples," processed herbs for grinding into water or sugar for liquid doses, pre-ground, or soaked in alcohol or water. Later, liquid extracts (alcoholic solutions of the herbs) became popular.

Shaker furniture, functional and long-lasting, often included practical innovations. Examples on display include sewing tables with drawers on the side so knees would not bump them, and chairs with "tilting boots" to prevent occupants from falling if they leaned backward.

In the laundry there are soapstone tubs, washing machines, a machine that extracts moisture from cloth, a clothes elevator, a mangle for pressing laundry, and unique drying-racks. Soapstone was used for tubs because it is soft, easy to carve, inert to acids and bases, and quarried in New England. The knitting machines used to make Shaker sweaters are also housed here.

The horse barn exhibits agricultural equipment such as plows and seed planters. The infirmary building is outfitted with a pharmacy and areas for dentistry and surgery. Walkways on the property reflect the Shakers' philosophy of building for endurance. Some walkways date back to 1793. They lead to the garden of culinary and medical herbs, the organic vegetable garden, the arboretum (the first in New Hampshire), and the apple orchard with its rare varieties of apple trees. A self-guided nature trail follows around the Turning Mill Pond, spillways, and mill foundation. Other trails lead to the carding mill and the meadow ponds.

To learn the story of the Shakers, take the Story Tour, which begins at the Visitor Center, and includes the meetinghouse, laundry, and school house. As an alternative for families with children aged five and up, the village provides a family tour, which lasts 45 minutes and examines the Shaker dwelling house. Other special-interest tours may be offered. Explore the vegetable, flower, and medicinal-herb garden with a one-hour seasonal tour.

Address: Canterbury Shaker Village, 288 Shaker Road, Canterbury, NH 03224.
Phone: (603) 783-9511.
Days and hours: Vary throughout the year. Check in advance.
Fee: Yes.
Tours: Guided and self-guided.
Public transportation: None available; possible taxi or limousine-service arrangements.
Handicapped accessible: Limited.
Food served on premises: Restaurant serves Shaker-inspired food; farm stand.
Website: www.shakers.org/

NORTH SALEM

America's Stonehenge

In the beautiful woodlands of New Hampshire is a megalithic site covering about 105 acres (42 hectares). Archaeologists believe that the structures here were built before Europeans arrived in the New World in 1492. Radiocarbon-dating of charcoal found here indicates the site was inhabited by humans over 4000 years ago. This site is probably the oldest human-made construction in the United States.

America's Stonehenge is a complex of chambers, walls, and what are believed to be ceremonial meeting places. The people who built these structures were knowledgeable about stone-construction techniques and astronomy. The stones are large, shaped, and in a standing position, some weighing as much as four tons. Some of the stones are arranged to form walls and chambers. The material is native schist containing mica and feldspar, which likely came from a nearby quarry. Although some of the stones have drill marks on them, many do not, nor do they show evidence of modern or historic quarrying techniques. Experts are unsure how they were extracted from the quarry and moved to the site.

Marks on the material indicate that the stone was probably quarried by percussion techniques typically used by indigenous stone workers. Colonial people did not use these methods, for the stones were shaped with stone hand tools rather than the metal tools typically used by European settlers. Given the weight of the stones and their distance from the quarry, the people who built the site clearly knew the principles of the fulcrum and the lever.

The exact purpose of most of the site's structures is unknown. Some stones near the main site have markings made by humans. The meanings of the inscriptions have not been deciphered.

The site is an accurate astronomical calendar. It was, and still can be, used to determine specific solar and lunar events of the year. These stones, placed in astronomical alignments, accurately predict North American solar and lunar events. If people are properly positioned on December 21 (the first day of winter), the sun sets directly over the Winter Solstice Sunset Monolith. This same monolith marks the 18.61-year nutation cycle of the moon's position of rising and setting. At the end of each cycle the moon appears to stand still for a night, relative to the monolith. On June 21 (the first day of

Sacrificial Table, America's Stonehenge.
Courtesy of America's Stonehenge.

summer), the longest day of the year can be marked by watching the sun rise over the Summer Solstice Sunrise Monolith.

The monoliths can also be used to detect other astronomical events, such as the equinoxes. There is a megalith that indicates the alignment of true north and true south. Yet another stone marks November 1, an important date to the ancient people that we celebrate on October 31—All Hallows' Eve or Halloween.

In addition to the path through the megalithic site, there is a nature trail through the beautiful New England woods with its wildlife. Several alpacas occupy a compound along the way. Because alpacas were once common in North America before the last ice age about 10 000 years ago, these animals, members of the camelid family, have been brought here. Alpacas originated on the Great Plains of North America about 40 million years ago and migrated to South America about 3 million years ago. They became extinct in North America about 10 000 to 12 000 years ago and now they have returned. Another point of interest is an Iroquois "Three Sisters Garden," a mounded garden in which the complementary vegetables beans, corn, and squash (the "three sisters") are grown together.

A ring of post molds 30 feet (9 m) in diameter are the remnants of a wigwam discovered near the parking lot, and are associated with two fire pits from 2000 years ago. A seven-foot (2.1-m) tall replica of this wigwam, along with a cooking rack, occupies a place along the trail. There is a reconstruction of 300-year-old canoe, constructed by a process of chopping and burning out the inside of a pine log. The original may be seen at the museum in the Visitor Center located near the entrance to the site.

Other artifacts displayed in the Visitor Center Museum which have been found on-site include slate and bone artifacts, pottery, stone tools, and other evidence of possible Native Americans habitation of this location at various prehistoric and historic times, with accompanying explanations.

Address: America's Stonehenge, 105 Haverhill Road, Salem, NH 03079.
Phone: (603) 893-8300.
Days and hours: Daily 9 a.m.–5 p.m.; last admission 4 p.m. Closed 12/25 and Thanksgiving.
Fee: Yes.
Tours: Self-guided.
Public transportation: Bus and train transportation are available to North Salem from Boston. There is taxi service from downtown Salem, New Hampshire.
Handicapped accessible: Limited.
Special note: Wear sturdy walking shoes to follow trails through the woods.
Food served on premises: None; picnic area is on-site.
Website: www.stonehengeusa.com/

Rhode Island

PAWTUCKET

Old Slater Mill National Historic Landmark

Old Slater Mill, the first successful water-powered mill in America, started producing textiles on the banks of the Blackstone River in 1793. The mill is named after its cofounder, Samuel Slater, known as New England's first industrial capitalist because of his emphasis on division of labor, tracking of supplies and production, and careful management of machines for top efficiency. This facility was declared a National Historic Landmark District and a National Historic Landmark within Blackstone River National Historic Park.

The original mill structure was only 29 feet by 42 feet (8.8 m × 13 m) and 2½ stories tall. Later additions have changed these dimensions. It originally looked very much like the farms, barns, and churches in the surrounding area. Its long narrow shape used the power from the waterwheel very effectively and made the best use of natural light. This design was then used throughout the Blackstone River Valley. Today visitors can tour the mill and nearby buildings to see historic machinery in operation, learn about the development of the factory system, and see pre-industrial techniques.

Slater's industrial career had its roots in his native England, where he was a member of the landed gentry. In 1782, at age 14, Slater became an apprentice in a textile mill and spent seven years in England learning how to manufacture cotton and manage a mill. By this time, the Industrial Revolution was well underway in England and the new cotton-spinning machinery invented by Richard Arkwright, who was Slater's employer, was well established. One of Slater's key lessons he learned was that more than the right equipment was necessary for a successful spinning business: the facility must be run properly.

After the American Revolution, the United States still had to depend on England for most manufactured goods. The new nation had citizens skilled in building machinery but did not have the advanced technology available in England. Industry was attracted to the Pawtucket area because the Blackstone River had waterfalls to provide waterpower and well-trained craftspeople living nearby. Although most of these craftspeople were shipbuilders, they could easily transfer their skills to other industries. Despite high hopes, however, industrial success did not come to the region until after Slater arrived.

In 1789 Slater left the mill in England and immigrated to America. He was working in a small textile mill in New York City when, in 1790, Moses Brown hired him to improve the fortunes of his mill in Rhode Island. Slater set about applying the lessons he learned and the ideas he conceived as an apprentice to change the operations of Brown's mill.

Slater and Brown went into partnership in 1792 and a year later built a new mill, known today as Old Slater Mill. Slater carefully planned the organization of the waterwheel-powered machinery, instituted record-keeping to track supplies and production, initiated division of labor, and introduced other practices to control efficiency and minimize waste.

Traditionally, manufacturers produced only goods that were already ordered, or "bespoke." Slater realized that idle machines were not profitable—the equipment should work at maximum capacity. He instituted a system of producing goods continuously even if they had not been ordered. To successfully manage the operation he had to thoroughly understand the machines and their potential. He had to keep a constant flow of raw materials and set the timing of each step so that no bottlenecks would occur. His hardest task was to make his partners realize that his approach was the best way to bring about industrial success. Once they were convinced, other manufacturers followed suit and the fledgling nation's textile industry became more successful.

Slater Mill Historic Site is made up of the Sylvanus Brown House, the Wilkinson Mill, and the Old Slater Mill. The house, built in 1758, was inhabited from 1784 to 1824 by Sylvanus Brown, a pattern-maker for Slater's early textile machines. It was moved to this spot in 1971. To re-create the home of a 19th-century artisan, historians used the probate inventory made at Brown's death as a guide. The inventory included a loom, spinning wheel, and various other equipment to make cloth by hand. Much of this type of work was done by women and children. This work was tedious, slow, and included cleaning and carding wool, spinning yarn, and weaving cloth. These pre-industrial tasks of weaving flax and wool are demonstrated in the Brown house. Much of the machinery built for Slater's Mill looks very much like the equipment in the Sylvanus Brown House.

The rubblestone Wilkinson Mill was built between 1810 and 1811 by Samuel Slater's father-in-law, noted ironworker Oziel Wilkinson. This mill incorporated 20 years of industrial experience into the new structure and is much larger than Slater's Mill. By making the exterior wall of stone, the chance of fire was significantly reduced. Oziel and his son David made nails and anchors at their mill and eventually did work for Slater. David contributed to the design of Slater's carding machine and was also credited with inventing the screw-cutting lathe in 1806. The 8-ton (7.3 tonne) waterwheel demonstrates the power of water in action.

When touring the Wilkinson Mill, the visitor sees how a 19th-century machine shop functioned, with its water-powered drills, lathes, planers, and milling machines. The original waterwheel still powers the machinery. Exhibits on textile production and fiber art are displayed in a gallery at the Wilkinson Mill. The brick tower was added in 1840 and the belfry was a modern addition recreated from a photograph taken in 1870.

Between the Sylvanus Brown House, built in 1758, and Wilkinson Mill is a kitchen garden, where historically correct plants provide flax, vegetables, and flowers. The Brown

House is a typical artisan cottage of the period. Brown was a woodworker, pattern-maker, millwright, and dam-builder, and used this house from 1784 to 1824.

A tour of Slater Mill introduces the visitor to the workings of the first fully mechanized cotton-spinning facility in the United States. With the help of Brown, David Wilkinson, and others, Slater duplicated Arkwright's spinning system. Visitors can watch as historic machinery demonstrates Slater's processes of converting raw cotton into finished cloth. On the tour, the development of the factory system and its effect on workers' lives are presented.

The Museum explores the history of textile manufacturing in America with 24 machines built between 1776 and 1922. This equipment demonstrates the how cotton was made into cloth.

Address: Old Slater Mill National Historic Landmark, 67 Roosevelt Avenue. Pawtucket, RI 02860.
Phone: (401) 725-8638.
Days and hours: Mar.–Apr.: Sat. and Sun. 11 a.m.–3 p.m.; May–Jun.: Tue.–Sun. 10 a.m.–4 p.m.; Jul.–Oct.: daily 10 a.m.–4 p.m.; Nov.: Sat.–Sun. 10 p.m.–4 p.m.; Dec.–Feb.: open for group tours by appointment.
Fee: Yes.
Tours: Guided tours offered on a regular schedule.
Food served on premises: None; restaurants are in Pawtucket.
Public transportation: Local bus stops about two blocks from site.
Handicapped accessible: Yes.
Website: www.slatermill.org

Vermont

RUTLAND

New England Maple Museum

Probably Vermont's best-known product is maple syrup. In this small, homegrown museum tucked behind a gift shop in the foothills of the Green Mountains, maple-sugaring, past and present, is explored.

The state has just the right conditions for growing sugar maple trees (*Acer saccharum*), which produce sap for maple syrup. These ideal growing conditions include the right amount of calcium in the soil, as well as cold nights and warm days in spring, and thus the sugar maple trees grow well here and provide sap for maple syrup. Thus Vermont is the main producer of maple syrup in the United States. Maple syrup itself is composed largely of sucrose plus water, with traces of other sugars. Potassium and calcium can be found, but it also contains significant amounts of manganese and zinc. Maple syrup is acidic because it also contains organic compounds such malic acid. There are trace amounts of amino acids as well as volatile organic compounds, such as hydroxybutanone, vanillin, and propionaldehyde.

A well-researched mural more than 100 feet (30 m) long, painted by Grace Brigham, explains the history of maple syrup. The museum displays its holdings of a very large collection of tools and other items related to maple-syrup production. The process of making the syrup is simulated with evaporators. There are dioramas and carved figures illustrating the process. All of this is documented with historical photographs.

The Iroquois first made what they referred to as "sugar water." They caught the sap in troughs and transferred it to wooden or clay vessels into which they dropped hot stones to make the sap boil. The final process was letting the syrup crystallize into sugar. On display are wooden troughs the Iroquois used hundreds of years ago.

When Europeans came to the area around 1750, the Native Americans told them of this wonderful, sweet syrup. The settlers tapped the trees; caught the sap in buckets of hand-shaped bark or wood; put it into brass, iron, or copper pots, which were not perishable; and boiled it over open fires. The resulting syrup was like the maple syrup we know today. The Europeans introduced these new syrup-making methods to the Native Americans.

The museum is an oversized replica of a sugar house, which at one time was part of most New England farms. Antique equipment, covering more than 200 years, is one of the most complete collections of this equipment in existence. These artifacts include an ancient block of wood with a sap-collecting gap made by Native Americans, syrup-making utensils used from about 1790 to 1930, gathering pails from 1830 to 1850, rock-maple hinge pins, and modern plastic pipelines. Also displayed are an oxen yoke from about 1875 with the capacity to carry 200 gallons (760 L), which weighs about 2000 pounds (910 kg) and a people yoke with an 8-gallon (30 L) capacity (75 lb., 34 kg). Equipment once hauled by oxen is displayed alongside modern tractors and boiling tanks.

The museum has a complete display of modern sugar-making equipment and gives live demonstrations of maple candy-making and wooden sap-bucket construction. Today, maple sap is collected by a network of interconnected plastic tubes that run from hundreds of trees to a central tank. The sap is picked up by trucks and taken to a central sugar house, where it is boiled and processed. A slide show describes the entire process, which takes place on a Vermont family farm. Samples of different grades of syrup are available to try at the tasting table.

The museum's forest display identifies different types of trees in the area and explains old and new tapping techniques. Though the traditional maple species for tapping is the sugar maple, other species can be used for syrup, including the black maple (*Acer nigrum*, very similar to the sugar maple) and the red maple (less-sweet sap, and timing for tapping is more constrained). To be tapped effectively a sugar maple should be about 40 years old, and tapped when temperatures rise above freezing by day, but below freezing at night.

Trees growing in open fields or other unforested settings generally provide about double the amount of forested maples. Typically one acre (0.4 hectare) of trees yields 25 to 35 gallons (95 to 130 L) of sap. The sap usually has a 30% sugar content. It takes 40 gallons (150 L) of sap to make 1 gallon (4 L) of maple sugar. The museum claims that "on a good day, the sap will run about two drops per heartbeat, and will fill a 16-quart (15 L) bucket in eight hours." The final syrup product is about two-thirds sugar. If the syrup is over-boiled beyond 67% sugar, crystals may precipitate out.

Address: New England Maple Museum, 4578 U.S. Route 7, Pittsford, VT 05763.
Phone: (802) 483-9414.
Days and hours: Mid-May–Oct.: 8:30 a.m.–5:30 p.m.; Nov.–Dec. 23 and mid-March–mid-May: 10 a.m.–4 p.m. Closed 12/25 through Feb.

Fee: Yes.

Tours: Self-guided. Guided tours by prior arrangement.

Public transportation: Bus to Rutland, then walk 0.7 miles (1.1 km) to site.

Handicapped accessible: Yes.

Food served on premises: None. Restaurants are in nearby Rutland.

Special note: Regional trees flower in the spring; vivid colors appear in the fall. Call to find out the best dates.

Website: www.maplemuseum.com

ST. JOHNSBURY

The Fairbanks Museum and Planetarium

The beautiful 1891 Romanesque building of carved red sandstone that houses this museum is on the National Register of Historic Places for good reason. Step through the entrance and enter the Victorian 130-foot (40 m) great hall covered with a golden, quartered oak barrel-vaulted ceiling and Tiffany stained-glass windows. The oak and cherry cases are filled with mounted animals and birds—some familiar, some exotic. Spiral staircases tucked into the corners wind up to the balcony for fine views of the hall below and ceiling above.

Industrialist, scientist, and naturalist Franklin Fairbanks, an avid collector of wildlife and natural-science specimens, donated this building that now has over 21 500 square feet (2000 m²) and the collections of more than 100 000 items to form a museum. The purpose of the museum, Fairbanks said, was to stimulate "moral, intellectual and scientific improvement in St. Johnsbury." Fairbanks's family and friends augmented the museum by donating their own cultural artifacts and natural specimens from around the world.

The Museum is organized into three different departments: Natural Science, Historical, and Ethnological. The Natural Science Department, which the museum considers its most important collection, includes extensive New England flora and fauna, with a vast collection of New England mammals and birds. Most native Vermont species are represented. There is also a noteworthy collection of North American fauna. The herbarium contains around 6500 New England plants, primarily from Vermont. Other significant collections include rocks, minerals, gems, coral, butterflies, Arctic insects, New England birds and eggs sets, and one of the worlds' most diverse mounted hummingbird collections.

Today the museum has a global bird collection with more than 3000 mounted specimens of nearly 1000 species. Some of these birds are extinct or endangered. The collection contains an almost complete set of Vermont birds and one of the world's best collections of hummingbirds and birds-of-paradise. The Philippine birds come from a well-documented 1874 expedition led by Michigan ornithologist Joseph Beal Steere.

Displays of extinct species like the passenger pigeon and Carolina parakeet, and the endangered Central American resplendent quetzal, support the museum's goal of conveying the need to protect the world's ecosystems. About 65% of the Natural Science collection is on display.

Nine unusual works of "bug art" were created by John Hampson, a mechanical engineer who worked for Thomas Edison in his Menlo Park, New Jersey, laboratory. (See

Fairbanks Museum and Planetarium.
Courtesy of the Fairbanks Museum and Planetarium.

Thomas Edison National Historical Park, West Orange, New Jersey.) A noted entomologist, Hampson created this bug art collection over a total period of 30 years using more than 70 000 insects. He fixed them in place with pins and fish-scale glue.

William Balch, one of the first taxidermists to use realistic naturalistic habitats, arranged many of the bird and animal dioramas. His diorama called "Muskrats in Summer and Winter" was completed in 1895. Balch gained national acclaim as the first to photograph a beaver at work and for his photography of New England orchids. Although he was an expert on Vermont wildlife, his knowledge of wildlife outside the state was limited, as demonstrated in his flamingo diorama. Balch shows the flamingo trying to eat a frog, a food not on its menu.

The Wildflower Table displays the diverse types of living flowers, grasses, berries, ferns, and evergreens growing in the northern part of Vermont. About 400 species are on view during the year, showing both the fruit and flowering stages.

The balcony exhibits include fossils, minerals, and ethnological artifacts from Eskimo, Navajo, Palestinian, Japanese, and other cultures. An introductory panel provides a history of the Fairbanks Company—now known as the Fairbanks Morse Company—a firm that originally manufactured stoves and plows. The company was formed by brothers Thaddeus and Erastus Fairbanks in 1824. In 1831 Thaddeus invented a compound lever platform scale, which revolutionized the weighing of heavy loads. The firm switched to scale production, which became a major industry in the region. Balancing Science and Wonder includes objects from the home of Franklin Fairbanks, along with artifacts from the founding and construction of the Museum.

The United States' official northern New England weather station has been at the site since 1894. One of the oldest continuously used weather stations in the country, it forecasts the weather for regional radio stations and newspapers.

This museum also displays temporary exhibits.

The Lyman Spitzer Jr. Planetarium is, at this writing, the only public planetarium in the state. Built in 1961 and refurbished in 2012, it has a Digitarium® Kappa system which projects a fisheye image on a 20-foot (6.1 m) dome and seats 45 people. Programs are presented to the public on a regular basis. An exhibit about outer space, which includes photos from the Hubble Space Telescope, and a meteorite from northern Argentina are next to the planetarium.

Address: The Fairbanks Museum and Planetarium, 1302 Main Street, St. Johnsbury, VT 05819.

Phone: (802) 748-2372.

Days and hours: Daily 9 a.m.–5 p.m. Closed 1/1, Easter, 7/4, Thanksgiving, and 12/25.

Fee: Yes.

Tours: Self-guided; audio tours are available.

Public transportation: Amtrak train to Waterbury/Stowe, then bus to St. Johnsbury, then walk 0.2 mile (0.4 km) to site. Alternatively, take Bus 84 from the Peoples Bank on State Street in Montpelier (nearest city) which goes to the St. Johnsbury Welcome Center, and walk 0.2 mile (0.4 km) to site.

Handicapped accessible: Partially.

Food served on premises: None. Restaurants of various styles and prices are in town.

Website: www.fairbanksmuseum.org

WINDSOR

American Precision Museum

The American Precision Museum holds a historically substantial collection of hand and machine tools and the products they made. This equipment is of major importance to the development of industry in the United States. The museum building is a three-story structure built in 1846 by Samuel Robbins, Nicanor Kendall, and Richard S. Lawrence, which served originally as an armory and machine shop. This structure is an outstanding example of New England mill architecture of mid-19th-century design. In 1966 it was placed on the National Register of Historic Places. The American Association of Mechanical Engineers has also placed the structure on its list of historic mechanical-engineering landmarks.

Robbins made his fortune in the lumber business; Kendall and Lawrence were experienced custom gunsmiths. These three men applied for and received a US War Department contract for the production of guns in 1846. Their contract called for delivery of the guns in 1850. Amazingly, they built this armory and fulfilled their contract by 1847.

Traditionally, gun production was slow and unreliable. Robbins and Lawrence designed and made interchangeable parts for the guns, using advanced production methods. These techniques speeded up the manufacturing process, enabling them to deliver their product early. When they exhibited their six precision-made rifles at the

Crystal Palace Exhibition in London in 1851, Robbins and Lawrence introduced the world to their revolutionary new manufacturing approach. This advance in gun production was quickly added to what the English called the "American System." These techniques were adapted for producing many other kinds of equipment.

The armory and the machine shop, constructed of handmade bricks and a slate roof, were built along the Connecticut River. In the long, narrow machine room, 24 tall windows capture as much sunlight as possible to illuminate the workers' area. Today it is a museum that displays a large variety of machines tools from its collections. This extensive collection is one of the largest in the world, which includes single- and multiple-spindle lathes, grinding machines, shapers, planers, and single and multiple spindle drills. These pieces range in age over more than 200 years of machine development up to the development of computerized systems.

The most important machine tools in the collection are those developed in this shop to mass-produce firearms with interchangeable parts. Quite possibly the museum has more historically significant machine tools than any site in the country. These tools changed manufacturing processes throughout the world. Some are important because of their role in the evolution of mechanical engineering; others show the ingenuity of inventors to meet particular needs. Older items are gradually being restored and rotated into the exhibits. Categories of machinery and tools in the collection include machine tools, measuring devices, sewing machines, typewriters, firearms, precision hand tools, and precision scale models.

An extensive gun collection illustrates the development of fire bicycle design in the Connecticut Valley, beginning with guns made by Nicanor Kendall, David Hilliard, and Asa Story. Every gun manufactured in this building is represented, including the Enfield Minie rifle, the 1841 Mississippi Rifle made by Robbins & Lawrence, the Model 1961 Special Musket manufactured by Lamson, Goodnow, & Yale, the Jennings rifle, the Palmer carbine, the Ball repeating carbine, and the seldom-seen Windsor Sharps rifle.

Measuring devices that assisted in fabrication of precision equipment are also on display, including comparators, calipers, and gauges used to produce gun locks, screws, and much more.

Sewing machines became an important product manufactured in factories after 1850 when the gun-making machines were reworked. Thus the Museum displays such items as Edwin Clark's Looper, made by Lamson and Goodnow. Many early sewing machines were very simple and are still in working order, incorporating detailed artistic elements. These include Howe, Singer, and Wilcox & Gibbs.

On view is an interesting typewriter collection with more than 50 items, such as Barlock, Remington, Oliver, and the Northern Typewriter Company. Unusual items are the Blickensderfer, introduced in 1893, which has a changeable type-wheel, allowing the use of different fonts or different languages, and the Hammond No. 12 that strikes the paper from behind, developed in the 1890s.

A variety of eclectic materials includes a gnome made by Maxfield Parrish—a machinist as well as an artist—and other items from his workshop; some working models built by John Aschauer; and John Keely's Etheric Force Machine from 1878 that used his so-called "etheric force" or "vibratory sympathy" to generate power.

The visitor can see a complete miniature machine shop with 40 working models of machines on a 1/16th scale, built by John Aschauer of Detroit. It took 24 years to complete.

A display floor that once held long rows of machines arranged for efficiency now showcases machines in clusters for easy viewing from all angles. The earliest piece is an engine lathe dating from about 1825, once in a textile mill in Massachusetts. An accompanying video shows how this piece of equipment works. The central working surface was usually made of metal, but this one is made of granite because it was cheaper and could stay flat when subjected to heat. Another lathe, built in 1911 in Hartford, Connecticut, is automatic. It has multiple spindles that allowed a single worker to do several operations at the same time, thus increasing productivity while cutting costs.

An 1895 beveled-gear grinder on display made gears for chainless bicycles—popular with women because chains could catch their skirts. A power rifling machine for drilling the interior of a gun barrel, devised by a Robbins & Lawrence employee in the 1850s, was in production until the end of World War II. An 1893 Edison-General Electric dynamo, one of the earliest generators, produced direct current. The ultimate result of mass production is a 1914 Model T Ford.

Special events, programs, and temporary exhibits occur. The Museum also sponsors a Machine Tool Hall of Fame, honoring those who have contributed significantly to the American machine tool industry.

For access to the library and archives, speak with the museum staff.

Address: American Precision Museum, 196 Main Street, Windsor, VT 05089.
Phone: (802) 674-5781.
Days and hours: Memorial Day through Oct. 31: 10 a.m.–5 p.m.; closed the rest of the year.
Fee: Yes. Extra fee for guided tours.
Tours: Self-guided. Guided tours are by prior arrangement only. Behind the Scenes tour is offered about once a month by prior arrangement and is not handicapped accessible.
Public transportation: Amtrak train to Windsor, VT, then walk about half a mile (1 km) or take taxi to the Museum (contact in advance).
Handicapped accessible: Yes, except for Behind the Scenes tour.
Food served on premises: None. Restaurants are in town, mostly a few blocks north of the site.
Website: www.americanprecision.org

WOODSTOCK

Billings Farm and Museum and Marsh-Billings-Rockefeller National Historical Park

The partnership of the Billings Farm and Museum and the Marsh-Billings-Rockefeller National Historical Park explores the story of conservation history and the evolution of the stewardship of the land in North America. This partnership is the first national park to tell this story. The 555 acres (225 hectares) of this national historic park encompass the reforested Mount Tom and rolling hills and pastures typical of eastern Vermont. There are 50 different stands of trees, 11 of which date to the time of Frederick Billings. These woods, one of the oldest continuously managed woodlands in the North America, climb Mount Tom. Trails and rivers traverse the property. At the foot of the mountain are the more than 200 acres (81 hectares) of pasture, meadows, and cropland that comprise the Billings Farm and Museum.

This working dairy farm nestled in Vermont's rolling hills is a living-history museum examining Vermont's agricultural heritage. Using four of the farm's original barns, rural late-19th-century Vermont life and the development of this dairy state are showcased. A tour of the modern dairy operation demonstrates how the farm changed between the 1890s and the present. In an 1890 Queen Anne–style farmhouse, see demonstrations of the era's domestic crafts.

Mary Rockefeller, the granddaughter of the railroad builder and agricultural innovator Frederick Billings, grew up on this farm and lived in the mansion across the road, which is now part of the National Park. The mansion overlooks Elm and River Streets in the town of Woodstock. Mary Billings Rockefeller and her husband, Laurance Rockefeller, created this farm and museum as a public venue and donated the land for the National Historical Park to the federal government.

Frederick Billings was a native of Woodstock, Vermont. He was a lawyer, conservationist, and pioneer in reforestation and scientific farm management. He had financial interests across the country, such as the Minnesota & Montana Land and Improvement Company that he organized in 1881, for the purpose of establishing a railroad center for shipping mined ore, wheat, and cattle. His influence was so great that Montana's popular industrial center and tourist city, Billings, was named after him.

Billings purchased his initial 250 acres (100 hectares) in 1869 from George Perkins Marsh, an early environmentalist and also a native of Woodstock. Marsh was disturbed by the rapid deforestation and the harm this process did to the land and to agriculture. Thus he argued for human stewardship of the land in his book *Man and Nature*, published in 1864. He was also interested in modern technology, installing the latest improvements on the farm, which was of prime importance to Billings too.

When Billings purchased the farm, he then increased its area to more than 2000 acres (810 hectares), a very large farm for its day in this part of the country. His main interests were breeding top-quality Jersey dairy cattle and reforesting the land. Many Vermont

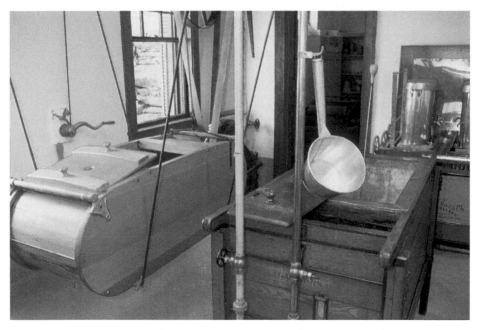

Creamery Equipment (Swing butter churn, separator, cooler, motor, water line).
Courtesy of Billings Farm and Museum.

hillsides had been clear-cut for farmland, lumber, and fuel, weakening their ability to hold moisture and destroying the land's agricultural value. Using Marsh's ideas, Billings planted more than 10 000 trees of many varieties on the denuded Mount Tom behind the estate.

Billings brought dairy cattle from England's Isle of Jersey to Vermont and used scientific breeding methods to develop a champion herd that produced milk with an improved butterfat content. The farm still reflects Billings's progressive agricultural philosophies.

Under the supervision of George Aitken, the resident manager from 1884 to 1910, Billings's Jersey cattle won many awards at the 1893 Colombian Exposition in Chicago. At its peak the farm employed 50 men, required a half-dozen teams of oxen, and was a leader in introducing new breeds to the area. Aitken brought Southdown sheep to the site. Billings bred the robust Vermont Morgan horse with the Hambletonian trotter to combine sturdiness and speed in one animal.

Some of the farm's contemporary operations are open to visitors, such as the modern cow barn where prized Jersey cows are milked in late afternoon, the calf nursery, and the trophy room. Next door in the horse barn there are Belgian and Percheron horses.

The museum has artifacts illustrating facets of Vermont farm life around the year 1890. The themed exhibits include Making the Land Produce, with plows and other agricultural implements; the Farm Workshop, displaying hand tools of the period; Sugaring, concerning artifacts and processes in the maple syrup industry; and the General Store, showing the types of products farmers bought from the outside world.

Audiovisuals, including the orientation film "A Place in the Land," and demonstrations help the visitor understand 19th-century agriculture in this part of the nation.

The farmhouse, restored to its 1890 state, has gas fixtures for lighting, an indoor toilet, hot and cold running water, and central heating—modern amenities for that time. It served as a business office and living quarters. The basement contains the ice house with its cold-storage room and creamery. The space was designed for efficiency and hygiene with its coolers, cream separators, and mechanized churn. A Chicago Ware motor, connected to a water line, drove the overhead shaft and pulleys that powered a Davis swing churn. The churn could produce butter in 30-pound (14 kg) batches.

Across the way, in the National Historical Park, there are tours of the 33 rooms of the Queen Anne–style mansion house, built in 1805–1807 by Charles Marsh, Sr., the father of George Perkins Marsh. The house features landscape paintings by Thomas Cole, Albert Bierstadt, Asher B. Durand, important photographic works, and other artifacts, as well as a 3000-volume library. The artworks illustrate the influence of art on the public and, in turn, on the conservation movement. The mansion is a national historic landmark in its own right. The tour continues through the gardens of azalea, rhododendron, and rock ledge. Hemlock and hedgerow decorate the gardens; some of these plants date from the Billings era. There is also a stand of Norway spruce. Parts of these gardens originated in 1869, and were augmented in 1954. They provide a historic view of landscape architecture.

The Carriage Barn, built in 1895, has several uses. It serves as the park's visitor center and has a permanent conservation exhibit. Horse-drawn carriages and sleighs in the collections are showcased in the Wood Barn. A walk along the carriage roads of Mount Tom provides a good way to examine the reforestation and the resulting forests.

Addresses: 69 Old River Road (Route 12 and Old River Road), Woodstock, VT 05091 (Museum); 54 Elm St, Woodstock, VT 05091 (park). **Phone:** (802) 457-2355 (Museum) and (802) 457-3368 ext. 222 (park).

Days and hours: Museum, Apr.–Oct.: daily 10 a.m.–5 p.m.; Nov.–Feb.: weekends 10 a.m.–4 p.m. Park, Memorial Day–Oct.: daily 10 a.m.–5 p.m.

Fee: Yes.

Tours: Self-guided; guided tours also available at regular intervals.

Public transportation: Not available.

Handicapped accessible: Farm accessible; house partially accessible; rest rooms accessible; no pets, except guide dogs permitted.

Food served on premises: Snack bar and country store at the Visitor Center.

Special note: Check schedules for milkings, special demonstrations, other events, and special exhibits.

Websites: www.billingsfarm.org, www.nps.gov/mabi/index.htm

2

Mid-Atlantic

Delaware

Johnson Victrola Museum

Just a very short distance from the 1792 State House, Delaware's seat of government until 1934, is the Johnson Victrola Museum, created to celebrate the work of Eldridge Reeves Johnson, the inventor and businessman who founded the Victor Talking Machine Company. He was known as a man who could fix anything that was broken. When Emile Berliner invented a flat record to replace the cylinder devised by Thomas Edison, he asked Johnson, a machinist, to make a machine that would not require the constant cranking of phonographs. In 1895–1896, Johnson invented a spring-driven motor able to play a disc without stopping. Eventually Johnson and Berliner merged companies, and Johnson continued to improve the recording system for the Berliner recordings. In 1901 Johnson formed the Victor Talking Machine Company and bought Berliner out. Johnson sold the company to the Radio Corporation of America (RCA) in 1929 (see The Sarnoff Collection, Ewing, NJ).

This small gem of a museum is designed to resemble a Victrola dealer's shop of the 1920s. An extensive collection of talking machines, gramophones, early recordings, and Johnson memorabilia fills the exhibit area. Although the items on display are labeled, it's a good idea to take a docent-directed tour. Docents add significant information and play some of the old machines, often using a few of the more than 20 000 historic Victor recordings in the collection.

The museum exhibits the Victor models I–VI, made from 1902 to 1920, which are easily recognized by their external horns. Available for view are early Victrolas with the horn enclosed in a cabinet. From 1906 to 1918 the company made the Auxetophone, a machine that used compressed air to amplify sound for theaters and restaurants. One of these is on display.

With the introduction of commercially produced electricity in the early 20th century, major changes hit the recording industry, such as the introduction of motorized record players and radio broadcasting. Take note of a 1925 radio and Victrola combination that plays an electrically produced recording from 1925. The device uses a small wooden needle, which makes a mellow sound. Also on view are a variety of wooden needles and the first record-changing machine, introduced in 1927.

His Master's Voice, one of 12 original paintings done in 1899 by Francis Barraud, hangs on the wall. This painting, which once adorned the RCA boardroom and was the property of Johnson, inspired the famous RCA Victor symbol of the dog (Nipper) and horn. On view is an assortment of items bearing Nipper's likeness, from radiator caps to clothing and salt and pepper shakers.

A bust of Enrico Caruso, the famous opera singer, is displayed—one of only 10 of these sculptures produced. Caruso presented it to Johnson in gratitude for the technology that brought Caruso's voice to more than a million listeners.

The renovated third floor, not handicapped accessible, also displays a variety of items related to the Victrola.

Address: Johnson Victrola Museum, Museum Square, 375 South New Street, Dover, DE 19901.
Phone: (302) 744-5055.
Days and hours: Wed.–Sat. 9 a.m.–4:30 p.m.
Fee: Free.
Tours: Self-guided; guided tours are also available at regular intervals. To schedule a tour, call in advance.
Public transportation: DART buses stop in front of the site.
Handicapped accessible: Only on the first floor.
Food served on premises: None; restaurants are within a few blocks of the site.
Website: history.delaware.gov/museums/jvm/jvm_main.shtml

WILMINGTON

Delaware Museum of Natural History

This fascinating museum was founded in 1957 by John Eleuthère du Pont (1938–2010), and opened in 1972 not far from Winterthur, Delaware. Du Pont began collecting natural history materials as a child, and his collection grew to be rather extensive. As an adult he obtained a doctorate in natural science and wrote several books on birds.

Delaware Museum of Natural History.
Courtesy of the Delaware Museum of Natural History.

Today 10 acres (4 hectares) of beautiful countryside are home to this museum that is well known for its wide-ranging collections of seashells, birds, and bird eggs. The bird-egg collection is one of the largest in North America. Some of the artifacts were collected during du Pont's expeditions to such places as the South Pacific and the Philippine Islands. Exhibits include the familiar, the exotic, and the extinct. Many of the once-living specimens appear in dioramas resembling natural habitats and labeled with brief explanations.

At the entrance to the museum—which emphasizes ecology—is a replica of a Giant Squid, *Architeuthis dux*, looming overhead. There are several permanent exhibits, along with temporary and traveling exhibits.

The Shell Gallery recreates the Australian Great African Barrier Reef with scallops, nautilus, and other mollusks. Within the Please Touch area are shells selected from a collection of over two million items that ranks among the top 12 collections of this type in the United States.

In the ornithological exhibit are mounted birds from around the world, selected from the museum's collection of 112 000 bird and nest specimens. These specimens come from a variety of places throughout the world, such as the Philippines and New Zealand. Mounted birds include the New Zealand owl, parrot, and kakapo, which are rare or endangered. Some extinct species are also on view: for example, the heath hen and passenger pigeon died out in the first half of the 20th century. Several of the dioramas have dome covers that allow for close-up viewing. The extensive collection of preserved bird eggs is particularly noteworthy; those on display come from one of the world's largest collections.

The Hall of Mammals Galleries include exhibits representing North and South America, Antarctica, and Africa. The African Water Hole has animals such as the gnu, warthog, lion, impala, and more. There are also displays of exotic flora and fauna of Mount Kenya, an extinct volcano. In these galleries the polar bear and South American jaguar are among the animals displayed. A sampling of native Delaware's flora and fauna are also on view.

The Dinosaur Gallery includes *Tuojiangosaurus* and *Yangchuanosaurus*, Asian relatives of North American species such as stegosaurus and allosaurus. In addition there is a *Parasaurolophus* head and *Archaeopteryx* on display. An exhibit about the legacy of Charles Darwin showcases the biologist's life and research.

Behind the museum in the countryside, the Larry F. Scott Nature Trail winds a mile (1.6 km) through the woodlands and wetlands. Included are two paths: one loop has a birdhouse that visitors can see through to observe the wildlife. This pathway has benches and is handicapped-accessible. The other, longer loop winds though wooded areas and into wetlands. The trails feature interpretive signs along the trail. Also behind the museum is a Wildflower and Native Grass Meadow, which focuses on the habitat of grasslands. A Pollinator Garden includes Delaware native flora such as milkweed and purple coneflowers to attract butterflies and other pollinating creatures. Fourteen boulders near the parking lots teach visitors about local geology, with samples of gneiss (metamorphic granules in layers), quartzite (metamorphic non-layered rock originating as quartz sandstone), sillimanite (official state mineral, an aluminosilicate), and more.

The museum is well known for its extensive research in natural history.

Address: Delaware Museum of Natural History, 4840 Kennett Pike (Route 52), Wilmington, DE 19807.

Phone: (302) 658-9111.

Days and hours: Mon.–Sat. 9:30 a.m.–4:30 p.m.; Sun. noon–4:30 p.m; early closings 12/24 (close at 1 p.m.), 12/31 (close at 3 p.m.). Closed Thanksgiving Day, 12/25.

Fee: Yes.

Tours: Self-guided.

Public transportation: DART bus stops next to site.

Handicapped accessible: Yes.

Food served on premises: None.

Special note: Theater runs nature films regularly.

Website: www.delmnh.org/

Hagley Museum and Library

This open-air museum and library traces the history of the DuPont Company from its humble beginnings to the industrial giant it is today. Located along the Brandywine River, the museum was the site of the first E. I. du Pont black-powder works, built in 1802. The company—once the nation's largest manufacturer of black powder—later diversified into other explosives and then expanded into chemical production. The 235 acres (95 hectares) of museum property are covered with trees, shrubs, restorations, and exhibits.

The industrial potential of the Brandywine River region was recognized by early Swedish settlers who held the land during the early colonial period. The last Swedish governor, John Rising, said of the area in 1651, "There are waterfalls; and at the most important, called the great fall … could be placed … a flour mill, a saw mill and a

Interior of machine shop at the Hagley Museum.
Courtesy of the Hagley Museum and Library.

chamois-dressing mill.... Besides this ... if we could here establish powder-mills it would bring us great profit." Rising foresaw what would come to pass: a major industrial area on the Brandywine River.

The story of what was originally known as E. I. du Pont de Nemours and Company goes back to the 18th century in France. There, Eleuthère Irénée (E. I.) du Pont was trained—by the great chemist Antoine Lavoisier—in the manufacture of black powder for the French government. When Eleuthère Irénée du Pont, his brother Victor, and their father Pierre Samuel du Pont de Nemours came to America with their families, E. I. also brought this expertise. They chose to settle here because other French families where already in the Wilmington area, and it provided the raw materials required for the manufacture of black powder: waterpower, willow trees for charcoal, and water transportation to bring in saltpeter (potassium nitrate) and sulfur. By 1804, black-powder production began and, with the War of 1812, increased greatly.

The main exhibit building, called the Henry Clay Mill in the 1800s, was completed in 1814 as a cotton-spinning mill and was named for the Kentucky statesman who strongly supported American industry in Congress. By 1884 the DuPont Company converted this structure to produce black-powder containers. It is now a visitor center and museum with exhibits on three floors that explore the region's industrial development in the era of water-powered mills. Artifacts, dioramas, and models—some working—present other local industries in addition to black-powder production that helped make this area prosperous. Exhibits and demonstrations of the extensive collection of patent models are on display.

Control of the Delaware colony passed from the Swedish to the Dutch in 1655 and to the British in 1664. After Thomas Willing founded Wilmington in 1731, its waterpower, rich local farmland, and well-traveled trade routes attracted flour mills. By the early 19th century, there were 14 mills in the area, producing 50 000 bushels (1800 m^3) of grain per year that was sold across the nation and around the world.

In the visitor center, a model of a Delaware gristmill tracks the grinding process. The exhibit traces the beginning of automated flour production to Oliver Evans, a Delaware-born engineer and millwright, who received patents for his process in the late 18th century.

The tanning industry, which began in the mid-17th century, became important in the Wilmington area because of the availability of the necessary components —waterpower, oak bark, and cattle. The Hagley property was host to a tannery from 1816 to 1825. Although the standard processing time to cure a hide was two years, a new method used here shortened it to two months.

Merino sheep, known for their high-quality wool, were introduced into the Brandywine area by E. I. du Pont. The textile industry flourished here.

To visit the expansive museum grounds, walk or take a museum bus that follows a regular schedule with selected stops. See an overview of quarrying techniques at a small quarry, one of several on the property that produced Brandywine gneiss to construct the site's buildings. The quarry, within walking distance of the powder yard, is at the foot of Blacksmith Hill—the remains of one of a dozen workers' communities built by the DuPont Company. The community has several restored buildings, some of which are staffed by interpreters in period dress, where visitors learn about the lifestyle of the workers and families who lived there. There is a worker's house, the Gibbons house, the foreman's house, a schoolhouse, and more.

Hagley yard, the powder-production site, has several original buildings and the remains of others still standing. In the restored buildings are exhibits and working machinery.

In the 1858 millwright shop, machinery and work vehicles were constructed and repaired during the era of manufacturing black powder. Today it houses a working machine shop powered by leather belts driven by overhead shafts. Models illustrate the powder-making process, and demonstrations of the machinery occur regularly. In the powder yard are three-walled stone buildings constructed in the 1820s that are linked together with a waterwheel. Their blow-out curtain wall, facing the river, allowed pressure out when explosions occurred—which helped save the equipment. Personnel were not permitted in the mills while they were operating. Originally the waterwheel powered the mills, but the mills and machine shop were later updated and powered by a water turbine. The machine shop's turbine was found on the property and restored to operation. Here stands a narrow-gauge railroad track, a boxcar, and two handcars at the ready to move the powder.

Four different power sources used by the mill at different times are at the original locations today. A reproduction of a 16-foot (4.8-m) 1840s-era wooden breast wheel is installed in the black-powder mills. Waterwheels were gradually replaced by water turbines, on site at the Eagle Roll Mill, the only working roll mill in the United States. As the number of mills increased to match production demands, the water supply proved insufficient, so the turbines and steam-engines were introduced. The restored Engine House has a horizontal slide-valve, box-bed steam engine dating from the 1870s, and a boiler of similar age. In the 1880s, an electric dynamo provided current to light the mill.

On the upper property is the du Pont family's Georgian-style home, opened to visitors, Eleutherian Mills, built in 1802. For convenience, it was built close to the powder making, but the location exposed the house and family to occasional explosions from production mishaps. The barn displays antique vehicles, agricultural tools, weather vanes, and a cooper shop. A French garden planted in 1803 has been restored. The company's first office, located next to the house, is open to the public.

Lammont du Pont, grandson of E. I., was a talented inventor, engineer, architect, builder, and chemist who worked at the DuPont Company in the second half of the 19th century. Through his efforts, the firm moved into the chemical industry. The small frame house built in 1840 that was his workshop and laboratory has been moved to the grounds. The interior, although not original, re-creates his workplace—one typical of the time—and showcases Lammont's achievements.

The Hagley Library, on the upper property, contains 65 000 artifacts including an audiovisual collection, many items from the DuPont corporate archives, including manuscripts from DuPont, the Pennsylvania Railroad, Sun Oil, Sperry-UNIVAC, samples of Kevlar®, Nylon, and Pyralin (a type of celluloid), antique gunpowder flasks, and more, and the nearly 5000 patent models from 19th-century America. All of these collections are open to the public for research. Some pieces are on rotating display in the library.

Address: Hagley Museum, 200 Hagley Road, Wilmington, DE 19807. Hagley Library, 298 Buck Road, Wilmington, DE 19807.
Phone: Museum, (302) 658-2400. Library, (302) 658-2400 ext. 309.

Days and hours: Museum, mid-Mar. to early Nov.: 10 a.m.–5 p.m.; mid-Nov.–Mar.: 10 a.m.–4 p.m. Closed Thanksgiving and 12/25. Library, Mon.–Fri. 8:30 a.m.–4:30 p.m.; second Sat. of the month 9 a.m.–4:30 p.m. Closed Thanksgiving and 12/25.

Fee: Yes, some extra fees for tours.

Tours: Self-guided, or guided on a regular schedule. The Brandywine Bus tour covers the Powder Yards, gardens and grounds, and the family home. The Extended tour also includes the Visitor and Worker's Hill. The Hagley Library is not included in the tours.

Public transportation: DART buses stop by the road at the edge of the property, then walk 0.6 mi (1 km) to the visitor center.

Special Note: Check website for events.

Handicapped accessible: Most, but not all buildings accessible; make special arrangements ahead of time. Bus accommodates wheelchairs; call (302) 658-2400 ext. 259 for details.

Food: Lunch-time café.

Website: www.hagley.org/

Maryland

ANNAPOLIS JUNCTION

National Cryptologic Museum

Located on an outer corner of Fort Meade military base housing the headquarters of the National Security Agency (NSA) is the National Cryptologic Museum. It is the "principal gateway to the public" for the NSA. This museum, which opened in 1993, displays artifacts and information relating to codes, code-breaking, and machinery used to encode and decode secret messages from the beginnings of this nation to the present. The institution holds a priceless collection with thousands of historical cryptological materials. It is the first and only public museum in the intelligence community holding this type of material.

Exhibits begin with coded materials and literature from the Revolutionary War and Civil War. There is also a large collection of material devoted to World War II and the birth of electronic code-breaking. One of the oldest American devices on display is a Jefferson Cipher Device, similar to one used by President Thomas Jefferson. Cipher wheels of this sort date back to the late 18th century. This device is a cylindrical machine wheel with scrambled alphabets at the edges, probably for use in French. The encoder aligns the original text in one row, and selects another row as the enciphered form. An identically aligned device owned by the recipient can decode the enciphered text. One area is devoted to describing cipher letters that General George Washington intercepted from Dr. Benjamin Church, Chief Physician for the Continental Army, and sent to a British officer, revealing details of American capabilities. Church was jailed, then exiled for treason.

By the time of the Civil War, messages were being sent by telegraph, and therefore encrypting them was required. Both the Union and Confederacy were encoding telegraph messages, and were able to collect enemy messages in bulk. An exhibit describes the methods for telegraphy, and line-of-sight signal flags, using the Myer flag system, waving the flags in various ways representing letters of the alphabet. This was a method

SIGSALY, the first true electronic voice encryption system.
Courtesy of the National Cryptologic Museum.

that anyone could intercept merely by viewing the flags; hence these messages also became encrypted. A rare star flag from the Signal Corps is shown in the museum, as well as code books and a Confederate cipher cylinder.

World War I involved early radio messaging, and the birth of US Army radio interception. With listening in on radio communications, intelligence could spy on enemies' conversations without requiring wiretaps. A mock-up of an intercept radio shack in Souilly, France, is on display as an illustration.

One of the larger exhibits in the museum describes various efforts in World War II to decrypt enemy messages. Among the machinery on display is an Electric Cipher Machine (ECM), the only machine system unbroken during the war. Also on view is the German Enigma machine, the standard encryption device used by the Nazis. It was an electromechanical device incorporating a plugboard, three or four rotors, and a reflecting plate. The Polish Cipher Bureau worked from 1928 for 10 years to break Enigma's three-rotor coding scheme, then they passed the secret to Britain and France just before Poland's invasion in 1939. By late 1941 the Germans added a fourth rotor, which complicated matters for the Allied codebreakers. The US Navy spent several years working on solving the new Enigma code, finally solving it in May 1943. Two of the few still-operating Enigmas are on view here and available for visitors to try.

World War II is when true electronic voice-encryption began, invented by Bell Telephone in 1943. The speech encipherment system SIGSALY was installed in Washington, DC, and London. A mock-up of only one-third of a complete station is on display; the complete version weighed 55 tons (50 tonnes) of 80 racks of equipment, needing 13 operators and 15 minutes to set up one phone call. Prominently displayed is the Cryptanalytic Bombe, an electromechanical forerunner to modern computers, built by mathematicians and engineers. The Bombe was used to solve the four-rotor Enigma encoding scheme for communication between German U-boats; the system was run by National Cash Register Company employees, sailors, and WAVES. Other exhibits describe the Navajo code-talkers, and devices used to decode Japanese messages.

A second hall is devoted to the Cold War, with the rise of modern digital computers and satellite spy networks. The museum has two Cray supercomputers, one from 1983–1993 (running at 420 megaflops) and one from 1993 (2.67 gigaflops). Other items on display include a 1962 HARVEST and TRACTOR system, the first fully automated storage and retrieval system; a RISSMAN telemetry-processing computer from the early 1980s; circuit boards and memory modules from a CDC 7600 from the 1970s; telephones with voice encryption, punch cards, and more. Exhibits also detail incidents from the Korean War and the 1960 U-2 spy plane over the Soviet Union.

The museum also runs a reference library open to the public, although visitors should call ahead to ensure that assistance is available. Historic cryptanalysis books dating back to the 16th century are in its collection, including the extremely rare *Polygraphiae Libri Sex*, compiled by German monk and mystic Johannes Trithemius, and published in 1518, the first printed book on this subject.

Address: National Cryptologic Museum, 8290 Colony Seven Road, Annapolis Junction, MD 20701.
Phone: Museum, (301) 688-5849; Library, (301) 688-5845.
Days and hours: Mon–Fri. 9 a.m.–4 p.m; 1st and 3rd Saturdays of the month 10 a.m.–2 p.m. Closed Sundays and federal holidays.
Fee: Free.
Tours: Self-guided; for guided tours, call in advance.
Public transportation: None.
Handicapped accessible: Yes.
Food served on premises: None.
Website: https://www.nsa.gov/about/cryptologic-heritage/museum/

BALTIMORE

Baltimore Museum of Industry

As a major American port city, Baltimore has a long industrial and urban history which has affected places well beyond the local region. The Baltimore Museum of Industry examines Baltimore's industrial developments and the effect on the city, the country, and the world. With so much of the city's commerce dependent on the waterways, the museum is appropriately set on a wharf in Baltimore's restored Inner Harbor. The museum is housed in the 1865 Platt Cannery Building, last of a dozen former cannery buildings in Baltimore.

The museum's theme is "the Industrial Age of America." During this period—1830 to 1950—America moved from an agrarian to an industrial society, setting the stage for today's high technology.

Cannery Exhibit at the Baltimore Museum of Industry. *Photo* by Aaron Clamage. *Courtesy* of the Baltimore Museum of Industry.

A Bethlehem Steel Company shipyard crane, 100 feet (30 m) tall, is the first thing visible upon entering the museum grounds. The restored 1906 steam tugboat *Baltimore* is docked at the museum's pier. One of the oldest surviving vessels of its type in the United States, the Baltimore worked the port until 1963. Also on display is an 1850 shipyard bell used to summon workers.

The diverse assortment of important industries in Baltimore's history includes canning, printing, metalworking, ready-to-wear garment manufacturing, and cargo handling. The museum, which considers itself "Baltimore's industrial attic," displays tools, machinery, instruments, furniture, equipment, models, and product samples. Its collection continues to grow.

In more than a dozen life-size exhibits, the visitor gets a good sense of the types of work performed in the 19th and 20th centuries. The technology, labor, and sociology of the times are explained using authentic factory and port artifacts. One exhibit called Clippers to Containers: A History of the Port of Baltimore displays cargo-handling equipment, ship models, a portion of a modern ship container, and even a section of the museum's old pier. Another exhibit titled Fruits of Labor: A History of Food Processing in Maryland explores canning, brewing, meatpacking, sugar refining, bottling, ice cream making, distilling, and related industries. Exhibits replicating a Baltimore oyster cannery of the 1880s explain the workings of the former Platt Cannery. In winter, oysters were canned; in summer, the machinery processed fruits and vegetables. The Shriver family of Baltimore designed the world's first commercial pressure cooker, which changed the way foods were preserved.

Among "firsts" in the US food industry that helped bring fame to Baltimore are the tin can and packaged ice cream. One of many interesting items in the museum is a small, hand-cranked horseradish grinding machine labeled "Tulkoff's 100% pure 'Hot' Horse-Radish."

In three working shops filled with authentic artifacts are demonstrations of machinery, as guides explain how modernization changed workers and the workplace. The belt-driven machine shop contains such pieces as an early 20th-century stamping press, wooden foundry patterns used in casting massive gears for the Bethlehem Steel Company, and a White and Middleton single-piston engine made in Baltimore circa 1897. The garment loft is set between 1870 and 1920, at a time when Baltimore was the second-largest producer of ready-made clothes, after New York City. The ready-made garment industry began in Baltimore when tailors realized that sailors landing in port for only a day or two needed clothes delivered almost immediately. Because there was no time for fitting and tailoring, the clothes were prepared ahead of time. Printing, another of the city's major industries, is demonstrated in the print shop, which contains thousands of pieces of metal type in various fonts and sizes. Ottmar Mergenthaler invented the Linotype machine in Baltimore, which was able to set type in whole lines. The shop contains a 1905 Linotype as well as an 1828 Hoe Acorn press and an 1871 Franklin Gordon platen press.

Among the modern businesses the museum illustrates are communications, electronics, and broadcasting. Telephone-related equipment, including switchboards, is on exhibit. A broadcasting theater has seating for radio shows and guest lectures.

The exhibit Turning on the Power: A History of Gas and Electricity in Baltimore contains working gas and electric apparatus. The country's oldest power company, Baltimore Gas & Electric, was launched in 1816 as the Gas Light Company of Baltimore. Here are shown an early wooden gas pipe, a control panel from the Limited Electric Railroad Company, and a replica of the 1816 "ring of fire" invented by Rembrandt Peale, the first indoor gas lighting used in a museum.

The Museum of Industry also houses an archive for local businesses such as Bethlehem Steel, Amoco Oil Company, Crown Central Petroleum, Kirk-Steiff, McCormick, Rukert Terminals, and Bendix Radio. There is a library of around 5000 books and journals on Baltimore-area industry. Researchers may visit the archives with prior appointment.

Address: Baltimore Museum of Industry, 1415 Key Highway, Baltimore, MD 21230.
Phone: (410) 727-4808.
Days and hours: Tue.–Sun. 10 a.m.–4 p.m. Closed Mondays, Memorial Day, Thanksgiving, and 12/25.
Fee: Yes.
Tours: Self-guided and docent tours are available. Group tours are by prior arrangement.
Public transportation: Local buses stop within 0.2 mile (0.3 km) of the museum, then walk to site.
Handicapped accessible: Yes.
Food served on premises: None. Food is available along the pier.
Website: www.thebmi.org/

B&O Railroad Museum

American railroading began in the heart of Baltimore, where this railroading museum—the world's first—was established in 1953. The collection at the core of the museum was started by the Baltimore and Ohio (B&O) Railroad in 1892. The museum preserves and displays rare documents, models, equipment, and other railroad-related materials from the Western Hemisphere, particularly North America.

The B&O Railroad was formed in 1828, the tracks were laid in 1829, and the first tickets for regularly scheduled service were sold in 1830. Travelers headed west to Ellicott's Mills, Maryland, using horse-drawn trains. By 1852 the railroad's tracks went from the Atlantic tidewater to the Ohio River.

Our visit begins at the historic Mt. Clare Station, built in 1851. In 1844 Samuel F. B. Morse used the B&O's right-of-way to string wire for a demonstration of his telegraph, which went from Washington, DC, to the town of Mt. Clare, now part of Baltimore. (See Historic Speedwell, Morristown, New Jersey.) "What hath God wrought?" was the world's first long-distance telegraph message—a message that changed how people saw communications and added a new role to railroading. In isolated towns across the United States, the railroad telegraph was the main means of keeping in touch with the outside world. At the entrance to the Mt. Clare structure is a plaque engraved with this historic sentence. Within Mt. Clare Station is an exhibit on the work of Morse.

In a grand Victorian structure added in 1884, known as the roundhouse, was the repair shop for passenger cars. Architect Ephraim Francis Baldwin modeled this building after a 13th-century cathedral. The 22-sided structure, more than 200 feet (61 m) in diameter and 120 feet (37 m) tall, still has its original wooden turntable. Extensive use of glass allows natural light to fill the building. These and other structures, such as the annex building—originally the headquarters for the printing department—house additional railroad museum exhibits. The displays include artifacts, models, and more. One section houses a working HO-gauge model layout that re-creates the time when both steam and diesel shared the rails. Topics covered include the development of the railroad tie and rail as well as the importance of railroad time. (See American Clock and Watch Museum, Bristol, Connecticut.) A small theater, designed to look like a railroad car, shows historic films about trains.

Twenty-two stalls in the roundhouse display the museum's oldest and finest steam, diesel, and electric locomotives. Some 19th-century passenger and freight equipment is on view as well. In the front and back yards are other cars and engines, some of historic significance; this is where shops were once used for designing and constructing thousands of them.

Included among unique pieces in the museum's extensive collection of cars and engines is the grasshopper-style engine built by John Hancock in 1886; the 1875 J. C. Davis, a 2-6-0 Mogul type engine; the 1869 Davis Camel engine designed by John C. Davis; a 2-8-2 Mikado-type standard locomotive; a 1905 three-truck Shay-type engine built by Lima Loco for the Greenbriar, Cheat, and Elk line; the 1888 B&O Consolidation-type 2-8-0 A. J. Cromwell freight locomotive; the 1941 C&O Allegheny, considered by many to be the most powerful engine ever built; and the 1925 Alco-GE-Ingersoll Rand 300-hp (224 kW) switcher, the first commercially successful diesel used on mainline America. (See Carillon Historical Park, Dayton, Ohio.)

Another one of the museum's prize possessions is a 1927 replica of the first steel locomotive ever built in the United States, the Tom Thumb by Peter Cooper. The Tom Thumb was tested at this site on August 28, 1830. A few times a year the replica travels the same route Cooper used when he raced his engine against a horse-drawn train. On weekends, visitors can board a passenger train from the 1940s and ride a mile and a half (2.5 km) to the spot where the first stone was laid for the nation's first railroad.

Address: B&O Railroad Museum, 901 West Pratt Street, Baltimore, MD, 21223.
Phone: (410) 752-2490.
Days and hours: Mon.–Sat. 10 a.m.–4 p.m.; Sun. 11 a.m.–4 p.m. Closed Easter Sunday, Memorial Day, 7/4, Labor Day, Thanksgiving, 12/24, 12/25, 1/1.
Fee: Yes. The train ride requires an additional fee.
Tours: Self-guided.
Public transportation: Local buses stop about 2.5 blocks away, then walk 0.1 mile (150 m) to site.
Handicapped accessible: Yes.
Food served on premises: Beverages and snacks year-round; café seasonally.
Website: www.borail.org/

National Aquarium

Sitting majestically at the edge of Baltimore's Inner Harbor is one of the largest and most sophisticated aquariums in the nation. This organization is a center for marine research and public education. The Aquarium consists of two structures, each capped by a glass pyramid; together these structures house a total of more than 20 000 aquatic animals living in natural habitat exhibits.

The larger structure, known as Pier 3, has five levels or floors, accessible by both escalator and elevator. Visitors may take a carefully directed, self-guided tour through the displays of sea life in habitats from all over the world. Displays range from the cold Icelandic coast to the hot and humid tropical rainforest.

Level 1 introduces the Indo-Pacific Reef, with nearly 70 species, mostly fish, including black-tip reef sharks. This exhibit is one of the largest for rays in the nation, with 50 southern stingrays, cownose rays, and bluntnose rays gathered off the Atlantic Coast. The huge 260 000 gallon (980 000 liter) saltwater ray pool sits at the base of the escalator, which provides an interesting view as you ascend. Through underwater viewing windows around the tank, watch rays as they feed.

Level 2, Maryland: Mountains to the Sea, displays four natural habitats. This exhibit follows the path of rainwater as it flows from a mountain pond to the sea. Residents of Allegheny mountain ponds are bullfrogs, softshell turtles, sunfish, crappies, and catfish. The tidal marsh exhibit is a typical Eastern Shore habitat with mummichogs, diamondback terrapins, and blue crabs. Coastal beach has a variety of sea life that includes needlefish, killdeer, and flounder. Tautogs, striped bass, and sea trout live in the Atlantic shelf habitat.

Level 3 displays some 22 exhibits that explain the importance of adaptation and illustrate the adaptive strategies that help animals survive, such as the defensive spikes of the lionfish, the bioluminescent winks of the deep-sea pinecone fish, the brilliant colors of the sea anemones, as well as electric eel, chambered nautilus, and giant Pacific octopus.

Level 4, the North Atlantic to Pacific gallery, provides visitors with a look at three diverse aquatic communities: puffins from the Icelandic coast, a California kelp forest, and a Pacific coral reef. Also located here is Children's Cove, also appealing to adults, which contains a touch tank where visitors can handle intertidal marine animals.

Level 5, the top level, has the lovely glass pyramid, 64 feet (20 m) high, covering the warm and humid South American Rain Forest. This stratified natural habitat houses 30 species of tropical fish and more than 600 species of tropical plants, as well as reptiles, free-flying tropical birds, and a pair of roaming two-toed sloths. As visitors go through this total-immersion jungle, watch for the Hidden Life exhibit with its poison dart frogs and other rare, small animals of the rain forest. Specimens from the poison dart frog collection, one of the most extensive in the United States, have been sent to other zoos and aquaria for breeding.

From the top level, descend by walking six levels down a ramp. The first stop, the 13-foot (4-m) deep ring tank, holds the Atlantic Coral Reef. This exhibit allows people to see a tropical reef as a diver sees it—except that the visitor stays dry. The tank holds bonnethead sharks, tarpon, endangered hawksbill turtles, and much more. Next the visitor finds the Open Ocean, an exhibit containing sharks, other large game fish, and two species of rays.

An enclosed bridge goes from the aquarium on Pier 3 to the Marine Mammal Pavilion on Pier 4 with its 1 300 000 gallon (4 900 000 liter) pool, as well as a variety of temporary exhibits. Listen for the songs and sounds of the humpback whale while walking through the bridge. Upon entering the pavilion, note the life-size model of a humpback whale, Scylla, which fills two levels of the atrium. By looking through the pavilion's viewscopes, the whale's anatomical features are visible up close.

In the 1300-seat amphitheater, mammalogists explain the natural behaviors of bottle-nosed dolphins and beluga whales. Videos of the mammals, augmented with animation, enhance the shows. At the lower level of the atrium, beluga whales and Atlantic blue-nose dolphins live side by side in a pool, which the visitor can see through underwater viewing windows.

Hands-on experiments in an educational arcade explain the behavior of whales and dolphins. The discovery room contains shark's teeth and jaws, whale vertebrae, and other objects to handle and examine up close.

Address: National Aquarium, 501 East Pratt Street (at Pier 3 Pavilion), Baltimore, MD 21202.
Phone: (410) 576-3800.
Days and hours: Sun.–Thu. 9 a.m.–5 p.m.; Fri. 9 a.m.–8 p.m.; Sat. 9 a.m.–6 p.m.
Fee: Yes.
Tours: Self-guided.
Public transportation: Metro Subway, MARC train, local bus, free Charm City Circulator, and light rail all stop nearby.
Handicapped accessible: Yes.
Food served on premises: Café.
Special note: Check schedule for times of amphitheater shows and other special events. Docking in the Inner Harbor is possible: call (410) 396-3174 in advance.
Website: www.aqua.org/

LINTHICUM

National Electronics Museum

Housed in a non-descript building reminiscent of an industrial or office park, the National Electronics Museum focuses on the history of the electronics defense of the United States. Displays are packed full of interactive and original examples of the electronics that have helped protect this country over the past centuries. There is something for everyone, from the electronics novice to the electrical engineer. First opened in the 1980s, the site has expanded as more and more materials have been be added to its collection.

Upon entering the facility, visitors see some half a dozen radar antennas of various eras stationed outside on the lawn. Indoor exhibits begin with the basics of electricity and magnetism along with hands-on displays that describe electromagnetic theory. Storage batteries, Samuel F. B. Morse's telegraph, and Alexander Graham Bell's telephone are demonstrated (see Historic Speedwell, Morristown, New Jersey), then early radio communications pioneered by Marconi (see InfoAge Science History Learning Center, Wall Township, New Jersey; and The Sarnoff Collection at The College of New Jersey, Ewing, New Jersey.). A reproduction amateur radio shack, incorporating a "spark gap" transmitter and an operational amateur telecommunications station, is also on display. Other countries are not ignored as one moves on to the rapid development of British radar systems in the late 1930s. There are examples of radar-controlled anti-aircraft guns to destroy German bombs over London in World War II.

Museum exhibits also examine Cold War Radar, with a demonstration of microwaves and how Doppler radar measures speeds of automobiles, as well as an array of Air Force and Navy systems. Side-looking radar explains how ground-mapping became possible under adverse weather conditions. Modern radar systems are also presented. This includes AWACS air-based and Joint Stars ground-based methods, an AN/APG-63 radar system for the F-15 aircraft and the AN/APG-68 for the F-16 aircraft, plus an AWACS antenna.

Self-defense methods to shield this country against enemies are examined. Rapid deployment of countermeasures is necessary, and the development of such systems is shown here. Chaff and decoy methods for fooling radar are displayed. The undersea world is not ignored. There are electronic measures, used by submarines, with underwater sound transmission, sonobuoys, and sonar methods that are all explored. Learn how optical detectors convert images into electronic signals. Visitors can compare their image in visible and infrared, and see the Pave Spike system for McDonald Douglas F-4 Phantom airplanes, used to deliver rockets and bombs to precise targets in total darkness.

Any modern defense electronics system includes satellites, which are exhibited. On display are the Lunar TV Camera from the Apollo missions to the moon, early weather satellite sensors, navigation methods from early furniture-sized satellites to today's modern GPS systems, and the AIMP and Ariel II satellites for scientific research.

Many of the docents in the museum are retired engineers who have worked with defense electronics systems, and so are very knowledgeable about the exhibits and materials.

The library serves as a research center and is opened to the public. Included in its collection are over 10 000 books, journals, magazines, video recordings, and technical documents related to Cold War–era communications, radar, defense, amateur radio, and military electronics history. There are works of art and photographs in the collection as well.

Address: National Electronics Museum, 1745 West Nursery Road, Linthicum, MD 21090.
Phone: (410) 765-0230.
Days and hours: Mon.–Fri. 9 a.m.–4 p.m.; Sat. 10 a.m.–2 p.m. Closed Memorial Day, 7/4, Labor Day, Thanksgiving Day and the following Friday, and 12/24 through 1/1.
Fee: Yes.
Tours: Self-guided; group tours are often available with two-weeks' advance notice.

Handicapped accessible: Yes.
Public transportation: Maryland Transit Authority buses stop next to the site.
Food served on premises: None.
Special note: Licensed ham-radio operators may use the amateur station with advance notice. Bring your license.
Website: www.nationalelectronicsmuseum.org/

SILVER SPRING

National Museum of Health and Medicine

During the Civil War, more soldiers died from diseases than from enemy bullets. In 1862, as the war raged, the Army Medical Museum was created to study and try to conquer such illnesses. At that time, Surgeon General William Hammond directed his officers to gather "specimens of morbid anatomy... together with projectiles and foreign bodies removed" from wounded soldiers for research. This collection is still at the museum, now called the National Museum of Health and Medicine (NMHM), and paved the way for the investigation of the illnesses and injuries of both the soldiers and the public as well. The museum's exhibits focus predominantly on trends in military medicine since 1862. Its mission is to inspire an interest in and understanding of the past, present, and future of medicine in the US military.

For many years the museum was located on the National Mall, within walking distance of the Capitol, but moved in the 1970s to Walter Reed Army Medical Center's

Advances in Military Medicine exhibit at the National Museum of Health and Medicine. *Courtesy* of the National Museum of Health and Medicine.

campus at the northern edge of the city. By 2011 the museum moved again, this time to the suburban site in Silver Spring.

Currently the museum is under the auspices of the Department of Defense as an element of the Research and Development Directorate of the Defense Heath Agency, with over 24 million artifacts and specimens. Materials of importance include photomicrography techniques, a library and cataloging system, the basis of National Library of Medical Research which investigates infectious diseases and vaccinations such as yellow fever and typhoid fever. During World War I this institution investigated ways to combat sexually transmitted diseases.

Over the years the museum has gained a national reputation for responding to the ever-changing health needs of the country. It has conducted research in trauma and wound-healing that led to important breakthroughs. Before the census of 1890 was conducted, curator John Shaw Billings helped devise a new data-reporting scheme to count the population. Billings worked with inventor Herman Hollerith to develop a way to record this information on paper cards; Hollerith later formed Tabulating Machines Company, the forerunner of IBM Corporation.

Exhibits are devoted to the medical traumas of the Civil War, during which dysentery and diarrhea were particular problems. An array of interesting topics is presented, including traumatic brain injury, forensic identification, anatomy and pathology of the human body, and innovation in military medicine. The museum narrates the story of how physician and surgeon Walter Reed, who was curator of the museum in the late 19th century, helped conquer yellow fever. The museum played a key role in bringing typhoid fever under control when curator Frederick Russell and his staff tested a typhoid vaccine early in the 20th century.

A display on the death of President Abraham Lincoln includes the bullet and the gun that killed him, bone fragments from his skull, and other materials taken at his autopsy. This Lincoln exhibit also includes a life mask.

During World War II, with a turn toward pathology, the museum was an important collector, analyzer, and disseminator of information about what were then called venereal diseases.

Among the important and diverse collections at this institution include 5 000 skeletal remains, 8 000 preserved organs, 12 000 items of medical equipment, as well as some remains of Albert Einstein's brain and a Robert Hooke microscope ca. 1660 (one of the earliest objects in the museum's collection).

Displays range from 19th-century stethoscopes to objects people swallowed that have been removed by modern techniques, to the world's most comprehensive microscope collection.

Researchers may visit the five NMHM collections (historical, anatomical, the Otis Historical Archives, Human Development, and neuroanatomical) by appointment only. These Archives were founded in 1968 to preserve the historical records of the Museum and collect documents, photographs, books, and manuscripts. Features of the collection include anatomical illustrations, photomicrographs, and descriptions of surgery on the battlefield.

Address: National Museum of Health and Medicine, 2500 Linden Lane, Silver Spring, MD 20910.
Phone: (301) 319-3300.
Days and hours: Daily 10 a.m.–5:30 p.m., including weekends and holidays. Closed 12/25.

Fee: Free.

Tours: Self-guided.

Public transportation: Montgomery County buses stop within 0.3 mile (0.5 km) of the site, then walk.

Food served on premises: None. In downtown Silver Spring, about 1 mile (1.6 km) southeast are restaurants.

Handicapped accessible: Yes.

Website: www.medicalmuseum.mil/

St. Michaels

Chesapeake Bay Maritime Museum

The name Chesapeake derives from the Indian word for the bay, *Chesepiook*, which some say means "great shellfish bay." The name is still appropriate; after more than 350 years, Maryland continues to depend heavily on the bay for its livelihood. Harvesting and processing crabs and shellfish and building boats are the important industries in the state. This waterside museum and 18-acre (7.3 hectare) complex on the Bay Hundred Peninsula contain restored and new structures tracing the history of boat-building, commercial fishing, seafood harvesting, yachting, waterfowl, and navigation on the Chesapeake Bay. The facility includes working shipwrights and apprentices in its own shipyard. It is possibly the largest shipyard of this type in existence with a small-boat collection of crabbing skiffs, workboats, and log canoes.

At the Bay History Building, there are boat models, paintings, a sailmaker's studio, and artifacts that interpret the bay's cultural history. Displays trace prehistoric times,

Hooper Strait Lighthouse, Chesapeake Bay Maritime Museum.
Courtesy of the Chesapeake Bay Maritime Museum.

Native American societies, the colonial era, the War of 1812, the Civil War, and the contemporary period.

During the American Revolution and the War of 1812, the town of St. Michaels was very active in shipbuilding. Although the importance of St. Michaels's harbor declined as the fortunes of Baltimore harbor increased, the town remained important as a boat-building center and a crabbing, shell fishing, and waterfowl area.

The Waterfowling Building features an extensive collection of working decoys and a decoy carver's workshop. The word "decoy," used by the Native Americans, comes from the Dutch word *kooi*, which means cage or trap. Also on exhibit are guns used in hunting waterfowl, including the punt gun and eight-barrel battery gun, which had been outlawed because they were destroying the wildfowl population.

The centerpiece and highlight of the museum is the 1879 Hooper Strait Lighthouse. This "screwpile" or cottage-type lighthouse, one of only three remaining on the bay, was built in 1879 and was moved 40 miles (64 km) to its present location in 1966. The screwpile method of construction means that it is built on seven screw-tipped iron pilings that were screwed into the bottom of the bay to at least 10 feet (3 m) deep. In years past, the lighthouse served as a navigation aid; to this day the lighthouse projects a continuous beam of white light. Besides providing a panoramic view of the grounds, the harbor, and Miles River, the lighthouse has exhibits about the lighthouse keeper's job.

The Small Boat Shed was built in 1933 from the former Claiborne steamboat and railroad terminal for Baltimore tourists on their way to the resort town of Ocean City, Maryland. The shed now displays more than 20 small work and pleasure boats typical of the bay. These include a Native American dugout canoe, a five-log oyster-tonging canoe, and several crabbing skiffs.

The process of building traditional wooden boats is demonstrated in the Boat Shop, where skilled craftspeople restore and maintain the historic boats and other vessels belonging to the museum. The seafood industry, a major part of Chesapeake Bay life, is explored in depth in the Oystering on the Chesapeake exhibit. Visitors can board the historic skipjack *E. C. Collier*.

Tied up along the docks is the museum's own fleet. These traditional Chesapeake Bay fishing boats include *Edna E. Lockwood*, the last historic log-built bugeye that still sails on the bay; *Rosie Parks*, one of the bay's fastest skipjacks; *Old Point*, a Virginia crab dredger; and *Winnie Estelle*, a buyboat.

The Museum collects and preserves the nation's most "comprehensive assembly of material culture relating to the Chesapeake's tidewater region." This collection includes working and decorative decoys, maritime paintings, and over 70 000 other objects documenting this region for over 200 years. Access to the archives is by appointment only.

Address: Chesapeake Bay Maritime Museum, 213 North Talbot Street, St. Michaels, MD 21663.

Phone: (410) 745-2916.

Days and hours: May–Oct.: 9 a.m.–5 p.m.; Nov.–Apr.: 10 a.m.—4 p.m. Closed Thanksgiving Day, 12/25, and 1/1.

Fee: Yes.

Tours: Self-guided; guided tours daily, group tours.

Public transportation: None.

Handicapped accessible: Mostly. Hooper Strait Lighthouse and the upper decks of the Boat Shop are not; call in advance for details.

Food served on premises: None. Restaurants are available in St. Michaels.
Special note: Contact the museum in advance for information on docking privileges.
Website: www.cbmm.org/

New Jersey

EWING

The Sarnoff Collection

David Sarnoff, born in 1891 in a small town in what is now Belarus, came to the United States with his family in 1900. The young Sarnoff began his career in 1906 at the Marconi Wireless Telegraph Company of America in electronic communications. Sarnoff was involved in electronic communication from its very inception from General Electric Company to Radio Corporation of America Company (RCA), from ship (the *Titanic*) to shore, from the telegraph key to radio and TV. (See InfoAge Science History Learning Center, Wall, New Jersey.)

RCA established the Sarnoff Collection in 1967 as the David Sarnoff Library. The focus of this library is the major developments in communications during the 20th century. Materials cover Sarnoff's papers, memorabilia, photos, notebooks, and RCA material—all of which number more than 6000 artifacts. There are various items concerning David Sarnoff, as well as RCA and RCA's predecessors and subsidiaries (Marconi Wireless Telegraph Company of America, Victor Talking Machine Company, and NBC). (See Johnson Victrola Museum, Dover, Delaware.) The collection covers the dawn of

The Sarnoff Collection.
Courtesy of The Sarnoff Collection.

radio starting before the turn of the 20th century, television and commercial broadcasting, recording of both audio and video information, as well as solid-state electronics and optical-display technology.

Though small, the museum in Roscoe L. West Hall at the College of New Jersey has a detailed exhibition hall with panels and displays in chronological order. Exhibits begin with a description of Sarnoff's early life, illustrated with photographs, showing his family in Belarus and later with Sarnoff at his telegraph operator's desk. Several painted portraits of Sarnoff are on view. There are some early phonographs, including a Columbia Graphophone from the late 1890s that plays cylinders, a 1935 portable Victrola in a briefcase, and even a CIA manual-cranked propaganda phonograph from the mid-1960s that played cardboard disks, designed for being dropped secretly into Eastern Europe to bring Western ideas behind the Iron Curtain.

The Age of Radio is illustrated by several RCA Radiolas in wooden cabinetry and a 13-inch (33-cm) diameter Radiola Model 100 speaker. A variety of televisions and associated electronic equipment is on exhibit, including Zworykin's iconoscope (the original modern cathode-ray tube, or CRT), the first color picture tube, and RCA's first color television set. A set of phosphors, which glow when struck by a beam of electrons, is on display as well.

With the invention of the transistor in 1947 at Bell Laboratories, RCA saw the potential in consumer electronics (radios, audio reproduction, and televisions) almost immediately. By 1952, RCA engineers built a 37-transistor prototype television that weighed only 27 pounds (12 kg), which is on view here in the collection. Within a decade, the race to miniaturize solid-state circuitry was on, and The Sarnoff Collection exhibits a variety of artifacts related to integrated circuits, all the way to a 1997 CMOS active pixel sensor ("camera on a chip").

Simultaneously with the revolution in consumer electronics, the digital revolution began after World War II. One interesting item is the handmade 256-bit core-memory, with magnetic parts strung on wires in a wooden frame, dating from 1950. Within a few years the fabrication process was mechanized, so nearby is a 10 000-bit memory dating from 1955, along with early RCA microcomputers from the early 1970s.

RCA was interested in new optical display technologies to improve upon the fragile and bulky CRT, which required dangerous high voltage to operate. In the mid-1950s, RCA built a prototype flat-panel display (10 × 10 matrix of bulbs) that visitors can see, but shortly thereafter RCA moved on to electroluminescent materials (materials that glow under the application of electric current) and the unusual optical properties of liquid crystals, materials that flowed, but had crystalline order in one or two dimensions, useful for liquid-crystal displays (LCDs). Another exhibit explores materials from the late 1960s on, such as an early electroluminescent panel, and LCD watches from 1971.

In the 1960s and 1970s, RCA worked on extending the idea of the phonograph to a video disc, in which both audio and video were recorded in microscopic grooves. This system was not used by CDs or DVDs, and eventually was discontinued. A 1980 SelectaVision VideoDisc Player, along with a demonstration model showing the needle and grooves from 1960, and associated advertisements, is on exhibit.

Besides consumer electronics, RCA was instrumental in developing the research-grade electron microscope. A rare EMB-4 electron microscope, a heavy, free-standing piece of equipment from 1944, is part of the collection. Electron microscopes revolutionized biology, making viruses and subcellular structures clearly visible for the first time. Outside the exhibition hall are a series of photographs documenting some milestones and equipment developed at RCA.

Exhibits do change, displaying other material in the collection.

The Sarnoff library material—papers, memorabilia, photographs, notebooks, and RCA materials—were placed at the Hagley Museum and Library in Wilmington, Delaware. (See Hagley Museum and Library in Wilmington, Delaware.)

Address: The Sarnoff Collection, 2nd floor, Roscoe L. West Hall, The College of New Jersey, 2000 Pennington Road, Ewing, NJ.
Phone: (609) 771-2654.
Days and hours: Wed. 1–5 p.m.; Sun. 1–3 p.m.
Fee: Free.
Tours: Self-guided. Guided tours and group visits are available; e-mail in advance to make arrangements at sarnoff@tcnj.edu.
Handicapped accessible: Yes.
Public transportation: Trains from Philadelphia to West Trenton via NJ Transit, then taxi; NJ Transit bus from Trenton stops on-campus at the College of New Jersey. Walk from the stop to Roscoe L. West Hall.
Food served on premises: Student food service, TCNJ Campus Town at the College entrance.
Special note: Check the website or call for lectures and events.
Website: davidsarnoff.tcnj.edu/

MORRISTOWN

Historic Speedwell

Historic Speedwell is a restored, 19th-century ironmaster's country estate, covering more than eight acres (3 hectares) with three original farm buildings, plus three 18th- and early 19th-century houses from Morristown that were moved to this site because they were in danger of destruction.

This site is a National Historic Landmark because of two important events. From 1818 to 1819, ironmaster Stephen Vail built the engine for the first steam-powered ship designed to cross the Atlantic, the S.S. *Savannah*. In the 1830s, Stephen's son Alfred and his partner, Samuel F. B. Morse, developed, then demonstrated on January 11, 1838, the commercially viable electromagnetic telegraph from this spot.

In 1815 Stephen Vail became the sole owner of the Speedwell Iron Works on the Whippany River. The foundry produced consumer goods from pig iron and made most of the machinery for the S.S. *Savannah*. Vail's marine engineering success firmly established his reputation, and he prospered. In 1829, just before becoming a county judge, he bought adjacent land that included a cotton-weaving mill and converted it to a gristmill.

The Vail House dates from the 1820s and was completely renovated by Stephen Vail in 1841 to add central heating, indoor plumbing, and other features for comfort and not for style. Much of the furniture and memorabilia on view are original. Period memorabilia have been returned to the house by descendants. Of major importance at the house are two paintings by Samuel F. B. Morse: portraits of Stephen Vail and his wife, Bethiah Vail.

Exterior of the Factory, Historic Speedwell.
Courtesy of Historic Speedwell.

Judge Vail, noting his eldest son's mechanical talent, wanted Alfred to join the business, but Alfred chose instead to study for the ministry at New York University. Samuel Morse, a trained artist, took an unpaid post as professor of literature and fine arts at the school. Later Alfred decided to explore the sciences as a route to financial success and became interested in the telegraph. In 1837, on a return visit to his alma mater, Alfred Vail saw Morse demonstrate the electromagnetic telegraph.

Alfred became fascinated by this magnificent machine and persuaded his father and brother at Speedwell to help Morse with mechanical advice, to provide a workshop with tools, and to offer financial assistance. The workshop was set up in the second story of the Speedwell gristmill.

Morse was a natural leader and organizer who was unafraid of controversy. Alfred Vail was a mechanical genius. A third member of the team, Leonard Gale, was a colleague of Morse and a friend of Vail. Gale's contribution was his chemical background and familiarity with the works of others in the field.

This gristmill, now known as the Factory, houses a small museum with documents, models, and instruments explaining the invention and development of the telegraph. In the beams, the nails used in 1838 to string two miles (3.2 km) of wire for demonstrations are still visible. The lower level of the gristmill houses the mill remains and a circular saw and other exhibits.

Other structures on the site include a wheelhouse with a 24-foot (7.3-m) overshot waterwheel and the 1830 granary, with a unique upstairs corn-storage area. The clapboard sides are louvered outward for good air circulation and to protect the grain from bad weather. Several Morristown houses built in the 18th and 19th centuries were in danger of being destroyed and so were moved to Speedwell. Buildings exhibit old-time industrial equipment and a collection of original wood patterns from the foundry, which were used to make cast-iron gears and parts for locomotives. One of the buildings moved to the property has a re-created 18th-century kitchen with an open hearth that is occasionally used for cooking demonstrations. Ask at the information desk for more information.

Exhibits also at Speedwell include information about the S.S. *Savannah*, Speedwell's ironworks and its place in American maritime history, as well its position in the Industrial Revolution.

Address: Historic Speedwell, 333 Speedwell Avenue, Morristown, NJ 07960.
Phone: (973) 285-6550.
Days and hours: Apr.–Oct.: Wed.–Fri. 10 a.m.–5 p.m.; Sat. 10 a.m.–6 p.m., Sun. noon–6 p.m.
Fee: Yes.
Tours: Self-guided and guided. Docents are available.
Public transportation: NJ Transit bus stops 0.4 mile (0.6 km) from site, then walk.
Handicapped accessible: Limited.
Food served on premises: None. Picnic area for bagged lunches.
Special Note: Call ahead for special events.
Website: 37.60.235.13/~morrispa/index.php/parks/historic-speedwell

OGDENSBURG AND FRANKLIN

The Sterling Hill Mining Museum and the Franklin Mineral Museum

Tucked up into the hills of northern New Jersey is an unusual confluence of zinc-, manganese-, and iron-based minerals that is duplicated nowhere else on earth. The minerals found here originated in the pre-Cambrian era, when the area was a seabed dotted with so-called black smokers. Later uplifting and metamorphic changes to the area, plus extensive weathering over millions of years, gave rise to rare fluorescence oxides, sulfides, and even arsenates.

The multiple kinds of minerals (at least 357 known types) are generally found within marble deposits. These minerals were lifted up and eroded around 100 million years ago. Ice-age glaciers transported many large marble boulders throughout the area. Of the types of minerals found here, 91 are fluorescent, and 35 are found nowhere else in the world. The Franklin area of New Jersey is the only place known on Earth where zinc and iron occur together.

Fluorescence itself is a physical property of a material, which occurs when light strikes the material, raising an electron from one orbital to a higher-energy orbital. The electron returns rapidly (usually within nanoseconds) back to its original state, while releasing its extra energy in the form of light. We can see fluorescence by using an ultraviolet lamp shining on materials, which fluoresce at lower-energy visible light. Mineral fluorescence was not discovered until the 19th century by Clark and Haüy, so the unusual optical properties of the local geology remained unknown until after the colonial period.

Together, the Sterling Hill Mining Museum and the Franklin Mineral Museum, just a short distance—about two miles (3 km)—from each other, explore the historic and scientific aspects of the area's notable zinc mines, and the surprising ores extracted from them.

The Sterling Hill Mining Museum

Here in Ogdensburg is one of the oldest mines in the United States. Mining probably began here in the mid-1600s, because the area was originally thought to have copper deposits (though this eventually was found not to be the case). Even so, by the 1730s this area was called the "Copper Mine tract." King George III of the United Kingdom granted the property to William Alexander, Lord Stirling, who was active in mining and farming, and built an estate in Basking Ridge. Lord Stirling sold the mine in 1765 to Robert Ogden.

The Franklin-Sterling Hill was dotted with several mines producing zinc and iron. These mines joined together to form the New Jersey Zinc Company in 1897. By 1986 this company closed the mine because low zinc prices made mining the ore uneconomical, all the while dealing with a tax dispute with the town of Ogdensburg. The town eventually sold the property at auction. Brothers Richard and Robert Hauck bought the mine and surrounding lands, and created the Sterling Hill Mining Museum in 1990.

Introducing visitors to the Mining Museum is a guided tour of much of the property, starting with the Zobel Exhibit Hall—originally the changing house for the miners. Lockers and hanging clothes baskets (in which to dry wet miners' clothes in the warmer air near the high ceilings) are still visible. Among the scientific and engineering artifacts on display in the Hall is an imposing periodic table of the elements, with samples of each element. The Oreck Mineral Gallery, lit by fiber optics, displays minerals from all over the world. Numerous chemical-analysis instruments such as analytical balances are on view as well, though they are not well documented or labeled. There is a generous selection of mining equipment, especially showing the development of lighting within mines from candles to electric lamps. Tour guides can be very informative.

After the Exhibit Hall, visitors walk 1300 feet (400 m) through part of the actual mine with displays of drilling equipment and the Rainbow Room, lit by ultraviolet lamps to show the fluorescence of the native minerals in the rock walls. Then the guide brings visitors to the Thomas S. Warren Museum of Fluorescence, displaying a variety of fluorescent minerals and a number of cases displaying the technological development of ultraviolet sources from iron-spark generators of a century ago through low-pressure mercury-vapor lights of today. Ask the guide for extra time to see these cases. Among the types of fluorescent minerals showcased are sphalerite and calcite, which glow in various colors because of impurities in the crystal structure, scheelite (blue to blue-white), fluorite (mostly blue to violet), and willemite (bright yellow-green). (See American Museum of Natural History, New York, New York.)

Local astronomy groups use the Ellis Astronomical Observatory on site, which contains a 20-inch (51-cm) reflector telescope, a 12.5-inch (32-cm) reflector, and a hydrogen-α telescope for viewing the Sun safely.

Address: The Sterling Hill Mining Museum, 30 Plant Street, Ogdensburg, NJ 07439.
Phone: (973) 209-7212.
Days and hours: Vary during the year. Check the website or call ahead for the current schedule.
Fee: Yes. Extra charges for the Rock Discovery Center and Fossil Discovery Center.
Tours: Guided tours are the only way to see this site. Ask many questions of the guides about the site and equipment on display.
Public transportation: None.
Handicapped accessible: Yes; seating areas in the mine.
Food served on premises: Indoor snack bar; outdoor picnic area.
Special note: Mine remains at a constant temperature of 56°F (13°C). Bring a jacket for the cool temperature in the mine, and wear sturdy walking shoes.
Website: www.sterlinghillminingmuseum.org/

Franklin Mineral Museum

Opened in 1964, the museum's mission is to preserve the mineral wealth, geology, knowledge, and history of "the greatest mineral locality on earth." The first documented mineral in the Franklin area, zincite, was discovered in 1810 by Archibald Bruce. Thousands of types of minerals from the local area and around the world are classified and displayed systematically, in the rooms of Local Minerals (over 4000), World Minerals (over 5000), and Fossils (including a slab of petrified wood over four feet [1.2 m] in diameter). A large hall (33 ft/10 m long) of fluorescent minerals is open to view. In the World Minerals collection there are a number of antique microscopes on loan, originally used by the Leidy Microscopical Society.

Important minerals include roeblingite (named for John Roebling, the New Jersey civil engineer who designed the Brooklyn Bridge; see the Roebling Museum, Roebling, New Jersey), which contains significant amounts of lead and calcium, and glows under ultraviolet and blue irradiation; willemite (zinc silicate); and three specimens of the rare lennilenapeite, a manganese-iron mineral named for a local tribe of Native Americans.

Visitors also may tour through a two-level mine replica, showing the tools and techniques used in northern New Jersey to extract ores. Among the interesting artifacts is a selection of miner's lamps, an Edison "battery jar," miners' paychecks and other memorabilia, as well as a reconstructed chemist's laboratory from the early 1900s. One room of the museum houses a collection of Native American artifacts. Rock collectors can scour the Buckwheat dump to find new treasures for their collections.

Address: Franklin Mineral Museum, 32 Evans Street, Franklin, NJ 07416.
Phone: (973) 827-3481.
Days and hours: Mar.: open weekends; Apr.–Nov.: Mon.–Fri. 10 a.m.–4 p.m.; Sat. 10 a.m.–5 p.m.; Sun. 11 a.m.–5 p.m.
Fee: Yes. Extra fee for visiting the Buckwheat dump area for rock collecting.
Tours: Self-guided, and guided tours hourly (when available)—check in advance. Because museum labeling is minimal, we recommend a guided tour.
Public transportation: None.
Handicapped accessible: Mostly, except for a small room upstairs.
Food served on premises: None, except for outdoor picnic area.
Website: www.franklinmineralmuseum.com

ROEBLING

Roebling Museum

Johann August Röbling, born in 1806 in Mülhausen, Prussia (which was to become part of Germany), was educated as an architect and engineer at the Royal Building Academy in Berlin. By 1825 Röbling was in Amsberg, Westphalia, building government military roads and making sketches for suspension bridges. Because the political and economic conditions were unstable in this part of Europe, Johann, his older brother Carl, and a group of young people left Germany and immigrated to the United States in 1831.

These two brothers purchased 1582 acres (640 hectares) in Butler County, near Pittsburgh in the hills of western Pennsylvania, for the purpose of starting a farming community for German settlers. They named the community Saxonburg. By this time Johann called himself John A. Roebling and became a farmer.

After his brother Carl died unexpectedly, John Roebling returned to his engineering profession. He went to Harrisburg and obtained employment with the Pennsylvania Canal System as a surveyor.

At this time in the United States, transportation became more important because there was a need to bring the industrial East together with the farming frontier. He began improving river navigation and canal building, surveyed the Allegheny Mountains from Harrisburg to Pittsburgh, and designed a bridge for Philadelphia. In 1841 he offered his version of a twisted-wire rope as an improvement upon the existing hemp ropes that pulled canal boats. The new rope was a success, and he founded a wire-manufacturing business in Saxonburg, and produced wire rope there from 1840 to 1844.

Roebling's first big success using the wire-rope was the construction of a suspension bridge in Pittsburgh in 1844. He moved his family to Trenton, New Jersey, in 1849, which was a burgeoning small city known for its pottery industry and excellent access to the Delaware River. Some of the bridges he and his family built over the years that are of major importance include a railroad suspension bridge over Niagara Falls and the Niagara Gorge (with the help of his wife, Emily Warren Roebling) and the Cincinnati-Covington Bridge. He designed the world-famous Brooklyn Bridge, spanning the water from the borough of Manhattan in New York City to (the then independent city of) Brooklyn. John Roebling died while the bridge was in progress in 1869, and never lived to see the finished product; he left his factory to his four sons. Charles Roebling was perhaps the most famous son, for he was also an engineer by trade, and became president of the Roebling firm in 1876.

By 1905, with the price rising for steel needed to make wire, Charles built a new factory and company town at the village of Kinkora, a few miles south of Trenton, along the Delaware River. This town eventually took the name of Roebling. At its height, the Roebling works was the leading wire-manufacturer in the world, contributing to the Brooklyn Bridge, San Francisco's Golden Gate Bridge, the George Washington Bridge, linking Manhattan and New Jersey, and was even an early supplier for the coiled steel Slinky® toy. The Roebling Museum, founded in 2009, recounts the story of the Roeblings, their factory (officially called the Kinkora Works) and company town, and its products.

The Museum itself is housed in the old Roebling Factory's refurbished Gate House, a small structure built in 1907, and expanded gradually through 1947, in two portions separated by an external walkway. The US Army Corps of Engineers constructed a suspension walkway—in homage to the Roebling legacy—between the two halves.

Outside, in the Millyard, are some items the Environmental Protection Agency moved while cleaning the area (a Superfund site), such as railroad and steel equipment. Upon entering the museum, visitors should view the 18-minute video documentary about the Roeblings and their legacy, narrated by two Roebling family members. The rest of the Museum includes products made by the factory, tools and machinery from the factory, wire rope, machinery, some of the 7000 wooden patterns of machinery parts, and artifacts from the Roebling family and the town they built to house the workers' families.

The Roebling room describes the years of the engineering careers of John and Charles Roebling. The Roebling Company room shows artifacts from the factory, ranging from samples of wire products to an employee's locker with its contents abandoned when the factory closed in 1974. Other objects on display include identification badges, photographs, drawings, and catalogs.

The General John Pustay Hall describes the surrounding town of Roebling that the company constructed, using photographs and ephemera. The Museum offers self-guided tours of the town.

The Ferdinand Roebling III Museum Archives, associated with the museum, preserves records pertaining to the factory and environs. Such records include 2000 photographs of the factory and 5000 engineering drawings of the equipment and products. Researchers may call in advance for access.

The museum offers temporary exhibits and regularly scheduled lectures.

Address: Roebling Museum, 100 Second Avenue, Roebling, NJ 08554.
Phone: (609) 499-7200.
Days and hours: Mar.: Sat. 11 a.m.–4 p.m.; Apr.–Aug.: 11 a.m.–4 p.m.; Sep.–Dec.: Thu.–Sun. 11 a.m.–4 p.m; Jan.–Feb.: only for researchers and groups of 10 or more.
Fee: Yes.
Public transportation: River Line light rail stops a block away.
Handicapped accessible: Yes.
Food served on premises: None. Nearby Bordentown has restaurants.
Special note: Call or check the website for the lecture series. For large groups, call to make reservations for guided walking tours of the company town.
Website: roeblingmuseum.org/

WALL

InfoAge Science History Learning Center

After becoming famous for communicating with the sinking *Titanic* in April 1912, Marconi Wireless Telegraph Company of America chose Wall, New Jersey, for a new location because the site was near the coast and therefore it provided an electrical connection to the ocean for their wireless receiving station. It was at least 20 miles (32 km) away from their transmission station near New Brunswick, New Jersey; and was at right angles to the direction of trans-Atlantic transmission. In 1914, Edwin Armstrong demonstrated his regenerative circuit (the first true radio amplifier) to David Sarnoff, an engineer employed at the site. By 1917, the US Navy took over the Marconi receiving station as part of the war effort, used the site to transmit important messages, and then returned the campus to Marconi in 1919. Simultaneously, American Marconi joined several other companies to form the Radio Corporation of America (RCA). By 1925, RCA moved and sold the land. (See The Sarnoff Collection, Ewing, New Jersey.)

With World War II, Fort Monmouth's US Army Signal Corps took over the property, now renamed the Signal Corps Radar Laboratory, and changed again to Camp Evans Signal Laboratory in 1942. The Signal Corps initially worked on the SCR-270 radar system, used at Pearl Harbor, as well as other electronic devices such as the Handie-Talkie, proximity fuses, and countermeasures against enemy radar (which may have led to victory for the Allies). Initial work on the LED even began here.

During the Cold War (1945–1990), when the United States and its allies were actively hostile to the Eastern-bloc countries, electronics technology continued to thrive. In 1946, the very famous Project Diana, using a repurposed radar antenna dish to bounce signals off the Moon and detect them for the first time in history, was performed at this location. The project proved that radio signals could be transmitted and detected beyond the ionosphere, making possible radio communications with spacecraft and orbiting satellites. The existing TIROS radio dish dates from the late 1950s, when satellite tracking of hurricanes to warn local inhabitants of impending storms was pioneered here.

Among other technological developments at Camp Evans was the first mobile digital computer, MOBIDIC, in 1956, though, in truth, it required two 30-foot (9-m) trailers which broke down under testing (yet the computer itself worked fine). Night-vision goggles were invented at this site. One of the last major projects at the site was "Pulse Power," during the 1980s, possibly designed to disrupt enemy satellites by sending high-voltage, high-current electricity, before the camp was decommissioned.

In the early 2000s, after the closing of Fort Monmouth and the United States Army Communications Electronics Museum, the InfoAge Science History Learning Center was founded and was designated a National Historic District. InfoAge began with a variety of small buildings displaying exhibits that were originally in the earlier museum. In fact, InfoAge itself is now a collection of small museums devoted to military electronics, early computers, shipwrecks off the New Jersey coast, and general radio and television technology. (Exhibits may change over time.)

Upon entry into main museum building, originally the 1913 Marconi Hotel (built by the Marconi company as its industrial complex for the radio receiver), the first stop is the Apollo Guidance Center. On display is a section of an Apollo Guidance Computer that traveled to the moon and back in an Apollo Command Module. Other areas describe Marconi's first attempts at transatlantic radio communication, and extensive holdings of military electronics, primarily from World War II.

Behind the hotel was a military facility with a number of structures, now a museum annex displaying a large collection of radio and television sets and memorabilia in the Radio Technology Museum. Available are hands-on demonstrations of telegraphy and early radio technology. An important part of early radio technology, Harold Wheeler's original prototype of electronic Automatic Volume Control, invented in 1925, is showcased near the 1920s radios. There is an extensive array of portable radios from the 1930s through the 1960s, and early black-and-white televisions and high-fidelity stereo equipment, some of which is still in working order. Also displayed are vacuum tubes, 1920s radio batteries (used for powering the radios because reliable AC-to-DC rectification was not invented yet), as well as a 1970s RCA tube-tester. See a Vanguard I satellite, never launched, with solar cells still attached. There is a small science hands-on demonstration room to show basic electromagnetic principles behind electronic technology.

Continuing through the military annex are several rooms dedicated to early computer technology, from the 1950s through the 1980s, including UNIVAC 1219, DEC PDP-8, IBM 1130, Altair 8800, and IMSAI 8080 minicomputers, TRS-80, Apple, Commodore, and other microcomputers and accessories. One room is devoted to electronic warfare (radio and radar) devices, and another with radio-relay equipment from World War II. In the WWII Miniatures room, note the 1:72-scale V-2 model rocket used to study radar signatures. The New Jersey Shipwreck Museum occupies a couple of rooms, describing a number of famous shipwrecks off the New Jersey shore. Many artifacts recovered from the ocean are displayed, including a millstone with the original chisel-grooves from its fabrication still visible.

Possibly the only fallout shelter converted into a theater is here as well. Among the array of paraphernalia on display are items thought necessary for post-apocalyptic survival, from food to toilet paper, that were stored in case of a nuclear attack, plus memorabilia from the Cold War related to protection against radioactive fallout.

Visitors can walk into the room that Senator Joseph McCarthy visited for hearings about security leaks in October 1953.

The Project Diana site is a short distance down Marconi Road by car or foot, where visitors can see the refurbished TIROS (Television Infrared Observation Satellite) antenna, which is 60 feet (18 m) tall. Currently the system is used for astronomical research on pulsars, but during special events, visitors can bounce their voices off the Moon.

Docents are available in many of the rooms to describe the equipment and answer visitors' questions.

Address: InfoAge Science History Learning Center, 2201 Marconi Road, Wall, NJ 07719.
Phone: (732) 280-3000.
Days and hours: Wed., Sat., and Sun. 1–5 p.m.
Fee: Yes.
Tours: Self-guided; docents available in many rooms. Guided tours with prior arrangement.
Public transportation: NJ Transit train from New York to Belmar, then take a taxi to the site.
Handicapped accessible: Yes.
Food served on premises: None; dining is available in local towns.
Website: www.infoage.org

WEST ORANGE

Thomas Edison National Historical Park

Thomas Alva Edison's home, Glenmont, and his laboratory complex in West Orange make up the Thomas Edison National Historical Park. The two facilities, used by Edison until his death in 1931, are only about one mile (1.8 km) apart. The site includes a wide range of exhibits on Edison's contributions to science, technology, and industry.

During his lifetime Edison created ideas for almost 1100 patents. In 1876 he started the first private, broad-based research and development laboratory in the world in Menlo Park, New Jersey. He soon realized that he needed a larger, more modern facility where inventions could rapidly and cheaply be turned into commercial products. After marrying and moving to Glenmont, he selected nearby West Orange for the new complex, which opened in 1887. This move marked a complete break with his past, both professionally and personally. Parts of the original Menlo Park facility were later moved to Greenfield Village in Dearborn, Michigan, by Edison's close friend and admirer, Henry Ford. (See The Henry Ford, Dearborn, Michigan.)

The West Orange complex is more than a collection of buildings where many of Edison's inventions were developed. It represents one of Edison's greatest contributions to science: the group-project approach to problem-solving. Until Edison's time, most technical and scientific research was accomplished in a one-person workshop, often with no specific industrial goal. Edison employed up to 200 people at various times, among them scientists, mathematicians, and trained technicians, who worked together to solve a specific problem and produce a reliable, easily repaired, commercially viable product. The industrial team model is now standard in most scientific and technical firms around the world.

At the complex are Edison's workshops, offices, and library, as well as machine shops, stockrooms, and laboratories. The chemistry laboratory is especially well-preserved, re-creating the feeling of the Edison era. A double-tiered library and office combination used by Edison is located in the main laboratory building. The museum and visitor center is in the original Laboratory Complex Powerhouse.

Many of Edison's inventions are on exhibit. The tinfoil phonograph of 1877, his most original gadget, caused him to be dubbed "the Wizard of Menlo Park." On display is the Strip Kinetograph he invented in 1889 and used with celluloid film developed by George Eastman. He cut the film into continuous strips, added perforations along its sides, and used sprockets to move the film in the projector. Other motion-picture devices are also visible, as well as Edison's incandescent electric-light developments and power equipment. (See George Eastman Museum, Rochester, New York.) In 1942 an underground vault, closed to the public, was installed to protect Edison's papers and memorabilia.

When visiting West Orange, be sure to see the Black Maria, a 1954 re-creation of the world's first building constructed as a motion-picture studio. From 1893 to 1903 the studio was near the water tower, just outside the current historic site. Now it is located on the other side of the property. Experience cinematography made over a century ago by watching *The Great Train Robbery of 1903*, one of Edison's early movies.

Glenmont, a 23-room, red brick-and-wood mansion, is on a 15-acre (6 hectare) estate in Llewellyn Park. Although built in 1880, it is in Queen Anne style, which is much simpler than the then-popular Victorian style. The interior contains opulent Victorian features: wide staircases, stained-glass windows, chandeliers, carved wood paneling, and ceilings with frescoes. Edison brought his 20-year-old bride, Mina Miller, to this mansion in 1886, and over the years she transformed it into what we see now. Inside are both the original furnishings and an extensive array of gifts the Edisons received from dignitaries around the world. In the upstairs sitting room, Edison's favorite room, are his and Mina's desks. (See Edison and Ford Winter Estates, Fort Myers, Florida.)

Address: Thomas Edison National Historical Park, 211 Main Street, West Orange, NJ 07052.
Phone: (973) 736-0550 ext. 11.
Days and hours: Laboratory Complex, Wed.–Sun. 10 a.m.–4 p.m.; Glenmont Estate, Fri., Sat., and Sun. 11:30 a.m.–4 p.m. Closed 1/1, Thanksgiving Day, and 12/25.
Fee: Yes.
Tours: Self-guided in the laboratory complex; guided tours only at Glenmont. Guided tours of both are offered by US National Park Service Rangers on a ranger schedule. Tour reservations are sold at the Visitor Center. Buy tickets for The Laboratory Complex and Glenmont at the same time.
Handicapped accessible: Mostly accessible at the laboratory complex, call ahead for assistance 973-736-0550; not handicapped accessible at Glenmont.
Public transportation: NJ Transit bus stops next to site. From NYC, private bus service (DeCamp bus line) stops 1 mile (1.6 km) from site.
Food served on premises: Nearby in the town of West Orange.
Special note: Visit the Laboratory Complex Visitor Center for tickets to both Glenmont and the Laboratory Complex.
Website: www.nps.gov/edis/index.htm

New York

New York State Museum

Founded in 1836, well before the Smithsonian and the American Museum of Natural History, this museum is one of the oldest in the nation. Originally named the "State Cabinet of Natural History," the site is now a major research and educational institution, a pioneer in the study of New York State's natural resources, with 100 000 square feet (9300 m²) of exhibition area. Scientists from around the world in such fields as geology, biology, anthropology, and technology use the museum's millions of specimens for research.

The Adirondack Wilderness Hall represents a trip through more than 10 000 years of wildlife and human history. Here visitors learn about the effect of humans on the Adirondack region and its changing ecology. Dioramas portray the wilderness in prehistoric times, during a transition, and as it stands today.

The Native Peoples of New York Hall focuses on native peoples of New York State from ancient times to the present. The exhibit begins with two Ice Age life-size dioramas, plus cases of artifacts, which trace the relationship between people and the environment. One diorama re-creates the wetland landscape of Storm King Mountain in the mid-Hudson Valley of about 12 000 years ago, on which the Ice Age people eventually settled. Most mastodon fossils found in the State of New York came from this region. The other diorama, on Ice Age hunters, is set about 11 000 years ago and shows a mid-Hudson Valley camp inhabited by a family whose ancestors crossed the Bering Strait. These caribou

New York State Museum.
Courtesy of the New York State Museum.

hunters are preparing game and making stone tools. The latest techniques of forensic science were used to identify the family characteristics likely for that time and place.

Forest Foragers is a diorama that represents the Late Archaic Period, about 4500 years ago. Based on a museum excavation at Lamoka Lake in the Finger Lakes region, it shows a typical day in the life of this settlement and depicts the geology, fauna, and flora of that era.

The Three Sisters diorama takes a look at Iroquois farmers of the Late Woodland Period, about 500 years ago. This exhibit, based on both archeological research in the Mohawk Valley and written records of the 17th and 18th centuries, gives visitors a good view of how the native Iroquois cleared and planted the land. Iroquois farmers often referred to corn, beans, and squash—three food staples that usually were grown next to each other—as "our life supporters" or "the three sisters." A centerpiece of this Hall is the Mohawk Iroquois Longhouse, an extended-family dwelling popular among the peoples of the northeastern woodlands.

The New York Metropolis Hall explores stages in the history of New York City, including prehistory, early European settlements, and contemporary life. Each display in this extensive exhibit presents a slice of life from a different period. An Algonquin wigwam, a 1630 West India Company cannon—possibly the oldest piece of ordnance in North America—and a rare gridiron map used to lay out the survey of the region's transportation systems for recreation and trade; an 1865 Goold Calb; subway cars from the 1940s and 1950s; an 1855 sandbagger sloop built at City Island; the skeleton of a right whale; a complete, late-19th-century Spanish barber shop from Manhattan; and a collection of weaponry used by the European settlers, are among the items on display.

Other displays include Minerals of New York, Bird Hall, and an exhibit about the response to the attack on the World Trade Center in Manhattan.

Address: New York State Museum, Cultural Education Center, 222 Madison Avenue, Albany, NY 12230.
Phone: (518) 474-5877.
Days and hours: Tue.–Sun. 9:30 a.m.–5 p.m. Closed 1/1, 7/4, Thanksgiving, and 12/25.
Fee: Free.
Tours: Self-guided.
Public transportation: Local buses stop in front of the museum.
Handicapped accessible: Yes. Wheelchairs are available at the Main Lobby Desk.
Food available on site: None. A number of restaurants are a few blocks away in Albany.
Website: http://www.nysm.nysed.gov/

CORNING

Corning Museum of Glass

Corning Incorporated was originally headquartered in Brooklyn, New York, before moving to the beautiful Finger Lakes region of upstate New York in 1868—a move made possible by the Erie Canal. (See Erie Canal Museum, Syracuse, New York.) In its early days, the company made thermometer tubing and pharmaceutical glassware, which it continues to this day. In 1879 it was the first company to produce incandescent light-bulb blanks for Thomas Edison. (See The Henry Ford, Dearborn, Michigan.) Today, Corning manufactures more than 34 000 products used in science, health, medicine, transportation, communications, and the home.

The company was incorporated as the Corning Glass Works in 1875 and was renamed Corning Incorporated in 1989 to reflect its growing diversification. In 1951 the Corning Glass Center was opened to teach the public about the 3500-year history of glassmaking. This facility explores the art, history, science, and industry of glass and glassmaking.

More than 50 000 objects make up one of the finest collections of glass in the world, which cover 35 centuries of human experience with glass. The exhibits are arranged chronologically from 1500 B.C.E. to the present. The pieces on display range from tools and weapons made from obsidian by prehistoric people to great European and American art glass, Asian pieces, and Arabic items. Contemporary works are on display as well. The glass items include scientific and industrial pieces. Among the specimens of natural glass are obsidian, fulgurites (tubes of melted sand from lightning strikes), a silaceous sea sponge, and a giant piece of glass that formed in the Libyan desert from a meteor hit.

Some of the highlights of the museum's collections include the priceless medieval treasure known as the Corning Ewer—an eggshell-thin cameo glass probably made in Iran around 1000 C.E.—and the spectacular 11-foot (3.4-m) high glass window created in 1905 by Tiffany Studios, which was headed by Louis Comfort Tiffany, son of the famous jeweler Charles Tiffany. Louis developed innovative techniques while creating his stained glass: Rather than paint the windows to obtain depth, he perfected a process of layering stained glass to create intensities of color and depth. Also on display is an 18th-century Chinese vase from the Ch'ien Lung dynasty, a beautiful example of carved glass.

Among the early scientific and technological glassware exhibited in the museum are an 18th-century European chemical retort and Benjamin Franklin's musical invention, the

Corning Museum of Glass.
Courtesy of the Corning Museum of Glass.

glass armonica. Other curiosities include some technically demanding glass biological models created by Leopold Blaschka and his son Rudolf in the town of Hosterwitz, near Dresden, Germany (see Harvard Museums of Science and Culture, Warren Anatomical Museum, and Arnold Arboretum, Cambridge and Boston, Massachusetts): some flowers, a series of fruit and glass eyes, and some invertebrate models.

The museum encourages hands-on participation. Try your hand at bending glass, watch glass transmit and reflect light, and explore other properties of glass in these exhibits. In one of the demonstration areas, lamp-makers show how to melt glass rods and form them into glass animals.

In the Optics Gallery, an exhibit on optical fibers investigates 100 years of scientific problem-solving, and shows how total internal reflection of light is used to transmit information through optical fibers and fiber-optics communications technology. In the center of the gallery is the first casting for the 200-inch (508 cm) telescope on Mount Palomar.

The Vessels Gallery presents the mundane yet crucial aspect of glass as a container. Heat-resistant glass and Pyrex®, the mainstay of chemical laboratories and kitchens, are discussed, along with light bulbs, CRTs, and fiberglass insulation.

Windows are an old yet relevant use for glass, showcased in the Windows Gallery. Here are shown the technologies behind the fabrication of sheet glass and tempered glass used in automobile windshields.

The Steuben Glass Factory, a very fine art-glass manufacturer, was once part of the Corning complex. This glass product was a division started in 1903 by Frederick Carder and named after the county in which the Corning plant is located. It was the only factory to make the famed Steuben glass art pieces, and became part of the Corning Glass Works in 1918. The Factory closed in 2011, and was transformed into the Amphitheater Hot Shop, one of the largest places around the world in which to view glassblowing. A collection of Steuben Glass is on display at the museum.

Allied with the museum and on the same campus is the Juliette K. and Leonard S. Rakow Research Library. Started in New York City as the Steuben Glass reference library, the institution now contains nearly 500 000 items, ranging from books to microfilms to drawings to photographs, all related to the history, art, and technology of glass. No appointment is necessary to visit and browse this non-circulating collection. Tours are available by prior request. The collection's earliest holdings range from medieval manuscripts such as the 12th-century *Mappae Clavicula* (a compendium of recipes for materials used in decorative arts), continue to the Blaschka family archive, and on to the present day.

Museum Address: Corning Museum of Glass, One Museum Way, Corning, NY 14831. Library Address: Juliette K. and Leonard S. Rakow Library, Five Museum Way, Corning, NY 14831.
Museum phone: (800) 732-6845; library phone: (607) 438-5300.
Museum days and hours: Sep.–late May: daily 9 a.m.–5 p.m.; Memorial Day–Labor Day: 9 a.m.–8 p.m. Closed 1/1, Thanksgiving Day, 12/24, and 12/25. Library days and hours: Daily 9 a.m.–5 p.m. Closed Good Friday, Memorial Day, 7/4, Labor Day, Veteran's Day, Thanksgiving Thu. and Fri., 12/24, and 12/25.
Library hours: Daily 9 a.m.–5 p.m.

Fee: Yes for museum; additional fees for certain demonstrations, hands-on displays, and tours.

Tours: Self-guided, and arranged. Call in advance for library tours.

Public transportation: ShortLine buses run from New York City and several other cities; Museum offers a free shuttle bus from ShortLine stop.

Handicapped accessible: Yes.

Food served on premises: Café in Museum.

Website: www.cmog.org/.

HAMMONDSPORT

Glenn H. Curtiss Museum

Hammondsport was the home of Glenn Hammond Curtiss, one of America's most important pioneer aviators. Here he began his career as a bicycle and motorcycle builder and manufacturer of reliable, light, powerful motors. Curtiss's fame spread quickly when, in 1907, he drove his motorcycle at 136.4 mph (219.5 km/h) and became known as "the fastest man on earth."

When inventor Alexander Graham Bell became interested in the possibilities of manned flight, he went to Curtiss for two engines. In 1907 he asked Curtiss to visit his summer home in Nova Scotia and to join the Aerial Experiment Association as chief of experiments. Curtiss spent a year and a half at that job. Although the organization disbanded, Curtiss's interest in aviation continued. He worked with J.A.D. Jack McCurdy to set the *June Bug* on catamaran-style floats. This creation, which they called the *Loon*, led to the first successful flying boat in 1912.

On a flight down the Hudson River, Curtiss commented that it would be easy to drop a bomb on a boat from a plane. By 1910, Curtiss—working with far-sighted naval personnel—gave the US Navy aviation capabilities. The first US Navy airplane, the A-1 Triad, was an amphibian built by Curtiss. He also founded the first flying school in the United States, where the first Navy pilots were trained. Curtiss also worked in the private sector: In 1909 his Golden Flyer was the first plane sold commercially.

The Curtiss JN Series, also known as the Jenny, was the training plane for World War I pilots. Curtiss's large flying boats, developed by 1912, added to the country's military aviation capabilities. World War I turned Curtiss Aeroplane Company into the largest US aircraft manufacturer of its day.

After the war, Curtiss broadened his commercial interests and developed the camping trailer, another important US industry. Later Curtiss moved his facilities from Hammondsport to Garden City, Long Island, New York, where he opened the first experimental factory and developed his NC-class planes.

The museum has more than 30 000 artifacts demonstrating Curtiss's contribution to American aviation. On display are a full-scale reproduction of the first airplane he built, the 1908 *June Bug*; an original 1918 Jenny; and a 1927 Curtiss Robin. You can also see a collection of Curtiss motorcycles (including a reproduction of the one that carried him in 1907) and engines.

The museum displays local memorabilia as well, such as cameras, radios, woodworking tools, and toys, and it has a theater and restoration workshop.

Address: Glenn H. Curtiss Museum, 8419 State Route 54, Hammondsport, NY 14840.
Phone: (607) 569-2160.
Days and hours: May–Oct.: Mon.–Sat. 9 a.m.–5 p.m., Sun. 10 a.m.–5 p.m.; Nov.–Apr.: daily 10 a.m.–4 p.m. Closed Easter, Thanksgiving, 12/24, 12/25, and 1/1.
Fee: Yes.
Tours: Self-guided.
Public transportation: None.
Handicapped accessible: Yes.
Website: www.glennhcurtissmuseum.org/

MOUNT LEBANON

Shaker Museum and Library

The religious community founded during the mid-18th century by Ann Lee was first called the United Society of Believers in Christ's Second Appearing. It later became known as the Shakers because of the physical manifestations of members' fervent prayer—shaking—that took place at prayer meetings. This small religious community left Manchester, England, in 1744 to seek religious freedom and a better way of life in the Province of New York. The Shakers were at their peak just before the Civil War, when they had about 6000 members. They lived in self-contained communities scattered in seven states from Maine to Kentucky. Today, in part caused by their religious belief in celibacy, they have perhaps only a handful of surviving members in one community: Sabbathday Lake, Maine. (See Canterbury Shaker Village, Canterbury, New Hampshire.)

Religious principles guiding the members include purity of life, equality of the sexes, confession of sin, separation from the world, conservation of resources, and consecration of strength, time, and talent within a communal society. They tried to make their communities an "earthly heaven." Despite their small numbers, the Shakers' effect on the United States and the world is remarkable. Thirty-seven known patents are attributed to the Shakers; many more of their inventions or improvements to existing technology were never patented. Their influence extends from furniture-making to agriculture, pharmacy, tailoring, and science. Shaker members developed the flat-bottom broom and wooden clothespins. They received a patent for the first commercial washing machine in 1858. In the 1820s, a Shaker designed tilter buttons for the bottom of chair legs to prevent the chair from falling over as a seated person leaned backward.

The museum was founded by John S. Williams, Sr., of Old Chatham, New York. Williams was a man who, although not a Shaker, greatly admired the community. Between 1935 and 1965, with the help of the Shaker leadership, he gathered most of the artifacts now contained in the museum and library. He used his farm, which was never part of the Shaker community, as a repository for his burgeoning collection. The Museum combined with the North Family of Shakers of nearby New Lebanon in 2004—today the main site—and it presents a complete view of Shaker life and culture, with over 70 000 items.

Shakers made their living by farming and selling products they produced or manufactured, such as seeds, herbs and medicines, wooden boxes, chairs, cabinets, farm equipment, and brooms. Shaker industries were known for their quality of workmanship and simplicity of line. Among the items on display is a ball mill from the second half of the 19th century that uses cannonballs to crush herbs, which were then packaged and sold to the outside world. The Shakers are credited with inventing small packets for selling plant seeds to the public. In 1810, Brother Thomas Corbett of the Canterbury community built an electrostatic machine to help cure rheumatism, a concrete reminder of the community's interest in science and medicine.

At least a half-dozen of the Shakers' patents are related to the waterwheel. Among other inventions on display are a 7-ton (6.4 tonne), double-trip hammer from the Mount Lebanon, New York, community, and a horse-drawn, hand-pumper fire engine from the Canterbury community, both from the year 1820. The extensive collection of woodworking equipment includes the first tongue-and-groove machine, invented in 1828. In the laundry is a patented washing machine from the Canterbury community that won a gold medal at the Philadelphia Centennial Exposition of 1876.

Shaker Elder Frederick Evans began construction of the Great Stone Barn on the North Family property in October 1858 as the locus of the most up-to-date scientific agricultural methods. A fire in 1972 severely damaged the structure. Restoration of the 1859 Great Stone Barn of the North Family, the largest stone barn in the country when it was built, is opened to visitors. The barn is being stabilized, and is on the list of the World Monuments Fund.

Still located on Williams's farm is the Emma B. King Research Library, which contains reference materials related to the Shaker community. The Library is open to the public by appointment, though only members of the site may borrow items.

Address: Shaker Museum and Library, 202 Shaker Road, New Lebanon, NY 12125.
Phone: (518) 794-9100.
Days and hours: mid-Jun. through mid-Oct.: Fri.–Mon. 10 a.m.–4 p.m.
Fee: Yes.
Tours: Self-guided; guided tours are offered all year round several times per day. Reservations are recommended for groups of 10 or more.
Public transportation: None.
Handicapped accessible: Mostly. Tours may cover uneven ground and sloping paths.
Food: Restaurants and other food vendors are available nearby.
Website: www.shakermuseumandlibrary.org/

NEW YORK

American Museum of Natural History

Calvert Vaux, landscape architect who designed Central Park with Frederick Law Olmsted, helped plan this museum complex that now has 28 interconnected buildings. (See Frederick Law Olmsted National Historic Site, Brookline, Massachusetts.) The Museum, overlooking Central Park, is situated on what was once swampy farmland. Although this institution was founded in 1869, the oldest portion of the complex was completed in 1877. Newer wings have been added over the years.

This facility is one of the largest museums in the world, with 45 permanent exhibit halls, a planetarium, and a library, as well as over 33 million artifacts located in 28 buildings. It was established to help the public understand the Earth, the diversity of living things that inhabit it, and humanity's place in the natural world. Only a small portion of the museum's holdings are displayed at any one time. Most exhibits have hands-on, interactive components.

Theodore Roosevelt Rotunda is at the Central Park West entrance to the museum. In this area is a diorama with the skeleton of a *Barosaurus*, the tallest free-standing mount of a dinosaur in the world.

Occupying the entire fourth floor are the Fossil Halls that display 600 specimens, 85% of which are real rather than casts. The Museum holds the world's largest collection of vertebrate fossils. The thrust of these exhibits is that natural science is a study of vertebrate evolution in process. They explore the mystery of how life developed, review how species are distributed around the Earth, and explain current threats of massive extinction. Two of these halls contain dinosaurs. The fossil exhibits are arranged in an evolutionary tree that connects the six halls. Each major evolutionary event is placed in context with the earth's history. Diversity and adaptation within groups of animals is explored in these exhibits. Hundreds of original fossils are used in realistic exhibits, such as the 15-million-year-old, bear-like carnivore *Amphicyon* and its prey, and the antelope-like *Ramoceros*. Many old favorites have been remounted according to modern understanding, including the well-known *Tyrannosaurus rex*. It is hard to miss the cast of a Titanosaur from Patagonia in Argentina at 122 feet (37 m) long on the fourth floor. The Titanosaur is just too large to fit properly vertically, so the neck and head are extended outward through an opening in the wall and the feet are slightly bent to accommodate its height.

The Rose Center for Earth and Space opened in 2000. This 333 500 square-foot (30 980 m^2) structure houses research facilities, three exhibit halls, and the Hayden Planetarium. The Cullman Hall of the Universe's Galaxies Zone explores aspects of galaxies, their formation, and evolution. Star Zone examines supernovae and the elements they synthesize: the chemicals of the human body. Planets Zone looks at the formation and evolution of the planets.

Scales of the Universe explores the wide range of size around us along a 400-foot (122 m) long walkway next to the glass wall of the building. The central Hayden Sphere is the reference-standard for comparison with presentation areas along the walkway. Examined are the size of atoms, stars, planets, and galaxies. Suspended from the ceiling are planet models in proportionate size to the Hayden Sphere. Follow 13 billion years of cosmic evolution down the 360-foot (110 m) spiral ramp of the Heilbrunn Cosmic Pathway, which begins with a dramatic Big Bang recreation in the base of the Hayden Sphere. Displays include the earliest fossil bacteria on Earth and a giant carnivorous dinosaur's fossilized tooth. The centerpiece of the Rose Center is the Hayden Sphere, 87 feet (27 m) in diameter. The upper portion houses the Space Theater with a Zeiss Mark IX Star Projector and a digital Dome Projection System.

Step from outer space into the older sections of the museum complex and visit the Gottesman Hall of Planet Earth. Topics such as the evolution of the planet, ocean basins, continents, mountains, climate, and habitability are examined. Thousands of mineral samples (from over 100 000 specimens in the museum's collection) are on view in the Ross Hall of Meteorites and Mignone Hall of Gems and Minerals. A display wall highlights the classification of minerals. Several massive geodes (including a 12-foot-tall [3.6 m] sample from Uruguay) are mounted in the Crystal Garden exhibition area. The Hall has such renowned samples as the Star of India, the largest star sapphire in the world, and the Patricia Emerald, an uncut 632-carat gem. Specimens from almost half of all known meteorite falls are in the museum's collection. On display is Ahnighito, a 68 000-pound (30 800-kg) piece of iron and nickel, the world's largest known meteorite, found in 1894, near Cape York, Greenland, and brought to the museum by Robert E. Peary, the famed explorer. (See Peary-MacMillan Arctic Museum, Brunswick, Maine.) Other displays include the oldest specimen, an Australian zircon crystal about 4.3 billion years old. For local color, a cabinet displaying New York City's minerals, including the "subway garnet," weighing 9 pounds (4 kg) from an 1885 sewer excavation. Fluorescent and phosphorescent rocks from the Sterling Hill Mining Museum in northern New Jersey are showcased. (See The Sterling Hill Mining Museum and the Franklin Mineral Museum, Ogdensburg and Franklin, New Jersey.) Global events, such as earthquakes, volcanoes, and atmospheric conditions, are broadcast as they occur on the electronic Earth Event Wall.

Some 11 000 square feet (1000 m^2) make up the Hall of Biodiversity, which focuses on protection and preservation of Earth's biodiversity. The center of this hall is a diorama of the Dzanga-Sangha (Central African Republic) Rain Forest with over 160 species of flora and fauna, one of the most biodiverse ecosystems. Some of the Museum's extraordinary collection of over 19 million arthropod specimens resides here, including the world's most diverse collection of over one million spiders and termites. Three aspects of the rain forest are presented: pristine, altered by nonhuman forces, and human effects.

The evolution of *Homo sapiens* is examined in depth in the Spitzer Hall of Human Origins. Featured are four dioramas of human predecessors in their habitats: *Homo ergaster*, *Homo erectus*, Neandertal, and Cro-Magnon. A cast of the "Turkana Boy" who died at age eight or nine, dating to roughly 1.7 million years ago, is on display. This nearly complete skeleton was found by Richard Leakey in 1984 at Nariokotome in Kenya. Limestone carvings, perhaps the earliest human artwork, made by Ice Age *Homo sapiens* from the Dordogne region of southwestern France about 26 000 years ago, are also on display.

The suspended 94-foot (29 m) long model of a blue whale, the world's largest animal, dominates the Hall of Ocean Life. The Hall of Primates is arranged according to families, from the smallest pygmy marmosets to the orangutan and gorilla, and examines these animals' relationship to humans. The exhibits in the Hall of Reptiles and Amphibians are laid out according to themes such as anatomy, defense, locomotion, distribution, reproduction, and feeding. Spectacular reptiles are shown in full-size, realistic dioramas. The Halls of Birds examine a wide variety of avian life and their variety of ecosystems in which they live, including a historic diorama of the peregrine falcon, and a diorama of Andean condors.

The Warburg Hall of New York State Environment explores local soil conservation, geology, plant life, and agriculture. On display is a slice of an ancient sequoia tree with 1342 rings, which grew from a seed that germinated probably in the year 540 C.E. Several halls are devoted to Native Americans and Eskimo cultures and their development of hunting and fishing technologies, agriculture, and household tools. A diorama of the forest floor examines the process of decomposition in a scene enlarged to 24 times actual size.

Among the other areas worth exploring are the Mammal Halls. The Akeley Hall of African Mammals, one of the world's great museum exhibits and dating from 1936, was named for Carl Akeley, a naturalist and sculptor who created this type of exhibit and these specific displays using innovative taxidermy and mounting methods. (See Milwaukee Public Museum, Milwaukee, Wisconsin.) The Bernard Hall of North American Mammals has some of the best diorama artwork anywhere, captured by master artist James Perry Wilson, who paid strict attention to detail. The Hall of Asian Mammals was created also along Akeley's methods based on expeditions in the 1920s, and includes a comparison between Asian and African elephants, plus exhibits on threatened species.

Cultural halls include Hall of Asian Peoples and Hall of Human Origins, Hall of African Peoples, Hall of Mexico and Central America, Hall of Plains Indians, Hall of South American Peoples, Margaret Mead Hall of Pacific Peoples, Northwest Coast Hall, Hall of Plains Indians, and the Hall of Eastern Woodland Indians.

Address: American Museum of Natural History, Central Park West at 79th Street in Manhattan, New York City.
Phone: (212) 769-5100.
Days and hours: Daily 10 a.m.–5:45 pm. Closed Thanksgiving Day and 12/25.
Fee: Yes.
Tours: Self-guided; highlights tours (including in languages other than English) are given at selected times.
Public Transportation: Subway (station on the lower level of the Museum); bus; taxi-ride or bus from other train stations.
Handicapped accessible: Yes. Phone (212) 769-5250 for more information.
Food served on premises: Cafés and food court.
Special note: Check schedule of Planetarium, movies, and special programs; some have extra fees.
Website: www.amnh.org

Brooklyn Botanic Garden

In the heart of Brooklyn, amid a forest of apartment houses, is a 52-acre (21 hectare) oasis that opened in 1910: the Brooklyn Botanic Garden, bordered on one side by the impressive Brooklyn Museum. On the other side is rambling Prospect Park, a masterpiece of public space.

Seventeen thousand years ago, the Wisconsin glaciation that surrounded Long Island, Manhattan, and the Bronx created a landscape of small ponds and hills and carved depressions in the soil. Some of this pristine landscape still remained in the 19th century. Fredrick Law Olmsted and Calvert Vaux submitted plans to the City of Brooklyn in 1860 for a park, and the City of Brooklyn purchased the land in 1864 for this purpose. (See Frederick Law Olmsted National Historic Site, Brookline, Massachusetts.) Brothers Fredrick Olmsted, Jr., and John Charles Olmsted, sons of Fredrick Law Olmsted, continued the project. In 1897 Brooklyn reserved 39 acres (16 hectares) for the park itself, resulting in the area now called the Brooklyn Botanic Garden.

From its conception, the goals of the Garden were different from those of similar contemporary institutions because it was dedicated to both scientific research and the art of gardening, with an emphasis on teaching gardening skills to anyone who wanted to learn. In 1914 it was the first botanic garden in the world to have a children's garden, which still flourishes.

The Brooklyn Botanic Garden (BBG) is a collection of gardens tied together with lovely lawns and walkways. Over 14 000 species and cultivars from almost every country of the world grow here. All the plants are carefully labeled with common and scientific names and place of origin.

Here is one of the most important places at which to view cherry trees outside of Japan. Over 200 cherry trees of over 20 Asian species and cultivated varieties grow here. The Japanese government gave the first trees to the Garden after World War I. Two rows of cherry trees define the Cherry Esplanade. The trees usually begin blooming in late March and early April, and finish by mid-May. (These times vary from year to year according to the weather and the variety of the trees.) *Hanami*, or cherry-blossom season, lasts for one month. *Sakura Matsuri* (cherry-blossom festival) is held during one weekend at that time. There are trails and paths among the trees encouraging visitors to wander and enjoy the natural beauty. Cherry trees are also found elsewhere within the BBG, including the Japanese Hill-and-Pond Garden.

The Japanese Hill-and-Pond Garden, opened in 1915, is the first Japanese garden created within an American public garden. Japanese landscape designer Takeo Shiota blended the ancient hill-and-pond style and the modern stroll-garden style with several types of landscapes that are revealed along winding paths of this three-acre (1.2-hectare) garden, which include artificially created hills, a waterfall, and an island. Along the walk are wooden bridges, stone lanterns, a viewing pavilion, tori gate, a Shinto shrine, and a Japanese temple dedicated to the wolf spirits. Hundreds of koi swim in the pond formed by the Wisconsin glaciation. This garden is considered Takeo Shiota's masterpiece.

The 1.5-acre (0.6-hectare) Cranford Rose Garden (1927), enclosed by a lovely, wooden latticework fence, is one of the largest rose gardens in the country, and was designed by landscape architect Harold Caparn and Montague Free, the Garden's horticulturist. Almost 5000 plants, representing nearly every type of hardy rose that grows in this climate, are represented here: hybrid tea roses, floribundas, grandifloras, climbers, ramblers, and more.

Darrel Morrison's 2013 landscape design was put in place to expand the space for plants native to the greater New York area. The Native Flora Garden displays prairie grasses, dry meadows, pine barrens, and a kettle pond. A wooden bridge spans across the various habitats for easier viewing.

Designed as a typical English cottage garden, the Shakespeare Garden has more than 80 plants mentioned in William Shakespeare's works. Each plant is labeled with its common or Shakespearian name, botanical name, with relevant quotations and image of the plant as needed.

The Alice Recknagel Ireys Fragrance Garden has complete information in Braille for visitors with visual impairments, as well as raised beds for those in wheelchairs. This garden, created in 1955 by Recknagel Ireys, a landscape architect, was the first garden of this type in the United States. For a full experience, visitors are encouraged to rub textured leaves between their fingers. There are four sections: plants to touch; plants with scented leaves; fragrantly flowered plants, and kitchen herbs. There is a fountain with calming sounds, which can be used to wash hands after touching the plants.

Local glacial boulders and low-growing plants, including alpines, make up the Rock Garden. It also contains woodlands, an alpine meadow, rock debris like that at the base of cliffs, and habitats for acid-loving and drought-tolerant plants.

The Herb Garden features a variety of worldwide edible plants, as well as a 16th-century Elizabethan knot-woven foliage of many colors and textures, and more than 300 medicinal, culinary, fragrant, and ornamental herbs.

Several special collections and displays help to make this facility stand out from others. For example, an extensive "Plant Family Collection" contains flora arranged scientifically in evolutionary progression. Another collection emphasizes dwarf and slow-growing plants that are particularly suitable for urban gardens. There is also a conifer collection.

The Louisa Clark Spenser Lilac collection has 150 fragrant lilac species and cultivars in shades of lavender, pink, and white, and magnolias, well as four varieties of trumpet daffodils.

Inside the Garden's Steinhardt Conservatory is an extensive collection of indoor plants from all over the world. This 1988 Conservatory has multiple pavilions: the desert, tropical, and temperate. The plants are arranged in naturalistic habitats. The Conservatory emphasizes the effect that plants have on the world and its people over time, as well as conservation and ecosystems.

A highlight of the Conservatory is the C. V. Starr Bonsai Museum, with one of the oldest collections of dwarfed, potted trees in the country. Some of these hundreds of trees are over a century old. Another collection on display here is the Robert W. Wilson Aquatic Collection of tropical water plants, insect-eating plants, and orchids. The Stephen K-M. Tim Trail of Evolution traces the history of plant evolution and the effects of climate change over 3.5 billion years.

The Children's Garden is the oldest in the world, operating as a community garden for children. Each youngster receives a small portion of the one-acre plot (0.4 ha) on which to grow and harvest vegetables and flowers for the season.

The Garden does research and has a herbarium (collection of preserved plants) and library with reference services to professionals and home gardeners. Please contact the Garden for more information.

Address: Brooklyn Botanic Garden, 990 Washington Avenue, Brooklyn, NY 11225.
Phone: (718) 623-7200.
Entrances: 150 Eastern Parkway; 455 Flatbush Ave.; 990 Washington Ave.
Days and hours: Mar.–Oct.: Tue.–Fri. 8 a.m.–6 p.m., Sat.–Sun. 10 a.m.–6 p.m.; Nov.–Feb.: Tue.–Fri. 8 a.m.–4:30 p.m., Sat.–Sun. 10 a.m.–4:30 p.m.; Dec.–Feb.: Tue.–Fri. 10 a.m.–4:30 p.m., Sat.–Sun. 10 a.m.–4:30 p.m. Closed Thanksgiving Day, 12/25, and 1/1. Closed Mondays, except for Martin Luther King, Jr., Day, Presidents' Days, Memorial Day, and Columbus Day, when open from 10 a.m–4:30 p.m.
Fee: Yes; additional fees may apply for events.
Tours: Self-guided. Guided tours are also available; check with the staff.
Handicapped accessible: Yes.
Food: Café and seasonal outdoor canteen.
Public transportation: Subway to the Brooklyn Museum Station or Prospect Park Station; bus; taxi.
Website: www.bbg.org/

Intrepid *Sea, Air & Space Museum*

The fourth ship named USS *Intrepid*, launched in 1943 and in service for 37 years, is now a floating museum in New York City harbor. (The first *Intrepid* was built for Napoleon Bonaparte in 1798 and was later transferred to US control.) This 20th-century ship served around the world as part of the US World War II effort, played a role in military operations in the Philippines and in the Vietnam War, and helped to recover Mercury and Gemini space capsules. Over its years of service it was struck by seven bombs, five kamikaze planes, and one torpedo, then was decommissioned in 1974. Fittingly, the US Department of the Interior declared the ship a National Historic Landmark in 1986.

The museum's exhibits, audiovisuals, models, dioramas, and original equipment give an overview of America's sea, air, and space history. Entrance is onto the cavernous hangar deck, once the service and parking area for planes, and now the museum's main exhibit area. The Navy Hall, located in the bow, displays a Grumman A-6E Intruder attack bomber and other aircraft, some of which are still in use. The Allison & Howard Lutnick Theater, occupying space that was once a flight elevator, regularly runs a film about life aboard a modern supercarrier. Intrepid Hall has replicas of period aircraft and a multimedia presentation that takes you back in time to World War II and operations performed by the Intrepid in the Pacific. Numerous models and samples of ancillary apparatus are available to view. For example, a cross-section of the flight deck reveals that it is made of teak over pine over steel plate. Displays include one of the four huge bronze propellers, uniforms, patches, and other memorabilia. Note the three conflags (conflagration stations), bunkers with small rectangular windows used to direct firefighting.

For an earlier history of flying, the years 1890 to 1914, head for Pioneer Hall. The displays of antique aircraft include a replica of a Curtiss Pusher biplane, the first aircraft to land on a carrier, and an original Voisin biplane, the first aircraft to fly in a circle.

Technology Hall, in the stern of the ship, contains a full-size spacecraft model explaining space flight. The nose section of a Boeing 707 provides a close-up view of the instruments and controls that keep this plane in the air. Another piece of 20th-century technology on display is the first US-built steam catapult, built in 1954 to launch aircraft from the carrier's flight deck.

When the USS *Intrepid* was in operation, pilots would land on the moving flight deck, which is the length of three football fields, and come to a stop within 150 feet (46 m). Now this area displays jets, propeller-driven planes, helicopters, and other items. A McDonnell Douglas X-A3 D, a Grumman F-11/F-1 Tiger, a Sikorsky UH-34 Seahorse, a North American FJ-3 Fury, and a Soviet-built MiG-21 are among the planes on view. A highlight of the display is a Lockheed A-12 Blackbird reconnaissance plane, one of the fastest planes ever built. It could travel at a speed of mach 3.6 and reach an altitude of 95 000 feet (29 000 m), one of the fastest and highest-flying aircraft in history. It is also the world's largest single-seat aircraft. There is also an AH-1 Sea Cobra, a US Marine helicopter used in Operation Desert Storm. The Douglas A4 Skyhawk, rolled out in 1956, carries 9000 pounds (4100 kg) of bombs, and has a top speed of over 600 mph (97 km/h).

The area used by the admiral to command the whole fleet, the chart room, and the admiral's and navigator's sea cabins are open to the public. Up one deck is the navigation bridge with its compasses, radarscopes, and other equipment. The original steering wheel is still in place.

Below the flight deck is the captain's stateroom, radio central, and the Combat Information Center, the USS *Intrepid*'s nerve center. It was from here that enemy craft were tracked and the ship's aircraft received their instructions.

Among other seacraft that are part of the museum is the 3000-ton (2700 tonne), 317-foot (97-m) long USS *Growler*, active from 1958 to 1964, a forerunner to the Polaris nuclear submarine. It is the only guided nuclear missile submarine open to the public. Crew size ranged from 88 to 105, and it was one of the first submarines with an inertial navigation system, which uses a computer, accelerometers, and gyroscopes to calculate the vessel's position, orientation, and velocity. The submarine was powered by three Fairbanks-Morse diesel engines running at high-frequency to avoid audible transmission through seawater. The USS *Edison*, a destroyer that was in action during the Vietnam War, and the lightship #112 *Nantucket* are also exhibited.

On display is the Space Shuttle Pavilion, first unveiled in 2012. In it, the space shuttle prototype *Enterprise* is featured, allowing visitors to view both the underside and from above. It is 137 feet (42 m) long, with a 78-foot (24-m) wingspan, and weighs 150 000 pounds (68 000 kg). Nearby is a small Soviet Soyuz TMA-6 descent module for comparison. On Pier 86 is a Concorde supersonic aircraft, serial number 100-010 (G-BOAD), which flew its maiden voyage in 1976, on loan from British Airways. The Concorde flew with a crew of nine at up to 1350 mph (2150 km/h) at an altitude of 60 000 feet (18 000 m), which is high enough for passengers to detect the curvature of the Earth. (See Steven F. Udvar-Hazy Center, Chantilly, Virginia.)

Often there are special temporary exhibits on display.

Address: *Intrepid* Sea, Air & Space Museum, Pier 86, 12th Avenue & West 46th Street, New York, NY 10036.
Phone: (212) 245-0072.
Days and hours: Apr.–Oct.: Mon.–Fri. 10 a.m.–5 p.m., Sat., Sun., and holidays 10 a.m.–6 p.m.; Nov.–Mar.: daily (including holidays) 10 a.m.–5 p.m. Closed Thanksgiving and 12/25.
Fee: Yes.
Tours: *Intrepid* self-guided; other ships guided tours only.
Handicapped accessible: Limited. Not handicapped accessible: The *Growler*, Foc'sle/Gallery Deck of the *Intrepid*.
Public transportation: Bus and subway (either 42nd Street or 34th Street Station) to near the site, then walk. Also available are a water-taxi to Pier 84, and a ferry to West 39th Street and 12th Avenue.
Food served on premises: Café.
Special note: Call for information on talks, presentations, and demonstrations.
Website: www.intrepidmuseum.org

Metropolitan Museum of Art

Why is the Metropolitan Museum of Art, one of the world's finest art museums, included in a science and technology tour book? Because it's a collection of collections—with some of the highest-quality artifacts in the world—and many exhibits have items and discussions of key scientific interest.

Housed in this museum is one the most extensive collections of arms and armor in the Western Hemisphere. The museum describes the collection primarily in terms of art. The technology exhibited is notable and much of it is described. Nine thematic galleries contain the best examples from the museum's 14 000 artifacts, items from the fifth through the 19th centuries collected in Europe, North America, the Middle East, and Asia, everything from small arrowheads to full suits of armor. The Equestrian Court displays armor from English royal workshops made from Elizabethan courtiers, Renaissance parade pieces including a 1543 helmet made by the Milanese armorer Filippo Negroli, armor made for King Henry II of France about 1555, and other armor for tournaments and parades.

The Japanese collection includes complete armor, swords, and elaborately worked sword guards. The most impressive piece is the 14th-century armor made for the Japanese military leader Ashikaga Takauji. Also exhibited are rare woven textiles made into a warrior's surcoat (a tunic worn over armor) and sashimonos (banners carried into a battle).

Among other intriguing items are European rapiers and court swords, civilian side-arms for self-defense, gentlemen's jewelry such as rings and brooches, pieces from the arms cabinet of Louis XIII of France, and Napoleonic firearms made in Versailles. Of special technological note is a double-barreled wheel-lock pistol made around the year 1540 by watchmaker and gunsmith Peter Peck of Munich for Holy Roman Emperor Charles V. It has an automatic wheel-lock ignition system invented early in the 16th century that combines two locks in one mechanism; even though the two barrels are attached to each other, each fires separately. An interesting hunting knife and wheel-lock pistol combination made in Germany in 1546 has a calendar on the blade.

A German hunting crossbow dated 1460 is an example of the most powerful missile weapon in use before firearms were introduced. It is made of horn and whalebone and covered in waterproof birch bark. Because it cannot be armed by hand it is spanned by a cranaquin, a mechanical device that operates like a modern automobile jack. When it releases an arrow there is no recoil and no noise, making it an ideal hunting weapon.

The North American arms gallery includes swords, colonial muskets, powder horns engraved with maps and cityscapes, Kentucky rifles, and Colt revolvers. It pays tribute to the technical inventiveness of Eli Whitney, Ethan Allen, Samuel Colt, and others who made and improved arms.

The 50 000 artifacts that make up the Egyptian collection—the largest outside Cairo—are arranged chronologically from 3100 B.C.E., beginning with the Predynastic Period, to the eighth century C.E., and ending with the Coptic Period.

Examined is the development of architecture, from temples to tombs to the pyramids, including an entire temple constructed of sandstone—the Temple of Dendur—transported from Egypt to this museum in order to save it from rising Nile River water. It stands proudly in an exhibit room of its own, overlooking the gardens outside through a glass wall. Much of what is discussed involves engineering and construction. A significant amount of what we know about ancient Egyptian life comes from finds in tombs, such as that of Chancellor Mekutra of the Middle Kingdom. Thirteen miniature models made of painted wood that were found in his tomb show stables, storerooms, kitchens, craftsmen's shops, a brewery, a bakery, a granary, a slaughterhouse, and fishing boats.

The museum holds a large collection of Egyptian textiles, particularly coffin sheets and fabrics. This fabric, Wah's linen, is a feather-light, transparent textile that has never been duplicated by modern techniques. The collection includes agricultural products such as dried figs, raisins, pomegranates, and nuts—as well a mummy or two. An exhibit on Mummification examines the process and why it was done.

A small exhibit on clocks and watches in the European Sculpture and Decorative Arts collection give people the opportunity to admire the beauty of these pieces, learn how they work, and understand their historical significance. One interesting piece contains a celestial globe with clockwork made in Vienna, Austria, in 1579 by Gerhard Emmoser.

Science-related pieces are scattered among other collections, such as an Etruscan chariot in the Greek and Roman Art gallery and a ninth-century bowl from Iraq in the Islamic Art collection. The technique of glazing and luster-painting earthenware, first employed in Iraq in the ninth century, had a permanent influence on the pottery industry all over the world.

The collections of musical instruments include some descriptions of how these instruments are made and how they work.

In the European Painting collection there is a 1788 portrait of the groundbreaking French chemist Antoine Lavoisier and his wife, Marie, painted by Jacques-Louis David. Marie was probably a student of David, and illustrated her husband's 1789 book, *Traité Elémentaire de Chimie*. This painting is sometimes on display.

Because exhibits change, all of these exhibits may not be on display at any one time.

Address: Metropolitan Museum of Art, 1000 Fifth Avenue (at 82nd Street), New York, NY 10028.
Phone: (212) 535-7710.
Days and hours: Sun.–Thu. 10 a.m.–5:30 p.m.; Fri. and Sat. 10 a.m.–9 p.m. Closed Thanksgiving Day, 12/25, 1/1, and the first Monday in May.
Fee: Yes.
Tours, lectures, and other events: Self-guided, as well as daily tours of selected sections of the museum. Other languages are available in written materials, lectures, etc. Check at the Information Desk located in the Great Hall.
Handicapped accessible: Yes.
Public transportation: Bus in front of the museum; subway (86th Street Station) to within three blocks of the site. Taxi stand is in front of the Museum entrance.
Food served on premises: Cafés, dining room.
Website: www.metmuseum.org/

The New York Botanical Garden

In the heart of the urban Bronx is this 250-acre (101-hectare) museum of living plants. When New York botanist Nathaniel Lord Britton and wife Elizabeth visited London's Kew Gardens in 1888 and returned home, they began a campaign to found a similar garden in New York City. Founded in 1891, the New York Botanical Garden is one of the oldest and largest gardens in the United States and a National Historic Landmark. Its site was chosen because the land has a rock-cut gorge and an old-growth forest. Calvert Vaux drew initial designs, and the Olmsted Brothers (Frederick Jr. and John Charles) continued the project by adding roadways and a circulation plan. (See Frederick Law Olmsted National Historic Site, Brookline, Massachusetts.)

This Garden features historic and valuable collections (over one million living plants), in more than 50 different gardens with an extensive collection of topical, temperate, and desert flora, ranging from 300-year-old trees to rare ferns, orchids, and palms from around the world. It is a sanctuary for plants and people, and a place to learn about biodiversity, the function of plants, and the use of plants for medicines and other economic products.

The Enid A. Haupt Conservatory is a Victorian greenhouse completed in 1902. The Palm House at Kew Gardens near London was the inspiration for the structure. It contains 11 galleries with different themes, climates, and plants. The building's glass is held in place in typical 19th-century fashion, by metal glazing bars lined with wood. This one-acre (0.4-hectare) building was renovated in the late 20th century to replace wood liners and glass panels, and to modernize irrigation, heating, electrical wiring, and environmental controls. Recent exhibits have examined multidisciplinary subjects such as the role of plants in Darwin's scientific research and the history of botanical gardens in the study of medicinal plants.

A historic and important feature of the Garden is the 50-acre (20-hectare) forest, the Thain Family Forest. This forest has never been cleared or developed, and is the largest remaining tract of the forest that once covered the land on which New York City stands. This area was originally part of the estate of tobacco merchant Pierre Lorillard. Here stand the renovated 1840 snuff mill and dam that Lorillard placed in the Bronx River, the only fresh-water river in New York City.

The forest has changed significantly over the past 100 years. A mature hemlock-and-oak forest established before the 17th century, it is going through a transition to a forest

Conservatory at the New York Botanical Garden.
Courtesy of the New York Botanical Garden.

of red maple, black cherry, and American beech. Researchers are trying to preserve the old-growth forest. A stroll through the forest is a wonderful escape from the urban setting that surrounds the gardens.

Specialty gardens include a world-famous collection of ferns, begonias, rock-garden plants, perennials, native wild species, orchids, daylilies, and bulbs. The Peggy Rockefeller Rose Garden has nearly 700 varieties. In the Native Plant Garden are shrubs that attract birds and butterflies. The Maple Collection has a worldwide selection of over 160 maple trees comprising 135 species, hybrids, and cultivars.

Much of the research done by the garden's scientific staff is carried out at the Garden's Pfizer Plant Research Laboratory, where scientists study plant DNA to better understand plant evolution and the genetic relationships between plant species. The Garden also has an Institute of Systematic Botany, whose research focuses on plants for food, fuel, fiber, and medicine.

The William & Lynda Steele Herbarium, the largest in the Western Hemisphere, holds nearly 8 million dried plant specimens from all over the world. About 50 000 new specimens are added each year. Stored here are specimens collected by Captain Cook on his 1768 voyage to the South Pacific, and John C. Frémont during his expeditions in the mid-1800s. (See Fort Clatsop & Visitor Center, Astoria, Oregon.)

Of special note is the LuEsther T. Mertz Library, founded in 1899, which has the largest and most comprehensive botanical library in the world: botany, horticulture, history, anthropology, landscape and building design, environmental design, and rare books from medieval times to the 17th century, as well as depictions of the magnificent gardens of Europe. There are accounts of botanical explorations, 18th-century writings of Carl Linnaeus, and 19th-century accounts by Charles Darwin. Access to the Library is by prior appointment, only to researchers.

Address: The New York Botanical Garden, 2900 Southern Boulevard, Bronx, NY 10458-5126.

Phone: (718) 817-8700.

Days and hours: Tue.–Sun. 10 a.m.–6 p.m.; Jan–Feb.: 10 a.m.–5 p.m.; open many holiday Mondays.

Fee: Yes.

Tours: Self-guided; guided tours of the grounds are available daily. Special events are scheduled throughout the year. Trams take visitors to various points in the Garden, where they can disembark and re-embark.

Handicapped accessible: Yes (though limited portions may not be because of rough terrain). Wheelchairs are available.

Food: Café and grill.

Public transportation: Metro-North railroad stops across the street; bus stops at the entrance; subway (Bedford Park Blvd. Station or Allerton Ave. Station) is about a 20-minute walk away.

Website: www.nybg.org

Staten Island Museum

Staten Island, probably the least known but certainly least populated borough of New York City, has a wonderful natural-history collection housed in the Staten Island Museum. In 1881, a small group of local naturalists pooled their resources because of the destruction of local habitats through population growth. This collection was then incorporated, and first opened to the public in 1908. The Museum—sometimes termed a mini-Smithsonian for its depth and breadth—is especially known for participating in bird counts. (See Smithsonian Institution, Washington, DC.) By 1919, the Museum took on its present role of preserving the local arts and sciences with a collection that now numbers over half a million items. The Museum is housed in multiple buildings: the original structure in St. George, two blocks from the Staten Island Ferry, serving as storage, and a second complex in the Greek Revival style at Snug Harbor, for the exhibitions and displays.

Of interest to the science-lover is the Natural Science collection. Here the fields of zoology, botany, entomology, geology, paleontology, and archaeology are well-represented. The William T. Davis Collection of cicadas contains 35 000 specimens, many of which were collected by Davis (a local naturalist, 1862–1945) himself. These insects are known for their loud calls produced by rapidly moving their tymbrals. Even larger is the Howard Notman collection of 75 000 beetles. Note the awe-inspiring Wall of Insects, which displays over 150 beetles, cicadas, butterflies, and moths.

The herbarium has a complete record of Staten Island flora from the 1860s to the present, and has notable specimens of hybrid oaks and hybrid ferns. There is a collection of alcohol-preserved local fish, frogs, toads, salamanders, and various reptiles, including some from Florida and Texas. A number of birds and mammals (as well as their skeletons) have been preserved through taxidermy.

Over 10 000 shells are housed in the Staten Island Museum, mostly based on local fauna, but global specimens are available as well.

Rock-based collections include the 7000-specimen mineral collection, including a variety of fluorescent rocks in a darkened mineral room, and especially the Zabinski Collection of microminerals with over 3200 micromounts from around the world. These microminerals are specially mounted in boxes for viewing under a microscope. Also collected are variety of fossils, especially Staten Island–based plants.

The History Archives includes books, maps, documents, and scrapbooks on the natural history and geology of Staten Island (by appointment only). A quarantine station that occupied a corner of Staten Island during the 19th and early 20th centuries is represented in the Quarantine Collection in the Archives.

The Museum offers a variety of lectures on scientific topics, plus nature and bird walks. Call in advance or check the website for schedules.

Address: Staten Island Museum at Snug Harbor, 1000 Richmond Terrace, Building A, Staten Island, New York 10301. Staten Island Museum History Center & Archives, 1000 Richmond Terrace, Snug Harbor Campus, Building H, Staten Island, New York 10301.
Phone: Snug Harbor, (718) 727-1135; Archives in Snug Harbor, (718) 727-1135 ext. 122.
Days and hours: Wed.–Sun. 12 p.m.–5 p.m. Closed 1/1, Memorial Day, 7/4, Thanksgiving Day, and 12/25. History Archives & Library, Tue., Thur., and Fri. 10 a.m.–4 p.m., by appointment only.
Fee: Yes. (Free on Tue. 12–2 p.m.)
Tours: None.
Handicapped accessible: Yes.
Food served on premises: None.
Public transportation: Staten Island Railway and bus, or Staten Island Ferry to Staten Island, and take Railway (Snug Harbor Stop) or bus to Archives.
Special Note: Check website or call for special lectures and nature walks.
Website: www.statenislandmuseum.org/

Wildlife Conservation Society

The Wildlife Conservation Society was chartered in 1895 as the New York Zoological Society to advance wildlife conservation, promote the study of zoology, and establish a prime zoological park. In 1993 the society assumed its current name, the Wildlife Conservation Society (WCS), to reflect its role as a sanctuary for protecting wildlife from ecological destruction. WCS was one of the first conservation organizations in this country. Over the past century this organization has established a long-term conservation presence in the last wild places in the Americas, Africa, Asia, and Oceania. These places are intact, biodiverse, mostly resilient to change, and have large, iconic wildlife species.

The Conservancy also operates five facilities in four of the five boroughs of New York City: the Bronx Zoo, the Central Park Zoo (Manhattan), the Queens Zoo, and the Prospect Park Zoo and the New York Aquarium (both in Brooklyn). All these facilities showcase the animal wonders of the world.

Bronx Zoo

The Bronx Zoo, the home base of the Conservancy, is next door to the New York Botanical Garden. (See The New York Botanical Garden, New York, New York.) This zoo is among the largest urban zoos in the world, situated on 265 acres (107 hectares) of woodlands, meadows, ponds, and streams that are populated with more than 4000 animals of 650 species from around the world. Some of these animals are endangered or threatened. When the zoo opened to the public in 1899 it held 843 animals in 22 exhibits.

On view, usually in native settings, are African animals, American Bison, a Baboon Reserve, Bears, Birds of Prey, the Carter Giraffe Building, a Butterfly Garden, Aquatic Birdhouse, Congo Gorilla, Himalayan Highlands, and much more.

Address: Bronx Zoo, 2300 Southern Boulevard, Bronx, NY 10460 (southern entrance); intersection of Bronx River Parkway and Boston Road, Bronx, NY 10460 (Bronx River entrance).
Phone: (718) 220-5100.
Days and hours: Nov.–Mar.: daily 10 a.m.–4:30 p.m.; Apr.–Oct.: Mon.–Fri. 10 a.m.–5 p.m., weekends and holidays 10 a.m.–5:30 p.m. Closed Thanksgiving Day, 12/25, 1/1, and Martin Luther King, Jr. Day.
Fee: Yes. Additional fees for some exhibits and entertainments, monorail, and zoo shuttle. Free all day on Wed.
Tours: Two-hour tour of the most popular exhibits; feeding demonstrations and educational programs; check with information.
Handicapped accessible: Mostly. Certain exhibits have rough terrain. Wheelchairs available but limited in number.
Public transportation: Bus; Express bus from Manhattan to Bronx River Entrance; subway; Metro North Harlem line to Fordham, then take bus.
Food served on premises: Café and seasonal food-stands.

Central Park Zoo
The Central Park Zoo covers 5.5 acres (2.2 hectares) in Frederick Law Olmsted and Calvert Vaux's masterpiece, Central Park. (See Frederick Law Olmsted National Historic Site, Brookline, Massachusetts.) This intimate facility focuses on climatic zones, the animals that inhabit them, and ways in which animals depend on each other.

The Tropic Zone is one of three habitats encircling a central garden and sea lion pool. The zone re-creates a tropical rain forest and is designed to house smaller creatures, such as monkeys, crocodilians, snakes, bats, free-flying birds, and leafcutter ants at work on their trails. The Temperate Territory contains Asian and North American animals, and the Polar Circle includes animals from the Arctic and Antarctic, such as Gentoo penguins, puffins, polar bears, harbor seals, and Arctic foxes. The Allison Maher Stern Snow Leopard Exhibit houses these endangered wild cats from Central Asia.

Address: Central Park Zoo, 64th Street and Fifth Avenue, New York, New York 10021.
Phone: (212) 439-6500.
Days and hours: Nov.–Mar.: daily 10 a.m.–4:30 p.m.; Apr.–Oct.: Mon.–Fri. 10 a.m.–5 p.m., weekends and holidays 10 a.m.–5:30 p.m.
Fee: Yes.
Tours: Self-guided.
Handicapped accessible: Mostly. Some areas have limited accessibility. Wheelchairs available on a first-come-first-served basis.
Public transportation: About seven blocks from the subway; bus stops near entrance.
Food served on premises: Café.
Website: www.centralparkzoo.com/

Prospect Park Zoo

The Prospect Park Zoo is a facility specializing in reaching the needs of children. Located on 12 acres (4.9 hectares), it contains over 600 animals from around the world, and offers a petting zoo and special shows.

Address: Prospect Park Zoo, 450 Flatbush Avenue, Brooklyn, New York 11225.
Phone: (718) 399-7339.
Days and hours: Nov.–Mar.: daily 10 a.m.–4:30 p.m.; Apr.–Oct.: Mon.–Fri. 10 a.m.–5 p.m., weekends and holidays 10 a.m.–5:30 p.m.
Fee: Yes.
Tours: Self-guided. Check with information on sea-lion feeding times.
Handicapped accessible: Yes.
Public transportation: Bus, subway.
Food served on premises: Vending machines; picnic tables for bagged lunches.
Website: www.prospectparkzoo.com/

New York Aquarium

The New York Aquarium, on 14 acres (5.7 hectares) alongside the Coney Island shoreline in Brooklyn, opened in 1896 as the first public aquarium in the United States. It works with the adjacent Osborn Laboratories of Marine Science to do research in fish diseases and fish genetics. The aquarium houses 10 000 specimens from around the world, representing 300 species. Such exotic creatures as the Pacific nautilus, the flashlight fish, and the giant Japanese spider crab populate indoor and outdoor exhibits.

The Main Gallery contains exhibits of coral and a wide variety of fish life. Species indigenous to the Red Sea live in one exhibit that has an intricate vertical reef. Fish from the Bermuda Triangle inhabit coral caves and rocky reefs. Other exhibits include sharks and freshwater fish. An authentic bathysphere used in the 1934 deep-sea dive by William Beebe and Otis Barton is also on view.

Discovery Cove has live aquatic animals, dioramas, graphics, interactive devices, and videos to teach visitors about coastal ecosystems, adaptation of marine mammals, and how people and the sea are interrelated. This section of the facility features three coastal ecosystems. Sandy Shore re-creates a 45-foot (14-m) section of the local coastline and contains striped bass, scup, sea robins, and kingfish. The Salt Marsh, a walk-in diorama, has a touch tank where you can examine hermit crabs and horseshoe crabs. Rocky Coasts has waves crashing against rock formations and touch tanks with sea stars, urchins, and periwinkles.

There are more than a dozen adaptation alcoves, including a kelp tank and a coral reef, which focus on how fish evolve to adapt to biological ecosystems. Some exhibits explain how sea creatures form schools, breathe, eat, protect themselves, move, reproduce, see, and hear. Others explore how people interrelate with the sea. You can see an authentic New England lobster boat, pharmaceuticals from the sea, and a ship chandlery. Sea Cliffs, which re-creates 300 feet (91 m) of the rocky Pacific coastline, lets you view sea life from above and below the waterline. On display in this naturalistic habitat are walruses, harbor seals, sea otters, and a breeding colony of black-footed penguins.

Address: New York Aquarium, Surf Avenue & West 8th Street, Brooklyn, New York 11224.
Phone: (718) 265-3474.
Days and hours: Sep.–May: 10 a.m.–4:30 p.m.; Memorial Day to Labor Day: 10 a.m.–6 p.m.
Fee: Yes.
Tours: Self-guided.
Handicapped accessible: Yes.
Public Transportation: Short walk from subway and bus stops.
Food served on premises: Food court and picnic area.
Website: www.nyaquarium.com/

Queens Zoo

The Queens Zoo is an 18-acre (7.3 hectare) site populated by familiar and lesser-known animals and birds indigenous to the Americas. Habitat exhibits are arranged off a pathway that encircles the zoo. The Buckminster Fuller geodesic dome, a remnant of the 1964 World's Fair, houses a large aviary. 175 feet (53 m) in diameter and originally covered in solid material, the dome was disassembled after the World's Fair and rebuilt at the Zoo with netting as the cover. The marsh is home to ducks, sandhill cranes, swans, herons, egrets, and Canada geese. The zoo also has black bears, Andean bears, puma, bobcats, prairie dogs, and wild turkey. Among animals in the zoo's Species Survival Plan* is the endangered red wolf, which is soon to be reintroduced into the wild of the southern United States.

Address: Queens Zoo, 53-51 111th Street, Flushing Meadows Corona Park, Queens, New York, 11368.
Phone: (718) 271-1500.
Days and hours: Nov.–Mar.: daily 10 a.m.–4:30 p.m.; Apr.–Oct.: Mon.–Fri. 10 a.m.–5 p.m., weekends and holidays 10 a.m.–5:30 p.m.
Fee: Yes.
Tours: Self-guided.
Handicapped accessible: Yes.
Public transportation: Subway and bus to near the site.
Food served on premises: Vending machines; picnic tables for bagged lunches.
Website: www.queenszoo.com/

OYSTER BAY

Planting Fields Arboretum State Historic Park

Located at Oyster Bay, Long Island, near the Long Island Sound, this arboretum benefits from both of these bodies of water because they keep the regional temperature relatively stable. The Matinecock Indians, who recognized the value of consistent temperatures and fertile land, planted corn and other grains here almost a thousand years ago. "Planting Fields" is a translation from the Native American name for this area. The land eventually became Planting Fields Arboretum, which was part of a large parcel of land bought by European settlers in 1653.

Oyster Bay remained a community of watermen and farmers for more than 200 years. Because the soil is sandy, the area was particularly suited to growing asparagus. Thus these fields became a major supplier of asparagus to the northeastern United States. In the 20th century the farm communities gave way to vast estates with large houses and splendid gardens. The area became known as the Gold Coast. In 1904 the Byrne family bought the land and later sold both the land and the houses on the property to William Robertson Coe, who hired several people to lay out gardens and create a horticultural showcase. These landscape gardeners included Guy Lowell, A. R. Sargent, and the Olmsted brothers. (See Frederick Law Olmsted National Historic Site, Brookline, Massachusetts.)

Coe deeded Planting Fields to the State of New York in 1949 as an educational and horticultural institution. This nature sanctuary of cultivated gardens, rolling lawns, and natural woodlands stretches over 400 acres (160 hectares). More than 800 species and hybrids of rhododendrons and azaleas form one of the arboretum's premier collections, the core of which was original Olmsted plantings.

A camellia greenhouse, built by Coe, displays over 300 plants, an extensive and expanding collection of camellias, one of the largest collections of camellias under glass in the northeastern United States. The main room displays hibiscus, orchids, cacti, succulents, ferns, begonias, and more. Seasonal displays are mounted regularly.

The Dwarf Conifer Garden displays plantings that are usable for the small gardens of local homeowners. At the entrance is an *Opuntia oppressa*, a native Long Island cactus. Other interesting areas are the Wildflower Garden, Formal Garden, Rose Arbor, Daylily Garden, Bird Sanctuary, Heather Garden, magnolia and maple collections, 175 varieties of holly, Native Tree and Shrub Walk, Conifer Trail, and remnants of a beech-oak forest from the pre-European period. The 5-acre (2-hectare) Synoptic Garden contains over 500 types of ornamental shrubs and trees, including beech, linden, elm, tulip, and cedar, that grow particularly well on Long Island. It is arranged alphabetically by botanical name. This helps visitors understand the local flora and shows local homeowners the species they can grow most successfully. The herbarium has more than 10 000 specimens native to Long Island.

A Sensory Garden boasts large-print and Braille signs, which allows people with a variety of differing abilities to better appreciate the smells, tastes, and textures of plants. People who visit this garden are encouraged to see, touch, taste, smell, and listen to all that goes on in the garden.

In addition to the land is the Tudor Revival Residence of the Coe family, known as Coe Hall. The original structure, which burned down in 1918, was replaced by the current Coe Hall, built between 1918 and 1921. This building is open to the public from April to September, with a guided tour. The house is arranged and appointed in the style of the 1920s. Visitors can see the craftsmanship, furnishings, painting, stained glass, and decorative arts of the time period.

Address: Planting Fields Arboretum, 1395 Planting Fields Road, Oyster Bay, NY 11771.
Phone: (516) 922-8605.
Days and hours: Daily 9 a.m.–5 p.m. Closed 12/25.
Fee: Yes.
Tours: Self-guided; guided tours may be arranged in advance during the summer season. Call or check the website.
Handicapped accessible: Yes.

Public transportation: Long Island Railroad train stops at Oyster Bay, then walk or take a taxi 1.7 mi (2.7 km) to the site.

Food served on premises: Café on weekends only. There are several restaurants in the town of Oyster Bay.

Website: www.plantingfields.org/

RHINEBECK

Old Rhinebeck Aerodrome

As a young man, Cole Palen fell in love with vintage aircraft, and in 1951 he began his own collection. In 1956 he purchased an abandoned farm and planned a small airfield in Rhinebeck, New York. He soon found, however, that his interest in antique airplanes overshadowed his original plans for an airport; thus the Aerodrome came into being.

The Aerodrome is a living museum in a pleasant, park-like setting. It is filled with original and re-created aircraft, land vehicles, motorcycles, early engines, and other memorabilia which date from 1896 to the 1940s. The vintage aeroplanes are divided into three periods: the pioneer era of 1900–1913; World War I, dating from 1914–1918; and 1919 and later, the golden era of barnstorming. There is also a collection of ground vehicles. On display is a 1909 Blériot XI, salvaged from a junkyard, one of the oldest airplanes in the world and the oldest flying airplane in the United States. This airplane has no flying instruments, so the Aerodrome estimates its flying speed at around 30 mph (48 km/h). Roll control is via wing-warping which is only slightly effective at this speed. A Blériot XI from 1910 was displayed at the Expo '67 World's Fair in Canada. Another Blériot XI, from 1911, was saved from a fire in the barn where it was stored. There is also a reproduction of Octave Chanute's 1896 glider, along with reproductions of some of the Wright Brothers' early flying machines.

The US-built 1909 Voisin on display has a French superstructure and American wood and fabric. The 1912 Thomas Pusher Model E was recovered from a barn in upstate New York and the 1917 Nieuport 28, originally owned by Paramount Studios, was used in a number of films.

At the end of World War I, Germany agreed to destroy all of its aircraft. In an effort to stay in manufacturing, Anthony Fokker smuggled 60 planes out of Germany into Holland, where he set up business. The Fokker D.VII, seen in this museum, was probably one of those planes. The 1918 SPAD 13 was flown into Roosevelt Field in 1930 by Colonel Benjamin Kelsey, who never returned for it. All his mail was returned unanswered, and he was never heard from again—an unsolved mystery that persists to this day. The 1918 Thomas-Morse S4 B is one of 497 originally built as part of the US effort to produce a fighter plane for World War I.

Some planes are reproductions. Many aircraft, whether original or reproduction, are in flying condition. The collection is displayed in a series of wooden hangars and barns, which re-create the ambience of the barnstorming days.

One of the barns is now a small museum that houses engines, automobiles, motorcycles, and other motor-powered items. The artifacts displayed include a 1903 Grout Steamer, a 1922 Buick touring car, a 1914 Packard moving van, a three-wheel Morgan from 1936, and a 1918 ambulance built by Columbia, predecessor to the General Motors Corporation. One of the displayed engines is a 9-cylinder air-cooled Wright JC-5

Whirlwind, boasting 223 horsepower (166 kW), the same type Lindbergh had on his famous *Spirit of St. Louis* when he flew to Paris in 1927.

A small exhibit honors Harriet Quimby for her 1912 feat: she was the first woman to pilot a plane across the English Channel. She flew in a 1911 Blériot XI similar to the ones on display here. Also on the grounds is a Model Airplane Museum displaying period scale models.

During the summer the Aerodrome has weekend shows where pilots and ground staff, in period costume, show the equipment in action. This is one of the few places in the United States where antique planes fly regularly.

Address: Old Rhinebeck Aerodrome, 9 Norton Road, Red Hook, NY 12571.
Phone: (845) 752-3200.
Days and hours: Daily 10 a.m.–5 p.m., from May to October.
Fee: Yes.
Tours: Self-guided; guided tours by prior arrangement.
Handicapped accessible: Limited.
Public transportation: Taxi or rental car from Rhinecliff-Kingston train station (Amtrak train) 6.7 miles (11 km) away.
Food served on premises: Snack stand, open on air-show weekends, cash only.
Special notes: Check times of air shows, which run on weekends from mid-June through Mid-October. Note that certain aircraft are being refurbished or in storage, not visible to the public; check in advance to be certain that particular aircraft are on display.
Website: www.oldrhinebeck.org/

Rochester

George Eastman Museum

When George Eastman's turn-of-the-century urban estate was bequeathed to the University of Rochester, it was used as the home of the university president for several years before becoming a photographic museum in 1949. Two years later it was opened to the public. Inside are prints, negatives, films, cameras, cinematographic and photographic devices, and literature. The mansion and the gardens are on the National Register of Historic Places.

In 1878, as a young accountant, George Eastman bought a view camera and became entranced with the photography process. Noting the bulky and unwieldy early photographic equipment, Eastman saw a potential for profit by simplifying it. In 1880 he founded a new company that produced dry plates, a simplified method for developing negatives. In 1888 he announced the availability of his box camera and flexible film in premade cartridges. This innovation brought photography to the masses. He came up with the name Kodak for a camera he was developing, and his company became the Eastman Kodak company.

By 1902 Eastman began to build his dream house, this 50-room colonial revival mansion, completed in 1905. It was one of the first houses in the region to have central heating and electrical service, and it was built of reinforced concrete to be fireproof. An extensive museum building underground replaced the historic gardens. This facility houses a large photographic collection. The historic garden was re-created above the photographic collections. Both the house and the garden have been restored, and most of the furnishings are original. The museum is a member of the Association of Botanical Gardens and Arboreta.

The museum's collection has grown considerably since the first photographs were donated. This facility includes one of the largest collections of photographic and cinematographic equipment, collections of photographic apparatus catalogs, and a lot more. Collections include the Medicus collection of Civil War photographs by Alexander Gardner, Eastman Kodak's historical collection, and the huge Gabriel Cromer collection from France. This fine collection of 19th-century photos contains daguerreotypes, albumin prints, and gelatin-silver prints.

The Mees Gallery, opened in 1962, is devoted to the science of photography and is named after the director of Kodak Research from 1912 to 1947. A photographic pioneer and industrial researcher, Dr. C. E. Kenneth Mees helped to create sophisticated photographic technology. The Mees Gallery houses a permanent exhibit entitled Enhancing the Illusion: The Origin and Progress of Photography, which traces the developments in photographic technology from the earliest experiments with light and optics in the 17th century to today's commercial imaging industry. It has interactive exhibits, video displays, and more than 1000 artifacts from the museum's extensive collections, including an 1839 Giroux Daguerre camera—the first camera to be marketed commercially—and daguerreotype accessories. It is estimated that by the 1850s three million daguerreotypes were produced each year.

See how John Herschel (1792–1871), the famed British astronomer, responded to the realization that photography and astronomy could be used together. He was able to record on film what he saw in the heavens. Explanations of the principles of light, optics, and image-formation accompany a room-sized camera obscura, a darkened chamber in which the image of an object passes through a small opening and is focused in color on a surface.

An additional building was put in place in 1989 to accommodate the growing collection. The museum examines photography from a sociological, cultural, and technological perspective, and provides step-by-step demonstrations of photographic processes. Special attractions include an exhibit of magic lanterns, a model of the Paris Diorama Theater, a 19th-century portrait studio, and home movie clips of the 1920s, 1930s, and 1940s. An interactive station provides detailed information on all the cameras exhibited and biographical sketches of important people in photography's history.

This museum has one of the largest collections of motion pictures in the world. It has one of three archival theaters in the United States equipped to exhibit nitrate films, the 500-seat Dryden Theatre.

Major collections here include Ansel Adams's early pictures, 19th-century photographs of the American West, and glass negatives from French photo-press agency Charles Chusseau-Flaviens, dating from the turn of the 20th century. Also included is one of the largest collections of daguerreotypes outside of France, as well as an important collection of some work of Alfred Stieglitz, Edward Steichen, and other notable photographers.

Nearby, in Chili, New York, is one of very few film-conservation centers. This facility houses the museum's 35-mm prints made of cellulose nitrate and a film-preservation school.

Address: George Eastman Museum, 900 East Avenue, Rochester, NY 14607.
Phone: (585) 327-4800.
Days and hours: Tue.–Sat. 10 a.m.–5 p.m., Sun. 11 a.m.–5 p.m. Closed Mon.

Fee: Yes.

Tours: Self-guided; check with museum for various guided tours of the museum.

Handicapped accessible: Yes (with some limitations: check with the museum; sign-language tour available by appointment).

Public transportation: City buses stop about 0.3 mile (0.5 km) from site, then walk.

Food service on premises: Café.

Website: www.eastmanhouse.org/

SYRACUSE

Erie Canal Museum

Once the American colonies were independent of the British, settlers began moving westward. For generations, travelers struggled to cross the Appalachian Mountains, which had only one land and water pass: the Mohawk River Valley in the State of New York, which was hindered by waterfalls and rapids. In 1825 the Erie Canal was completed to provide this much-needed link to the West. It became a vital transportation route between New York City and the Great Lakes, bringing coal, lumber, and grain to the East, and European immigrants to the West.

The design of the canal followed the model of the European canals, specifically those in England. The size of the Erie Canal and the challenges in building it were unmatched by any previous canal project. Eighty-three locks were installed and 18 aqueducts were added to lift barges 676 feet (206 m) as they moved 363 miles (584 km) from the Hudson River, through Buffalo, and on to the Great Lakes.

The Erie Canal became the world's most successful canal, shaping the economic and social fiber of this part of the nation. Shipping costs dropped dramatically, New York City became the most important port in America, and industries developed all along the canal's route.

The time span from the establishment of the New York State Commission, in 1817, to the closing of the canal in 1917 was a hundred years. Advances in technology, such as steam power on the canal, and steel boat construction, eliminated the need for a waterway as protected as the old Erie Canal. However, there are small pieces of the canal still in use.

The Erie Canal Museum, opened in 1962, works to advance public interest in the canal and preserve its history. The Weighlock Building, probably the most significant artifact of the museum, is listed on the National Register of Historic Places. Here, much of the museum collections are housed, and it is the only surviving structure in the country that was designed to weigh canal boats and their cargo. This restored Greek Revival building was constructed in 1850 at the junction of the Erie and Oswego Canals. Boats up to 100 feet (30 m) long could be weighed, and the weight of the cargo could be determined. Weighmasters would then collect tolls based on the weight of the cargo. Tolls were collected here for 33 years.

The Weighlock Building now displays a variety of exhibits with artifacts from the collections. A 14-minute video on the history of the canal and its effect on the region is shown regularly in the Orientation Theater in this building. A re-created weighmaster's office shows visitors how the business aspects of the site were managed.

The museum holds not only one of the largest collections of canal-boat models and canal patent models in the world, it also displays more than 600 implements used in canal and boat construction and related industries. Other canal-related memorabilia, such as commemorative china, paintings, lithographs, drawings, and etchings, are displayed in both permanent and rotating exhibits. The archive and library contain a broad collection of materials related to the Erie Canal, the New York State Barge Canal, American and European canals, and the Panama Canal. The museum also houses the Syracuse Urban Cultural Park Visitor Center and features exhibits on the business and cultural prosperity fostered by the Erie Canal in Syracuse.

Outside and next to the Weighlock Building is a typical lock-tender's garden. Gardening was one way that lock-tenders used their free time between jobs.

A 65-foot (20-m) canal boat, the *Frank B. Thomson*, was constructed for the museum so that visitors can view life and work aboard a canal boat. The crew quarters are where the captain, cook, steersman, and mule drivers ate, slept, and relaxed. The cargo hold is the place that raw goods and manufactured materials were stored and are now displayed. These products were moved east and west on the canal. The forward cabin has an exhibit on the various immigrant groups that traveled these waterways.

Address: Erie Canal Museum, 318 Erie Boulevard East, Syracuse, NY 13202.
Phone: (315) 471-0593.
Days and hours: Mon.–Sat. 10 a.m.–5 p.m.; Sun. 10 a.m.–3 p.m. Closed 1/1, Easter Sunday, 7.4, Thanksgiving, and 12/25.
Fee: Free.
Tours: Extra fee, Mon. through Fri.; arrange in advance.
Handicapped accessible: Yes.
Public transportation: Amtrak train and Greyhound bus service to Syracuse: Local bus service, some hotels have shuttle-bus service, taxi service goes to the site. Convention Center nearby.
Food on premises: None; food service within walking distance.
Website: eriecanalmuseum.org/

Pennsylvania

COLUMBIA

National Watch & Clock Museum

Starting with fewer than 1000 specimens, the National Association of Watch and Clock Collectors opened this museum and library in 1977 as a scientific and educational institution to preserve the craft and heritage of timepieces. In 1999 the museum expanded this facility with a two-story addition providing additional exhibition space. Now more than 12 000 items from more than four centuries of horological development make this facility one of the nation's largest collections of precision watches, clocks, tools, and related items. Everything is arranged in galleries according to type and period, which helps visitors understand the historical context and scientific growth of horology.

People have devised ways of calculating and measuring the passage of time for more than 5000 years, as the museum's historical overview explains. The Romans used the word *horologium* to identify timepieces. Today "horology" means the study of time and timekeeping devices. The earliest clocks were primarily mechanisms to ring bells. Eventually mounted in towers, these devices signaled the start of religious services and public announcements. In Latin, *clocca* means "bell." In France, towers with bell-ringing devices became known as *cloches*. For almost 300 years, during the Middle Ages, public clocks had no hands but struck the hours with a bell, calling monks to prayer and signaling time to the region.

The museum's collection comes from all over the world, including Japan, China, Europe, and the United States. The glass bell musical clock, made in Germany circa 1790, is driven by wooden gears and plays one of six different tunes every hour on its glass bells. A late 18th-century tall clock, made in Haarlem, Netherlands, not only tells the time but also gives the days of the week, presenting each day's deity and an astrological symbol for the planet that governs the day. A dial indicates the month and day number. The Yagura-Dokei, a Japanese lantern clock made between 1710 and 1725, is regulated so the length of the hour is longer in summer than in winter. Thus the number of hours of daylight is always the same.

The chronometer, a highly precise timepiece, was developed in the 18th century when ocean trade increased. Navigators needed to track time accurately to chart the ship's location, but could not use devices with traditional pendulums because they were unable to keep accurate beats on a rocking ship.

One of America's unique contributions to clock technology was mass-production, which clockmaker Eli Terry pioneered in 1812. Terry developed a small clock that could sit on a shelf, a major change from the traditional tall clocks that stood on the floor. He used interchangeable parts, which made shelf clocks affordable by the masses. The pillar-and-scroll clock Terry made between 1818 and 1822 in Bristol, Connecticut, introduced interesting design that could be made cheaply. (See American Clock and Watch Museum, Bristol, Connecticut.)

Leaders of American railroads, frustrated by their inability to maintain reasonable train schedules because each town set its own time, initiated the idea of standard national time. A joint agreement between the United States and Canada created separate time zones in 1883, with five zones at intervals of approximately 15 degrees longitude. The railroad's need for accurate timepieces made the term "railroad watch" synonymous with high quality. The "regulator" was invented in the late 19th century to help railroads, banks, and stock exchanges operate on precise schedules. This clock did not lose time while being wound or change speed according to the room's temperature.

The first electric clock is credited to Alexander Bain of Scotland who, in 1840, used the earth's magnetic field to help power a clock. Modern battery-powered clocks were mass-produced in the early 20th century, as demonstrated by the museum's 1915 clock made by the Rempe Manufacturing Company of Danville, Pennsylvania. By the 1930s, small synchronous motors were being installed to run clocks.

Although wristwatches were first built in the 1500s, the term "wristwatch" was not coined until the 20th century. The museum displays many modern wristwatches and pocket watches, including a 1947 pocket chronograph made by Patek Phillipe and Company of Geneva, Switzerland. It has a calendar, moon-phase indicator, five-minute chime, and hour chime. A masterpiece in the museum's collection is Stephen Engles's 19th-century monumental clock, which has 48 moving figures and two pipe-organ movements. The museum also has exhibits on the process of making clocks and watches.

Address: National Watch and Clock Museum, 514 Poplar St., Columbia, PA 17512.
Phone: (717) 684-8261.
Days and hours: Dec.–Mar.: Tue.–Sat. 10 a.m.–4 p.m., closed Sundays and Mondays; Apr.–Nov.: Tue.–Sat. 10 a.m.–5 p.m., Sun. 10 a.m.—4 p.m., closed Mondays. Also closed major holidays except Memorial Day and Labor Day throughout the year.
Fee: Yes.
Tours: Self-guided; group tours by prior arrangement only.
Public transportation: Local bus to within 0.3 mile (0.5 km) of site, then walk.
Handicapped accessible: Yes.
Food served on premises: None. There are several restaurants about 6–8 blocks to the southeast of the site.
Website: www.nawcc.org/index.php/museum

CORNWALL

Cornwall Iron Furnace

Cornwall was one of 80 blast furnaces in the American colonies during the Revolutionary War. It was built by Peter Grubb, who discovered a deposit of magnetite, or iron ore, in 1734 during a prospecting expedition. He purchased 400 acres (162 hectares) of the land, built the furnace, and named the site Cornwall after the rich mining area of Great Britain, home of Peter Grubb's father. One of very few furnaces of this type left in the world today, Cornwall is now part of a National Historic Landmark District by the National Park Service, is a National Historic Landmark chosen by the American Society of Metals, and a National Historic Mechanical Engineering Landmark.

By the start of the American Revolution, one-seventh of the world's iron was produced in America; only Russia and Sweden produced more. Because of the colonies' iron supply and production capacity, the colonists were able to make the ammunition and ordnance they needed to win a war with England. Pennsylvania, with its rich iron deposits, abundant forests, large limestone reserves, and plentiful water supply, became a natural center of iron production. It had 24 furnaces, more than any other single colony.

By the mid-19th century, the Cornwall facility was outdated and the furnace was refurbished. The owners added a 20-horsepower (15 kW) steam engine to replace the waterwheel and a vertical blast engine with two wooden blowing cylinders. Steam was created by placing the boilers directly in the path of the hot furnace gases. The furnace prospered for more than three decades thereafter, but became obsolete as others came on the scene using the newest hot-blast production techniques and anthracite coal.

What makes this site unique among the hundreds of iron furnaces from the 18th and 19th centuries scattered around the Pennsylvania countryside is that it has survived essentially intact. The Coleman family, which took over its operation in 1798, kept it going until it closed in 1883. They continued to operate other, more modern furnaces in the area and maintained the Cornwall site as a monument to early Coleman ironmasters. This "iron plantation," an entire working community, was deeded to the Commonwealth of Pennsylvania as a historic site in 1932. It remained in continuous operation for 141 years.

Cornwall Iron Furnace.
Courtesy of Craig A. Benner and the Pennsylvania Historical & Museum Commission.

The tour begins at the Visitor Center, originally a 19th-century charcoal house. On display are exhibits explaining mining, charcoal-making, and iron production, thus providing a sense of the vast amounts of charcoal needed for fuel during the smelting process.

The entrance into the Furnace Building is through a shed that once protected the charcoal from inclement weather. The building, a cluster of structures around the furnace, has an elegant Gothic Revival façade that reflects the owner's taste and wealth when it was remodeled and enlarged in the 19th century. On the upper level is the charging area where raw materials were introduced into the furnace. A middle level holds the blast equipment that supplied air to the furnace, perhaps the only surviving example of this type of machinery. At the lowest level is the Casting Room, where iron was let out of the furnace to be cast into 100-pound (45-kg) pigs (blocks of metal) or formed directly into finished products.

The furnace produced about 21 tons (19 tonnes) of iron a week, which was made into 12- and 18-pound (5.4- and 8.2-kg) shot, 8- and 10-inch (20- and 25-cm) cannon shells, and 10-inch (25-cm) plate stoves. Records indicate that during 1777 some Hessians— German mercenaries—taken prisoner were employed as laborers at the prevailing wage. A cannon and several cannonballs produced at the furnace during this period are on display. One view is a roasting oven, circa 1825, in which layers of charcoal and iron were alternated. This process removed sulfur from the lower-grade ore that came from the mine during this period.

Several auxiliary structures that were needed to keep the furnace operating are still on the site. The Wagon Shop built and repaired wagons for use in and around the mines. The Blacksmith Shop made and repaired tools for mining, iron-making, and household use. The smokehouse, the stable, and the manager's house and office are among other privately occupied structures on the site and therefore closed to visitors.

Minersvillage, begun in 1865, is one of the many nearby villages that housed the mine and furnace workers. It is still privately occupied. The ironmaster lived in a mansion built about 1773 that is now part of a local retirement community. It was totally remodeled in 1865 in a Victorian style. Although visitors cannot go inside these dwellings, the exteriors provide a feel for how the workers lived.

The drop open-pit mine that operated from 1735 to 1973 is not open to the public, but is visible from a distance. More than 100 million tons (91 million tonnes) of ore were removed over the years, leaving a pit 440 feet (134 m) deep. This was one of the great iron deposits in the world. (See Lake Vermilion–Soudan Underground Mine State Park, Minnesota Discovery Center, and the Hull-Rust Mahoning Mine, Soudan, Chisholm, and Hibbing, Minnesota; and Michigan Iron Industry Museum, Negaunee, Michigan.)

Address: Cornwall Iron Furnace, 94 Rexmont Road, Cornwall, PA 17016.
Phone: (717) 272-9711.
Days and hours: Thu.–Sat. 9 a.m.–5 p.m., Sun. noon–5 p.m. (including Memorial Day and Labor Day), and additional hours by appointment.
Fee: Yes.
Tours: Guided tours only.
Public transportation: None.
Handicapped accessible: Visitor Center only; furnace is not accessible.
Food served on premises: None on site. There is a picnic area, and restaurants exist in the town of Cornwall.
Note: Neither heating nor air conditioning is on the site tour; wear comfortable shoes.
Website: http://www.cornwallironfurnace.org/

DOYLESTOWN

The Mercer Mile: Mercer Museum, Moravian Pottery and Tile Works, and Fonthill

Located about a mile (1.5 km) apart are three remarkable structures created by Henry Chapman Mercer, an archaeologist, architect, historian, and tile-maker. The buildings, all National Historic Landmarks, are unique both in their designs and in the objects they contain.

Fonthill, which was Mercer's home, looks something like a medieval castle. He designed and built it to display his tile and worldwide print collection. The castle-like Mercer Museum, topped with towers and terraces, is where Mercer exhibited his encyclopedic collection of Americana. In the Moravian Pottery and Tile Works, which resembles a cloister, Mercer manufactured tiles and pottery for sale around the world.

Mercer was graduated from Harvard University in 1879. By family tradition he studied to be a lawyer, but he was unhappy and turned to archaeology. He soon became curator of American and prehistoric archaeology at the University of Pennsylvania's Free Museum of Science and Art.

In 1897 the whole direction of his life changed when, searching for fireplace tools in a junk dealer's barn, he found obsolete objects left to decay because they had been replaced by power machinery. As an archaeologist he realized the importance of these objects, and thus began his lifelong task of finding and saving artifacts that help us understand the lifestyle of 18th- and 19th-century Americans.

His collection of pre-industrial tools became so enormous that in 1913 he began constructing a museum. He designed a seven-story poured-concrete castle with asymmetrical styling and unmatched windows arranged for maximum light and dramatic effect. He said the building "was made for the collection … the collection was not made for it." Running the full height of the Mercer Museum is a central court with galleries surrounding it. Each tier is lined with small, glass-enclosed rooms that vary in size and shape. Each room displays the tools and products of a different craft or trade.

Mercer developed his own classification system for the artifacts. Those fulfilling primary needs, things essential for survival, include tools for building shelters, cooking food, and making clothing. Objects for learning, amusement, transportation, and law are categorized as secondary. Small items are displayed in cubicles; large ones, such a Conestoga wagon, a fire engine, and a whaleboat, are suspended from the ceiling of the central court or are left freestanding.

This extraordinary collection has grown to more than 50 000 objects, representing 60 trades and crafts of pre-industrial America. An extensive collection of lamps, candle molds, and wick-trimmers, ranging from an Etruscan lamp of about 200 B.C.E. to a local iron lard lamp of 1820, demonstrates the historical development of artificial lighting. Kitchen tools on display include stove plates and firebacks, often used as patterns for the tiles Mercer designed at the Moravian Pottery and Tile Works. There is even a gallows for executing criminals.

Mercer was entranced with Pennsylvania German pottery-making and its designs, which were fast disappearing. He learned the art of tile-making and built the Moravian Pottery and Tile Works, styled somewhat like California's Spanish mission churches. He

Fonthill.
Courtesy of Nic Barlow.

named the tile works after his first designs, which were taken from 18th-century stove plates made by Moravians, a Pennsylvania German religious sect. During its years as a commercial enterprise, the Tile Works supplied handmade decorative tiles for projects such as Pennsylvania State Capitol, the Gardner Museum in Boston, and the John D. Rockefeller Estate in New York. Today the Tile Works is a working history center where you can watch the entire process of tile-making, and even purchase the end results in the Tile Shop. The techniques are similar to those Mercer used, and the molds are copies of Mercer's originals.

Fonthill, built by Mercer between 1908 and 1912, served both as his living quarters and as a museum illustrating the history of tiles. Although concrete was a common building material at the time, it was most often used for constructing barns and factories. Mercer chose concrete for the structure because of its plasticity during construction and its fire-resistance. The core of Fonthill is a small stone farmhouse built around 1810. Mercer incorporated this building into his larger edifice of more than 40 rooms. Architectural features of the farmhouse are still visible.

In Mercer's unique style, his collections were integrated into the building's construction and décor. For example, some of his tiles are set in glass-covered niches; others, such as his antique Dutch tiles, outline windows; and still others, such as his Spanish and Persian tiles, panel the walls. He pressed stove-plate designs into one of the ceilings.

Addresses: Mercer Museum, 84 South Pine Street, Doylestown, PA 18901; Fonthill Castle, East Court Street & Route 313, Doylestown, PA 18901; Moravian Pottery and Tile Works, 55 East Court Street, Doylestown, PA 18901.

Phones: Mercer Museum, (215) 345-0210; Fonthill, (215) 348-9461; Moravian Pottery and Tile Works, (215) 348-6098.

Days and hours: Mercer Museum and Fonthill, Mon.–Sat. 10 a.m.–5 p.m., Sun. noon–5 p.m.; Moravian Pottery and Tile Works, daily 10 a.m.–4:45 p.m.

Fee: Yes.

Tours: Self-guided for the museum and tile works. To visit Fonthill, reservations in advance are required.

Public transportation: Local bus to within a mile (1.6 km) of sites, then walk to and between each; SEPTA train stop is about 0.3 mile (0.5 km) from the station to the Mercer Museum (about a seven-minute walk).

Handicapped accessible: Limited, call in advance.

Food served on premises: None; restaurants are in Doylestown.

Special note: Ask at the museum for directions to the other buildings.

Websites: Mercer Museum and Fonthill, www.mercermuseum.org/; Moravian Pottery and Tile Works, www.buckscounty.org/mptw

ELVERSON

Hopewell Furnace National Historic Site

Mark Bird, son of an ironmaster, selected this convenient site in 1771 as the home of Hopewell Furnace. Once a highly productive producer of iron goods, the furnace is now a living-history museum that reflects the life and technology of a mid-18th-century iron plantation.

Bird situated Hopewell Furnace near roads that provided access to markets and to raw materials needed for iron production: iron ore, limestone to free iron from iron oxide, and forests for charcoal. Bird, brother-in-law of James Wilson, a signer of the Declaration of Independence, was upset with British control of the iron industry and became a political activist. His furnace produced cannons, shot, and shells for the patriots during the Revolutionary War. The furnace drew from three local mines: Old Hopewell Mine (or Birdtown Mine), on the Stockton geologic formation; Jones Good Luck Mine, in Caernarvon Township; and Warwick Mine, known for magnetic hematite. Later, ownership passed through several hands, and finally Matthew Brooke, Thomas Brooke, and Daniel Buckley purchased the site. They and their descendants made iron stoves and other items until 1883. Louise Brooke sold the site to the federal government in 1935.

This early American iron plantation is typical of hundreds once found in the eastern part of the country, forerunners of today's iron and steel industry. As anthracite coal replaced charcoal and was in turn replaced by coke, the iron plantation became obsolete.

Hopewell, the most completely restored of the country's old iron plantations, is set in its peak period of prosperity, the 1820s to the 1840s. It produced kettles, pots, sash weights, machinery castings, and prison bars, and was particularly well-known for its high-quality tin-plate stoves.

The Visitor Center's illustrated program, exhibits, and examples of cast-iron products introduce you to the industry, technology, industry, history, and lifestyle of this company town. Your next stop, the highlight of the tour, is the restored furnace, built in 1771. It is a charcoal-fueled, cold-blast furnace typical of the day, with an impressive chimney 32.5 feet (9.9 m) high. The waterwheel, the power source for the furnace, is a reconstruction of the 1822 wheel.

Originally the molten iron was poured into sand impressions in the cast house, where it hardened into rectangular blocks called pigs (hence "pig iron") or other shapes for direct use. In the late 18th century, the more efficient flask-casting method was adopted. To form a mold, damp sand would be packed around a pattern in a box, and then the pattern would be removed. The iron was poured into the mold and allowed to cool and harden. Then the casting was taken out of the mold and taken to the cleaning shed to be prepared for market.

The furnace was fired by charcoal made from wood collected in the surrounding forests. The wood was coaled—burned to form charcoal—in one of several charcoal hearths on the property, one of which you can see today. The charcoal house and the cooling shed stored 5000 to 6000 cords (18000 to 22000 m^3) of wood for making charcoal. About 90000 bushels (3200 m^3) might have been found here at any one time. Coaling, however, did not cause deforestation, because the best wood for this purpose was 25 to 30 years old; old-growth forests were not as efficient.

In 1853, in an attempt to keep up with new technology of the day, an anthracite furnace complex was installed. The remains of this facility, including the coal storage house, heating furnace, and bridge house to the furnace top, are still visible. Unfortunately, the venture wasn't profitable. The costs of shipping anthracite coal, the poor quality of the iron produced, and structural defects in the furnace design all probably contributed to its demise in 1855.

The facilities where the workers lived are restored and staffed. The company built two- to four-room tenant houses for some workers. Others lived off the company grounds in private homes, and some lived in the boardinghouse. The schoolhouse was built in 1837 by the company and funded by the local school district. A church built in the late 18th century was also part of the community. The Big House, built between 1800 and 1828, was the home and office for the ironmaster and family. This four-story, 19-room mansion provided quarters for guests visiting the facility as well.

A completely restored company store that operated on the barter system is open to visitors; so is the restored blacksmith shop. Coaling is demonstrated. The barn that once housed Hopewell's draft horses now displays period horse-drawn wagons, carriages, and sleighs. You can visit the springhouse, through which cold spring-water flowed to refrigerate perishables kept in submerged crocks. It also served as a wash facility and a place to render lard, make soap and candles, and cook apple butter. Other buildings the public can visit include the smokehouse, bake oven, and molder's kitchen. During the summer, docents in period dress give demonstrations of typical tasks the inhabitants performed at work and at home.

The site has 12 miles (19 km) of hiking trails that link to French Creek State Park.

Address: Hopewell Furnace National Historic Site, 2 Mark Bird Lane, Elverson, PA 19520.
Phone: (610) 582-8773.
Days and hours: Wed.–Sun. 9.a.m. to 5 p.m. Closed Thanksgiving Day, 12/25, 1/1, Martin Luther King, Jr., Day, and President's Day.
Fee: Free.
Tours: Self-guided.
Public transportation: None; Greyhound bus service comes to Reading Pennsylvania, and then taxi 15 miles (24 km) to park.
Handicapped accessible: Visitor's Center, yes; limited in some historic buildings. The site is on a hillside.
Food: Vending machines in Visitor Center. Picnic area is available; apple-picking in season in the apple orchard (extra fee).
Special note: Check schedule of demonstrations during summer months.
Website: www.nps.gov/hofu/index.htm

NORTHUMBERLAND

Joseph Priestley House

This beautiful, Federal-style house, once the home and laboratory for the theologian and scientist Joseph Priestley, sits on a hill overlooking the Susquehanna River. It is an elegant house located in a peaceful setting, which showcases Priestley's accomplishments and wide interests, and belies his difficult path from England to America.

Priestley, born in Yorkshire, England, in 1733, studied and wrote about many subjects, including history, education, literary criticism, politics, and religion. Although today he is best remembered as a scientist, he thought of himself as a theologian. Priestley was a minister of a religious group that was dissenting with the Church of England, the Unitarians. His scientific experiments, which resulted in the discovery of oxygen and seven other gases, were religious investigations to help him better understand the natural system devised by the Creator.

Priestley's interest in science and his friendship with Benjamin Franklin earned him an invitation to join the Royal Society of London, a prestigious scientific organization. Shortly after meeting Franklin, Priestley published the results of several experiments with electricity, including the first detailed account of Franklin's kite-and-key research.

After living in various places in England, Priestley moved to a ministry in Birmingham. There he became a member of the Lunar Society, an informal science-oriented group whose members were often referred to as "Lunatics." Among the society's members were luminaries such as James Watt, designer of the steam engine; industrialist Matthew Boulton; William Small, teacher of Thomas Jefferson; Erasmus Darwin, well-known physician and grandfather of scientist Charles Darwin; iron industrialist John Wilkinson, who was Priestley's brother-in-law; and Josiah Wedgwood, the father of English pottery. Benjamin Franklin visited the group often. The present-day relationship between science and industry stems, in part, from the example set by this society.

Members of the Lunar Society who were sympathetic to the French Revolution held a dinner to celebrate the second anniversary of the fall of the Bastille. Upon hearing of the party, a mob supporting the British king and the Church of England gathered and burned Priestley's Unitarian meeting house and his house and laboratory. With the encouragement of Priestley's son, Priestley and his family quietly escaped to America in 1794.

Priestley's reputation preceded him in America. A New York newspaper wrote, "There is no doubt that England will one day regret her ungrateful treatment to this venerable and illustrious man." In Philadelphia he was welcomed by the American Philosophical Society and was offered a faculty position in chemistry by the University of Pennsylvania, which he declined. He chose to settle in Northumberland, a five-day journey from Philadelphia. He was urged to open an academy in Northumberland, and plans were discussed for a colony of English dissenters, but neither ever materialized.

In 1874 a group of leading chemists met at Priestley's house to commemorate the centennial of Priestley's discovery of oxygen. This meeting inspired the idea of the American Chemical Society, which today has more than 150 000 members devoted to the study of basic chemical principles and the industrial use of chemicals.

Joseph Priestley House and Laboratory.
Courtesy of the Friends of Joseph Priestley House.

Owned by the Pennsylvania Historical and Museum Commission, Priestley's house has been restored to its original 1798 appearance. The house is furnished with period furniture; the laboratory contains reproductions of his equipment. In 2011 an upgraded laboratory exhibit was completed, based on archaeological research plus an architectural drawing done in 1800. In the laboratory are two furnaces with chimneys, a fume hood, glassware, and ceramic retorts. It was in this laboratory that Priestley discovered carbon monoxide. Because he had difficulty obtaining certain important pieces for his laboratory, he wrote to Josiah Wedgwood back in England, who made them to Priestley's specifications and sent them to America. Archaeologists working on the site have uncovered remains of this equipment. The Franklin memorial at the Franklin Institute in Philadelphia also displays some of Priestley's apparatus. (See The Franklin Institute, Philadelphia, Pennsylvania.) Dickinson College in Carlisle, Pennsylvania, owns original glassware from the laboratory, as does the Smithsonian Museum in Washington, DC. (See Smithsonian Institution, Washington, DC.) The glassware for this exhibit was commissioned in 1974 by the American Chemical Society, and given to the Priestley House in 1976. Priestley's achievements are enumerated in the Priestley Timeline exhibit, displayed in the adjacent Pond Building.

Priestley lived in this house, designed by his wife, Mary, for only six years and published more than 30 papers on scientific experiments during that time. The site is on the National Register of Historic Places in Pennsylvania and is designated as a National Historic Chemical Landmark.

Priestley conducted Unitarian services in his home every Sunday, and gave lectures on religion when he traveled to Philadelphia. Descendants of Priestley, together with the local Unitarian congregation, built a Memorial Chapel in 1834 in Northumberland that is open to the public by appointment only. In Northumberland he wrote a dozen theological books, including the last four volumes of his General History of the Christian Church and his four volumes of notes on the books of the scriptures.

Priestley's library held about 1600 volumes at the time of his death. It was one of the largest libraries in America at the beginning of the 19th century.

Riverview Cemetery, established in 1854 in Northumberland by Priestley's descendants, holds the grave sites of both Joseph and Mary, as well as many members of their family.

Address: Joseph Priestley House, 472 Priestley Avenue, Northumberland, PA 17857.
Phone: (570) 473-9474.
Days and hours: Sat. and Sun. 1–4 pm from early Mar.–mid-Nov. Closed Easter, Mother's Day.
Fee: Yes (no credit cards).
Tours: Docent-guided tours only. Additional private tours may be arranged with advance notice.
Handicapped accessible: Yes.
Public transportation: Susquehanna Trailways bus to Sunbury, then take a taxi to the site.
Food served on premises: None; restaurants are in Northumberland.
Special Notes: Check calendar for special events including Oxygen Day in August.
Website: www.josephpriestleyhouse.org/

PHILADELPHIA

Bartram's Garden

America's first great botanist, naturalist, and plant explorer, John Bartram, established this garden in 1731. The oldest living botanic garden in the United States, it is a National Historic Landmark and is part of the National Recreation Trail System.

Quaker farmer John Bartram loved to stop and admire the beauty of the native plants around him while he worked on his farm near Philadelphia. He became a self-taught botanist who spent his life collecting plant species indigenous to the American colonies. He traveled from Lake Ontario to Florida and west to the Ohio River searching for specimens to add to his garden and to share with other collectors in America and Europe.

The prevailing 18th-century view of natural science assumed that the universe had a finite number of plant and animal species, most of them already known. In Bartram's travels he discovered many new species, which contradicted standing scientific theories. This led Swedish botanist Carl Linnaeus to call Bartram "the greatest natural botanist in the world" and to devise a system for classifying plants. Bartram also was a friend of Benjamin Franklin; together they founded the American Philosophical Society in 1743. Bartram's international reputation was thus established, and he was appointed Royal Botanist to King George III.

William Bartram, John's son, continued the family tradition. From 1773 to 1776 he traveled through eight Southern colonies making drawings and taking meticulous notes on plants, animals, and Native Americans. His drawings now reside at the British Museum, the American Philosophical Society, and the Earl of Derby's library. His journals, *Travels Through North and South Carolina, Georgia, East and West Florida*, were published in 1791 and are still considered classics in American natural history. William's excellent literary style reached far beyond the scientific community: the English philosopher and poet Samuel Taylor Coleridge quoted *Travels* in several of his writings, including his important work *Kubla Khan*. William was the first naturalist to keep a journal of bird migration in America. He and his father identified and introduced into cultivation more than 200 native plants.

The garden became a profitable commercial nursery for the Bartram family. In 1798 brothers William and John Jr. published the first plant catalog in America, and supplied plants to such notables as George Washington, Thomas Jefferson, and E. I. du Pont. (See Monticello, Charlottesville, Virginia; and Hagley Museum and Library, Wilmington, Delaware.) Ann Bartram Carr, of the next generation of Bartrams, continued the family business until 1850. The family raised more than 4000 species of native and exotic plants in the garden and greenhouses.

Bartram's Garden, located a stone's throw from Philadelphia International Airport, contains 44 acres (18 hectares) of the original farm and is located on the West Bank of the Schuylkill River. Some of the old, historic specimens are still standing. Descendants of the Franklin tree, *Franklinia alatamaha*—discovered in 1765 and named after Benjamin Franklin—were saved from extinction by the Bartrams. French plant explorer André Michaux gave William a yellowwood that still stands on the property. Also here is America's oldest ginkgo tree, imported to the United States from China.

John Bartram built the first version of his home in 1731, but continued to enlarge the structure throughout his life. He espoused a deist philosophy, which caused the Quaker community to disown him in 1758. Not constrained by Quaker ideals of plainness, he added some architectural elements, among them a classical façade with three ionic columns by 1770. The 13-room house, now restored, contains Bartram family artifacts and local Philadelphia furniture of the era.

On site is an archive and research library with books including copies of John Parkinson's English herbals, *Paradise in Sole*, Alexander Wilson's 1813 *American Ornithology of the Natural History of the Birds of the United States*, the family's copy of Linnaeus, and editions of William Bartram's journal of his Southern travels.

Some outbuildings still on the property date back to the colonial period, including John Bartram's greenhouse from 1760. The purpose of some of the buildings changed over the years, and only the foundation remains of several.

The foundation of John Bartram's cider press, carved in bedrock, is still here. Apples were placed in the circular trench, a revolving wooden wheel crushed them, and the juice was drawn through a small hole into a reservoir.

The garden is a unique microclimate for this region. The stone house, sitting on a bluff above the Schuylkill River, captures the heat of the sun and reflects it back into the nearby garden. The river below the bluff helps moderate the temperature, too. This unusual climate allowed Bartram to grow such exotics as figs, pomegranates, and Southern magnolias.

Next to the house is an arbor planted with native wisteria where the Bartrams entertained guests such as Washington, Jefferson, and Franklin.

The restored upper garden demonstrates the design and contents of local 18th-century kitchen gardens. It is divided into three parts: raised beds of herbs and vegetables for culinary, medicinal, and domestic use; a common flower garden with herbaceous plants and bulbs grown for exchange with other collectors; and the new flower garden where seeds, roots, and cuttings gathered on trips and donated by collectors were grown. Period plants from John and William Bartram's time are featured in the lower garden. There is a re-created pond and water garden. On the rear side of the house is a restored garden from Anne Bartram Carr's era.

In recent times a 15-acre (6 hectare) meadow was installed on a part of the property that was once an industrial site. It contains wildflowers and meadow grasses reminiscent of what grew here in John Bartram's day and sparked his love of the natural world. In 2012, a Community Farm and Food Resource Center covering 3.5 acres (1.4 hectares) was added at the southern part of the property to create farming connections to the local community. By 2017 the Bartram's Mile Trail was opened along the riverbank, connecting parkland surrounding the site.

Address: Bartram's Garden, 5400 Lindbergh Boulevard, Philadelphia, PA 19143.
Phone: (215) 729-5281.
Days and hours: Gardens are open daily dawn to dusk. Welcome Center is open 10 a.m.–4 p.m., Tue.–Sun; closed Monday. Entire site is closed for Philadelphia-observed holidays. The research library is available by prior appointment.
Fee: Free for gardens; fee for guided tours.
Tours: Self-guided and guided tours of house and garden.
Public transportation: SEPTA trolley stops across the street from the property.

Handicapped accessible: Mostly; only the first floor of the Bartram house is handicapped accessible.

Food served on premises: Snacks and drinks in Welcome Center; picnic tables on site.

Special note: Check schedule for guided tours of house and garden, and other programs.

Website: bartramsgarden.org/

College of Physicians of Philadelphia and the Mütter Museum

In 1787, 24 physicians, members of the American Philosophical Society, a scientific organization with a worldwide reputation, saw a need for a new group focusing on medicine. They formed the College of Physicians of Philadelphia, which remains an important private medical society, the oldest in continuous existence in the United States. The purpose of the College is to promote high ethical standards in medicine, exchange medical information, and attend to the public's welfare. It serves the medical community and public through its programs.

The Historical Medical Library of the College of Physicians of Philadelphia, founded in 1788, is a premier research collection focusing on the history of medicine. This facility has grown into one of the largest collections of medical history in the country. The book collection contains about 120 000 books and journals published before 1966, of which some 400 of these publications were printed before 1500, and 12 000 were printed between 1500 and 1801. The manuscript collection includes illuminated manuscripts, hundreds of students' lecture notes, and papers by leaders in American medicine. This library served as Philadelphia's central medical library for medical schools, hospitals, physicians, other medical professionals, and medical students, as well as professional writers, for more than 150 years. The library can only be visited by appointment.

The Mütter Museum, just off the entrance hall of the College, contains more than 20000 antique instruments, preserved pathological specimens, examples of rare anatomical anomalies, and other items. A large part of the collection came from Dr. Thomas Dent Mütter, a professor of surgery at Jefferson Medical College, who gave his fascinating acquisitions to the College of Physicians of Philadelphia in 1858. The collection is displayed on two floors of galleries paneled in wood and lined with glass cabinets, evoking the Victorian era. Also displayed here is the Hyrtl Skull Collection. The 139 skulls of this collection were donated by Joseph Hyrtl (1810–1894) in 1874. Hyrtl, a Viennese anatomist, collected the skulls from European Caucasians to attempt to rebut phrenologists who tried to prove intelligence and personality differences between European Caucasians and other races. The Mütter Museum was originally intended to teach medical students. Today it is still used for research and also remains open to anyone who is interested in these absorbing collections.

One of many unusual items the museum holds is a cancerous tumor removed from President Glover Cleveland's jaw in 1893. Cleveland was operated on secretly on the yacht Oneida while it moored in the middle of the East River in New York City. The president recovered and lived another 15 years. Pathologists continue to re-examine the tumor for additional clues about cancer.

The livers of the conjoined twins from Siam, Chang and Eng Bunker, who died in 1874, are also in the holdings. Their autopsy was performed at this location. Another artifact is a megacolon, which we now know to be caused by Hirschsprung's disease. Sections of Albert Einstein's brain, sliced to a thickness of 20 μm, are here as well; the Museum is one of only two sites in the world to display Einstein's brain.

Among other artifacts in the museum's collections include an antique tonsil-guillotine used before anesthesia, when speed was of the essence. The optometry equipment includes a collection of spectacles, a pair of early Chinese glasses, tortoiseshell frames, some dating from the 19th century, and a Hebrew-alphabet eye-chart used in New York City in the early 20th century. On the walls of the upper gallery are 18th- and 19th-century portraits of physicians who were fellows of the college. The Chevalier Jackson collection of objects that the otolaryngologist extracted from patients' lungs, esophaguses, and throats numbers over 2000, and includes detailed notes on each extraction. The specimens are stored in drawers in cabinets, and most are available for visitors to see.

Outside the college headquarters is the small Benjamin Rush Medicinal Plant Garden, first proposed in 1787 but not added until 1914. Benjamin Rush was a signer of the Declaration of independence, a colonial physician, and a civic leader. Rush was also one of the founders of the College of Physicians of Philadelphia. The purpose of the garden was to urge the College Fellows to maintain this medicinal garden as a way to replenish their medical chests. The original garden included herbs as well as trees and shrubs to shade the south side of the building. In 1937 the garden was finally converted to match the plan proposed 150 years earlier, and now grows more 60 kinds of herbs with historical importance, some of which still have contemporary medical value. The plants are labeled with common names, scientific names, and use. This small garden is arranged into four symmetrical beds and benches shaded by a magnolia tree. A tree from the Island of Kos in the Aegean Sea, where Hippocrates taught, is also located here.

The permanent exhibits now on view include A Stitch in Spine Saves Nine: Innovations in Spinal Surgery, exploring the evolution of spinal medicine using items donated by Dr. Purviz Kambin, an expert in minimally invasive surgery, which traces the treatment and technology over the years. Another exhibit, The Soap Lady, revolves around a young female body found in Philadelphia in 1875, who died no earlier than the 1830s based on a 1986 examination of buttons and pins on her clothing. The body was encased in a fatty substance, adipocere, giving rise to the nickname. Broken Bodies, Suffering Spirits: Injury, Death, and Healing in Philadelphia during the Civil War, examines the medical care given during the Civil War. (See National Museum of Health and Medicine, Silver Spring, Maryland.)

Address: College of Physicians of Philadelphia and the Mütter Museum, 19 South 22nd Street, Philadelphia, PA 19103.
Phone: (215) 560-8654.
Days and hours: Daily 10 a.m.–5 p.m. Closed on Thanksgiving, 12/24, 12/25, and 1/1.
Fee: Yes.
Tours: Self-guided; group tours by appointment.
Public transportation: SEPTA trolley line is ½ block from the site.
Handicapped accessibility: Yes.
Food served on premises: None.
Website: muttermuseum.org/

The Franklin Institute

A majestic four-story building, housing The Franklin Institute, stands on Logan Circle, along the Benjamin Franklin Parkway. Founded in 1824, it is one of the oldest museums in the nation. Originally known as the Franklin Institute of the State of Pennsylvania for the Promotion of the Mechanic Arts, it was first located in the county courthouse building, now part of the Independence Hall National Historic Site. The institute's original missions—teaching and exhibition—remain today. In 1826 it opened Franklin Institute High School, which has an innovative curriculum of theoretical and practical sciences blended with liberal arts. The Committee on Inventions, formed during the institute's first year of operation, still issues the prestigious Franklin Institute Medals for scientific achievements.

In the 1930s the Institute and Poor Richard's Club of Philadelphia joined forces to raise money for a new facility and memorial to Franklin. They gathered $5 million in only 12 days and opened the current building on January 1, 1934. Soap magnate Samuel S. Fels donated the funds for the Fels Planetarium. Interestingly, Franklin's father was a soap maker.

The main entrance of the Institute opens into the Benjamin Franklin Memorial Hall. This grand octagonal room, modeled after the Pantheon in Rome, houses a huge statue of Franklin on a 92-ton (83 tonne) Seravezza marble pedestal. The Benjamin Franklin Center of the Franklin Institute manages the memorial hall and selects the artifacts displayed from the Institute's collection.

Near the memorial hall is the Franklin Gallery, put in place for the 175th-anniversary celebration of Benjamin Franklin. Here the extraordinary scientific achievements of Franklin in such areas as music, meteorology, electricity, aquatics, and optics are explored. Works of other scientists who were influenced by the great man are also featured. Items inspired by Franklin include bifocals, a glass armonica, and a lightning rod.

A four-story-tall Foucault's Pendulum hanging in a wide stairway demonstrates the rotation of the earth. Restarted each morning, the pendulum knocks down a peg every 20 to 25 minutes, from a circular array.

The model walk-through Giant Heart, with sound and lighting effects—the largest in the United States—opened in 1954. It is the centerpiece of a bioscience exhibit, The Giant Heart. Some 5000 square feet (460 m^2) of interactive devices and displays discuss Anatomy and Physiology, Health and Wellness, Blood, and Diagnostics and Treatment. Highlights include a mock surgical theater with video of open-heart surgery and latest technologies used in human-heart treatment. A 3D monitor creates the experience for those not going through the heart.

Your Brain explores the physiology and neurology of this organ with an 18-foot (5.5 m) tall climbing structure recreating neural pathways that send messages. Neuroscience, bioethics, and more are also discussed.

The Franklin Air Show, a 5000 square-foot (460 m^2) aviation exhibit using the Wright Brothers Aeronautical Engineering Collection, recreates the environment of an air show with the history, science, and technology of powered flight. More than 20 interactive displays in three sections represent the grounds of an air show: an aircraft hangar, a midway, and a pilot-training area. See a restored 1911 Wright Model B Flyer—one of the first mass-produced aircraft—and a 1948 T-33 jet-trainer for would-be pilots. Several training stations allow visitors to try maneuvers. Explore the history of and physics of aviation, learn aspects of aeronautical technologies, and examine the history of some aviation pioneers. Artifacts used by the Wright Brothers, such as wind-tunnel observations and original airfoils, are on view. (See Carillon Historic Park, Dayton, Ohio; Wright Brothers National Memorial, Kill Devil Hills, North Carolina; The Henry Ford, Dearborn, Michigan; Old Rhinebeck Aerodrome, Rhinebeck, NY.)

Electricity showcases the physical phenomenon most associated with Benjamin Franklin—lightning—his invention, the lightning rod (the first practical application dealing with electricity), and several hundred years of pure and applied research into electrical effects, electronics, and electrostatics.

The planetarium is equipped with a computerized Digistar® projector. The seamless aluminum dome is 60 feet (18 m) in diameter for visualizing astronomy in three-dimensional space. Accommodations for the hearing-impaired are available. The Joel N. Bloom Observatory has a 10-inch (25 cm) Zeiss refractor and four Celestron® CPC 800 GPS computerized 8-inch (20 cm) Schmidt-Cassegrain telescopes.

Space Command has 30 interactive stations, such as Remote Command showing robotic techniques used by scientists to manipulate devices on distant planets, with a computer-simulated mission. A home-tracking device allows visitors to view their own homes and neighborhoods from space via satellite and aerial photography. Outer Space Outfitters examines clothing that accommodates for the harsh conditions of outer space, such as a Soviet KVO-9 suit for training on the former Soviet Union's manned lunar program, and an A7-L prototype Nomex® suit for the Apollo missions, and more.

The Train Factory is an authentic train factory circa 1925. The centerpiece of this exhibit is the huge 350-ton Baldwin 60000 locomotive, the major teaching tool, as well as an assembly shop and a locomotive engine mechanics facility. Explore how steam and diesel power are used, the purpose of the boiler, combustion engines, and types of wheels. Learn the role of magnets and electricity in the operation of MagLev and electric trains. Take a short ride on the Baldwin engine.

Sir Isaac's Loft, with some 3600 square feet (300 m²) of displays, demonstrates a variety of physics ideas using experiments to explore these principles. Displays include Combining Motions, Changing Light, Illusions, Physics Feats of Strength, Chain Reactions, and Energy Transfer. Among the exhibits are energy-transfer demonstrations showing how energy is transmitted between objects, including pneumatic action; a giant kinetic sculpture reveals chain reactions; optical effects explore stroboscopic principles, optical illusions, color-theory, and moiré patterns.

Other exhibits and attractions including a Flight Simulator, Observatory access, Sky Bike, Sports Zone exploring the science behind sports with interactive physics and technology, and Amazing Machine, which includes Maillardet's automation (built around 1800, with possibly the largest memory based on cams of its day).

Location: The Franklin Institute, 222 North 20th Street, Philadelphia, PA 19103.
Phone: (215) 448-1200.
Public Transportation: Bus; short taxi-ride or bus from train stations.
Hours: General admission 9:30 a.m.–5 p.m. Science after hours (held about once per month, advance tickets required): 5 p.m.–10 p.m; last admission 7 p.m. Some events, particularly in the evening, are often geared to adults. Closed 12/24 and 12/25, 1/1.
Handicapped accessible: Yes.
Fee: Yes. Extra fee for some events and programs.
Food served on premises: Café and fast-food shop.
Special note: Check schedule of planetarium, movies, and special programs in theaters and IMAX®, with interactive science demonstrations from weather to chemistry and machines.
Website: www.fi.edu

Pennsylvania Hospital

Pennsylvania Hospital, founded in 1751, was the nation's first private hospital. Over the years it has grown significantly. Today it has more than 500 beds, offers a full range of diagnostic and therapeutic services, and serves as a teaching and research facility. To inform the public of its long and illustrious history, this facility has opened its doors to visitors. Tours of the facility provide visitors with a chance to admire its beautiful architecture and learn about the story of medicine in Philadelphia and the United States.

The hospital was established after Thomas Bond, a respected Philadelphia physician, convinced Benjamin Franklin and other prominent Philadelphia citizens that a facility was needed for the sick and insane. The Pine building, named for the street on which it was built, is a National Historic Landmark. Although the entire colonial structure was designed at one time, lack of funds caused it to be built in three phases. The East Wing, constructed in 1755, is the oldest section. The West Wing was not added until 1797, and in 1804 the two structures were connected by the Center House.

Over the years this hospital has initiated many medical firsts. It started the first outpatient clinic and the first occupational therapy program between 1752 and 1753, and the first hospital apothecary in 1755. It is the only hospital that has cared for sick and wounded military personnel from all the wars fought on US soil. The nation's first medical apprentices, now called residents, served here. The hospital also kept systematic meteorological records each day from 1766 to 1929.

Upon entering Center House, visitors are greeted by a fire engine, purchased in 1803, which was placed there to use in the event of a fire and thus preserve the building.

The old apothecary is no longer used as a pharmacy. It is now the office of the hospital's president and available to tour. On display are an antique musical planetarium clock made by David Rittenhouse, the famous astronomer and clockmaker, and a chair in which William Penn sat when he negotiated a treaty with Native Americans at his home in Pennsbury, Pennsylvania.

Among the eminent staff members who worked at the hospital is Dr. Phillip Syng Physick, father of American surgery, who invented a stomach pump at this facility in 1812. (His Philadelphia home, located nearby, is open to the public.) Dr. Benjamin Rush, another hospital staffer and America's first psychiatrist, wrote the first textbook on psychiatry in America that same year.

The Historic Library, opened in 1762, is the nation's first medical library, housing 13000 medical books such as herbal and horticultural volumes, anatomy, surgery, internal medicine, natural history, science, and botany. Today it houses the most complete collection of medical books printed from 1750 to 1850. Some books in the collection date to the 16th century. Here on display is a set of three 18th-century plaster casts of a uterine dissection and a seven-pound (3-kg) tumor once used to teach anatomy to medical students. The library may only be visited by appointment.

On the third floor of Center House is America's oldest existing surgical amphitheater. It was used from 1804 to 1868. Because a skylight and candles were the only sources of light, operations were scheduled for bright, clear days at midday, whenever possible. Medical students were required to purchase tickets to sit in the banked seats surrounding the operating table below.

Pennsylvania Hospital is the first institution in America to care for the mentally ill. The Public Hospital in Williamsburg, Virginia, built some 20 years later, was the first public hospital to care exclusively for the mentally ill. (See Jamestown Settlement and Historic Jamestowne, Williamsburg, Virginia.) Remnants of the dry moat once used to exercise the patients and separate them from visitors are still visible on the hospital grounds.

The hospital opened its first nurses' training program in 1875 and continued to train nurses until 1974. It has a small museum dedicated to the history of nursing.

Here, as in many institutions, it often takes years to complete capital improvements. A request placed in 1774 for a botanical garden at the hospital was finally granted in 1976 as part of the bicentennial celebration of America. The garden contains plants used for medicines in the 18th century.

Artworks are on display throughout the building. These include portraits by Thomas Sully and Thomas Eakins.

Address: Pennsylvania Hospital, 800 Spruce Street, Philadelphia, PA 19107.
Phone: (215) 829-3000.
Days and hours: Mon.–Fri. 9 a.m.–4 p.m.
Fee: Free.
Tours: Self-guided for five people or fewer; guided tours (required for six or more people) with 48-hour advance notice.
Handicapped accessible: Limited.
Food service on premises: Café.
Public transportation: Subway to 8th and Market Streets, then walk about 0.5 mile (0.8 km) to site.
Special note: Enter the Hospital using the main entrance on 8th Street (between Spruce and Pine Streets). Check in at the welcome desk upon arrival and receive a visitor badge. Self-guided tour brochures can be purchased at the gift shop for a donation.
Website: www.uphs.upenn.edu/paharc/

Philadelphia Zoo

During the colonial period, animals collected in the wild were often put on display for public entertainment. The first exotic animal on view in the colonies was a lion brought to Philadelphia in 1729. During the late 18th century the Peale Museum of Philadelphia, a place to see wild animals in captivity, drew many visitors. When London established its zoo in the 1840s, Philadelphia began exploring the possibility of opening one, too.

In 1859 the Zoological Society of Philadelphia, which would oversee the zoo, defined the zoo's role as "the purchase and collection of living wild and other animals, for public exhibition ... [and] for the instruction and recreation of the public." The Civil War delayed the Philadelphia Zoo's opening, but finally, in 1874, it became the first zoo in America. From its earliest days, when it had 282 animals living on 42 acres (17 hectares), research and education have been top priorities.

At the turn of the 20th century, the Penrose Research Laboratory opened at the zoo—the world's first living laboratory for studying animal health, reproduction, and social behavior. Research carried out here has led to breakthroughs affecting zoos around the world. For example, an early study of the effects of fresh air and low temperatures (about 40°F or 4°C) on tropical and subtropical animals showed that the animals remained healthy. As a result, zoos changed their methods of animal care. The lab studied tuberculosis, a major killer of animals, especially primates, and in 1911 developed a tuberculin test for both animals and humans.

Zoocake, a formulated food for captive wild animals, was developed here and introduced to the market in the 1930s. The new diet, still in use today, resulted in significantly longer lives for zoo creatures.

Over the years the zoo's role expanded to safeguarding the future of endangered and protected species. The zoo participates in 21 Species Survival Plans® (SSP) for such animals as the Siberian tiger, Indian rhinoceros, African and Asian elephants, Western lowland gorillas, and the orangutan. SSP is a scientifically managed, nationwide program for breeding wildlife and reintroducing them to their natural habitat. This institution started a new conservation program, One With Nature, to save threatened species in the wild and in captivity. The zoo also works with a group called Conservation International to keep species alive and ensure that breeding animals have diverse genetic backgrounds. Today this institution is known worldwide as a premier zoo for breeding animals that are difficult to breed in captivity.

The zoo now has more than 1300 animals on its well-planted grounds, a Victorian garden setting using over 500 plant species. Many of the exhibits have been redesigned into natural habitats. Carnivore Kingdom encompasses 1.5 acres (0.6 hectare) for unusual animals that live on the forest's edge: snow leopards, jaguars, red pandas, jaguarundis, coatimundis, and waterfowl.

The McNeil Avian Center (or the Bird House), a neoclassical structure built in 1916, is an architectural treasure. Its waterfalls and abundant foliage are the tropical backdrop for more than 100 species of exotic birds from all parts of the world. The rain forest, the first exhibit you see as you enter the building, has free-flying birds. The Bird House places particular emphasis on Pacific Island birds. The Philadelphia Zoo has won awards for its pioneering efforts for saving endangered birds such as the Micronesian kingfisher and the Guam rail.

A re-creation of the African veldt, an open area for grazing with a sloped savanna, allows you to see giraffes, zebras, sable antelopes, ostriches, and other compatible animals and birds living together. The Rare Animal Conservation Center has naked mole rats and tree kangaroos on view. In the Small Mammal House there is a wing for nocturnal creatures like leaf-nosed bats, Egyptian bats, vampire bats, sugar gliders, and bush babies. In another section of this building are meerkats and aardvarks. The Reptile and Amphibian House contains one of the finest collections of reptiles and amphibians in the world.

A waterfowl pond at the zoo is home to many unusual species, including South American crested screamers, ruddy-headed geese, comb ducks, and Chiloé wigeons. The pond is adjacent to the Solitude, the oldest building on the grounds, which was built by William Penn's grandson John in 1784. Tours of the building must be pre-arranged.

A beautiful retreat within the zoo is Penn's Woodland Trail—a replica of a typical Philadelphia-area woods of the 1600s. The trail features native plants and animals in enclosures that complement the forest-like ecology. The PECO Primate Reserve, built after a disastrous fire in 1995 destroyed the World of Primates, includes 2.5 acres (1 hectare) of outdoor and indoor areas devoted to primates, such as lemurs, gorillas, aye-ayes, and gibbons. Zoo360 is a network of see-through mesh trails throughout much of the site, in which the animals roam around the zoo grounds as visitors roam outside of the mesh and observe the animals.

Address: Philadelphia Zoo, 3400 West Girard Avenue, Philadelphia, PA 19104.
Phone: (215) 243-1100.
Days and hours: Mar.–Oct.: 9:30 a.m.–5 p.m.; Nov.–Feb.: 9:30 a.m.–4 p.m.
Fee: Yes. Some special programs may have additional fees.
Tours: Self-guided; group tours by prior arrangement; chat with the keepers and check exact times at the information booth for more details.
Public transportation: SEPTA trolleys and buses stop near the site.
Handicapped accessible: Yes. Wheelchair and electric-scooter rentals are available. Call ahead for details.
Food served on premises: Cafés.
Website: www.philadelphiazoo.org/

Science History Institute

Center-city Philadelphia is home to one of the few—and certainly the largest—museums in the United States devoted largely to the chemical sciences, the Science History Institute. As its stated goal, the organization "focus[es] on the sciences and technologies of matter and materials and their effect on our modern world, in territory ranging from the physical sciences and industries, through the chemical sciences and engineering, to the life sciences and technologies."

The Science History Institute, within walking distance from historic Independence Hall, is housed within the First National Bank, built in 1865. A joint effort in 1982 by the University of Pennsylvania and the American Chemical Society (see Joseph Priestley House, Northumberland, Pennsylvania), with founder Arnold Thackray, resulted in the Center for the History of Chemistry. Two years later, the American Institute of Chemical Engineers became the third sponsor for the organization. The Center

Interior of Science History Institute. *Photo* by Conrad Erb.
Courtesy of the Science History Institute.

became an independent nonprofit in 1987, renamed the National Foundation for the History of Chemistry. By 1992, to reflect its multidisciplinary activities, it became the Chemical Heritage Foundation (CHF), and moved to its present location in center-city Philadelphia in 1996. By 2018, after merging with the Life Sciences Foundation (also founded by Thackray), the organization became the Science History Institute.

Displays at the Institute introduce the public to several areas of chemistry, chemical biology, and related subjects. Many of the exhibits are permanent and rotate artifacts from the Institute's extensive collections into the displays.

Masao Horiba Gallery houses the permanent exhibition, Making Modernity. Exhibits here explore how chemistry has affected the modern world, by covering many topics that are integral part of our current civilization and use artifacts from the museum's extensive collections. Displays explore the revolution in chemical instrumentation, chemical education, electrochemistry, and the science of color. Among the collection's highlights are a Beckman IR-1 spectrophotometer and Model G pH meter, John Fenn's electrospray mass spectrometer (for which he shared the 2002 Nobel Prize) and Bruce Merrifield's solid-phase peptide synthesizer (assisting Merrifield in winning the 1984 Nobel Prize), as well as a 1960s Varian A-60 NMR spectrometer from Stanford University. Artifacts include commercial samples and products of Bakelite (a popular plastic invented in 1907). Laboratory glassware and smaller apparatus are on display. There are numerous examples of chemical-related ephemera, such as posters, advertisements, and a selection of celluloid trading cards from the first decade of the 20th century. Exhibits explore synthetic materials, the revolution in chemical instrumentation, chemical education, electrochemistry, commercial chemistry sets by the Chemcraft and Gilbert companies from the 20th century, and the science of color. Instruments and other artifacts from the Institute's extensive collection are rotated in and out of this exhibit on an ongoing basis.

The Whole of Nature and the Mirror of Art, also explored in the DuPont Gallery, present a number of chemistry-related artworks as photo-reproductions from chemistry and alchemistry texts in the Science History Institute's collection. A particular strength of the collection is its 17th- through 19th-century depictions of iatrochemistry, the idea that chemical principles can explain disease and health. There are numerous artworks illustrating chemical transformations such as distillation, metallurgy, and textile production, and illustrations from books such as Robert Fludd's *Utriusque cosmi maioris scilicet et minoris metaphysica, physica atque technica historia*, published in 1617–1618, describing the philosopher's stone and alchemical methods of transformation.

Over a thousand photographs of important chemists, images related to the development of nylon, and informal snapshots of chemical-related personalities in the Institute's collections are sometimes used in displays.

Temporary exhibits in the Hach Gallery often include the subject of the intersection of art and science. The Institute also holds regular events and lectures related to the history of science; check the website or call for details.

Much of the Institute's collection is not on display, but members of the public can, with prior application, review and research within the foundation's archives.

An important part of the Institute is the Donald F. and Mildred Topp Othmer Library of Chemical History, founded in 1988, which collects materials connected with the history of science, technology, medicine and its relationship with chemistry, and chemical engineering from ancient times to the present. The collection holds 160 000 prints and microfilm volumes, rare books and manuscripts, significant archival materials, and historic photographs covering six miles of shelves. The library is open only by prior appointment.

For academic research into the history of chemistry, the Foundation sponsors the Arnold and Mabel Beckman Center for the History of Chemistry, which supports the largest private fellowship program dealing with the history of science in this country.

Address: Science History Institute, 315 Chestnut Street, Philadelphia, PA 19106.
Phone: (215) 925-2222.
Days and hours: Tue.–Sat. 10 a.m.–5 p.m.; Mar.–Dec.: first Friday of the month 10 a.m.–8 p.m. Closed Memorial Day, Labor Day, Thanksgiving Day, the Friday after Thanksgiving, 12/25, and 1/1; open on certain holiday Mondays.
Fee: Free.
Tours: Self-guided; guided tours by appointment.
Public transportation: Market-Frankford subway stops 2½ blocks away at 2nd Street, then walk to site. SEPTA buses also stop near the site.
Handicapped accessible: Yes.
Food served on premises: None. Numerous local restaurants are nearby.
Special note: Call or check the website for lectures and events.
Website: www.Sciencehistory.org

Wagner Free Institute of Science

This little-known treasure in the heart of Philadelphia, just a stone's throw from Temple University, is often referred to as "a museum of a museum." The original display cases and systematic collections in this three-story Victorian structure have barely been touched by time.

In 1847 William Wagner, a Philadelphia merchant, philanthropist, and amateur scientist who wanted to make education available to the public, started giving free lectures on the natural sciences at his home. These lectures were one of the earliest free adult science-education programs in the United States. He augmented his lectures with the large collection of specimens he had gathered from around the world in pursuit of his hobby, mineralogy. By 1855 his lectures had become so popular that he moved them to the center of the city and formally incorporated the Wagner Free Institute of Science. Faculty members at the institute gave six lectures per week on the natural science. In 1865 the institute moved into a new building—now a National Historic Landmark.

Professor Wagner, as he became known, continued to lecture until his death in 1886. The institute then turned to Joseph Leidy, a nationally renowned biologist and natural scientist, to direct its programs. Under Leidy's leadership the organization expanded its programs to include scientific research and hired leading scientists and explorers for its faculty. The institute's first expedition went to the west coast of Florida, until then unexplored. The vast collection of fossils the expedition brought back included the first known fossil of a saber-toothed tiger. Leidy identified it and placed it on display, where it remains today.

To accommodate the materials from the expedition and Leidy's purchases, the exhibit hall was removed. The artifacts were reorganized to reflect the theory of evolution, and in 1891 the building reopened as the museum of the institute. Many displays and specimen arrangements that Leidy supervised remain intact; it's perhaps the only place where you can still see his personal touch.

Original cherry-wood and glass cabinets from the 1880s display the collections. The exhibits are arranged in the original Victorian "systematic" scheme of the period. William Wagner's personal mineral collection, fossil collection with specimens from Europe and America, and mounted skeletons, skulls, and skins are on display. The entomology collection, with its original arrangement and labels, are displayed and used as a tool for education and resource for research.

Samuel Wagner, the second president of the institute and a leader in public education, was very active in getting a public library system for the city of Philadelphia. In 1892 several of the institute's classrooms were altered to make room for the public library's first branch. In 1902 the west wing of the building was completed to house the Philadelphia Free Library. Very little has changed, and much of the original library furniture is still in place.

This library has a collection of materials of primary scientific works that date from the late 17th century to the early 20th century. These materials include natural and physical sciences, education, medicine, archaeology and anthropology, pseudo-sciences, instrument building, and engineering, as well as monographs, serials, slides, and much more. The business records of William Wagner are also in the collection, as well as international science journals dating back to the 19th century. The library now serves as a research center for materials on the history of American science.

The institute still gives adult science courses in accordance with William Wagner's goal of free science education for the public. It maintains a natural history museum with more than 100 000 minerals, fossils, archeological materials, freshwater specimens, shells, and stuffed and mounted birds and animals. It also has an extensive entomological collection.

A prize display exhibits the dinosaur bones gathered by Edward Drinker Cope, a student of Leidy's, a lecturer at the institute, and one of the first major paleontologists in America. (Although Cope was a Philadelphian, most of the dinosaur bones he collected are now housed at the American Museum of Natural History in New York City.) (See American Museum of Natural History, New York, New York.)

Ninety percent of the institute's specimens are on display in their original cases, some with glass-covered study drawers. The institute has a large microscope-slide collection and a Victorian auditorium little altered from Leidy's days. The auditorium's turn-of-the-century lantern slide projector is still used with the old lantern slides.

Address: Wagner Free Institute of Science, 1700 West Montgomery Avenue, Philadelphia, PA 19121.

Phone: (215) 763-6529.

Days and hours: Tue.–Fri. 9 a.m.–4 p.m. Closed major holidays. Usually closed for two weeks in Aug.

Fee: Free; fee for guided tours.

Tours: Self-guided; guided tours by appointment. Children under age 18 must be accompanied by an adult.

Public transportation: SEPTA bus stops 2½ blocks away, then walk to site.

Handicapped accessibility: No elevator to second floor. Sign interpretation for some programs: contact museum.

Food served on premises: None. There are nearby restaurants.

Special programs: Contact museum for schedule.

Website: www.wagnerfreeinstitute.org

Carnegie Museums of Pittsburgh

Industrialist and philanthropist Andrew Carnegie established this foundation in 1896. According to his plan, the complex would revolve around four disciplines, which he called "the noble quartet" of literature, art, music, and science. His aim was to create a major educational and cultural complex for the people of Pittsburgh. Until the present, Carnegie Museums of Pittsburgh has maintained Andrew Carnegie's goal.

Still called "the Carnegie" by locals, this institution has expanded significantly since it was founded, and now has two campuses. The Carnegie Museum of Natural History and the Carnegie Museum of Art remain in their original locations and are part of the Carnegie Institute and Library Complex in the Oakland section of Pittsburgh. Both the Carnegie Science Center and the Andy Warhol Museum are located on a newer site in the North Shore section of Pittsburgh on the Ohio River.

For the purpose of this book we review the two science museums: the Carnegie Museum of Natural History and the Carnegie Science Center.

Carnegie Museum of Natural History

The Carnegie Museum of Natural History has extensive, varied displays, in a multilevel complex with almost 15 acres (6 hectares) of floor space. It continues to evolve and grow, exploring the natural world and world cultures, past and present.

Exhibits explore animals from the miniscule to gigantic, plants from forests to meadows, geological instruments and specimens, and ecosystems from equatorial to polar. More than 10 000 artifacts and specimens from the museum's collection of over 20 million objects are used in these exhibits. The museum is well known for its dinosaur expeditions during the first quarter of the 20th century and is still involved in them. It has more than 500 catalogued dinosaur specimens. Dinosaurs in Their Time, re-designed in the first decade of the 2000s, contains many dinosaurs in the most accurate poses and environments according to the latest paleontological research. At the turn of the 20th century, the museum purchased a collection of Mesozoic fossils belonging to the private collector Baron Ernst de Bayet of Brussels. Bayet's fossils of vertebrates and invertebrates that lived at the same time as the dinosaurs are displayed in Dinosaurs in Their Time.

Benedum Hall of Geology looks at the ever-changing nature of Earth, and reflects the discipline's latest thinking. Primarily using Pennsylvania, West Virginia, and Ohio as examples, the exhibits explain plate-tectonic theory of the movement of the earth's crust. Exciting interactive exhibits investigate coal, oil, and gas: how they are formed and retrieved from below the earth's surface. The Stratavator simulates an elevator ride down 16 000 ft. (4900 m) into the bedrock beneath Pittsburgh, and then back to the surface.

Hillman Hall of Minerals and Gems has extensive displays, including minerals from Pennsylvania and a collection of rare and aesthetic masterpieces, such as Powellite from India and a fine example of golden topaz on quartz.

Polar World documents 4500 years of humans inhabiting the Arctic. Artifacts, pictures, and dioramas explore the environmental challenges the region has faced, the special flora and fauna that live there, the existence of the Inuit people who occupied the region before Europeans arrived, and the impact of explorers and modern technology.

The Walton Hall of Ancient Egypt displays more than 600 objects (from a collection of 2500) arranged according to seven themes related to ancient Egypt's social system, technology, and beliefs. At video kiosks and computer stations, explore the Pharaoh Tutankhamen, the Great Sphinx, and other specialized topics. One of the hall's most important artifacts is a rare, 3800-year-old royal funerary boat, probably from a pyramid complex at Dashur, Egypt. The cedar boat, which originally was painted, reveals important information about the ancient technology of boat-building in Egypt. A life-size diorama of artisans at work shows how the ancient Egyptians crafted jewelry.

The museum also oversees the Powdermill Nature Reserve, a 2250-acre (911-hectare) environmental research center in Rector, Pennsylvania, with a bird-banding program to study bird migrations, climate change, and more.

Address: Carnegie Museum of Natural History, 4400 Forbes Avenue, Pittsburgh, PA 15213.
Phone: (412) 622-3131.
Days and hours: Mon., Wed., Fri., Sat. 10 a.m.–5 p.m.; Thu. 10 a.m.–8 p.m.; Sun. noon–5 p.m. Also open Election Day, Veterans Day. Closed Tue., 1/1, Easter, Thanksgiving Day, 12/25.
Fee: Yes.
Tours: Self-guided; free guided mini-tours of specific sections on weekends.
Public transportation: Local bus stops about 400 feet (120 m) from museum.
Handicapped accessible: Yes.
Food served on premises: Cafés.
Special note: Check schedule of special shows, events, and OMNIMAX theater productions.
Website: www.carnegiemnh.org/

Carnegie Science Center

First opened in the 1930s, and moved to its current location in the 1990s, the Carnegie Science Center, with its interactive exhibits on the physical and life sciences and technology, is billed as an "amusement park for the mind."

The H2Oh! exhibit presents the ecology and properties of water, why it is important in our daily lives, and fluid dynamics. SpacePlace atrium shows artifacts and models as it teaches about outer space. A life-size replica of the International Space Station is available for exploring, the original Zeiss II planetarium projector (1939–1991), a 21-foot (6.4-m) climbing wall to simulate zero-gravity, and more are on display for visitors. SportsWorks studies the human body, including the physics of sports (e.g., momentum and center of gravity) and healthy foods.

Roboworld is one of the world's largest exhibits on robotics. There are a number of robots on display, including some that interact with visitors. Three general themes of interaction are presented: Sensing (collecting data); Thinking (processing data); and Acting (performing actions). The Exploration Station presents activities related to energy-usage and natural forces.

The USS *Requin*, docked next to the museum, is a World War II diesel-electric submarine that uses ingenious voice, visual, and electronic communications equipment to explain to visitors the science and history of water vehicles. Once used for both scientific and military purposes, it is 312 feet (95 m) long, housing 80 crew members.

The Carnegie Science Center also includes the Buhl Planetarium, observatory, and OMNIMAX theater. The Buhl Planetarium was established by local businessman Henry Buhl in 1939, and moved in 1991 to a new facility, which now is 50 feet (15 m) in diameter, and seats 150. By 2006 a fully digital projection system was installed.

Address: Carnegie Science Center, One Allegheny Avenue, Pittsburgh, PA 15212.
Phone: (412) 237-3400.
Days and hours: Sun.–Fri. 10 a.m.–5 p.m.; Sat. 10 a.m.–7 p.m. Closed 12/25.
Fee: Yes.
Tours: Self-guided.
Public transportation: Buses or T-Plus subway to Allegheny Station, next door to the site.
Handicapped accessible: Yes, except for the USS *Requin*.
Food served on premises: Café.
Website: www.carnegiesciencecenter.org/

Pittsburgh Zoo & PPG Aquarium

One of the older zoos in the United States, this facility opened in 1898 as the Highland Park Zoo, for it was in the Highland Park area of Pittsburgh. As was typical of the period, the zoo was organized as a holding area for exotic animals. Beginning in the 1930s and faster through the 1980s, the site was redesigned gradually in a naturalistic fashion, and the goal of this facility became focused on conservation.

The Pittsburgh Zoo is one of very few combination facilities in the United States that include both zoo and aquarium. This site, on 77 acres (31 hectares), houses more than 9000 animals of 475 species that include 24 threatened or endangered species. These carefully landscaped exhibits have concealed barriers and special viewing positions that permit visitors to watch the animals in environments similar to their native habitats. The Pittsburgh Zoo participates in Species Survival Plans®, special programs to protect and breed endangered animals, for more than a dozen species.

A highlight of the zoo is the 5-acre (2 hectare) Tropical Forest, an indoor facility landscaped with a river, waterfalls, pools, and thousands of plants representing more than 50 tropical species. This natural-looking jungle is home to 16 species of endangered primates. The tour begins in Madagascar and weaves through fog-shrouded jungles of South America, Southeast Asia, and Africa. Each primate species lives in a riverside clearing designed especially for it. The topography and flora of the habitats reflect the species' typical feeding habits and exercise needs. Signs with graphics give information about each animal and its lifestyle.

Visitors learn, for example, that pacas are ground-dwelling rodents that sometimes bury fallen fruit for future use, behavior that helps to seed the forest. Black howler monkeys, very rare in captivity, have a unique call produced by their egg-shaped larynxes. Sloths are slow-moving mammals that can take up to a month to digest a meal. Orangutans, or "men of the woods" in the Malay language, are the only species to have the same number of ribs as human beings. At times they cover their heads with leafy branches when it rains and have been seen using leaves to scoop water to drink. Although the rock hyrax looks like a rodent and is only 18 inches (46 cm) long, its nearest relative is probably the elephant.

The Tropical Rain Forest has a 1-acre (0.4-hectare) outdoor exhibit for a gorilla troupe. It contains a grassy meadow with dead trees and shrubs, a stream, and a sheltered area with a heated rock floor for cooler days.

The PPG Aquarium, built in 1967 and renovated in 2000, is a two-story 45 000 foot aquarium with several aquatic habitats designed to portray different ecosystems: a tropical rainforest and a lowland equatorial evergreen rainforest. Habitats include a living coral reef and 800 specimens of marine animals and plants. The technology used to replicate the reef's original site of Mayaguana Island in the Bahamas was developed by the Smithsonian Institution. The water's movement is similar to that of natural waves, and the light level is similar to that of a Caribbean day. The plants and animals in the PPG Aquarium create a natural food chain. Among the occupants are octopi, penguins, and South American freshwater dolphins. The PPG Aquarium also includes an insect exhibit, one of the few in the nation.

The Asian Forest has a Siberian tiger, a waterfowl pond with northern Asian plants, and a snow leopard exhibit. The African Savannah has seven major exhibits that include wild African plants. One of the exhibits displays cheetahs. The Niches of the World Amphibian Building has small nocturnal mammals, reptiles, and amphibians.

Address: The Pittsburgh Zoo & PPG Aquarium, 7340 Butler Street, Pittsburgh, PA 15206.
Phone: (800) 474-4966.
Days and hours: Spring: 9 a.m.–5 p.m.; summer: 9:30 a.m.–6 p.m.; fall: 9 a.m.–5 p.m.; winter: 9 a.m.–4 p.m. Closed Thanksgiving, 12/25, 1/1.
Fee: Yes.
Tours: Self-guided.
Public transportation: Local buses stop 0.4 mile (0.6 km) from zoo, then walk to site.
Handicapped accessible: Yes. Wagons, wheelchairs and electric scooters are available with fee, as is a tram that stops around the site (primarily during the summer). Call ahead; without a reservation is first come, first served.
Food served on premises: Café and picnic area.
Website: http://www.pittsburghzoo.org/

SCRANTON

Pennsylvania Anthracite Heritage Museum and Scranton Iron Furnaces, and the Lackawanna Coal Mine

Together, the Pennsylvania Anthracite Heritage Museum and Scranton Iron Furnaces and the Lackawanna Coal Mine illustrate that for more than a century, coal dominated the lives of the people who lived in and around Scranton.

To explore the Lackawanna Coal Mine—once a working mine—visitors can take an underground railcar 250 feet (76 m) below the earth's surface on a track one-fifth of a mile (0.3 km) long. This mine, now operated by Lackawanna County, and which produced anthracite coal until 1966, looks much as it did when it closed. Below, where it is always cool (53°F, 12°C), damp, and drafty, there is a one-hour walking tour led by a coal miner. Sturdy shoes are a must; hard hats are provided. The tour focuses on the peak production period of 1890 to 1925 and explores the coal veins, the mining process, and the workings of the mine equipment. Above ground, in the Shifting Shanty, are exhibits, artifacts, and continuous-play video on mining techniques in the early 20th century.

Under the auspices of the Pennsylvania Historical and Museum Commission, the Pennsylvania Anthracite Heritage Museum, with 22 000 square feet (2000 m²) of space, is adjacent to the mine. This facility explains the role of coal and its effect on the lives of the local population. See how coal was extracted from the mines and brought to market, examine equipment used in the mines, and gain an overview of mining techniques from early methods to the most modern technology. Among the items on display are an electric mine locomotive and a model of a coal-breaker, a machine that breaks large coal chunks into pieces.

Several exhibits explore industries attracted to the area by the anthracite coal deposits. Nottingham-style lace was manufactured in Scranton. The silk industry moved here from Paterson, New Jersey. Both trades employed the wives and daughters of miners. Displayed are the tools used, such as a two-harness loom built by John Martinkas around 1900, and an 1880 Nottingham Lace loom from the Scranton Lace Curtain Company.

Exhibits examine Immigration and History in Pennsylvania's Hard Coal Region. Here the lives of the people associated with coal mining are extensively explored. Welsh immigrants, the first to settle in the region, were experienced miners in their native country. It is thought that the phrase "keeping up with the Joneses" may have originated among the miners: Jones is a common Welsh name, and the Welsh had some of the best jobs at the mines because many of these people had extensive training in the mines of Wales. They had good incomes to match the work they did. Soon after came other ethnic groups; the museum traces the immigration patterns to the Anthracite Region around Scranton.

The museum has an exhibit of a miner's village and a recreation of a textile factory that produced lace and some garments. Local women and girls worked in these factories.

The Scranton Iron Furnaces, on the banks of Roaring Brook, were fueled by coal from local area mines. Four towering, stone-blast furnace stacks still stand near the city's railyards as a monument to Scranton's early industrial prowess. In 1840 William Henry and George and Seldon Scranton came to the wilderness town of Slocum Hollow, which

Interior of Anthracite Heritage Museum.
Courtesy of the Anthracite Heritage Museum.

eventually was renamed Scranton. Because all the necessary resources were in easy reach—coal, iron ore, limestone, and waterpower—they built a furnace. When the railroad came through Scranton in the late 1840s, the Lackawanna Ironworks was chosen to produce the rails. It made such high-quality rails that business boomed and other industries came to town. By the 1860s the Lackawanna Iron and Coal Company was one of the nation's major iron-producing complexes. In the 1880s it was the second-largest iron producer and made T-rails for the railroad until 1902.

Today the Lackawanna Iron and Coal Company complex looks much as it did in 1860. The four stacks from the massive stone furnaces are connected by 30-foot (9 m) high brick arches constructed between 1846 and 1857. The remains of the casting house, where iron was formed into oblong blocks is on site; the blast engine house that contained blowers for the furnace—now the site of the visitor's center—and the rolling mill, where the iron was rolled into sheets, are also here. On Shanty Hill, where the miners lived, many houses still stand and are occupied.

Addresses: Pennsylvania Anthracite Heritage Museum and Lackawanna Coal Mine, R.D. #1, McDade Park, Bald Mountain Road, Scranton, PA 18504; Scranton Iron Furnaces, 159 Cedar Avenue, Scranton, PA 18504.

Telephones: Museum and Furnaces, (570) 963-4804; Coal Mine: (570) 963-6463.

Days and hours: Museum, Mon.–Sat. 9 a.m. to 5 p.m., Sun. noon–5 pm; closed 1/1, Thanksgiving, day after Thanksgiving, 12/25, and many other holidays. Coal Mine: Apr. 30–Nov. 30; closed Easter and Thanksgiving. Furnaces: Open daily.

Fee: Yes.

Tours: Self-guided in the museum; mine by guided tour only.

Public transportation: COLTS runs bus service to McDade Park seasonally; taxis are available. Handicapped accessible: Yes, in the museum, wheelchairs and walkers are available, call in advance; limited accessibility in the mine.

Food served on premises: Snack shop at the mine entrance, restaurant off-site, 0.9 mile (1.4 km) away.

Special note: Wear sturdy shoes and jacket in the mine.

Websites: Museum and Furnaces, www.anthracitemuseum.org/; Coal Mine, www.lackawannacounty.org/index.php/attractions/coal-mine

Steamtown National Historic Site

This reconstructed railroad with a collection of vintage locomotives represents an important aspect of the industrial past of both Pennsylvania and the nation. About 1800, in a section of Scranton known as Slocum Hollow, the Slocum brothers started surface-mining iron and operating a bloomery, a small smelting furnace. The surface iron soon played out, and the bloomery closed. Thus, in the 1840s William Henry started mining iron underground and called on Seldon and George Scranton and Sanford Grant for the finances to build a large iron furnace at the original bloomery site. Because anthracite coal was so abundant locally, in 1847 the furnace started using this coal to make T-rails for the railroad. (See Pennsylvania Anthracite Heritage Museum and Scranton Iron Furnaces, and the Lackawanna Coal Mine, Scranton, Pennsylvania.)

One of the greatest problems for the foundry was moving the T-rails to market. In 1849 the Scranton family decided that the solution was a railroad that would serve their region. Soon after, the Lackawanna and Western Railroad was built, with financing by the Phelps family. With cheap transportation now available, both the furnace and the anthracite coal mines prospered. The ability of the railroad to move coal efficiently encouraged mining expansion, which, in turn, influenced the growth of the railroad. By the 1860s and 1870s, almost 90% of the freight hauled by the railroad was anthracite coal.

The Lackawanna and Western Railroad continued to thrive for the next 100 years. At its peak it operated about 1000 miles (1600 km) of track from Hoboken, New Jersey, to Buffalo, New York. In the late 1940s and 1950s the railroad changed from steam power to diesel, which was cheaper and cleaner and fueled faster, more powerful engines. Industrial competition from other parts of the world brought hard times to the Lackawanna and Western Railroad. By 1972 it was in bankruptcy, and in 1976 it was absorbed into the Conrail system.

F. Nelson Blount, a wealthy New England seafood processor, saved from the scrap heap many standard-gauge steam locomotives that were being replaced by diesel engines, and preserved them along with freight and passenger cars. In 1967 Blount displayed his collection in Bellows Falls, Vermont, as Steamtown USA. In 1983 the foundation that maintained the collection moved it to a more central spot, the Scranton yards of the Delaware, Lackawanna & Western Railroad. In 1986 Congress designated the new Steamtown in downtown Scranton a National Historic Site.

Steamtown National Historic Site RR Museum and Heritage Railroad occupies 62.48 acres (25.3 ha). It incorporates the working turntable and parts of the 1902 DL&W roundhouse, a circular building where locomotives were repaired, the switch yards, a locomotive shop, associated buildings, track, and equipment. Reconstructions of buildings or parts of buildings are carefully done in period style.

On display are locomotives, freight and passenger cars, mail cars, dining-room and sleeping cars, lounge cars, box cars, cabooses, and a reconstructed DL&W station. Among the 29 locomotives in the collection is one from the 1913 Main Central Railroad; the Boston & Maine 3713, built in 1934 and named the Constitution; the Union Pacific Big Boy 4012, one of the largest and most powerful steam locomotives ever built; and the Delaware, Lackawanna & Western 565, one of only two DL&W steam locomotives still in existence.

Some of the interesting passenger and freight cars in the collection are the Mountaineer, a 1930 Pullman heavyweight dining car from the Boston & Maine Railroad; the Louisville & Nashville 1100, which is a 1913 heavyweight railway Post Office car; and a Long Island Railway rotary snowplow, built in 1898 and displayed at the New York World's Fair.

A guided walking tour shows off the roundhouse; restoration shops where mechanics repair and care for locomotives; the gas house where coal was used to generate power and heat; iron furnaces; and the Lackawanna railroad station. During the summer, a steam-powered train brings visitors on a short ride to the roundhouse. Also available is one of the round-trip summer excursions to Moscow, Pennsylvania, a 2½-hour tour of the countryside. Some of the site's buildings are under restoration, and new ones are being constructed. The Visitor Center's exhibits explain the relationship among the economy, people, and lifestyles of the region, and touch upon social history, ethnology, immigration patterns, the labor movement, and union history. Artifacts from the railroad's peak period of 1850 to 1950 are on display, and the visitor can learn about coal mining and iron manufacturing during the steam era.

Address: Steamtown National Historic Site, 350 Cliff Street, Scranton, PA 18503 (address of parking lot). With GPS also use the coordinates +41.410730 (n-latitude),—75.671329 (w-longitude).

Phone: (570) 340-5200.

Days and hours: Regular hours 9 a.m.–5 p.m.; Jan.–Mar.: 10 a.m.–4 p.m. Closed Thanksgiving, 12/25, 1/1.

Fee: Yes. Extra fee for excursion trains.

Tours: Self-guided at Visitor Center. Ranger tours offered at regular intervals daily. Excursion train rides during the summer on a varied schedule; call for information.

Public transportation: Lackawanna County buses stop nearby.

Handicapped accessible: Yes.

Food served on premises: None. Food is available nearby.

Website: www.nps.gov/stea/index.htm

TITUSVILLE

Drake Well Museum and Park

This museum is in a 240-acre (97-hectare) park on the spot where the petroleum industry was born. In August 1859, Edwin Laurentin Drake drilled for oil and demonstrated that petroleum could be collected in large quantities by drilling, a pivotal discovery with far-reaching effects on modern civilization.

For thousands of years, people around the world collected oil that trickled from springs and oozed from seeps. Native Americans used oil to waterproof objects, to make war paint, and as a medicinal salve. Early settlers bottled it and sold it as a cure-all called Seneca Oil, and used it to lubricate sawmills. For many other tasks, whale oil was used. As the number of whales declined from overfishing, oil that flowed naturally from the ground, known as rock oil, started to replace whale oil for lamp fuel. In 1850 Samuel Kier, who sold homemade medicines, found a way to refine crude oil into a new material with improved burning characteristics. The demand for this new fuel, kerosene, exceeded supplies.

The Seneca Oil Company of New Haven, Connecticut, sent Drake to examine the oil spring the company leased near Titusville and try to improve the local oil output. In his own newly formed company, Drake dug for oil, but he was unsuccessful. Next he brought a salt well driller from Pittsburgh to try drilling for oil, an approach never attempted before. The loose glacial till of the valley was a problem, however, because sand and gravel would slide back into the hole. Drake tried placing an outer tube around the drill, and it worked. The well started producing about 20 barrels (3.2 m³) of oil a day, more than double the production of any facility known at that time.

By 1861 wells had sprung up all along the Allegheny River and Oil Creek. Thousands of barrels (hundreds of m³) of oil were produced daily, which drove the price of oil down. Exports to Europe began in 1860, and to Japan in 1862. Drake and his partners finally were forced out of business, and the original well was shut down in that year. By the 1870s wells in other parts of the country surpassed the output in the Titusville area.

In 1934 the American Petroleum Institute built a museum on the site of the original well and gave it to the Commonwealth of Pennsylvania. E. C. Bell, a late 19th-century collector of oil-industry memorabilia, donated the core of the museum's collection. The

Drake Well.
Courtesy of Drake Well Museum and Park.

museum presents the importance of petroleum in current life by showcasing the oil industry's history, growth, and modern effect on society.

The museum, now in a modern park setting, has both indoor and outdoor exhibits which explore the modern oil industry using orientation videos, exhibits, demonstrations, operating oil-field machinery, and historic buildings. The museum building contains more than 6000 items, including machinery, tools, gas and steam engines, and drilling rigs. Manuscripts, maps, and other printed specimens document the rise of the petroleum and gas industry.

Here the museum exhibits explore the region's geology, how oil is formed, and how other industries supported oil production. Coopers, for example, made wooden casks for storing oil, and blacksmiths made and repaired machine parts. An interesting vehicle on display is the nitroglycerin wagon, which transported explosive torpedoes to the wells, where they were used to fracture oil-bearing rock. On display are artifacts explaining how wealth was created via the oil industry, and how this wealth helped build new communities. Exhibits reveal how the Pennsylvania oil industry spread around the globe.

The research library contains about 3500 books, 1500 periodicals, over 10 000 images, many maps, and more, for use by appointment only.

Outdoors there are exhibits about the technical side of early drilling, such as how power was moved from the central gas engine throughout the oil field. On view are oil-collecting pits dug by prehistoric people. Sights, sounds, and smells of Pennsylvania's early oil industry are now re-created by the museum's oil-field buildings and machinery. A reconstruction of Drake's derrick and engine house with a working steam engine is open to visitors. The original oil rig was a wooden structure 35 feet (11 m) tall, designed to drill to 2000 feet (610 m). Later, rigs were built of steel, reached 85 feet (26 m) high, and could drill down to 10 000 feet (3050 m) faster and deeper. The Pennsylvania design for oil rigs became the industry standard around the world. Exhibits, both indoors and outdoors, explain new technologies developed to speed the pumping process, how natural gas is captured from the oil, ways of powering the machinery, and how pipelines from oil fields to refineries came to replace barrels of oil hauled by railroads. Interpreters are on hand seasonally to answer visitors' questions about the people who lived and worked in local oil fields.

South of the park, off State Route 8 south of Rouseville, is McClintock Well #1, the oldest continuously working oil well, was begun in 1861 on Hamilton McClintock's property by Joel Angier, and produced 175 barrels (6680 L) of petroleum per day initially, but gradually dropped to half a barrel (80 L) per day 60 years later. The Pennsylvania Historical and Museum Commission acquired the well in 2000, and uses the proceeds from sale of the oil to maintain the well site. There is a public unpaved road from the historical marker on Route 8 to the well.

Visitors can take a narrated, 2½-hour trip on the Oil Creek and Titusville Railroad, which stops in Titusville at the restored Perry Street Station (a freight house built in 1893) and in Drake Well Park.

Address: Drake Well Museum and Park, 202 Museum Lane, Titusville, PA 16354.
Phone: (814) 827-2797.
Days and hours: Tue.–Sat. 9 a.m.–5 p.m.; Sun. noon–5 p.m. Closed Mondays and holidays except Memorial Day, 7/4, and Labor Day. Closed Mon.–Fri. in Jan.
Fee: Yes.
Tours: Self-guided; call in advance for guided tours.
Food on site: None, but there is a picnic area. Titusville, 1.25 miles (2 km) away, has several eateries.
Public transportation: None.
Handicapped accessible: Yes. Wheelchairs are available; call in advance.
Website: www.drakewell.org/index.html

YORK

Harley-Davidson, Inc.

The motorcycles manufactured by Harley-Davidson have become icons of the American lifestyle and spawned a unique culture of motorcycle enthusiasts. Visitors can tour Harley-Davidson's York plant to see how these vehicles are built and learn about their history.

Although motorcycles were first developed around 1868, when steam power was a popular curiosity, it wasn't until 1901 that the E. R. Thomas Motor Company produced the first commercially viable motorcycles. The Harley-Davidson Company, founded in 1903, was one of 150 companies that tried to market motorcycles in the beginning of the 20th century. By the end of World War II, only two—Indian and Harley-Davidson—were still in business. Harley-Davidson has been the sole manufacturer of American motorcycles since 1953.

The company was launched in the Davidson family's backyard shed in Milwaukee, Wisconsin. William S. Harley (a draftsman), Arthur Davidson (a pattern-maker), and Arthur's brothers Walter and William—who knew about engines and mechanics—pooled efforts to build their first engine. It was a leather-belt, direct-drive, 2⅛-inch by 2⅞-inch (5.4 cm × 7.3 cm) bore and stroke capable of three horsepower (2.2 kW).

As their designs improved, standard bicycles became too flimsy to hold the engines, and they needed to develop a stronger frame. Harley and Davidson got help from Ole Evinrude (who later opened his own outboard engine company) and started making motorcycles commercially the same year that Henry Ford produced his Model T. The company grew, faltered during the Depression of the 1930s, and regained the market. American Machine and Foundry Corporation (AMF) bought the family-run business in 1969. In 1981 Willie G. Davidson, grandson of one of the original owners, joined with other investors to buy back the company.

Today there are three plants: an engine division in Milwaukee, Wisconsin; a fiberglass plant in Tomahawk, Wisconsin; and an assembly factory in York, Pennsylvania. All three locations give tours to the public.

At the York plant, established in 1973, the first stop in the "Steel Toe Tour" is the Antique Motorcycle Museum. More than 40 historic vehicles are arranged chronologically to show the development of motorcycle technology. The 1906 model was a single-cylinder, four-stroke, belt-drive machine with an idler pedal and coaster brake. By 1914 a footrest, kick-starter, and free luggage rack had been added. The 1915 model had three speeds and an enclosed chain. The 1918 Army model had carbide lights because Harley-Davidson didn't think electricity would become popular on these vehicles. The 1923 model, the first made by Harley-Davidson for police departments, included a key innovation: a motorcycle speedometer. It averaged 87 miles per gallon (2.7 L per 100 km) of gasoline. The 1926 model had the first teardrop-shaped gas tank, which has become famous over the years. In 1936 the company started making motorcycles in colors other than the standard gray or olive. In 1941 the company produced motorcycles for only three months because of World War II. One of these rare models is on exhibit. In addition, many different motorcycle engines, including the 74-cubic-inch (1200 cm^3) flat head, the 61-cubic-inch (1000 cm^3) knuckle head, the 74-cubic-inch (1200 cm^3) pan head, and the 80-cubic-inch (1300 cm^3) shovel head are displayed.

The tour follows a 2-mile (3.2 km) walk to see the bolt-by-bolt assembly of a motorcycle. See materials cut by laser to fabricate special models. Machines are tested at the end of the assembly line. Visit the new team production line where sport-model motorcycles are assembled by small groups of workers, who stay with a motorcycle from beginning to end and are responsible for the quality of each machine it produces. Clocks monitor the amount of time allotted for each production step.

Address: Harley-Davidson, Inc., 1425 Eden Road, York, PA 17402.
Phone: (877) 883-1450.
Days and hours: Museum only, Mon.–Fri. 8 a.m.–4 p.m., tours 9 a.m. through 2 p.m. Closed Good Friday, Memorial Day, 7/4, Labor Day, Thanksgiving Day and the day after Thanksgiving, the last week in December, and 1/1.
Fee: Yes.
Tours: Guided only; call ahead or purchase tickets online for reservations.
Public transportation: None.
Handicapped accessible: Yes, but extensive walking is required.
Special note: Tours may be canceled, so call ahead that day. Children over age 12 only. No photography, video recording, or camera bags are allowed in the plant, and visits to the plant require closed-toed shoes. A more limited, free tour is also offered.
Website: www.harley-davidson.com/content/h-d/en_US/home/events/factory-tours/yorkpa.html

Washington, DC

Bureau of Engraving and Printing

Billions of dollars of paper money for the federal government are printed at the US Department of the Treasury's Bureau of Engraving and Printing each year. Here visitors watch currency come off the press and learn about the history of banknotes in the United States.

During the colonial period, English shillings, French *louis d'or*, and Spanish doubloons were the media of exchange in commerce. Having no uniform money system caused confusion, and trade slowed. In 1775, before releasing the Declaration of Independence, the Continental Congress issued the nation's first money, continental currency. The famous silversmith Paul Revere, known for his midnight ride, engraved the plates for these notes. Because this money had little value, "not worth a continental" became a common saying.

In 1785 the country adopted a monetary system based on the decimal system, which became the dollar we know today. Four years later, the same year the US Constitution was adopted, paper money came into use—not as federal currency, but as promissory notes issued by banks, payable on demand.

Coins have been produced since 1792 at the US Mint, a separate entity from the Bureau of Engraving and Printing. Today's system of paper money did not come about until the 1860s, during the Civil War. In 1861 the Department of the Treasury issued Demand Notes as a way finance the Civil War. These notes were government IOUs, printed by a private firm, and sent to the Treasury, where dozens of clerks and laborers signed, separated, and trimmed them by hand. The following year United States Notes, also called legal tender notes, were issued. In 1862 the Bureau of Engraving and Printing was launched; it consisted of several people in the basement of the Treasury Building separating, bundling, and sealing notes printed by private banks. By 1877 the bureau was printing all paper currency. In the early part of the 20th century, the basic design of the paper money we use now was established. Federal Reserve notes, the only US currency banknote now printed, were issued in 1914.

Besides banknotes, the Bureau at one time printed securities and obligations. Postage stamps were printed at the Bureau from 1894 through 2005, when private firms were contracted to print US postage stamps.

The manufacture of currency is a complex process. The self-guided tour through the Washington branch of the bureau provides visitors with a view of each of four basic stages involved in producing paper money: Offset (in which the background images are printed); intaglio (where the backs and faces of notes are added); mechanical examination; and currency overprinting equipment and packaging (in which seals and serial numbers are added, and the sheets of currency are cut into single notes).

The bureau can print more than 8000 sheets per hour. Banknotes are printed in sheets of 32 and 52 notes. The faces of $1 notes are printed in black and the backs in green (hence the term "greenback" for US paper money). Some special anti-counterfeiting measures are also put in place (depending on the denomination). Banknotes are first shipped to the Federal Reserve, which then issues the notes to one of 12 Federal Reserve Banks, which then distributes them to local banks.

Intaglio printing press.
Courtesy of the Bureau of Engraving and Printing.

More than $200 billion in banknotes (about 7 billion notes on average) are printed annually in denominations of $1, $2, $5, $10, $20, $50, and $100, primarily to replace worn or mutilated currency that has been removed from circulation. The largest note ever printed was a $100,000 Gold Certificate bearing President Woodrow Wilson's picture, and it was only used for transactions between Federal Reserve Banks. No denomination larger than $100 has been printed since 1945.

The Washington facility, authorized by Congress in 1878, was completed in 1880. To accommodate more production space, the Main Building was constructed in 1914 in a neoclassical style spanning a 505-foot (154-m) façade, incorporating fireproof concrete; an annex was added in 1938. The Fort Worth, Texas, facility was added in 1991.

Address: Bureau of Engraving and Printing, 14th and C Streets, S.W., Washington, DC 20228.
Phone: (202) 874-2330.
Days and hours: Mon.–Fri. 9 a.m.–2 p.m. Evening tours run during the summer from 2 p.m.–6 p.m. Visitor Center is open Mon.–Fri. 8:30 a.m.–2:45 p.m. Closed weekends, federal holidays, and 12/25–1/1.
Fee: Free, although same-day tickets (the booth is on Raoul Wallenberg Place, S.W., and opens at 8 a.m.; tickets for that day are usually gone by 9 a.m.) are required from Mar.–Aug.
Tours: Partially self-guided; for guided tours call (202) 874-2330. Evening tours during spring and summer 2:00–6:00 p.m. every 15 min.
Public transportation: Local buses stop less than 0.1 mile (160 m) away, then walk to the site. Metro station (Smithsonian Station) is one block away.

Handicapped accessible: Yes.
Food served on premises: None.
Website: www.bep.gov/services/takeatour.html

Smithsonian Institution

Established in 1846, the Smithsonian is the world's largest museum complex. It consists of 19 museums and galleries, plus the National Zoo; 10 of the museums are along the mile-long (1.6-km) National Mall between the Capitol building and the Washington Monument. The facility holds 150 million objects, only a tiny percentage of which are on display at any given time.

The Smithsonian was launched when English scientist James Smithson (1765–1829) bequeathed a large sum of money to the United States to establish "at Washington, under the name of the Smithsonian Institution, an establishment for the increase and diffusion of knowledge among men." Besides being dedicated to public education, the Smithsonian is a world leader in scientific research, which is conducted at sites across the United States and in 140 countries around the world.

The Smithsonian's oldest building, known as "the Castle," was designed by architect James Renwick, Jr., in Norman style and was completed in 1855. It houses James Smithson's crypt and the Smithsonian Visitors Center, which offers orientation films, an information desk, interactive maps, and an exhibit showcasing artifacts from the various subsidiary museums. Originally it contained a public exhibition area, offices, laboratories, and sleeping quarters for the Secretary's family and scientists. Today, the Smithsonian has 10 museums devoted to science and technology, many located along the Mall: the National Museum of Natural History, the National Museum of American History, and the National Air and Space Museum.

At the Natural History Museum are permanent and temporary exhibits such as the Kenneth E. Behring Family Hall of Mammals, which features more than 200 preserved specimens in their natural habitats. Many of the mounted mammals on display were collected by President Theodore Roosevelt. The Hall of Geology, Gems, and Minerals showcases an extensive collection of meteorites, moon rocks, and gems, including the famous 45.5-carat Hope Diamond. The Hall of Human Origins delves into the millions of years of human evolution.

In the Insect Zoo, where visitors can watch ant colonies at work, hissing cockroaches, and tarantula feedings, is a Butterfly Pavilion housing butterflies and exotic plants from all around the world. The Q?rius center is an interactive space that lets visitors examine and handle over 6000 objects, including fossils, shells, and minerals. The Living Coral Reef provides the opportunity to study the interrelationship between coral and fish. In the Discovery Room are fossil footprints, seashells, and snake skins available to touch, plus other objects that can be studied at close range. The Naturalist Center is for teens and adults who want to examine biological specimens at their leisure.

The National Museum of American History illuminates the scientific, technological, and cultural heritage of the United States. Here the visitor can learn about the Industrial Revolution, agriculture in America, printing and graphics, maritime history, and musical instruments, among other topics. Items that are especially interesting are the gunboat *Philadelphia*, part of Benedict Arnold's flotilla that fought the British on Lake Champlain in 1776, and the John Bull locomotive. Historic inventions on display in the American Enterprise Exhibit include the Morse-Vail telegraph, experimental telephones designed

Aerial view of the Castle (foreground) and National Museum of Natural History (domed structure).
Courtesy of the Smithsonian Institution.

and used by Alexander Graham Bell, and light bulbs made by Thomas Edison. (See B&O Railroad Museum, Baltimore, Maryland; Historic Speedwell, Morristown, New Jersey; and Thomas Edison National Historical Park, West Orange, New Jersey.)

Visitors can take a journey through Places of Invention and learn about cardiac innovations during the 1950s in Minnesota's Medical Alley and the development of the personal computer in the 1970s and 1980s in Silicon Valley. (See Computer History Museum, Mountain View, California.) Modern Medicine and the Great War discusses advances in medicine, medical technology, and medical professionals during the First World War. (See National Museum of Health and Medicine, Silver Spring, Maryland.)

The National Air and Space Museum is one of the most popular attractions in Washington. It explores the technology used to develop air- and spacecraft, traces the history of space flight from the 13th century to the present, and displays flight equipment, engines, propellers, instruments, memorabilia, and documents. The exhibit includes the 1903 Wright brothers Flyer that made the first powered, manned, controlled flight of a heavier-than-air craft near Kitty Hawk, North Carolina. (See Carillon Historical Park, Dayton, Ohio; and Wright Brothers National Memorial, Kill Devil Hills, North Carolina.) Here on display are also the Viking Mars Lander, used to send information back to Earth from the atmosphere of Mars; the Spirit of St. Louis, flown across the Atlantic Ocean in 1927 by Charles Lindbergh; and the Mercury capsule launched in 1959 to learn about flight dynamics and spacecraft procedures.

The museum reviews the history of astronomy from Stonehenge, believed to be an ancient religious and astronomical site, to the Hubble space telescope, which was used to take an unprecedented look at outer space. The museum's Sea–Air Operations section has a reproduction of a carrier hangar deck, a wooden-winged Boeing F-4 B-4 from the 1930s, a Grumman FM-1 Wildcat used in World War II, and other important pieces.

(See Steven F. Udvar-Hazy Center, Chantilly, Virginia; and *Intrepid* Sea, Air & Space Museum, New York, New York.)

Each museum of the Smithsonian Institution regularly has special events and exhibits, and shows films. The National Air and Space Museum offers a series of films daily on its five-story-high movie screen.

Address: Smithsonian Institution, 1000 Jefferson Drive, S.W., Washington, DC 20560
Phone: (202) 633-1000.
Days and hours: Daily 10 a.m.–5:30 p.m. Closed 12/25.
Fee: Free.
Tours: Self-guided; guided tours are offered at selected times at each facility.
Public transportation: The Smithsonian Metro stop is only 0.2 mile (0.3 km) from the Castle. Walk from the station to the site.
Handicapped accessible: Yes.
Food served on premises: Cafés.
Website: www.si.edu/

Smithsonian's National Zoo and Conservation Biology Institute

In the late 19th century, people noticed that the North American bison was almost extinct, and some species of deer, elk, and other North American wildlife faced a similar fate. Samuel Pierpont Langley, secretary of the Smithsonian Institution, and William Hornaday, naturalist and Smithsonian taxidermist, asked Congress for a reserve in the District of Columbia to save these species and display native and exotic animals. In 1889 President Grover Cleveland signed a bill establishing a National Zoological Park "for the advancement of science and the instruction and recreation of the people." In 1890 it became part of the Smithsonian.

Langley and Hornaday designed this large, urban park with the internationally renowned landscape architect Frederick Law Olmsted. (See Frederick Law Olmsted National Historic Site, Brookline, Massachusetts.) The zoo is made up of 163 acres (66 hectares) of hard rock on the Piedmont plateau in the heart of Washington, DC. The land here weathers into claylike soil that supports deciduous trees such as oak, hickory, and tulip.

Today, 15 000 animals from more than 300 species live here in natural social groups. The park places equal importance on animals and plants, seeking to help visitors understand the biological world's diversity, intricacy, and interdependencies. Over 70 species of grasses (including 36 species of bamboo) contribute to the diverse landscape. While the zoo is best known for giant pandas, visitors also can see birds, great Apes, big cats, Asian elephants, insects, amphibians, reptiles, aquatic animals, and more. Integrated into the park are art displays and cultural artifacts.

Finished in 2013, Elephant Trails is the exhibit hosting up to 10 elephants. Included are a pool for swimming and bathing, a trail .25 mile (0.4 km) long through woods, and an Outpost for visitors to view these large beasts. There is also the Elephant Community Center, where visitors learn about elephants.

The reptile house, built in 1929, has been refurbished and is now the Reptile Discovery Center, where visitors are permitted to touch these cold-blooded vertebrates. Available for observation are more than 600 reptiles and amphibians, where human explorers learn about their life cycles, and find out how they adapt to their environments. At 12

Asian elephants at the National Zoo. *Photo* by Jen Zoon.
Courtesy of the Smithsonian Institution.

strategically placed learning modules, discover facts about reptiles and amphibians, and gain insights into why many people have misconceptions about these creatures. Zoo personnel observe as visitors go through the exhibit. This helps the personnel learn better ways of designing the zoo's facilities.

Amazonia, the zoo's Brazilian rain-forest exhibit, contains 100 animal species and more than 400 plant species. It is a total-immersion display designed to transport visitors into a rain forest's climate, geography, and plant and animal communities. There is a tropical river, a village hut and canoe, and many kinds of animals, birds, insects, reptiles, and amphibians. At a replica of a biologist's field station, visitors learn how scientists study the rain forest. The interactive educational gallery explores the interrelationship among animals, plants, and humans in the rain forest.

Most people associate this zoo with the giant panda Hsing-Hsing (a gift from China), which died in 1999. The zoo's giant pandas, plus six other Asian species (red pandas, Asian small-clawed otters, clouded leopards, fishing cats, a Japanese giant salamander, and sloth bears), live in the Asia Trail, opened in 2006. The zoo is also the first place outside of Indonesia to breed the endangered Komodo dragon. Scientific methods of animal husbandry were applied to bring about the birth of 13 Komodo dragons in October 1992, 237 days after zoo workers first noticed the eggs.

Modern animal conservation began at this zoo, where American bison were exhibited when they were almost extinct at the close of the 19th century. Now the species is recovering, with over half a million individuals either wild or raised as livestock. The American Bison exhibit, opened in 2014, displays a pair of American bison in order for the public to learn about this symbol of the American West.

Address: Smithsonian's National Zoo and Conservation Biology Institute, 3001 Connecticut Avenue N.W., Washington, DC 20008.

Phone: (202) 633-4888.

Days and hours: Exhibit buildings, 9 a.m. 4 p.m.; grounds, 8 a.m.–5 p.m. (last entrance 4 p.m.). Closed 12/25.

Fee: Free.

Tours: Self-guided and guided: Ask at information desk and call in advance: (202) 633-3025.

Public transportation: Local Metrobus stops directly in front of the zoo.

Handicapped accessible: Yes.

Food served on premises: Cafés.

Website: nationalzoo.si.edu/

3

Southeast

Alabama

Sloss Furnaces National Historic Landmark

The 30-acre (12-hectare) site where the Sloss Furnaces at one time produced iron now showcases the South's industrial heritage. Birmingham's birth can be attributed in large part to the iron industry. The city was established just after the Civil War at the junction of two railroads that were attracted to the area because of its rich supply of iron ore, coal, and limestone.

James Withers Sloss, a successful business and railroad executive, saw great potential in the Birmingham area because it had all the raw materials needed to produce iron. Thanks to Sloss Furnaces Company, founded in 1881, and other iron producers in the area, Birmingham prospered. By the early 20th century it had evolved from a small town to one of the South's major industrial, railroad, and iron-producing centers.

Sloss Furnaces grew rapidly, producing tons of pig iron for sale to distant foundries. After J. W. Sloss retired in 1886, the company was bought by a group of entrepreneurs from Virginia, who changed its name to Sloss Sheffield Steel and Iron in the hopes of making steel. Even though the company never produced steel, the name remained. Sloss Sheffield Steel and Iron was modernized between 1927 and 1931, then merged with U.S. Pipe in 1952, and continued operating until 1971, after which the facility was donated to the city of Birmingham. Since then, Sloss Furnaces has been restored and preserved for public use. In 1981 the site became a National Historic Landmark and was opened to tourists. On display are the technological changes that took place at the complex between 1927 and 1931. Here visitors feel the character of the iron industry and get information about Birmingham's iron-industry heritage.

The 16 000 square-foot (1500 m²) Visitor Center, which vaguely resembles an industrial container lying on its side, houses a historical introduction. A 15-minute slide show and small museum provide background. A self-guided walking tour begins here. This tour follows the same path as the raw materials took when they were turned into pig iron, passing hulking smokestacks, stoves, and cast sheds. Eight enormous air compressors run by steam engines once pumped blasts of air into the hot blast stoves. Nearby, in the blower building, the factory grew and new, more modern equipment was added. On display is the more recent equipment that made the old steam engines obsolete.

Railcars fitted with bottom doors transported raw materials to the elevated stock trestles, dumping them into bins below the trestles. The skip hoist moved these raw materials to the top of the furnace in the cast shed.

Of the two original furnaces in the Cast Shed exhibit, only the No. 1 blast furnace is open to the public. During the years when it was a functioning furnace, the liquid iron was released every four hours through the runner into ladle cars. The ladle car, a steel container lined with heat-resistant bricks and mounted on wheels, acted like a large kettle as it moved the molten metal to the pig caster. Unfortunately, not much is left of the pig caster, which cast molten iron into pigs—large bars of solid metal. Railroad cars that hauled the pigs out of the plant are on display.

The Pyrometer House contains instruments called pyrometers, which measured the interior temperatures of the furnace, thus giving the name to this facility. The fireproof pyrometer house doubled as a safe haven for workers if something went dangerously wrong with the furnace. It also contained a millwright shop for furnace repairmen, now a blacksmith shop that helps maintain the Sloss Furnace site.

Boilers generated the steam on which the plant depended. By 1929 a powerhouse generated electricity from the steam to power motors and lights and run the plant. Because some of the machinery was installed as late as the 1950s, we have a sense of how iron production changed over time. Thus we follow the evolution of this industry for nine decades, from 1881 to 1971.

This facility hosts a nationally recognized metal arts program which preserves the forging and fabrication of metallic items. Concerts, weddings, and festivals are also held here.

Address: Sloss Furnaces National Historic Landmark, 20 32nd Street North, Birmingham, AL 35222.
Phone: (205) 254-2025.
Days and hours: Tue.–Sat. 10 a.m.–4 p.m.; Sun. 12–4 p.m. Closed Mondays, Veteran's Day, Thanksgiving Day, the day after Thanksgiving, 12/25, and 1/1.
Fee: Free, except fee for group tours.
Tours: Self-guided; group tours by advance appointment only. Wear weather-appropriate clothing appropriate for outdoor exploration. Bring water when the weather is hot.
Public transportation: Buses stop several blocks away from the Visitor Center.
Handicapped accessible: Limited.
Food served on premises: None. Restaurants are in the area.
Website: slossfurnaces.com/

TUSKEGEE

Tuskegee Institute National Historic Site

Founded in 1881 with a class of 30 students, Tuskegee Institute's initial purpose was to offer African-American students secondary education and teacher training. As it grew, it expanded its educational goals to include many areas of training, such as agriculture and practical industrial training. It is now a university and is designated a National Historic Site.

There are several places of interest to visitors to the campus. These sites include a museum dedicated to the school's most famous faculty member, the scientist George Washington Carver, and the beautiful Victorian home where the founder and driving force behind the school, Booker T. Washington, lived. Washington's house, called The Oaks, is an attractive structure built by the students of Tuskegee Institute. It has been restored to look as it did when Washington and his third wife, Margaret, lived here.

Booker T. Washington's extraordinary abilities and his skill at working within the system made the school financially self-sufficient. He wished to maximize the students' potential by having them learn a trade as they undertook their academic studies. These students built the school buildings with the bricks they made, and worked in campus facilities to keep the school operating. Some of the original student-constructed buildings (including structures designed by Robert Robinson Taylor, the first African-American graduate of the Massachusetts Institute of Technology) are still in use.

This extraordinary man was able to persuade some of the best people in their fields to join the faculty, including George Washington Carver (see George Washington Carver National Monument, Diamond, Missouri), who was hired to head the school's new Department of Agriculture. Carver left a faculty position at Iowa State College, where he had earned a Master's Degree of Agriculture in 1896 and was involved in botany and bacteriology research.

For 47 years, Carver taught and researched agricultural and mineral products at Tuskegee Institute. Most of his work focused on peanuts and sweet potatoes, which grew well in the soil of the South and easily could be stored to provide winter food for the poor Southern farmers. His aim was to develop new products to replace the faltering cotton economy, badly hurt by depleted soil and the boll weevil. Carver became a strong advocate of crop rotation to save the soil.

Carver's work on products derived from peanuts and sweet potatoes made him a self-taught chemist knowledgeable about dyes, paints, stains, talc, cosmetics, medicines, and foods such as coffee substitutes, mock coconut, starch, sugar, and flour. He created more than 300 useful products from peanuts and 175 from sweet potatoes. Together, Washington and Carver developed the idea of extension programs for local farmers. They set up courses focusing on agriculture in order to teach farmers better ways to use the land and how to produce plentiful, superior crops. A mule-driven wagon converted into a classroom traveled the countryside, stopping at various locations. This wagon-classroom, known as the Jesup Wagon, was replaced by a bus called the Movable School, now on display in the George Washington Carver Museum.

The museum, established in 1938 and thoroughly refurbished in 1980, is in what was the laundry building of Tuskegee Institute's original student-built school. Displays include exhibits and interpretive programs as well as virtual displays, Civil War flags, a Civil War collection, the USS *Cairo*, and the Vicksburg siege. Exhibits also discuss American presidents and the Revolutionary War. However, most exhibits focus, in large measure, on the life and career of Carver, who died in 1943 and is buried on the Tuskegee campus. He was originally trained as a musician and artist, but became a scientist out of his love of nature, his basic curiosity, and his need to help his fellow human beings. On display are Carver's extensive collection of native plants, minerals, birds, and vegetables, as well as many of the products he developed from the peanut and sweet potato. Displays of the clay ores Carver used in some of his research on stains, dyes, and pigments give insight into his work. Some pieces of Carver's laboratory equipment and some of his paintings, drawings, and textile art are also displayed. Additionally, information exhibits explore Tuskegee Institute's growth into Tuskegee University. Two 30-minute films are shown: one on Carver, and one on Washington.

Address: Tuskegee Institute National Historic Site, 1212 West Montgomery Road, Tuskegee, AL 36088.
Phone: (334) 727-3200.
Days and hours: Mon.–Sat. 9 a.m.–4:30 p.m. Closed Thanksgiving Day, 12/25, and 1/1.
Fee: Free.

Tours: Self-guided in museum; wheelchairs are available on first-come first-served basis.
Public transportation: None.
Handicapped accessible: Yes.
Food served on premises: None. Restaurants are in town.
Website: www.nps.gov/tuin/index.htm

Florida

FORT MYERS

Edison and Ford Winter Estates

During a vacation in 1885, the great American inventor Thomas Edison became entranced with Fort Myers, a small cattle town located along the Caloosahatchee River on the Gulf Coast of Florida. He purchased 13.5 waterfront acres (5.5 hectares), on which two houses, which he designed, were assembled a year later. The structures were built in prefabricated sections in Fairfield, Maine, and shipped to Fort Myers on two schooners. Henry Ford and his wife, frequent visitors to Edison's Fort Myers estate, bought the adjoining property in 1916. The two homes have been authentically restored to their 1929 appearance. Together these two properties now consist of 20 acres (8.9 hectares) of historic buildings and historic gardens.

One of America's first prefabricated houses, Edison's home is a gracious, rambling residence well-suited to the tropics. Large overhanging porches surround the main house

The winter homes of Thomas Edison and Henry Ford sit along the Caloosahatchee River on more than 20 acres (8 hectares) of botanical gardens.
Courtesy of Edison & Ford Winter Estates.

and the guest house, which are connected by a wooden pergola. French doors on the ground floor can be opened to admit cooling breezes. The only access to the estate was by boat, or horse and buggy, for the railroad had not yet been extended to Fort Myers. (See Thomas Edison National Historical Park, West Orange, New Jersey.)

Edison's genius is evident throughout the house. For example, the living room and other rooms have brass chandeliers that he designed and made in his own workshops. These electric chandeliers (or "electroliers," as he termed them) are still in use.

Although Edison is best known for his inventions connected with electricity, he did a significant amount of work in other fields, such as horticulture. With the help of naturalist John Burroughs and other friends from around the world, Edison developed one of the most complete tropical gardens in the United States at the time. These gardens contain hundreds of varieties of plants from many places around the world.

In addition to useful plants such as cinnamon, coffee, tea, vanilla orchid, camphor, allspice, bay, and bamboo, the gardens have unusual trees such as the frangipani with its exotic perfume, and a host of flowering shrubs and citrus, ficus, and palm trees. The 1929 landscaping has been kept throughout the property, including the Moonlight Garden.

The orchid collection includes Florida's native species and specimens from around the globe. This collection began when Edison and two friends, John Burroughs and Henry Ford, went on a camping trip into the Florida Everglades and returned with wild orchid plants.

Early in the 1920s Edison grew concerned about the possibility of war or disease, and the consequent loss of access to natural rubber from the Pacific. Edison, Ford, and Harvey Firestone formed the Edison Botanic Research Corporation to pursue alternate sources of natural rubber. Edison realized that the existing rubber plantations could not grow enough latex to meet an immediate crisis. He had rubber plants sent to Fort Myers from around the world, and tested over 17 000 plants. After extensive experimentation, Edison found that a common native plant, goldenrod, produced a sap high in latex. He crossbred goldenrod until he developed a gigantic strain, *Solidago edisoniana*, which could be harvested quickly and efficiently. To convert the goldenrod sap into natural rubber, Edison constructed a large chemical laboratory, the Edison Botanic Research Laboratory, on the premises, employing many assistants. Samples of rubber were sent to Firestone, who made prototype goldenrod tires for Edison's Model T Ford. During World War II the US government found many uses for goldenrod rubber. Today, the laboratory still looks as it did in the 1920s.

Edison's most important contributions, the research laboratory and research teams, are exemplified here. The tour goes through the laboratory and stops at the office where Edison worked and often took catnaps. The Edison Botanic Research Laboratory has been designated a National Historic Chemical Landmark by the American Chemical Society.

The Edison Museum houses a huge collection of Edison's inventions, representing just a portion of the more than 1000 patents he received over his lifetime. Here on view are hundreds of Edison artifacts: phonographs, batteries, movie projectors, light bulbs, an early talking doll, family memorabilia, a Model T presented as a gift from Ford to Edison, and an early-model Cadillac.

The Henry Ford Winter Home has been authentically restored to its 1929 appearance. Visitors can tour the house and the grounds, on which several automobiles are displayed. For example, there is a 1923 Model T with a hand-crank and a 1929 Model A with a trunk attached to the rear of the car.

The lifelong friendship between Edison and Ford began when Ford was the chief engineer of the Detroit Edison Illuminating Company and met Edison at an employees' convention. Ford explained his gasoline-powered-car ideas to Edison, who gave Ford great encouragement for the project. (See The Henry Ford, Dearborn, Michigan.)

Cruises on the Caloosahatchee River offer views of the island rookery. Guides explain the history and ecology of the river, along with the history of the town of Fort Myers. Visitors may see river life, including manatees, dolphins, and various birds of the region.

Address: Edison & Ford Winter Estates, 2350 McGregor Boulevard, Fort Myers, FL 33901.
Phone: (239) 334-7419.
Days and hours: Daily 9 a.m.–5:30 p.m. Closed Thanksgiving Day and 12/25.
Fee: Yes.
Tours: A variety of guided tours are conducted on a regular basis; self-guided tours include audio; boat cruises on select days. Check schedule online in advance.
Public transportation: None.
Handicapped accessible: Yes, service dogs permitted. Wheelchairs are available; check in advance.
Food served on premises: Light refreshments and snacks; picnic areas with tables.
Website: http://www.edisonfordwinterestates.org/

HOMESTEAD

Everglades National Park

The largest subtropical wilderness in the nation, Everglades National Park covers 1.5 million acres (610 000 hectares) and is the United States' biggest national park after Yellowstone. Because of its vast size, the park maintains three separate, unconnected entrances in very different locations of southern Florida: in Miami Dade County, Monroe County, and Collier County.

Congress recognized the unique qualities of this ecosystem at the southern end of Florida by 1929. In 1946 the state moved to save its flora, fauna, geology, and hydrology by establishing Royal Palm State Park. A year later, President Harry S Truman, acting under a 1934 congressional mandate to the Secretary of the Interior, created Everglades National Park to preserve "a wilderness, where no development would take place and for the entertainment of visitors ... preserving intact ... the flora and fauna and the essential primitive natural conditions now prevailing in the area."

Although the Everglades is located in the temperate zone, the winds, rain, and temperature make it subtropical. Flora from both zones coexist here, a unique situation that prompted the United Nations to declare the park an International Biosphere Reserve and a World Heritage Site, with wetlands of international importance and an ecosystem to be preserved for all time.

This park has three significantly different ecosystems. The freshwater habitats contain sawgrass—not a grass, but a sedge—and many fish, such as Florida gar and largemouth bass, as well as zebra butterflies and black vultures. Hardwood hammocks (a local term for hummock) support air plants, barred owls, royal palms, strap ferns, and slash pines. Saltwater habitats house great white herons, loggerhead turtles, mangrove snappers, blue crabs, coon oysters, brown pelicans, and southern bald eagles.

The Everglades watershed is a network creating a slow-moving, freshwater, shallow river averaging about 6 inches (15 cm) deep, 40 miles (64 km) wide, and 100 miles (160 km) long. It is fed by Lake Okeechobee and the Kissimmee River, 65 miles (100 km) north of the park. Ultimately the system flows into the Gulf of Mexico and Florida Bay. Its porous limestone base and thin mantle of marl and peat support fragile and unique life forms.

Climate in this park consists of two seasons. The summer wet season goes from the end of April until the beginning of December and is punctuated with heavy downpours, lightning, temperatures in the 90s (mid-30s Celsius) and biting insects such as mosquitoes and sand flies at their most active. Best for visitors is the dry season, from mid-December through mid-April, because it generally has milder weather, more visible wildlife, and a less problematic insect population. Mosquito repellent, head covering, long sleeves, and long pants are suggested all year long. Much of the park can only be reached by hiking, canoe, or boat.

More than 300 species of birds, including 12 that are endangered, migrate through the park or live here year-round. Animals that breed in the park are the American crocodile, the snail kite, the southern bald eagle, the wood stork, the Florida panther, and the West Indian manatee. (See John Pennekamp Coral Reef State Park, Key Largo, Florida.)

Each of the Visitor Centers offers videos, slide shows, exhibits, and park information. The main entrance is the Ernest F. Cole Visitor Center at Royal Palm, Flamingo, open 365 days a year, at the eastern park entrance near Homestead. There the visitor can find information about the excellent naturalist-led walks and illustrated talks scheduled during the tourist season, as well as displays which may include works of local artists. The restrooms for this park area are located here.

Along the 38 miles (61 km) of road that curve through the park are self-guided walking paths leading to mangroves, waterways, sawgrass marshes, areas of hardwood hammocks, and salt prairies. From these vantage points visitors can safely observe alligators, anhingas, gallinules, and other wildlife.

Just past the Ernest Coe Visitor Center, at the Royal Palm area, is a trail through a tropical hammock with an alligator hole, a viewing tower, royal palms, live oaks, strangler figs, and red-barked gumbo-limbo trees—the only trees native to this part of Florida. Watch for liguus tree snails wrapped around the smooth-barked lysiloma tree.

The Flamingo Visitor Center is the southernmost point in the park accessible by road, with additional walking trails that wander past mahogany hammocks and diving anhinga birds. Ranger talks discuss the wading birds that roost on nearby mangrove islands as well as the crocodiles, mosquitoes, bobcats, and raccoons that live in the park.

Gulf Coast Visitor Center, on the Gulf Coast side of the park, cannot be accessed from Flamingo or any other spot on the Atlantic Coast. It offers ranger-led nature talks and boar tours through the Ten Thousand Islands of the mangrove forests. The drive between Flamingo and Everglades City, the western entrance, is more than 128 miles (206 km).

Along the drive on US 41, visitors pass Big Cypress National Preserve, which the federal government created on adjacent lands in 1974. Visitors can stop at the Oasis Ranger Station to see a 20-minute video describing the preserve, and exhibits illustrating its flora and fauna. Visitors can drive a loop road through the preserve and stop at an interpretive center surrounded by hardwood hammocks. Visitors can take a two-hour tram ride into the heart of the sawgrass Everglades while park rangers explain the hydrology, geology, vegetation, and wildlife of the region. Visitors can also climb a 65-foot (20-m) tower for a spectacular view.

Among several other areas in the Everglades ecological system are Biscayne National Park and Florida Keys National Marine Sanctuary, which includes Fort Jefferson National Monument, a Civil War fort accessible only by boat or seaplane.

In the 20th century, 600 000 acres (240 000 hectares) of nearby wetlands were drained and turned into productive farmland. The flow of water and the grass in the Everglades' shallow river, the birds and animals that rely on the water, and the national park are all threatened by this change in land use. Scientists believe that the ecosystem can be saved, however, with appropriate reclamation and restoration. To re-create natural conditions, the government is in the process of purchasing more than 100 000 acres (40 000 hectares) of land to the east and turning it over to the park.

Address: Everglades National Park; Ernest Coe Visitor Center, 40001 State Road 9336, Homestead, FL 33034; open 365 days a year. Flamingo Visitor Center is at 25°08′28.96″ N 80°55′25.73″ W; Shark Valley Visitor Center, 36000 SW 8th Street, Miami, FL 33194; Gulf Coast Visitor Center, 815 Oyster Bar Lane, Everglades City, FL 34139.
Phone: (305) 242-7700.
Days and hours: Open 24 hours. Visitor Centers' hours vary throughout the year.
Fee: Yes, per car.
Tours: Self-guided; ranger-guided walks.
Public transportation: None.
Handicapped accessible: Partial accessibility. Some trails are, others are not.
Food served on premises: Cafés, usually at the visitor centers.
Special note: Additional fees for boat tours and tram tours.
Website: www.nps.gov/ever/index.htm

KEY LARGO

John Pennekamp Coral Reef State Park

Some 200 separate islands and inlets, of 70 nautical square miles, make up the Florida Keys. Among the keys are a group of living coral-reef formations along the 221 miles of coast and three miles into the Atlantic Ocean. This is the only coral reef formation in the continental United States. These reefs form a fragile, complex ecosystem consisting of skeletal remains of corals and other animals and plants typically found in shallow, warm marine waters.

While studying marine species, Dr. Gilbert Voss, American conservationist, oceanographer, and biologist at the University of Miami, found damage to the reef and, in 1957, at a conference, he expressed his concerns about the future of the coral reefs along the Florida coast. He then organized a group of environmentalists, including journalist and newspaper editor John Pennekamp, in order to save this very special place. Pennekamp succeeded in the fight to protect this delicate marine ecosystem in the Florida Keys. In 1960 the area became America's first underwater park, John Pennekamp Coral Reef State Park and the adjacent Key Largo Coral Reef National Marine Sanctuary, which cover almost 190 nautical square miles (650 km^2) and together extend 21 miles (34 km) along the Florida coast, beginning about an hour's drive south of Miami. Because of its unique status, this park was listed on the National Register of Historic Places in 1972.

Coral, an invertebrate related to jellies, hydra, and the sea anemone, has a hard, stony skeleton and soft body. Corals are polyps living in a colony. A thin surface of living tissue, the mesoglea, gives them their color. Coral reefs form in water $75 \pm 2°F$ ($24 \pm 1°C$) when the skeletal remains of coral are glued together by limestone and calcareous algae. Algae and plankton are brought to this region by the seawater eddies of the Gulf Stream. Saltwater and the strong light of subtropical and tropical latitudes add to the slow process of reef-formation. It takes a thousand years for 1–16 feet (0.3–4.8 m) of coral reef to form. Hard coral is formed from polyp colonies; soft coral is formed from sea fans, sea whips, and sea plumes, which move gently in the tropical sea currents. Key Largo is the northern limit of reef formation.

The reefs form a habitat for thousands of animals and plants, some large and some microscopic. The ecosystem includes seagrass beds, mangroves, or marshy areas filled with trees that send down air roots, and tropical hardwood hammock communities. Hammock, a Native American word meaning "shady place," in English has come to mean a small island covered in forest trees.

The most popular sights at John Pennekamp Coral Reef State Park are the more than 40 coral formations, 300 species of fish including wrasses, tangs, parrotfish, damselfish, and angelfish, and 27 species of gorgonians, a marine life form similar to anemones. The park's waterways are populated with rare, endangered West Indian manatees, or sea cows, as they are commonly called, and sea turtles. (See Miami Seaquarium, Miami, Florida; and Everglades National Park, Homestead, Florida.)

There are several ways to view underwater life at the park. Several times a day, glass-bottomed boats take passengers on a guided tour of Molasses Reef with high coral ridges, tunnels, many types of coral formations, and a wide variety of reef fish. Scuba divers and snorkelers can rent equipment and arrange on-board rides for visiting the reefs. At French Reef, divers can investigate the remains of a deep cove and a 1942 shipwreck where large fish, rays, and moray eels are found.

A reconstruction of an early Spanish shipwreck, located 130 feet (40 m) offshore, provides a glimpse of a time when galleons sailed the Atlantic Coast, often carrying treasure. On display in the reconstruction are artifacts recovered from shipwrecks, including 14 cannons from ships that sank in a severe storm in 1715, an anchor from a 1733 shipwreck, and many ballast stones.

Pennekamp Park's 2300-plus acres (930+ hectares) of limestone uplands have many rare and endangered plants and large numbers of birds. The bird population is made up mostly of perching birds, particularly in the higher wooded areas of the Keys. Visitors can watch some of the nocturnal birds while they are active at dusk, night, and dawn. Among the hammock birds are warblers, bobolinks, and buntings. Follow the park's nature trails through mangrove swamps and sawgrass to see West Indian mahogany, strangler figs, and tropical vegetation such as gumbo-limbo, the only native Florida tree.

The Visitor Center has natural-history exhibits explaining the land and its wildlife, aquariums displaying local sea life, and slide shows. A 30 000-gallon (110 000-L) saltwater aquarium and several smaller displays with local sea life are interesting. The theater shows nature videos at regular intervals. Guided tours along nature trails are offered depending on the season.

Address: John Pennekamp Coral Reef State Park, 102601 Overseas Highway (mile marker 102.5), Key Largo, FL 33037.
Phone: (305) 451-6300.

Days and hours: Daily 8 a.m.–sunset. All park buildings are open 8 a.m.–5 p.m.

Fee: Yes. Fee for boat tours and other amenities.

Tours: Self-guided; ranger-led seasonally: contact the Visitor Center for schedules. Boat tours by reservation; call (305) 451-6300.

Public transportation: Miami-Dade Transit bus to Key Largo, then take a taxi 4.4 mi (7 km) to site.

Handicapped accessible: Mostly; contact the park for specific issues. Wheelchairs are available without cost; call ahead to reserve.

Food served on premises: Snack bar, picnic area.

Special note: Scuba diving available to certified divers by reservation. Snorkeling with training by reservation. For information on boat, scuba, or snorkeling tours, see the website.

Website: pennekamppark.com

MIAMI

Phillip and Patricia Frost Museum of Science

Founded in 1949 by the Junior League of Miami (a women's educational and charitable organization) as the Junior Museum of Miami, this museum now occupies its third home, a five-story complex opened in 2017 that is the only combined planetarium, aquarium, and science museum on the East Coast of the United States. The structure is one of several cultural facilities, including an art museum, concert hall, and sports arena, all situated near the waterfront with a picturesque view of Biscayne Bay.

The complex, designed by Grimshaw Architects, is composed of four structures: an aquarium, a planetarium, a North Wing, and a West Wing. Much of the structure is open to the air to allow sea breezes to cool the open-air sections of the aquarium and outdoor walkways between the indoor exhibits. The site works particularly hard to stress

Oculus at Frost Museum of Science.
Courtesy of Frost Museum of Science.

sustainability in its architecture, such as rainwater and gray water collection to water the rooftop gardens, as well as solar panels to power much of the building. Scattered throughout the museum are plaques (which visitors can follow as a self-guided tour) describing the environmental features of the structure.

The centerpiece of the architecture is the Gulf Stream Aquarium, an awesome display of various fish that populate the Gulf Stream in the Atlantic Ocean. The exhibit is a conical 100 000-gal (380 000-L) tank 100 feet (30 m) wide at the upper brim, constructed of 1200 cubic yards (920 m³) of concrete poured continuously over 25 hours. From underneath, visitors can view multiple species of simultaneously schooling fish, such as scalloped hammerhead, devil rays, silky sharks, mahi-mahi, rainbow runners, and blue runners, through a 31-foot (9.4-m) diameter Oculus, made of 13.5-inch (34.3-cm) thick acrylic. The circular viewport, slanted at a 21.5-degree angle, was crafted in Italy in eight segments that were baked together in a custom-designed oven.

There are other large outdoor aquarium tanks in the museum, primarily dealing with southern Florida marine ecosystems such as live coral reefs and Mangrove Forests. The Frost Museum is working with other institutions to grow coral reefs and deposit them in areas afflicted with bleaching to restore reefs. A touch tank allows visitors to gently handle stingrays. Also in the outdoor areas are an aviary and an exhibit where injured Eastern screech owls unable to be returned to the wild are shown. The museum operates a satellite facility, the Batchelor Environmental Center in East Greynolds Park in North Miami, where wild birds are rehabilitated after injury.

Indoor portions of the aquarium showcase a variety of marine life such as jellies, seahorses, urchins, and other creatures on a changing basis. A large *Xiphactinus* fossil from 85 million years ago, which lived in the seas that covered much of western North America, is on display. Several microscopes allow visitors to examine different types of sand from around the world up close for their structural and compositional differences. Other interactive stations demonstrate and teach marine ecology. An augmented-reality sandbox, or Topobox, includes a scanner and video projector that superimposes elevation contours and colors in real time on the sand. As visitors rearrange the miniature mountains and lakes by hand and learn about water flow and erosion, the topographic projection is updated.

Unique to the museum is a display describing the work of Dr. Roy Waldo Miner, a curator at the American Museum of Natural History, who sent a team in the 1920s to characterize the Andros Island reefs in the Bahamas. This group included sculptors in glass, model-makers, scientists, and painters. A display case of artifacts, models, early diving equipment, and biological drawings reveals much of what they found. (See American Museum of Natural History, New York City, New York.)

Feathers to the Stars is an exhibit showing how flying creatures (early insects, flying reptiles, and birds) eventually inspired humanity to build aircraft and rockets. Biological displays include a lifelike 30-foot (9.1-m) long reconstruction of *Yutyrannus huali*, the largest dinosaur known with fossil evidence of feathers. A computerized representation of airflow around various objects is projected on a long wall, a variety of aircraft (including an F-15 fighter) hang from the ceiling, and interactive activities explain lift and aerodynamics.

Design Lab: *Engineering* includes a two-part tour of the process of designing a tool or object. The first part compares early with modern iterations of everyday objects such as toasters, electric fans, or toys. The second part includes a Design Lab where participants attempt to solve engineering problems or puzzles. Within a health sciences area,

MeLaβ is an exhibit about human health and the mind. Five zones (relax, connect, eat, learn, and move) tell the story about the human body and its biological systems. There is a cabinet of curiosities displaying a variety of scientific instruments from the museum's collections.

The rooftop is an outdoor gallery with rooftop gardens (including milkweed and other plants to attract butterflies), two weather stations, and the Sun Spot exhibit that explains solar power. Note the quote by Thomas Edison on a rooftop structure: it is composed of reversed 10-inch (25-cm) aluminum letters that cast a shadow which varies according to the angle of the sun during the day. The words are attributed to Edison in a 1931 conversation with Harvey Firestone and Henry Ford. (See The Henry Ford, Dearborn, Michigan; and Edison and Ford Winter Estates, Fort Myers, Florida.) An observation deck provides a spectacular vista of Biscayne Bay and the Miami skyline.

A planetarium 67 feet (20 m) in diameter houses an 8K digital projection system presenting an array of space-based shows. In the planetarium's lobby is a touchable 240-pound (109-kg) iron meteorite from Argentina. Throughout the museum are changing exhibit galleries.

Address: Phillip and Patricia Frost Museum of Science, 1101 Biscayne Boulevard, Miami, FL 33132.
Phone: (305) 615-7092.
Days and hours: Daily 9:30 a.m.–6 p.m.
Fee: Yes. Behind-the-scenes tour with animal encounter costs extra.
Tours: Self-guided, and behind-the-scenes tour. Group tours (10 or more) with two-weeks' advance notice.
Handicapped accessible: Yes. Wheelchairs are available on a first-come, first-served basis.
Public transportation: Miami-Dade MetroMover and buses stop nearby.
Food served on premises: Café.
Special Note: The museum offers occasional adult evening programs. Call or see the website for details.
Website: frostscience.org

Miami Seaquarium

One of the first oceanariums, founded in 1955, is the Miami Seaquarium, located near downtown Miami. It is on Virginia Key along the Rickenbacker Causeway, a highway that connects Miami with Key Biscayne. On display are several types of dolphins, sea lions, a killer whale, endangered manatees, sharks, sea turtles, tropical fish, flamingos, an American crocodile, exotic birds, and other marine life. This spot was the setting for more than a hundred *Flipper* television episodes during the 1970s and for a motion picture about Flipper.

At the Golden Dome Arena, a geodesic structure designed by architect R. Buckminster Fuller, visitors watch sea lions perform while trainers discuss their typical behavior and natural environment. Visitors have a chance to feed the sea lions just outside the arena. The males are massive, more than 600 pounds (270 kg), while the females are usually a svelte 200 pounds (91 kg). (See Sea Lion Caves, Florence, Oregon.)

Whale Stadium presents shows in which killer whales, white-sided dolphins, and Risso's dolphins perform side by side. White-sided dolphins, which typically eat squid and small fish, can be identified by their playfulness and their fast, powerful swimming.

Risso's dolphins typically grow to 600 pounds (270 kg) or more and can be recognized by the characteristic crease in their foreheads. They are often found with scars on their backs from wounds inflicted by other dolphins. Killer whales, whose diet consists of seals, sea lions, penguins, other whales, and fish, can grow to 10 000 pounds (4 500 kg). They are highly intelligent animals that communicate with whistles, calls, and fin slaps.

Atlantic bluenose dolphins, very social, creative, bright animals, illustrate their native behaviors during shows at the Top Deck. These dolphins grow to an average of 10 feet (3 m) long and weigh more than 400 pounds (180 kg).

Other presentations that teach about marine life include daily shark-feeding accompanied by a narrator discussing sharks' features and habits. A display of harbor seals, 5-foot (1.5-m) long animals usually found in Northern Hemisphere waters, are on view. These animals have no external ears, and they have hind flippers behind rather than at the side of the body.

The main building houses more than 30 saltwater tanks containing Atlantic and Caribbean tropical species. A presentation at the Reef Aquarium describes the fragile coral reef environment and provides feeding demonstrations of fish and other marine animals.

The Seaquarium has acres (hectares) of lush landscaping, providing a "mini-Everglades" experience. (See Everglades National Park, Homestead, Florida.) Visitors can walk along a boardwalk through an ecosystem called Lost Islands Wildlife Habitat to explore mangrove forests complete with tropical birds, exotic fish, and turtles. Faces of the South American Rainforest, another ecosystem exhibit, displays such exotic creatures as iguanas and parrots.

The Life on the Edge exhibit has touch pools where visitors can handle starfish and other small sea animals that live in Caribbean reefs. Tidal Shallows exhibits stingrays, creatures that normally inhabit shallow waters. In the American Crocodile exhibit, visitors may sometimes see this reclusive endangered animal up close.

Among other special exhibits is the Sea Trek Reef Encounter, which allows visitors to don a diving helmet, and walk underwater through a 300 000-gallon (1.1 million-L) tropical reef to meet sting rays and various tropical fish. Penguin Isle is an 800-square-foot (74-m^2) dry spot combined with a 9000-gallon (34 000-L) pool for a colony of African penguins, which are adapted for warm weather.

The Seaquarium conducts research jointly with the Rosenstiel School of Marine and Atmospheric Science of the University of Miami and the National Marine Fisheries Service. (See Virginia Aquarium & Marine Science Center, Virginia Beach, Virginia.) Some of the research involves studies on the habits of turtles, dolphins, manatees, and whales. The Seaquarium is also part of the Marine Stranding Network and the US Department of Fish and Wildlife's efforts to rescue and rehabilitate marine mammals. (See John Pennekamp Coral Reef State Park, Key Largo, Florida.) The facility's rescue program includes a Manatee Halfway House for injured and orphaned manatees, where these gentle animals work to regain their strength before being released back into the wild. In conjunction with this rehabilitation program, the Endangered Manatee Exhibit provides educational presentations on manatees several times a day.

Address: Miami Seaquarium, 4400 Rickenbacker Causeway, Miami, FL 33149.
Phone: (305) 361-5705.
Days and hours: Daily 9:30 a.m.–6 p.m.

Fee: Yes. Many special tours cost extra.

Tours: Self-guided, and optional special tours.

Public transportation: Miami Dade Transit buses stop in front of the site.

Handicapped accessible: Mostly; certain special tours have accessibility restrictions.

Food served on premises: Cafés.

Special note: There are additional fees for the causeway toll and the monorail ride.

Website: www.miamiseaquarium.com/

TAMPA

Busch Gardens Tampa

Busch Gardens Tampa is one of a series of family theme parks originally owned and operated across the United States by the Anheuser-Busch Corporation, the well-known beer brewer. The Tampa Bay site is a 300-acre (121-hectare) entertainment center offering thrill rides, live entertainment, games, shops, restaurants, and, what is significant for this book, a major zoo collection with 12 000 animals of over 300 species. Originally opened in 1959, it is one of the first natural-habitat zoos in the country.

The park's predominant theme, carried through each of its eight parts, is Africa at the turn of the 20th century. There are also animal exhibits representing other parts of the world. The Bird Gardens area has the lush foliage favored by the 2000 birds that live here. More than 7000 birds have been hatched at this institution, which is famous for its captive breeding program of parrots. Prime attractions in this section are a walk-through free-flight aviary and Eagle Canyon, an exhibit with bald and golden eagles. The koala exhibit, Walkabout Way, features two kinds of koalas, Dama wallabies, rose-breasted cockatoos, and other Australian species. Flamingo Island has a flock of the birds most often associated with Florida fauna. Bird Garden Theater presents demonstrations of macaws, cockatoos, and birds of prey.

The Congo theme area contains Claw Island, rich with vegetation and surrounded by a river with a waterfall. This is the habitat for rare white and yellow Bengal tigers. Jungala features the Tiger Trail, including a glass turret to view the central area of the tiger enclosure.

The largest animal theme area, Serengeti Plain, has 65 acres (26 hectares) of grassy veldt where more than 500 African plains animals roam freely. These large animals include rhinoceroses, giraffes, many varieties of antelopes, elephants, and gazelles, all of which travel in herds. The site is a successful breeding and survival center for several endangered species, including black rhinos and Asian elephants. Visitors cannot walk through the veldt, but can take an air-conditioned monorail ride above it, view it from the sky ride that goes from one side of the park to the other, or tour it from the Trans-Veldt Railway, which is hauled by a steam locomotive. Each of these three rides provides informative commentary on the sights. For a special treat, take a behind-the-scenes tour that includes a safari truck ride through the Serengeti Plain. Visitors can have a leisurely view by having lunch or dinner at the Crown Colony House, which has a massive picture window overlooking the veldt. Visitors can also take a behind-the-scenes tour and learn how elephants are cared for.

The Crown Colony theme area includes Clydesdale Hamlet, where visitors can view these famous, massive horses. The Clydesdale breed was originally found along the Clyde

River Valley of Scotland, where these strong draft animals worked. Today Clydesdales are still prized for their size, strength, and beauty.

The Nairobi theme area contains Myombe Reserve: The Great Ape Domain, with three acres (1.2 hectares) of natural tropical habitat for gorillas and chimpanzees. At Nocturnal Mountain visit animals that normally wander about at night. Special lighting convinces these creatures that day is night, so you see them when they are most active. The area also includes a reptile exhibit with crocodiles. The Aldabra tortoise display features the last tortoise species left in the Indian Ocean. This species has a life span of about 100 years, can grow to 40 inches (102 cm) long, and typically weigh more than 500 pounds (230 kg). Here visitors see baby tortoises resulting from the park's breeding program. The animal nursery shows young animals ranging from fragile birds to gazelles, Asian elephants, and monkeys. Penguin Point showcases a colony of warm-adapted penguins from southern Africa.

In the Stanleyville area, visitors can see orangutans and warthogs, distinctive animals that have warts on their faces to protect them from injury during fights. At the Timbuktu Dolphin Theater, bottle-nosed dolphins display their intelligence and agility in the show Dolphins of the Deep.

Mobile carts staffed with knowledgeable personnel are strategically placed throughout the park providing visitors with information. The carts also have exhibits where visitors can touch live animals and other materials. A special walking tour of the veterinary facilities on-site is also available.

Address: Busch Gardens Tampa, 10165 N. McKinley Drive, Tampa, FL 33612.
Phone: (888) 800-5447.
Days and hours: Daily 10 a.m.–6 p.m. Open to 8 p.m. in the summer, although the park reserves the right to change hours as needed.
Fee: Yes.
Tours: Self-guided and guided. The site recommends that visitors book many tours in advance for guaranteed tickets.
Public transportation: Hillsborough Regional Area Transit stops about 0.4 mile (0.6 km) from the park, then walk. Mears Destination Services offers free (with purchase of park tickets) bus service from sites around Tampa to the park.
Handicapped accessibility: Yes.
Food served on premises: Restaurants and cafés.
Website: seaworldparks.com/en/buschgardens-tampa/

TITUSVILLE

Kennedy Space Center Visitor Complex, Merritt Island National Wildlife Refuge, and Canaveral National Seashore

These attractions are located on Merritt Island, and are sheltered from the Atlantic Ocean by Cape Canaveral, a peninsula along Florida's coast. Spanish sailors who landed here five centuries ago named the peninsula "Canaveral," the Spanish word for the cane or hollow reeds that grew on it so abundantly. Shortly after the 1963 assassination of President John F. Kennedy, the name was changed to Cape Kennedy; 10 years later it was changed back to Canaveral. Launches to space still take place from the cape.

Aerial view of Merritt Island National Wildlife Refuge.
Courtesy of Merritt Island National Wildlife Refuge.

Kennedy Space Center Visitor Complex

Today Merritt Island is home base for many of America's space flights. The Kennedy Space Center occupies the northern end of the island. A Visitor Complex has indoor and outdoor exhibits, demonstrations, movies, and interactive displays that explain the past, present, and future of America's space program. They are organized chronologically into Mission Zones.

Heroes & Legends traces the astronauts from the early years of NASA through the present. The Astronaut Hall of Fame honors the best space explorers from the Mercury missions to the Space Shuttle, as determined by the Astronaut Scholarship Foundation, including memorabilia from the early Mercury astronauts through the present day. Overhead is a Redstone rocket, named *Sigma 7*, and the *Gemini 9* capsules.

An outdoor Rocket Garden contains manned and unmanned rockets that once explored space, including a Saturn I B rocket similar to the one used on the first manned Apollo launch, as well as an access arm used by the Apollo 11 crew.

Bus tours go through areas of the working space center and launch pads. When launches are scheduled, for safety reasons an alternate tour visits the Vehicle Assembly Building and mobile launcher of NASA's Space Launch System. An extended tour also visits Launch Complex 39.

The Race to the Moon Zone showcases the Apollo era, using the Saturn V rocket. Visitors walk underneath a 363-foot (111-m) long Saturn V rocket, one of only three surviving examples. Firing Room Theater simulates the excitement of an Apollo 8 launch to orbit the Moon, using actual computer consoles. Lunar modules and rovers, a full-scale model of the 1970s Skylab, plus samples of Moon rocks, are displayed in this Zone. Lunar Theater recreates the final few minutes of the Apollo 11 mission landing on the Moon, using video clips and recordings from Mission Control. Other technological items of interest in this Zone are prototypes and training equipment for astronauts, the Apollo 14 capsule, Alan Shepard's space suit (covered with Moon dust), a Skylab mating adapter, and the "Astrovan," a van that brought astronauts to the launch pad from their quarters.

The Shuttle: A Ship Like No Other is the Zone describing the Space Shuttle reusable craft and other projects from the 1980s and 1990s. Displayed is the *Atlantis*, used for 30 years, and presented in the same way astronauts viewed her in orbit, tilted 43.21° with the payload doors open and robotic arm out. The Hubble Space Telescope Theater describes the orbiting telescope sent into orbit in 1990, and shows an exact full-scale model hanging from the ceiling. Astronaut simulators demonstrate some of the activities required by astronauts of that day, including piloting and docking the Space Shuttle. There is also a memorial to those astronauts killed during Space Shuttle missions, which includes some hardware recovered from those disasters. Modules of the International Space Station provides the visitor with a view of what life is like 300 miles (480 km) above planet Earth. On board, the space factory reveals how protein crystals are manufactured in space. The physiology laboratory documents the long-term effects of weightlessness on the human body. A salad machine grows vegetables in space. The space bath and toilet have a sealed shower and a vacuum to siphon away floating water droplets. Because the environment in the space station has only microgravity, all objects float, including water. A simulation of a Space Shuttle launch cements the experience of lift-off.

NASA Now & Next is the final Zone, displaying plans of current and future missions to the International Space Station and more. Scale models of rockets and space vehicles under construction are visible, planned missions to Mars, as well as images from current space observatories. There are mission status briefings in the Astronaut Encounter Theater. A six-foot (2-m) Science on a Sphere globe gives a spherical view of the planets in our solar system. An IMAX® theater shows 3D space-related movies daily.

Address: Kennedy Space Center, Route 405, Titusville, FL. Use GPS coordinates 28° 31′ 34.10″ N and 80° 40′ 45.12″ W.
Phone: (866) 737-5235.
Days and hours: Mar. 31–Sep. 1: 9 a.m.–7 p.m.; Sep. 2–Dec. 31: 9 a.m.–closing time varies.
Fee: Yes; Certain activities are extra.
Tours: Self-guided at the visitor center; guided bus tours of Space Center. There are many different activities, so plan in advance which to try during your visit.
Public transportation: None.
Handicapped accessible: Yes. Wheelchairs and scooters can be rented.
Food served on premises: Restaurants and snack stands.
Special note: Reserve early for bus tours.
Website: www.kennedyspacecenter.com/

Merritt Island National Wildlife Refuge and Canaveral National Seashore
Merritt Island, part of a chain of barrier islands, is located at the interface of temperate and tropical ecozones. It has plants and animals of both zones living side by side in the protected shelter of Merritt Island National Wildlife Refuge and Canaveral National Seashore.

Some of these creatures are unusual, and many are endangered and rare, including the Florida Scrub Jay, manatee, and loggerhead sea turtle. A number of migratory waterfowl winter here. Visitors can take a self-guided automobile tour of the wildlife refuge and follow walking trails with interpretive signs, as well as kayaking with local guides. The National Wildlife Refuge Visitor Center has exhibits and information for the public. Canaveral National Seashore's 24 miles (38 km) of coastal beach, dunes, and wooded areas are virtually unchanged since Ponce de Léon first saw this island in 1513. Visitors can drive along 12 miles (19 km) of the seashore; the other 12 miles (19 km) are accessible by foot.

Address: Canaveral National Seashore, 212 South Washington Avenue, Titusville, FL 32796.
Phone: (321) 267-1110.
Days and hours: Daily 9 a.m.–5 p.m.
Fee: Yes.
Tours: Self-guided.
Public transportation: None.
Handicapped accessible: Yes.
Food served on premises: None.
Special note: Summer visitors should expect biting insects, high temperatures, and frequent rain showers.
Website: www.nps.gov/cana/index.htm
Address: Merritt Island Wildlife Refuge, 1987 Scrub Jay Way, Titusville FL 32781. (This is not the true address, but from here go east over the Max Brewer Bridge. At .25 mile [0.4 km] past the bridge is the main entrance sign. GPS coordinates for the Visitor Center are Latitude 28° 38′ 29.28″ N, Longitude 80° 44′ 9.03″ W.)
Phone: (321) 861-0069.
Days and hours: Visitor Information Center is open daily 8 a.m.–4 p.m.; closed most federal holidays. Black Point Wildlife Drive and all other public areas, daily from sunrise to sunset.
Fee: Yes.
Tours: Self-guided. Local guides offer kayak eco-tours and fishing trips. See the Refuge website for the current listing of guides, and Facebook page for current schedules.
Public transportation: None.
Handicapped accessible: Yes.
Food served on premises: None.
Special note: Summer visitors should expect biting insects, high temperatures, and frequent rain showers.
Websites: www.fws.gov/refuge/Merritt_Island/ and www.facebook.com/MerrittIslandNWR

Georgia

ATLANTA

Atlanta Botanical Garden

Three very different aspects of the botanical world are featured at this 30-acre (12-hectare) oasis in the middle of Atlanta, next to Piedmont Park. Here visitors can enjoy nature in outdoor gardens, a hardwood forest, and a controlled-climate conservatory.

Northern Georgia's extended growing season allows a diverse collection to grow in the outdoor gardens, including herbs, perennials, roses, vegetables, annuals, trees, shrubs, and ornamental grasses. The display gardens are primarily designed to provide gardening ideas to homeowners. There is also the intimate Japanese Garden, and an Edible Garden including a kitchen set outside.

The Upper Woodland and Storza Woods together consist of 15 acres (6 hectares) of mature hardwood trees. As one of the few remaining forest areas in Atlanta, many of the trees are 100 years old. Recently restored walking trails provide visitors with the opportunity to enjoy the forest's sights and solitude. From the Botanical Garden's visitor center, take the Kendeda Canopy Walk for 600 feet (180 m) through the Storza Woods at an altitude of 40 feet (12 m) for a view of a woodland canopy.

The most spectacular part of the garden is the Dorothy Chapman Fugua Conservatory. This facility, opened in 1989 to honor the wife of local businessman J. B. Fugua, specializes in conserving rare and endangered plants from three climate zones: tropical, Mediterranean, and desert.

A lush, humid Tropical Rotunda has rare palms from Madagascar, the Mascarene Islands, and the Seychelles island group. The rotunda is a 50-foot (15-m) tall glass cylinder with a 14-foot (4.3-m) waterfall flowing over volcanic rock. Aquatic plants grow on the hardened lava; rare palms, endangered ferns, bromeliads, and other exotic species proliferate. Visible is a profusion of orchids, cycads, epiphytes, fragrant flowers, and flowering vines.

This facility's significant collection of tropical palms contains a 30-foot (9.1-m) *Bismarckia nobilis* specimen, many decades old, and three different species of *Neodypsis*,

Chihuly Fountain, Atlanta Botanical Garden.
Courtesy of Atlanta Botanical Garden.

including a triangle palm from Madagascar. The Mascarene Islands are represented by three species of *Hyophorbe*. *Verschaffeltia splendida* (a tropical palm with wide leaves and a spiny trunk) and *Lodoicea maldivica* (a double coconut palm seed) both come from the Seychelles. An endangered coconut palm seed is the heaviest, biggest seed in the plant kingdom and also one of the rarest and most difficult to grow.

The conservatory is proud of its massive orchid collection in the Fuqua Orchid Center. Some unusual specimens are the *Coryanthes speciosa* from Panama, which grows only in ant nests; the moth orchid *Phalaenopsis amabilis* from Borneo; the *Galendra batemanii*; and the *Cataseum pendulum* from Mexico. Cyclads are another important group in the collection. Among those worth special note are the *Encephalartos*, *Zamia* and *Macrozomia*, *Dioon*, and *Cycas circinalis*.

The conservatory's second climate zone, the Desert and Mediterranean House, contains botanicals from the Mediterranean Sea region; Africa; Australia; Chile; and San Francisco, California. Look for fragrant and culinary herbs, eucalyptus trees from Australia, *Metrosideros* from New Zealand, honey flower from Southern Africa, and conifers from Africa and the Canary Islands.

A third climate zone, the Desert Collection, displays two groups of succulents from a small desert on the island of Madagascar. Here are found all 11 known species of the Didieriaceae family of Madagascar succulents. Other species include tree-size *Euphorbia*, *Pachypodium*, and aloe. The section Old World Succulents includes such oddities as living stones, a group of plants that look like pebbles (an adaptation that helps protect them from grazing animals). The *Welwitschia* from the Namib Desert in Africa, another of the conservatory's very rare and endangered species, is a succulent that has remained unchanged for millennia.

The Specialty Exhibits area has more than 30 carnivorous plants. Predominant here are different forms of Nepenthes, the pitcher plant, which comes from Southeast Asia. *Nepenthes truncata*, the largest carnivorous pitcher plant, has been known to consume an entire lizard or frog. Sundews and Venus flytraps are also part of this group.

An Orangerie displays both rare and common tropical fruit trees: mango, papaya, star fruit, litchi, coffee, citrus, and many more.

Address: Atlanta Botanical Garden, 1345 Piedmont Avenue NE, Atlanta, GA 30309.
Phone: (404) 876-5859.
Days and hours: Apr.–Oct.: Tue.–Sun. 9 a.m.–7 p.m.; Nov.–Mar.: 9 a.m–5 p.m.; closed Mondays except for Monday holidays such as Labor Day, Memorial Day, and Columbus Day. Closed Thanksgiving, 12/25, and 1/1.
Fee: Yes.
Tours: Self-guided.
Public transportation: MARTA buses stop at the site.
Handicapped accessible: Yes.
Food served on premises: Restaurant, café, and snack bar.
Website: atlantabg.org

Fernbank Science Center and Fernbank Museum of Natural History

The Fernbank Forest was home of the Creek Native Americans, who lived on this land until giving up their rights to the land in an 1820 treaty. Today it is the location of two museums focusing on Georgia's geology and plant, animal, and sea life. It is extraordinary that this 65-acre (26-hectare) forest remained undisturbed as it went through a series of owners over the years. Colonel Z. D. Harrison, who died in 1935, was the last private owner of this property. The land first appears on official records as the Fernbank Forest in the early 20th century; it may have been named by Emily Harrison, the colonel's daughter. Harrison was a conservationist and environmentalist who dreamed of creating a "school in the woods." In 1939 she persuaded her family to protect and preserve the land as a living laboratory under the guidance of the nonprofit organization Fernbank, Inc. The Fernbank Science Center opened here in 1967 on four acres (1.6 ha) of land as an independent, supplementary facility run by the De Kalb County public school system, bringing teachers and students to a central location for science education. The Fernbank Museum of Natural History opened in 1992 by Fernbank, Inc., on the opposite side of the forest to educate the public about natural history.

Fernbank Museum of Natural History

After walking through Dinosaur Plaza toward the Museum of Natural History, with a bronze display of a family of *Lophorhothon atopus*, a variety of hadrosaur, the focal point of the museum building is the six-story, glass-covered atrium called the Great Hall. This atrium floor contains limestone tiles imported from Germany in which 150 million-year-old fossils are embedded, forming an exhibit in itself. From the hall, visitors have an excellent view across the Fernbank Forest. The Great Hall is filled with a number of casts of prehistoric reptiles (Argentinosaurus, Gigantosaurus, Pterodaustro, Anhanguera) from the Cretaceous period and fossils of other creatures that lived simultaneously. The originals are from Patagonia in Argentina.

The museum's theme is "A Walk Through Time in Georgia." Participatory exhibits illustrate Georgia's biodiversity and mountain-to-sea geology. Flora and fauna are set in dioramas exploring the state's geographic regions. Each of these dioramas represents a different season and time of day.

Jim Cherry Planetarium at the Fernbank Science Center. *Photo* by Tony Madden. *Courtesy* of the Fernbank Science Center.

A trip "across" the state begins in the Cosmos Theater with a movie on the big-bang theory of the world's birth some 15 billion years ago. Visitors then move on to exhibits emphasizing plants, animals, and dinosaurs from the Jekyll Island Pier to the Chattahoochee River. Subjects that the exhibits examine include how life began on the earth, the northern Georgia piedmont, the Appalachian Mountains, and ridges and valleys. The role of photosynthesis in algae and more complex plant life is discussed in depth.

Dioramas in the section called Life Develops in the Ancient Sea illustrate five geologic periods and describe the first simple life, how fossils are made, and the evolution of modern fish. Interactive computers offer additional information.

The Cumberland Plateau of northwestern Georgia is the setting for the Life Adapts to Land exhibit, which discusses the role of plants and insects. Triassic, Jurassic, and Cretaceous periods come to life in the Coastal Plain and Ruling Dinosaur Gallery. Here, models of long-extinct creatures fill a two-story area of the Okefenokee Swamp, which visitors "cross" via boardwalk. The scene begins in the early morning and finishes in the late evening.

The Rise of Birds and Mammals Gallery features a model of an 18-foot (5.5-m) tall megatherium, or giant ground sloth, browsing in the treetops. Human migration into Georgia, an important aspect of these exhibits, is illustrated on a large map.

Into the Future Gallery is a four-part re-creation of Georgia's coast and barrier islands. This exhibit ties together concepts from the other galleries and examines humankind's responsibility for the future of the earth.

Other highlights of the museum include the thousands of shells from around the planet that make up the World of Shells. Special pieces in this display are two examples of *Perotrochus maureri*, or Maurer's slit shell, discovered off the coast of North Carolina. A 1000-gal (3800-L) aquarium contains a living coral reef and specimens of giant clams, sea anemones, coral, and tropical fish.

This facility includes the Harris Naturalist Center, which encourages the public to bring in specimens for identification. Experts on site help visitors with technical questions. The Fernbank Science Center provides laboratory equipment, space for student research, experts with whom students can consult, and a planetarium, observatory, and library. For many years the science center has managed the official test rose gardens of the All-American Rose selection organization and the American Society Award of Excellence Gardens. They are part of the museum's landscaping, creating a beautiful, fragrant environment.

The Fernbank Forest is open to the public. Two miles (3.2 km) of paved walkways make it easy for the visitor to explore the forest and learn about forest life before human intrusion. Guide sheets help the layperson identify native animals and foliage.

Address: Fernbank Museum of Natural History, 767 Clifton Road NE, Atlanta, GA 30307.
Phone: (404) 929-6300.
Days and hours: Mon.–Sat. 10 a.m.–5 p.m., Sun. noon–5 p.m.
Fee: Yes.
Tours: Self-guided; group tours by prior arrangement.
Public transportation: MARTA buses stop about 0.2 mile (0.3 km) from the site, then walk.
Handicapped accessible: Yes.
Food served on premises: Café.
Special note: Extra fee for the Giant Screen Theater.
Website: www.fernbankmuseum.org/

Fernbank Science Center

In the Fernbank Science Center, an observatory, is open late Thursday and Friday evenings, weather permitting. An exhibit hall includes 21 of the 25 known meteorites to have fallen within the State of Georgia, along with an Apollo 6 Command Module dating from 1968. There are also galleries for temporary exhibits. Many of the center's 25 000 specimens are on display.

The observatory includes a 36-inch (0.9-m) Cassegrain reflector telescope. Inside a 70-foot (21-m) dome, the planetarium includes a classic Zeiss Mark V projector for sharp projections of the night sky, together with a Spitz SciDome 4K digital projection system for ancillary programs.

Address: Fernbank Science Center, 156 Heaton Park Dr., Atlanta, GA 30307.
Phone: (678) 874-7102.
Days and hours: Mon.–Wed. noon–5 p.m., Thu.–Fri. noon–9 p.m., Sat. 10 a.m.–5 p.m. Closed Sun. Observatory is open later on Thurs. and Fri. Planetarium programs run Thurs.–Sat. See website for detailed schedule.
Fee: Free. Fee for the planetarium.
Tours: Self-guided; group tours by prior arrangement.
Public transportation: MARTA buses stop 0.4 mi. (0.6 km) from site, then walk.
Handicapped accessible: Yes.
Food served on premises: None.
Website: www.fernbank.edu/

BRUNSWICK

Hofwyl-Broadfield Plantation State Historic Site

This plantation, in what used to be rice country, came into being when William Brailsford and his son-in-law, James M. Troup, purchased a parcel of virgin cypress swamp in about 1806 and named it Broadfield. Rice grew here successfully from 1806 to 1915. In the mid-19th century, Ophelia Troup and her husband, George Dent, built a house, the same one visible today, and added Hofwyl to the name in honor of the Swiss school George had attended. Visitors can tour the house and outbuildings, explore the grounds, and view exhibits in the Visitor Center.

The Hofwyl-Broadfield Plantation, on a narrow stretch of land along the Altahama River, is an ideal spot for rice propagation. The river carries fertile topsoil from the Piedmont to the floodplain, enriching the fields regularly. The water supply for irrigation comes from a tidal flow that reaches about nine miles (14 km) upstream. The local warm, rainy weather is also favorable for growing rice.

Rice cultivation during the Dents' day was labor-intensive and heavily dependent on slaves. Plantation owners preferred slaves from West Africa because many came with expertise in growing rice, a crop that fares well in West Africa's heat and humidity. Also, many of the West African slaves had sickle-cell anemia, which made them immune to malaria, a parasite carried by the mosquitoes that proliferated in the swamps. Back then, malaria was believed to be caused by poisonous swamp vapor or "miasma." Because the plantation owners and their families were susceptible to malaria, they would flee the tidal low country during mosquito season, late spring through the first frost, and leave the plantation's operation to the slaves and an overseer. Without these slaves the plantations would likely have failed.

Hofwyl-Broadfield grew rice steadily and successfully until the Civil War. During that period, the war, natural disasters, and loss of slave labor caused the rice empire to collapse. With the introduction of more efficient machinery better-suited to other areas, rice cultivation here became unprofitable. This site was last planted with rice in 1913. It was converted into a dairy farm, and by 1942 fell into disuse. With the death of the granddaughter of Ophelia Troup Dent in 1973, the site went to the State of Georgia, as specified in her will.

The Visitor Center has a 17-minute video that provides an excellent introduction to the plantation and the process used to cultivate rice on the grounds. The center also has exhibits, family memorabilia, tools, and objects that help illustrate the region's rice culture. Many of the tools were made by slaves and are based on West African equipment. An interesting piece of rice technology on display is a model of a floodgate, or trunk, that illustrates how water flowed in a controlled manner onto rice fields.

Before undertaking the one-mile (1.6-km) walk around the grounds, keep in mind that the hot, humid conditions that made rice cultivation so successful are not always kind to people. Insect repellent and sunscreen are suggested during the late spring and summer. Fire ants and snakes are in abundance, so keep a watchful eye.

Walking toward the plantation, the visitor passes the rice-field dikes, some of which were installed in the early part of the 19th century by William Brailsford. The dikes used by the plantation for both irrigation and transportation, and the floodgates that controlled water-flow to the rice fields, are visible. Now the area is a freshwater marsh and canal. The dikes, as well as the rice mill (now in ruins) and canals, were constructed by hundreds of slaves over many years.

The modest 1850 plantation house has been modernized over the years. In 1903, Dr. James Troup Dent, who then owned the site, added screens to the windows, porch, and fireplace to keep flies and mosquitoes out, and thus help prevent malaria. Interesting architectural features include a pointed chimney to keep rainwater out, and jib doors for better summer ventilation. Original furniture, some made in Philadelphia and South Carolina, dates from the 1790s.

Behind the house are outbuildings, including a milking barn and bottling house used during the dairy-farm days. At the peak of the dairy farm's prosperity, about 35 Jersey and Guernsey cows produced 100 to 150 bottles of milk daily. Electric milking machines were introduced in the 1920s. This dairy and several others in the area closed when tighter government controls on milk-processing went into effect.

Visible also are the pre–Civil War slaves' quarters, built in customary duplex style. When the slaves were granted their freedom, many stayed on the plantation and were paid in cash for their services. The pay shed and commissary, or plantation store, both built after the Civil War, still stand, as does the ice house. The brick fence near the ice house kept cattle from brushing against laundry hung in the sun to dry.

The plantation is now on a wildlife preserve. Cattle and some wild animals roam the property, helping to re-create the time when it was a dairy farm. Several marked nature trails cross the narrow piece of land between the saltwater marshes and the pinelands. The site (part of the Colonial Coast Birding Trail) is known for bird-watching (including painted buntings, ibis, herons, and egrets).

Address: Hofwyl-Broadfield Plantation Historic Site, 5556 US Hwy 17 N., Brunswick, GA 31525.

Phone: (912) 264-7333.

Days and hours: Wed.–Sun. 9 a.m.–5 p.m. Closed Thanksgiving, 12/25, and 1/1.

Fee: Yes.

Tours: Self-guided at Visitor Center, grounds, and outbuildings; only guided tours in the plantation house.

Public transportation: None.

Handicapped accessible: Limited.

Food served on premises: None; picnic area.

Website: gastateparks.org/HofwylBroadfieldPlantation

JEFFERSON

Crawford W. Long Museum

Jefferson, a town founded and laid out in the early 19th century, was formerly the Native American village of Thomacoggon. Because of its convenient location and four springs providing the area with a good water supply, it became the Jackson County seat and was renamed after the country's third president. The farming community prospered as a stop for travelers heading west through Georgia. Here a young country doctor, Crawford W. Long, set up his practice in 1832.

About 10 years later, on March 30, 1842, Dr. Long used ether to carry out what many people believe was the first painless surgical operation. (Some historians contend that nitrous oxide, or "laughing gas," was the first anesthesia used in surgery.) To commemorate Long's work, the citizens of Jefferson established this museum on the site where Dr. Long performed his history-making surgery. (See International Museum of Surgical Science, Chicago, Illinois.)

In addition to telling the story of Long's discovery, the museum presents the histories of Jackson County, 19th-century medicine, and the development of modern anesthesia. This museum occupies three interconnecting historic buildings, two of which were built in the 20th century. The entrance to this facility is through the building that once served as the medical office of physician James Stovall from 1934 to 1957. During the 1987 museum expansion, his small one-story brick building was incorporated into the museum. A visitor center and several displays are located here.

Exhibits in the adjoining two-story building constructed in 1879 explore Crawford W. Long's work. His idea for anesthesia evolved from the ether-inhalation parties popular during the early 19th century, commonly known as Ether Frolics. During one of these parties attended by Long, several party-goers, while in an ether stupor, were injured but seemed to feel no pain. This prompted Long to try using ether to deaden pain during surgery. He tested its effectiveness while removing a cyst from the neck of patient James Venable. This exhibit area includes a diorama depicting this first painless surgery, as well as Long's personal and family items.

The discovery of anesthesia was one of two key factors leading to modern surgery; the second was antiseptics. Before these techniques were used, surgery patients were likely to die of either shock caused by pain, or infection from unsterile procedures. Members of the Georgia Society of Anesthesiologists and the Medical Association of Georgia have joined forces with the museum to present an exhibit on the history of anesthesia from ancient times to the present. Several inhalation machines and historic equipment are on display, such as a 1914 Gwathmey Foregger, nitrous oxide-ether apparatus, and a 1930s Connell's anesthesiometer apparatus with a flow-meter. Karl A. Connell developed machinery to specifically measure and mix ether vapors with other gases in accurate measurements. An anesthesia machine of the late 1960s, displayed in cutaway fashion, shows its inner mechanisms. Also on exhibit is the Ohmeda Excel 2001 machine. This piece of equipment is of importance to the development of anesthesia because of its excellent ventilation capacities. To complete this exhibit is a video about anesthesia and Dr. Long.

The Pendergrass Store, from the 1840s, is the only pre–Civil War commercial building in Jefferson. Dr. Long's office stood near the rear of the store and was the site of the first painless surgery. Here there is a re-creation of an apothecary shop and Dr. Long's office, as well as several other period exhibits, including one on medicine using plants. The Store also houses temporary exhibits.

Address: Crawford W. Long Museum, 28 College Street, Jefferson GA 30549.
Phone: (706) 367-5307.
Days and hours: Tue.–Fri. 10 a.m.–5 p.m., Sat. 10 a.m.–4 p.m. Closed on major holidays.
Fee: Yes. Optional audio-tour rental available. Docent-led tours incur an extra fee.
Tours: Self-guided; guided tours are available with advanced booking.
Public transportation: None.
Handicapped accessible: Limited.
Food served on premises: None on site. There are restaurants in town nearby.
Website: www.crawfordlong.org/

North Carolina

ASHEBORO

North Carolina Zoological Park

Asheboro, a small city in the rural Appalachian Mountains, near the center of the state, has the largest natural-habitat zoo in the world. This 2200-acre (890-hectare) zoo, which is both a research site and a facility for educating the public about animal conservation, is part of the North Carolina Department of Natural and Cultural Resources, and is one of only two zoos in the country sponsored by a state. The complex opened in 1974. Presently the animal population consists of more than 1600 animals of more than 250 species, primarily representing Africa and North America, including the largest troop of baboons in the United States. This population will eventually expand to include more animals, representing other continents and seas of the world.

This zoo is designed with a "natural habitat" philosophy, one of the first in America to implement this scheme. It presents animals living together with plants resembling those they would live with in the wild, using a much larger expanse of land than typical zoos encompass. Such a philosophy seems to have reduced the behavior problems of the animals.

The Africa exhibits and the North America exhibits are located on different sides of the zoo, with a Desert and Aviary between the two. There are two parking lots, one at each location of these two major exhibits. Some five miles (8 km) of walking paths weave their way across the zoo. Trams and air-conditioned buses take visitors through the park, stopping at important sites along the way. The natural habitats contain either native plant species or local plant look-alikes.

The Forest Aviary, an 18 000-square-foot (1700-m^2) structure capped by an acrylic dome, encloses a carefully landscaped tropical forest of more than 3000 plants, including palms, mangroves, and ficus. The aviary has about 140 exotic birds from the world's equatorial areas, including the eclectus parrot, sunbittern, scarlet ibis, and Pekin robin.

The 3.5-acre (1.4-hectare) Forest Edge represents equatorial Africa's grasslands and savannas. Plant-eaters roam this habitat, such as reticulated giraffes and Grant's zebras. Rhinoceroses, lions, and the endangered African elephants live in separate grassy areas.

A band of lowland gorillas inhabit a forest clearing, which is where these animals tend to gather naturally. Lush vegetation provides food for foraging and allows gorillas some needed privacy. Kitera Forest, a chimpanzee habitat, a wooded half-acre (0.2 hectare) with tall grass, has a 25-foot (7.6-m) interactive climbing tree. The zoo has one of the largest troops of chimpanzees in the United States.

The Watani Grasslands Reserve resembles an African veldt. Twelve species of African hoof stock, including herds of greater kudus and gemsboks, graze on 40 acres (16 hectares) of rolling grasslands. African elephants share two large spaces in the Watani habitat. It is worth stopping at the hamadryas baboon island habitat to see the baboons' antics.

Elephant herd on the Watani Grasslands at the North Carolina Zoo.
Courtesy of the North Carolina Zoo.

North America is composed of numerous exhibits on landscaped with hundreds of plants. Here are animals from the Arctic Circle all the way south to Mexico's Sonora Desert.

A boardwalk meanders through the Cypress Swamp and its thick vegetation: bald cypress trees, hollyberries, and rhododendrons. The visitor passes an alligator pond, glass aquariums, and terrariums with turtles and other amphibians. The swamp also has wild ducks and cougars. In the carnivorous-plant garden there are insect-eating pitcher plants and Venus flytraps.

A natural extension of the swamp is the Marsh area, with lagoons containing wild rice, cattails, bulrushes, water lilies, and grass-covered islands. Bald eagles and native wildfowl live here.

The two-building complex, Streamside, has a re-created a North Carolina stream that flows from high in the mountains down to the Atlantic Coast. A bobcat lives on rocky ledges and game fish such as trout swim in their own tank.

The Rocky Coast exhibit has polar bears, big-beaked puffins, and California sea lions, as well as Arctic foxes scampering among the gray rocks that make up their lair. Peregrine falcons fly over the rocks and crevices. Polar bears live in a replica of the Arctic tundra landscaped with pines and junipers.

Desert Pavilion is filled with spiny cacti and slithering snakes. The dry world of the Sonora Desert exhibit is divided into four biomes: Desert Flats, displaying desert-floor plants and animals; the Uplands, filled with cacti and appropriate animal life; Desert Canyons, with streams, toads, and pupfish; and Desert Night, where nocturnal desert rovers such as the sand cat can be found. Desert tortoises and Gila monsters are among the many other animals seen here. Across from the Sonora Desert, the Honey Bee Garden teaches visitors about bees and beekeeping with a live honeybee hive.

The Prairie is a 10-acre (4-hectare) grass-filled area where deer, elk, and bison roam. Nearby are black and grizzly bears and gray wolves. Black and grizzly bears swim nearby in a series of pools in the midst of aspen and spruce trees. Endangered red wolves occupy their own lair and watering pond.

Throughout the zoo, visitors can find distinctive yellow "Smart Carts," staffed by well-trained docents, to help you understand the exhibits. Because the habitats at this zoo are so large and the walking so extensive, it is a good idea to take the free tram when traveling between the North America and Africa exhibit areas.

Address: North Carolina Zoological Park, 4401 Zoo Parkway, Asheboro, North Carolina 27205.

Phone: (800) 488-0444.

Days and hours: Apr.–Oct.: 9 a.m.–5 p.m.; Nov.–Mar.: 9 a.m.–4 p.m. Closed 12/25.

Fee: Yes.

Tours: Self-guided. Guided tours available through the North Carolina Zoo Society.

Public transportation: None.

Handicapped accessible: Yes (except for Zoofari and 4D theater).

Food served on premises: Restaurants and snack bars; picnic area.

Special note: Free tram crosses the site; regular shuttles run between the two parking lots. Extra fee for Zoofari tours.

Website: www.nczoo.org/

CHARLOTTE

Discovery Place

One of the nation's leading comprehensive science and technology museum organizations is found in Charlotte, the center of North Carolina's manufacturing base and the largest metropolitan area in the state. Discovery Place, a model for other science museums around the country, promotes the idea that learning about and understanding science is fun for people of all ages. It is one of the few science museum organizations that specifically reach out to senior citizens. The museum was selected by the Kellogg Foundation and the Smithsonian Institution to examine the educational role of museums in the community and to develop prototype programs to enhance education in schools and universities. The philosophy here is that people learn best by doing. Visitors are encouraged to touch and manipulate materials throughout the museums.

In 2016 the organization reconstituted itself as three "brands": Discovery Place Science, Discovery Place Nature, and Discovery Place Kids. This book concerns itself with the first two "brands," Discovery Place Science, and Discovery Place Nature.

Discovery Place Science

Materials and their properties are emphasized in the Thinker Space workshop and laboratory, where "doing" skills are emphasized, ranging from woodworking and sewing, to soldering, laser cutting, and 3D printers. The Explore More Stuff is a laboratory for people of all ages to explore forces of physics and chemistry dealing with matter. Unusual physical phenomena are discussed in the Cool Stuff gallery. Here are thermography cameras registering infrared waves, vacuum-powered technology, gyroscopes and rotational inertia, pulleys, wind-power, and more.

Being Me is a journey through anatomy and health of the human body, including exercise in The Big Wheel (a giant hamster wheel that displays distance, speed, and heart rate); the Human Mirror, which captures movement and creates an image of the body's systems; near-infrared light, used in the VeinViewer to see blood vessels in the human hand; and plastinated human bodies and organs.

Explore More Life is a laboratory devoted to exploring biodiversity, including a touch tank populated with horseshoe crabs and sea urchins; other displays show hissing cockroaches and albino axolotls. Often visitors can perform an actual dissection. An entire exhibition hall devoted solely to frogs is the Fantastic Frogs gallery, with live specimens, and wall- and panel-illustrations in graphic-novel style.

The museum's aquariums illustrate a variety of marine environments and include creatures such as eels, nurse sharks, and groupers. Regional marine habitats contain local fish life. A wave tank demonstrates the action of waves, and an ocean touch pool has starfish and other marine life.

Various habitats explore biodiversity in the World Alive hall. Here is an aquarium with a reconstructed Indo-Pacific coral reef populated with tropical fish, giant clams, and puffers. On display are jellies and cuttlefish. An exhibit of the North Carolina coast houses sea stars, oysters, and urchins. A rainforest includes birds, reptiles, frogs, turtles, and more.

A more generic exhibit with no directions offered is Think It Up. Here visitors must solve problems on their own, involving light sensors that emit noises when touched,

LED displays that vary depending on dials turned, and materials that may float or sink when dropped into Flutter Tubes. An exhibit dedicated to the various inventions that Leonardo da Vinci described is Da Vinci's Machines, in which more than 75 of these creations, ranging from cars to diving equipment to flying machines, are reconstructed.

The Explore More Collections Gallery has a very large collection of minerals, shells, insects, and anthropological artifacts, not all of which are displayed at any one time. About 300 different minerals come from North Carolina, 75 of commercial quality. North Carolina has one of the world's largest surface quarries for granite. It is also one of the few places in the United States that have rubies and other precious stones in the ground. North Carolina gemstones at the museum include the Reverend Bruce Owens quartz collection, each piece acquired and hand polished by Owens. At an observation area, visitors can watch mineral specimens being prepared for exhibit. Visitors bring sea shells and minerals to the Trading Post to exchange them with other collectors or to get expert help in identifying samples.

A three-story dome-covered wing contains an OMNIMAX theater. A variety of science-related live shows, from chemistry to biology and physics, occur regularly throughout the museum.

Address: Discovery Place Science, 301 North Tryon Street, Charlotte, NC 28202.
Phone: (704) 372-6261.
Days and hours: Sun. noon–5 p.m., Mon.–Fri. 9 a.m.–4 p.m., Sat. 10 a.m.–6 p.m. Closed 2/1, Easter Sunday, Thanksgiving, 12/24, and 12/25.
Fee: Yes.
Tours: Self-guided. Staff members give a variety of lectures and demonstrations at intervals; check schedule upon arrival.
Public transportation: Local bus stops 0.2 mile (0.3 km) away from site, or take the LYNX train to 7th Street Station, then walk.
Handicapped accessible: Yes.
Food served on premises: Café.
Special note: Additional fees for the OMNIMAX theater.
Website: www.discoveryplace.org/

Discovery Place Nature

Discovery Place Nature was the original home of Discovery Place, formerly called Charlotte Nature Museum, next to Freedom Park. Here are exhibits of small creatures of the world and naturalists who talk of their characteristics and habits. A Butterfly Pavilion houses species including white peacocks, monarchs, and swallowtails. The Naturalist Lab is a gallery displaying a variety of natural specimens, both biological and geological. Insect Alley features the life and diversity of insects.

The Charlotte A. Kelly Planetarium hosts regular shows using a Digitarium® Zeta projection system. There are also daily live animal feedings and encounters in the museum.

Outdoors is the Paw Paw Nature Trail, which winds through a century-old woods, where visitors may see Jack-in-the-pulpit or box turtles, and hear cicadas.

Address: Discovery Place Nature, 1658 Sterling Road, Charlotte, NC 28209.
Phone: (704) 372-6261.

Days and hours: Sun. noon–5 p.m., Mon.–Fri. 9 a.m.–4 p.m., Sat. 10 a.m.–6 p.m. Closed 2/1, Easter Sunday, Thanksgiving, 12/24, and 12/25.
Fee: Yes.
Tours: Self-guided. Staff members give a variety of lectures and demonstrations at intervals; check schedule upon arrival.
Public transportation: Local buses from the Charlotte Transportation Center stop within 0.5 mile (1 km) of the site.
Handicapped accessible: Yes.
Food served on premises: None, except for picnic area. Nearby restaurants are available.
Website: www.discoveryplace.org/

KILL DEVIL HILLS

Wright Brothers National Memorial

Wilbur and Orville Wright, who launched the world into the age of aviation, are memorialized at a site along the barrier islands along the North Carolina coast. This site is north of Cape Hatteras National Seashore and just south of Kitty Hawk. These long, thin stretches of beach, sand dunes, marshes, and woodlands, known as the Outer Banks, have been subject to the ravages of wind and sea over the ages. The Wright brothers, from Dayton, Ohio, chose to do their flying experiments in this area because, as they described it, the site is "absolutely bare of vegetation and flat as a floor from ocean to sound, for a distance of nearly five miles, except for a series of sandwich hills, the largest of which is one hundred and five feet high." They took off near the base of Big Hill of the Kill Devil Hills sand dunes, four miles (6.4 km) south of the tiny town of Kitty Hawk. (See Carillon Historical Park, Dayton, Ohio.) When the first efforts to honor the Wrights began in the 1920s, it was discovered that Big Hill had migrated about 50 yards (46 m) since the Wrights' first flight.

Wilbur and Orville had no formal training in engineering. They became interested in aviation while selling, repairing, and manufacturing bicycles. They read material available about the theories of flight and communicated with many experts in the field. As their knowledge grew, they devoted the rest of their work lives to aviation.

Because weather conditions in the Dayton area were not the best for flying, they studied national weather-bureau records and chose the Kitty Hawk area for their flying experiments. Originally they erected tents at Kill Devil Hills in which to work. Later they constructed more substantial sheds of wood, and eventually they built wooden living quarters.

In 1902 the Wright brothers undertook almost a thousand glider flights at the site. Although Wilbur had piloted gliders earlier, Orville had never tried flying until taking part in these tests. Together the brothers perfected the controls and the art of free, controlled flight—that is, flight directed by the pilot rather than the wind. In controlled flight, the pilot lay prone with his head forward, while his left hand operated the elevator lever. His hips were in a saddle, which could, with one side motion, control the wing tips (balance) and the rudder (steering). This three-axis control system, enabling the pilot to steer the aircraft and maintain equilibrium, is still in use today and remains the current standard for fixed-wing aircraft. On December 17, 1903, the brothers returned to Kitty Hawk and achieved the first powered flight of a heavier-than-air aircraft. Although

West View of Wright Brothers National Memorial.
Courtesy of the National Park Service.

others soon built aircraft, the Wright brothers chose only to pilot airplanes they designed themselves or for which they had built the engines.

The Visitor Center at the site contains a reproduction of the 1903 Wright Flyer—the "heavier-than-air" craft. The Wright Flyer, which reached an altitude of 120 feet (37 m) in its 12-second historic flight, weighed a mere 605 pounds (274 kg), had a wingspan of slightly more than 40 feet (12 m), and was 21 feet (6.4 m) long. The original airplane is in the National Air and Space Museum of the Smithsonian Institution. (See Smithsonian Institution, Washington, DC.)

Because the brothers could not a get a gasoline-powered motor manufactured to their specifications for the 1903 Flyer, their machinist, Charlie Taylor, built one, which is on display, and reconstructed here. They also designed and built the propellers. (See The Henry Ford, Dearborn, Michigan.) The visitor can also see a model of the first wind tunnel, designed and constructed by these two aviation pioneers.

Some of the exhibits at the center chronicle earlier ventures of the two brothers, such as their newspaper business and the bicycle shop. Others explain the principles of aeronautics that they discovered. A small exhibit honors other first flyers, such as Edwin Aldrin of the first lunar mission, and Theodore Gordon Ellyson, the first Navy aviator. (See Virginia Air and Space Center, Hampton, Virginia.)

The 60-foot (18-m) pylon on Kill Devil Hills memorializes the first flight. A short walk from the center across the dunes leads to a reconstruction of the Wrights' 1903 camp and flight area, where the reproduction wooden skids used to launch the plane are visible. One wooden building re-creates their hangar, and another their workshop and cramped living quarters.

While in the area, be sure to visit the lighthouses on each of the three islands within the national seashore: Bodie Island, Hattera, and Ocracoke. Pea Island National Wildlife

Refuge, between Bodie Island and Cape Hatteras, has observation platforms and walking trails from which to view waterfowl and other birds.

Address: Wright Brothers National Memorial, 1000 North Croatan Highway, Kill Devil Hills, NC 27948.
Phone: (252) 473-2111.
Days and hours: Daily 9 a.m.–5 p.m. Closed 12/25.
Fee: Yes.
Tours: Self-guided. There are also daily ranger-led tours of the facilities. Check the website or call for exact times and days.
Public transportation: Taxi service only.
Food served on premises: None. There are a number of restaurants in the Kill Devil Hills area, mostly to the northeast and southeast of the Memorial grounds.
Handicapped accessible: Yes.
Website: www.nps.gov/wrbr/index.htm

RALEIGH

North Carolina Museum of Natural Sciences

The iconic "Daily Planet" gigantic model of the Earth, on the external façade of the North Carolina Museum of Natural Sciences, marks the focus of this site: the natural history of the Earth, with its mission questions of "what do we know?" "how do we know?" "what's happening now?" and "how can you participate?"

The North Carolina State Legislature created the North Carolina State Museum in 1879 as a merging of the state's geology and agriculture collections. Shortly thereafter, two English immigrants, Herbert Hutchinson Brimley and Clement Samuel Brimley, came to North Carolina, and began work as chief collectors. Herbert Brimley was the curator for the Museum, and the brothers continued their efforts over the next six decades, until their deaths in 1946. The museum expanded its collection over the coming half-century against the confines of tight spaces. The museum was reopened in enlarged quarters in 2000 as the North Carolina Museum of Natural Sciences, billing itself as the largest natural history museum in the southeastern United States. A Nature Research Center was added as a new wing in 2012, rebranding the earlier museum building as the Nature Exploration Center.

Currently the Museum's world-spanning research collections include over 260 000 specimens of reptiles and amphibians, 7500 samples of rocks and minerals, 1.3 million fish, 56 000 paleontological invertebrates, 18 000 mammals, 500 000 meteorites, 250 000 non-mollusk invertebrates, 20 000 birds, 1600 ancient botanical objects, and 63 000 vertebrates of paleontological status. Research at the site includes astronomy, biodiversity, genomics, geology, and paleontology.

In the Nature Exploration Center, various exhibits discuss and display aspects of the North Carolina ecosystems, from the Atlantic shores to inland mountains, including wildlife, prehistory, and geology. Natural Treasures of North Carolina showcase the Venus flytrap, a carnivorous plant native to this state, along with the extinct Carolina parakeet. Coastal North Carolina explains to visitors about barrier islands, maritime forests, saltmarshes, and sand dunes, including tanks of native species of fish and

Exterior of North Carolina Museum of Natural Sciences. *Photo* by Eric Knisley.
Courtesy of the North Carolina Museum of Natural Sciences.

invertebrates. A Coastal North Carolina Overlook affords the visitor of views of skeletons of a blue whale, a True's beaked whale, and a sperm whale. Underground North Carolina features the geological and natural history of this state, from soil and creatures that live in it, to various gems and minerals found in the area. Mountain Cove is a display of various plants and animals—many nocturnal—that inhabit North Carolina. Many of the nearly 40 species of snakes native to the state are on exhibit in the Snakes of North Carolina section.

The Terror of the South is the sole real fossil skeleton of the *Acrocanthosaurus*, an early Cretaceous bipedal predator similar to the Allosaurus (see American Museum of Natural History in New York, New York, for a cast of this fossil). Other paleontological specimens (*Thescelosaurus*, a late Cretaceous ornithopod; and a giant ground sloth) are shown in the Prehistoric North Carolina exhibit.

North Carolina is also a transition area from the tropics to the more northerly temperate ecosystems, as detailed in the Tropical Connections area of the museum. Here visitors get a glimpse of various hummingbirds and butterflies that migrate, as well as native species of emerald tree boas and poison dart frogs. The Living Conservatory is a re-creation of a tropical dry forest, including butterflies, turtles, sloths, and fish. An Arthropod Zoo is an exhibit including live insects and horseshoe crabs, plus enlarged models of these living creatures.

The Nature Research Center's focus is on the tools and methods scientists use to interpret the natural world. The largest section of this wing is the Daily Planet, a three-story theater with a 42-foot-tall (13 m) screen displaying various items of information and daily shows. The Micro World Investigate Lab is a hands-on laboratory demonstrating microscopic techniques for learning about DNA, photosynthesis, bioluminescence, and the like. The Visual World Investigate Lab shows off how scientists visualize ideas, and includes 3-D printing plus virtual reality. The Natural World Investigate Lab offers visitors techniques for investigating the natural world.

A 10 000-gallon (38 000-L) aquarium is a model of a hardbottom habitat off the coast, displaying a variety of fish, including balloonfish, angelfish, ladyfish, and lookdowns. Other aquatic exhibits include the Exploring the Deep Sea, modeling the Atlantic sea bottom, and Investigating Right Whales. Researching Weather shows how meteorologists predict and analyze the weather. Ancient Fossils, New Discoveries highlights mostly non-dinosaur fossils. Unraveling DNA discusses the helical molecule central to biological inheritance.

The Museum has a satellite campus at Prairie Ridge (with 45 acres [0.18 km²] of a Piedmont prairie), wetlands, trails, and ponds for hiking and exploration.

Address: North Carolina Museum of Natural Sciences, 11 West Jones Street, Raleigh, NC 27601.
Phone: (919) 707-9800.
Days and hours: Mon.–Sat. 9 am–5 pm, Sun. noon–5 pm. Closed Thanksgiving Day, 12/24, 12/25, and 1/1. The Living Conservatory is closed on Mon. The Naturalist Center and iLabs close one hour before the rest of the museum. The Naturalist Center is closed on Mon.
Fee: Free.
Tours: Self-guided.
Public transportation: Amtrak trains reach Raleigh. Local buses and taxis service downtown.
Food served on premises: Cafés.
Handicapped accessible: Yes.
Special note: Many of the Labs have classes when they are closed to the general public. Call in advance for schedule.
Website: naturalsciences.org/index

South Carolina

CHARLESTON

Middleton Place

This colonial plantation, now a National Historic Landmark, remains in the hands of the original Middleton family. Mary Williams brought the property to Henry Middleton as her dowry upon her marriage in 1741 and together they named it Middleton Place. The plantation prospered during the era of indigo, rice, and cotton. (See The Rice Museum, Georgetown, South Carolina.) It served as not only the seat of the family but also the headquarters for a group of area plantations. Today the complex contains 65 acres (26 hectares) of formal gardens, the House Museum, and a working stable yard. (See Plimoth Plantation, Plymouth, Massachusetts.)

The Middletons were dedicated to public service. Henry Middleton, one of the richest men in colonial America, was the second president of the First Continental Congress. Arthur, his son, was a signer of the Declaration of Independence. The second Henry, son of Arthur and grandson to the first Henry, was governor of South Carolina and American minister to Russia. William, the first Henry's great-grandson, signed the Ordinance of Secession, the formal withdrawal of South Carolina from the federal union. This is one reason why General Sherman's army ransacked and burned the plantation in 1865. The main house, built before 1741, and the north wing, added in 1755 as a conservatory and library, were both destroyed. All that remained was the 1755 gentleman's quarters, now the House Museum.

The Great Charleston Earthquake of 1886, registering a minimum of 7 on the Richter scale, probably had its epicenter at Middleton Place. This catastrophe caused the ruins of the main house and the west wing to tumble to the ground, the terraces to break apart, and the water to drain from Butterfly Lake.

J. J. Pringle Smith, a direct descendant, inherited the property in 1916 and began restoring the gardens in the 1920s. The Smith heirs continue to restore and maintain this historically significant property. Middleton Place is now one of only six gardens in the United States recognized by the International Committee on Monuments and Sites, and is a winner of the Buckley Medal, the highest award from the Garden Club of America.

Charleston's gardens, well-known in colonial America, sparked Henry Middleton's interest in botany. In 1741 he ordered more than 250 species of plants from England for a garden on his property (see Bartram's Garden, Philadelphia, Pennsylvania). Today it is the oldest existing landscaped garden in America.

André Le Nôtre, the designer of the gardens at Versailles, seems to have influenced Henry Middleton's planning. Within the garden's right-triangle configuration are a variety of geometric plots, reflecting the principles of French philosopher Blaise Pascal and his spirit of geometry. Parallel and perpendicular paths and allées, woods, water, ornamental canals, and sundials are all major components of the plan. The 19th-century expansion of the gardens reflects a romantic English design, less influenced by geometry.

The gardens still contain one of four *Camellia japonicas* brought by French botanist André Michaux in 1786. These are the first camellias planted in any American garden. One of the largest known specimens of giant crape myrtle also graces the garden. Live oaks grow profusely here, chief among them Middleton Oak, believed to be 1000 years old. This tree, one of the oldest in America, served as a marker along a Native American trail before Columbus arrived in the New World. The gardens contain almost all the plant varieties found in the Southern United States.

The gardens were originally situated in the middle of a working rice plantation. From the colonial period until just after the Civil War, rice was the plantation's major agricultural product. Butterfly Lake was part of the waterway used to flood the rice fields. We see evidence of rice and sugar cultivation in the remains of the dikes, floodgates, and irrigation system. The rice mill, at one end of the gardens, now has exhibits on rice cultivation and art. The sugar mill still makes cane syrup in season. The spring house, originally for storing and keeping food cool, has exhibits about the history of the gardens and recent archaeology of the area.

Because the rivers served as highways for the Low Country plantations, most of the houses were placed facing the river. Middleton Place, 20 miles (32 km) from Charleston, faces the Ashley River. The 14-mile (23-km) land route to the property was part of the ancient Cherokee Trail.

The portion of the original mansion that still stands is the gentlemen's guest quarters, restored in 1870 for use as the family residence. Today it contains displays of furniture, paintings, books, and documents chronicling two centuries of the Middleton family. John Izard Middleton (1785–1849) is considered America's first classical archaeologist. His painting, *Grecian Remains in Italy* (1812), hanging in the front hall, reflects his interest in archaeology. Along the walls of the upstairs hallway hang the Bien edition chromolithographs of Carolina birds from John James Audubon's *Birds of America*. The library features materials on ornamental horticulture, landscape design, architecture, local and Southern history, and natural history, subjects of special interest to the Middleton family. Among the holdings is a 17th-century first edition of *The Natural History of Carolina* by Mark Catesby.

The Museum House portrays antebellum life. The winter and summer bedrooms show how sleeping rooms were altered periodically to make them comfortable in different seasons. The winter bedroom has heavy drapes and bed covers; the summer bedroom has mosquito netting and a bed—with headboard removed—in the center of the room to catch river breezes.

At the site's re-created Plantation Stableyards, docents in period dress demonstrate blacksmithing, coopering, pottery, spinning, weaving, and other skills once used on the plantation. On display are early farm tools, 19th-century vehicles, and farm animals typical of those present during the antebellum period. A tour of the historically significant animal breeds is available. Eliza's House (named for the last inhabitant of the dwelling, who died in 1986) captures the life of the African-American slaves and freedmen who worked on the plantation.

Address: Middleton Place, 4300 Ashley River Road, Charleston, SC 29414.
Phone: (843) 556-6020.
Days and hours: Daily 9 a.m.–5 p.m. Dec. 24 closing at 1:00; closed 12/25. Check ahead, the schedule for when various sections open and close varies.
Fee: Yes. Some guided tours have an additional fee.
Tours: Self-guided and guided tours; house tour is guided only.
Public transportation: None. From the train station in Charleston take taxi or other car service to site.
Handicapped accessible: Partially: check with site. Wheelchairs are available; call ahead.
Food served on premises: Restaurant.
Website: www.middletonplace.org/

GEORGETOWN

The Rice Museum

Colonists quickly recognized that the area that now includes southeastern North Carolina, South Carolina, and Georgia is ideal for cultivating rice and aquatic plants that grow well in warm, wet climates. Rice cultivation began sometime during the 1680s or 1690s, when the first rice seeds arrived here from Madagascar. Before the American Revolution, indigo was an important crop in the region as well, but later rice became the dominant agricultural product. This small museum in the Old Market Building conveys the local history of rice and indigo from about 1700 to 1900. The Old Market Building, constructed in 1842, together with its clock tower added in 1857, is on the National Registry of Historic Places.

At the museum the visitor learns that in 1741, colonial William Lucas of Charleston sent indigo seed from the West Indies to his daughter, Eliza, to plant on Wapoo, the family plantation. Word quickly spread of how well this plant thrived. Indigo, used commercially to dye fabric and in household laundering, soon rivaled rice as South Carolina's major agricultural product. The plant looks very much like asparagus and does well on dry land. It was used to produce three colors of dye: copper for wool, purple for linen, and blue for silk. Indigo's role in the local economy declined after the Revolution because the British could get higher-quality indigo plants from foreign markets and because insect infestations in the 1790s made production difficult. (See Middleton Place, Charleston, South Carolina.)

After indigo's failure, rice again became the key crop of South Carolina's Low Country, reaching its peak in the 1850s. When slavery was abolished, however, this labor-intensive crop was no longer economically worthwhile. The lumber industry's higher wages lured workers away, and new technology enabled other areas of the United States, such as Texas, Louisiana, and California, to produce rice more efficiently. Today, these other

Clock tower at The Rice Museum.
Courtesy of The Rice Museum.

regions produce about 95% of the nation's crop. South Carolina, nevertheless, continued to produce rice into the 20th century.

The museum's exhibits focus in part on technological advances in rice production. In 1758 McKewn Johnstone perfected a tidal flow method of rice cultivation, which worked well because of the five rivers flowing through the county on their way to the Atlantic Ocean: the Waccamaw, the Pee Dee, the Black, the Sampit, and the Santee. Land would be flooded with fresh water during high tide and drained during low tide. Dikes, floodgates, and canals were constructed to form a rice floodplain. Controlled flooding of the land enhanced the topsoil, ensuring that rice cultivation could be carried on for many years. This was very important in a region where other key crops—tobacco and cotton—depleted the soil. Some of the flood canals were also used to transport goods in and out of the area.

In this museum there is a permanent collection of dioramas, maps, paintings, artifacts, and other exhibits which tell the story of rice cultivation in this county, Georgetown County.

A highlight of the museum is a miniature working model of the pounding mill invented by Jonathan Lucas, Sr., in 1787. Separating hulls from the kernels of rice was originally done by hand using a mortar and pestle. Later, pounding mills were developed that were powered by tidal water. Still later, the Lucases, Jonathan Jr. and Sr., incorporated steam power into the machine, which used a series of drums and grinders to remove the hulls. Harvested rice was hauled to the third floor of the mill for hulling, then sacked on the ground floor and shipped. This invention was probably as important to the rice industry as Eli Whitney's cotton gin was to the cotton industry. Steam-driven pounding mills helped the region provide almost 50% of the rice consumed in the country before the Civil War.

Address: The Rice Museum, 633 Front Street, Georgetown, SC 29440.
Phone: (843) 546-7423.
Days and hours: Mon.–Sat. 10 a.m.–4:30 p.m., Sun. 11:30 a.m.–3:30 p.m.
Fee: Yes.
Tours: Self-guided. Guided tours are available throughout the day.
Public transportation: Buses from Charleston stop right at the museum.
Handicapped accessible: No.
Food served on premises: None. In the local area are restaurants.
Website: www.ricemuseum.org/

Virginia

ALEXANDRIA

Stabler-Leadbeater Apothecary Museum

More than 200 years ago, this apothecary shop and manufacturing facility opened just across the Potomac River from what is now Washington, DC, in Alexandria, Virginia. Edward Stabler founded the shop in 1792, and by the mid-1840s, his son-in-law, John Leadbeater, joined the business. The Leadbeater family took over ownership in the 1850s and the business remained in continuous operation for 141 years, serving such historic figures as George and Martha Washington and Robert E. Lee. The Great Depression and the rise of the chain drug store led the company to go into bankruptcy and close in 1933. This site, one of the best-preserved American apothecaries from the 1700s, is now a museum and is on the National Registry of Historic Places. (See New Orleans Pharmacy Museum, New Orleans, Louisiana.)

The shop's collection of pharmaceuticals, equipment, elixirs, and other items presents important information about the science of pharmaceutical manufacture in bygone days, those who made the medications, and those who bought and used them. On the floors above the apothecary shop are the remains of the only early 19th-century medical manufacturing facility still in existence.

The building's beautiful colonial architecture has been restored to its early 19th century appearance. In the mid-19th century, the shop's interior was modernized to the then-popular Gothic Revival style, covering the original shelving and drawers dating to 1805. Gothic arches frame the well-worn shelves lined with hundreds of hand-blown bottles. The original retail counter, store fixtures, and display equipment have all been left in place. On this floor, bottles, jars, boxes, drawers, and bins, most with original labels, are on display. The cases and shelves contain medical glassware and antique drugstore equipment, including mortars and pestles, old eyeglasses, weights, and balances, as well as temporary exhibitions covering topics relating the history of the business. Especially of note are several 18th-century hand-blown medicine bottles that Edward Stabler had imported from England for use in his shop.

At the front of the shop, near the rounded windows, stands the counter where Edward Stabler and his sons took full advantage of the daylight to compound prescriptions for his customers. The addition of gas, and later electric, lighting to the building allowed prescriptions to be prepared in the rear of the retail space and the second-floor manufacturing space. Additional lighting also allowed the pharmacist to do the daily bookkeeping at the rear of the store at the large desk.

Interior of Stabler-Leadbeater Apothecary Museum. *Photo* by Erik Patten.com.
Courtesy of Stabler-Leadbeater Apothecary Museum.

In the second-floor manufacturing area are drawers, tins, and boxes of herbal materials that have been left intact, showcasing the use of early plant-based materials in medicine and household products. In addition to patent medicines, the company also sold garden seeds, glass, hairbrushes, paints, and perfumes. The building next door, one of the two original buildings, housed the wholesale business. Today it contains the gift shop for the apothecary on the first floor, a period pharmacy display case, and a collection of antique crates and boxes.

Address: Stabler-Leadbeater Apothecary Museum, 105–107 South Fairfax Street, Alexandria, VA 22314.
Phone: (703) 746-3852.
Days and hours: Nov.–Mar.: Wed.–Sat. 11 a.m.–4 p.m., Sun. 1 p.m.–4 p.m., closed Mon. and Tue. Apr.–Oct.: Tue.–Sat. 10 a.m.–5 p.m., Sun. and Mon. 1–5 p.m. Closed 1/1, Thanksgiving, 12/25.
Fee: Yes.
Tours: Guided group tours of 10 or more only; call two weeks in advance. Extended and specialty-themed group tours are available.
Public transportation: Local buses stop four blocks (0.3 mi., 0.5 km) from site, then walk. The King Street Trolley stops at King and Royal Streets, near the site.
Handicapped accessible: First floor only.
Food served on premises: None. The City of Alexandria has a variety of restaurants nearby.
Special note: Check the website or call about special lectures and programs and museum closures (because of special events and tours).
Website: www.apothecarymuseum.org

CHANTILLY

Steven F. Udvar-Hazy Center

Despite the unrivaled size of the Smithsonian Institution (see Smithsonian Institution, Washington, DC), even the Smithsonian cannot display some of its large specimens from its Air and Space Museum on the National Mall. Hence, with a grant in 1999 from aircraft-leasing magnate and Hungarian immigrant Steven F. Udvar-Házy, construction began at the annex across the Potomac River in Fairfax County, Virginia. This site, at the Washington Dulles International Airport, which opened in 2003, encompasses 760 000 square feet (71 000 m²) with two hangars (Boeing Aviation and James S. McDonnell Space hangars) containing larger samples from the Smithsonian's aerospace collection.

Over 300 items are displayed at this site, divided up into 23 different types of craft, including vertical flight, business aviation, human spaceflight, rockets and missiles, and more. Among them is the famous Boeing B-29 Superfortress, model B-29-45-MO, called the *Enola Gay*, that dropped an atomic bomb on Hiroshima, Japan. This model of airplane, designed in 1940, was the first bomber with a pressurized compartment for its crew, and included radar. The *Enola Gay* itself was delivered to the US Army Air Forces in June 1945, and last flew in December 1952.

Another craft shown is the space shuttle *Discovery* (or officially OV-103), which began flying in the space shuttle fleet in 1984, and finished in 2011. With an entire year of flight, *Discovery* traveled almost 150 million miles (240 million km).

Also on display is a supersonic transport Concorde airplane (SST), a creation of the French firm Aérospatiale and the British Aviation Corporation. Work on the Concorde began in 1962, and the first version was tested in 1969. The Concorde aircraft could fly at a cruising altitude of 10.4 to 11.4 miles (16.8 to 18.3 km)—high enough to view the curvature of the Earth. This particular plane, a Concorde F-BVFA with a top speed of 1350 mph (2179 km/h), flew between New York City and Rio de Janeiro, Brazil, and was donated by Air France in 2003 to the Smithsonian.

The *Gemini VII* space capsule, nearly 11 feet (3.3 m) tall, was launched on December 4, 1965, carrying its two-man crew, Frank Borman and James Lovell, and stayed in Earth-orbit for 14 days, a record that lasted five years.

Interior of Hangar at Steven F. Udvar-Hazy Center. *Photo* by Dane Penland.
Courtesy of the Smithsonian Institution.

Preserving the delicate and myriad parts that make up air- and spacecraft requires much work. Hence the Mary Baker Engen Restoration Hangar allows visitors a view from a glassed-in mezzanine level of how restoration and preservation proceeds, with space for several aircraft simultaneously.

Visitors can also enter the Donald D. Engen Observation Tower, which provides a panoramic view of Dulles airport, with airplanes taking off and landing. Two levels of exhibits in the tower show how the US air-traffic control system operates.

Numerous science demonstrations are held daily in the museum. There is an IMAX® theater with regular shows. Researchers can visit the archives, which houses probably the preeminent collection of the history, science, and technology of flight, such as books, photographs, film, video, and technical drawings.

Address: Steven F. Udvar-Hazy Center, 14390 Air and Space Museum Pkwy, Chantilly, VA 20151.
Phone: (703) 572-4118.
Days and hours: Daily 10 a.m.–5:30 p.m. Closed 12/25.
Fee: Free. There is a fee for parking and IMAX® Theater with regular shows.
Tours: Self-guided; docent-led 90-minute free guided tours are offered twice daily. Reserve two weeks in advance for visitors with disabilities.
Public transportation: Fairfax Connector buses travel from the Wiehle-Reston East station on the Metro, and stop at the site.
Handicapped accessible: Yes. A few wheelchairs are available at the Security Desk. Reserve two weeks in advance for visitors with disabilities.
Food served on premises: Fast food.
Website: airandspace.si.edu/udvar-hazy-center

CHARLOTTESVILLE

Monticello

Thomas Jefferson is popularly known as a statesman and third president of the United States. Yet his intellectual diversity in such areas as architecture, horticulture, and science is not popularly understood. The building he designed and called home, along with the surrounding plantation, reflects these profound interests.

Monticello, the name that Jefferson selected for the estate, is Italian and means "little mountain." It is a very fitting name for the complex because it sits atop a hill 867 feet (264 m) high. Monticello is the only American home awarded a place on the UNESCO World Heritage List and is also a National Historic Site.

The Thomas Jefferson Visitor Center, at the entrance to Monticello Plantation, is composed of The David Rubin Visitor Center and Smith Education Center. This facility provides an excellent introduction to Thomas Jefferson, his life at Monticello, his work in the areas of architecture, gardening, farming, and science, and his ideas about liberty. Many of the artifacts displayed were excavated on the grounds of Monticello. Audiovisual presentations, architectural drawings, manuscripts Jefferson wrote, and models are all part of the exhibits. Mulberry Row, the living quarters for the house slaves at the plantation, as well as the working quarters of some of these people, is reviewed. The Smith Education Center is a two-story history education center with classrooms, an exhibition center, a multipurpose theater, a café, museum, and shops. An excellent film about Thomas Jefferson and his plantation is also shown.

Thomas Jefferson, a self-taught architect, began constructing his home in 1768. The early mansion design was typical of Georgian architecture then popular in Virginia. As he learned more about then-contemporary designs introduced into architecture, he changed the mansion to reflect these new European and American styles, as well as some of his own innovations. He chose a formal, classical look based on the work of 16th-century architect Andrea Palladio, known as Italian Renaissance design. In 1784, when Jefferson departed to serve as ambassador to France, the residence at Monticello was still unfinished. The architecture he saw abroad influenced him greatly, and after returning to the United States he spent the next 40 years altering the building and grounds, adding new features.

The building, Monticello, clearly reveals Jefferson's fascination with gadgets and science. Hanging above the entrance to the house is a seven-day clock, built in Philadelphia, to Jefferson's specifications. It indicates both the day and hour.

At one time the entranceway served as both a reception hall and a museum for Jefferson's collection of fossil bones and Native American artifacts. Some of these pieces were brought back by Meriwether Lewis and William Clark, the explorers Jefferson sent on an expedition to the American Northwest while he was president. (See Fort Clatsop & Visitor Center, Astoria, Oregon.) Because Jefferson bequeathed most of the Lewis and Clark artifacts to the Academy of Natural Sciences in Philadelphia, these artifacts are no longer on display at Monticello. What we see today are reproductions.

On display in the house are Jefferson's telescope, theodolite (surveying instrument), celestial globe, medicine bottles, a camera obscura, lenses, and other science-related items. His first collection of books was destroyed by fire in 1814. The second library, consisting of 7000 volumes, he sold to the federal government in 1815. These books form the core of what is now the Library of Congress. The books currently at Monticello are the third, replacement, collection. Off the library room is the greenhouse where Jefferson kept his workbench and tools.

He enclosed a dumbwaiter on each side of the dining-room fireplace to transport wine from the cellar. He designed a set of single-acting double-glass doors. These doors open or close simultaneously when either door is operated. Through these doors is the parlor.

The plantation consists of 5000 acres (2000 hectares) of land Jefferson inherited from his father. This mostly self-sufficient complex included four adjacent farms. Jefferson began as a tobacco farmer, but soon moved to wheat as his main cash crop. He also raised corn, potatoes, and small grains, experimenting with new agricultural techniques such as crop-rotation and contour plowing.

The plantation grounds became Jefferson's horticultural laboratory. His extensive writings on the subject reveal much about early American gardens and his scientific and creative abilities. He planted the Columbian lily discovered by Lewis and Clark, native cardinal flowers, and new European imports. One of the flower beds held the *Jeffersonia diphylla*, a native American wildflower named for Jefferson by botanist Benjamin Smith Barton during a meeting of the American Philosophical Society in 1792. (See College of Physicians of Philadelphia and the Mütter Museum, Philadelphia, Pennsylvania; and Joseph Priestley House, Northumberland, Pennsylvania.)

Among the many trees Jefferson planted—some still living today—were newly introduced species such as copper beech, mimosa, and gingko. He obtained many of his plants from Philadelphia botanist John Bartram. (See Bartram's Garden, Philadelphia,

Pennsylvania.) Jefferson studied useful and ornamental plants from all over the world, and tested more than 200 varieties of vegetables and herbs including sea kale, the relatively unknown tomato, and almost 20 types of English peas, probably his favorite vegetable. Many of the early varieties are on view in the restored gardens, vineyard, and orchard. These historic gardens, in which he experimented on foods and ornamental plants from around the world, were his laboratory.

Mulberry Row, the location of the cabins for some slaves, was a 1000-foot (300-m) avenue of horses, artisan shops, sheds, and stables, and acted as the center of many activities for the plantation. The gate to Jefferson's 2-acre (0.8-hectare) vegetable garden and 400-tree orchard once stood here. Today much of what was here is being refurbished and reconstructed.

Jefferson directed that at his death he should be buried on this estate. This area is now designated as Monticello Cemetery. Over the years many of his descendants also have been buried here. The cemetery and Jefferson's grave site are opened to visitors.

With time, the wealth of the Jefferson family declined. After Jefferson's death, Martha Jefferson Randolph sold the estate to pharmacist James Barclay, who did not keep up the property. Commodore Uriah Levy, a great admirer of Jefferson, bought Monticello from Barclay, preserved the estate, and thus we have this gem.

In addition to Monticello, Jefferson also designed the University of Virginia.

Address: Monticello, 931 Thomas Jefferson Parkway, Charlottesville, VA 22902.
Phone number: (434) 984-9800.
Days and hours: Daily 8:30 a.m.–5 p.m. Closed 12/25.
Fee: Yes.
Tours: Guided tours at regular intervals. Shuttle buses are available around the property; check at the Visitor Center.
Public transportation: Buses from Washington, DC, to Charlottesville. Taxis are available from Charlottesville to Monticello.
Handicapped accessible: Yes.
Food served on premises: Café in the Visitor Center.
Main website: www.monticello.org/;Visitor Center website: www.monticello.org/site/visit/thomas-jefferson-visitor-center-and-smith-education-center

FORT EUSTIS

U.S. Army Transportation Museum

The history of transportation development is an integral part of the history of the U.S. Army, as reflected in a quote on a plaque in the museum: "Transportation is and always has been the controlling factor in military operations with our Army, with all armies since the dawn of history." The U.S. Army Transportation Museum, on a 7-acre (2.8-hectare) site near Colonial Williamsburg, was first organized in the 1950s as the Transportation Corps Circus. It started with traveling exhibits and circus acts, but in 1976 put down roots in this permanent exhibition facility. The museum collects artifacts and documents related to the history of the United States Army Transportation Corps from the Revolutionary War to the present. Its indoor and outdoor exhibits include dioramas, scale models, and life-size displays.

In 1827, when the newly formed Baltimore & Ohio Railroad (see B&O Railroad Museum, Baltimore, Maryland) asked Secretary of State James Barbour for help, the government recognized the railroad's military importance and agreed to pay its surveying costs. This was the first time a private enterprise asked for and received federal aid because of its military implications.

The museum's holdings include prototype machines, equipment used by the Army, and Army memorabilia. Several cutaway engines help explain the equipment's inner workings. Videos provide additional background. Newer exhibits focus on the Desert Storm and Desert Shield operations in Iraq and Kuwait, especially the work of the 7th Transportation Group and the 22nd Support Command. A full-size diorama on the living conditions of men and women who participated in the operations includes a Bedouin tent they used as living quarters, uniforms from the U.S. military, chemical suits, and captured Iraqi uniforms and equipment. Work in Somalia, Haiti, and Afghanistan is also covered.

One unique piece at the museum is the experimental Flying Bicycle of Ground Effects Machine X-2, developed at Princeton University in 1960. This machine, with a saucer-shaped aluminum base attached to a seat, engine, and handlebars, could lift off the ground, hover, and move a 250-pound (113 kg) payload or person. The handlebars controlled and steered the 15-horsepower (11-kW) engine.

The outdoor displays are arranged in four exhibit areas. The 20 000-square-foot (1900-m^2) Aircraft Pavilion contains fixed- and rotary-winged aircraft from the Korean War to the present. Of note are two experimental vehicles designed for use in the Korean War, the flying Jeep and the VTOL (Vertical Takeoff and Landing) vehicle. The Caribou, an aircraft used by the Army Parachute Team, was the largest army aircraft used in Vietnam. Built by the De Haviland Aircraft Corporation to provide close support to ground forces in forward battle areas, the Caribou is typically used to transport up to 32 personnel or up to 3 tons (2.7 tonnes) of supplies or casualties. It has a ready loading door for aerial supply drops. Among other experimental machines on view are the Cybernetic Walking Machine, a device for traversing difficult terrain on earth or in space, and the Bell Rocket Belt, a jet pack strapped to the back that enables the wearer to fly like a helicopter.

Marine Park, the second outdoor exhibit area, shows landing and watercraft such as a tugboat with a 400-horsepower (300-kW) engine and the LARC (Lighter, Amphibian, Resupply, Cargo). The third area, the Truck Park, contains common and rare military transportation vehicles: trucks, trailers, Jeeps, and armored vehicles, some from Russia and other foreign countries. An armor-plated gun truck called The Eve of Destruction, the only one of its kind to return from Vietnam, has 50-caliber (12.7-mm) machine guns and grenade launchers. This truck and others like it provided escort for other vehicles delivering supplies. Here the visitor learns about the evolution of one of the best-known Army vehicles, the land Jeep or quarter-ton truck.

An exhibit on World War II explains the truck route called the Red Ball Route, which was created in August 1944 for military vehicles too large for the narrow European roads. It went from St. Lô, a French port along the English Channel, to La Loupe-Dreux-Chartres, west of Paris. The U.S. Army used it to move more than 800 000 gallons (3 million L) of gasoline and 100 000 tons (91 000 tonnes) of supplies inland as the Germans retreated. The success of Generals Patton and Hodges and the First and Third Armies rested, in great measure, on the logistical strategies and capabilities of the transportation unit.

Railroad transportation, so important to the U.S. Army, is a major exhibit area. It houses 13 pieces of railroad rolling stock from World War II and the Korean War, including ambulance cars, flat cars, a German tank car, a steam crane, and locomotives. In this area the chronology of rail transportation is explained.

Address: U.S. Army Transportation Museum, 300 Washington Boulevard, Besson Hall, Fort Eustis, VA 23604.
Phone: (757) 878-1115.
Days and hours: Tue.–Sat. 9 a.m.–4:30 p.m. Closed on Sun., Mon., and federal holidays.
Fee: Free.
Tours: Self-guided; guided tours for large groups with two weeks' advance notice.
Public transportation: Local buses stop about 0.4 mile (0.6 km) from the site, then walk.
Handicapped accessible: Yes.
Food served on premises: None. Some fast-food restaurants are within a few blocks of the fort.
Website: www.transportation.army.mil/museum/transportation%20museum/museum.htm

HAMPTON

Virginia Air and Space Center

When Langley Air Force Base closed its Visitor Center in the 1990s, it transferred all the exhibits to this specially constructed learning facility, which resembles a modern airplane hangar. Its seven-story-high observation deck overlooks Hampton's working harbor. Inside, the center traces America's sea and air history from the 17th century to the present.

Gosport Navy Yard opened in the coastal town of Hampton in the 1860s. The site's history includes both shipbuilding and military research. In the first of many military breakthroughs at the yard, military observation balloons flew over Yorktown in 1860. Here in 1910, an airplane took off from a ship for the first time. In 1928, using captured German ships that had been brought to Chesapeake Bay, the facility demonstrated that ships could be bombed from the air. In 1917 an aeronautic experimental station named Langley Field was built near here, which in 1948 became known as Langley Air Force Base. Today, work at the Langley research center focuses on airplane design and space programs like the Mercury, Gemini, Apollo, Viking, and Space Shuttle projects. The first astronaut training took place here in the 1950s.

At the Virginia Air and Space Center there are more than 30 historic aircraft for the public to see and a number of simulations to experience. Throughout the museum are many hands-on exhibits.

The theme of the museum is "From the Sea to the Stars." Visitors can climb a 60-foot (18 m) gantry to walk among a dozen air- and spacecraft hanging from the ceiling. The F-86 and the Vought A-7 E Corsair II, built between 1965 and 1976, were used by the U.S. Air Force and Navy as their standard light attack aircraft. The last military mission of the Vought A-7 E Corsair was Desert Storm in 1991. Many other pieces represent major events in aerospace history, such as the Apollo 12 command capsule, similar to the one that went to the moon in 1969, and a lunar sample collected by Apollo 17. To view aircraft suspended from the ceiling above, visitors walk the catwalk designed for just such an excursion.

Both a Grumman Yankee, used by NASA to test pilots' abilities to maintain control of an aircraft while it spins violently, and a mock-up of the International Space Station are on view. If you want to feel what it is like to fly a modern aircraft, climb aboard an F-16 fighter plane with a hands-on audiovisual flight simulation program.

There is an amateur radio exhibit with both modern and historic radio examples on display.

In the galleries are exhibits on astronomy, the history of space transportation, and the lunar landscape. The Gallery of Technological Advances has a wind tunnel that is a modern version of the one first developed by the Wright brothers (See Wright Brothers National Memorial, Kill Devil Hills, North Carolina.) Many of the center's exhibits are interactive.

Address: Virginia Air and Space Center, 600 Settlers Landing Road, Hampton, VA 23669.
Phone: (757) 727-0900
Days and hours: Mon.–Wed. 10 a.m.–5 p.m.; Thu.–Sat. 10 a.m.–6 p.m.; Sun. 12 p.m.–5 p.m.
Fee: Yes.
Tours: Self-guided.
Public transportation: Local bus stops 0.4 mile (0.6 km) away, then walk to site.
Handicapped accessible: Yes.
Food served on premises: Café.
Special note: Additional fee for the IMAX° theater.
Website: http://www.vasc.org/index.html

LURAY

Shenandoah National Park

Shenandoah, a Native American word meaning "daughter of the stars," aptly describes these beautiful mountains. Shenandoah National Park is part of the Blue Ridge Mountains of the Appalachian Mountain Range, one of the oldest ranges in the world. Much of the park's 195 000 acres (78 900 hectares) were cleared for farming long before the national park was created in 1936. When it became part of the national park system, the land was left to return to its natural forested state. Skyline Drive, the backbone of Shenandoah National Park, runs 105 miles (169 km)—the full length of the park—along the highest crest of the Blue Ridge Mountains. Parks of the Appalachian Trail parallel the drive.

Naturalist George Freeman Pollock first came to these mountains in 1886 when his father became owner of more than 5000 acres (2000 hectares) of land around Stony Man Mountain. The senior Pollock purchased this property as a copper-mining investment, which proved unsuccessful. At the end of the 19th century, George Pollock built Skyland resort at the highest point along Skyline Drive. At 3000 feet (914 m), Skyland offers a magnificent view of the region. Pollock worked tirelessly, lobbying influential people to make the region a national park.

In 1929, needing a spot within 100 miles (161 km) of Washington, DC, for weekend retreats, President Herbert Hoover purchased land in the Blue Ridge Mountains. The

United States Marines built 13 buildings with materials purchased by Hoover and created roads along the Rapidan River. The site became known as the Camp on the Rapidan. When Hoover left office in 1933, he donated the land and buildings to the government to make it part of Shenandoah National Park, which was officially dedicated by President Roosevelt in 1936.

For the sake of convenience the park is divided into five areas, each with its own services. Each area has a visitor center with exhibits and information for the public. At the north end of the park, near the town of Front Royal, is Shenandoah Valley Overlook. From here you can see Massanutten Mountain, which during the Civil War had a communications post known as Signal Knob. Dickey Ridge Visitor Center, a major information facility in the park, offers a movie that provides an overview of the park and its facilities. Byrd Visitor Center, at milepost 51 in the Hawkesbill Gap, near the entrance to Big Meadow, covers the area's human and natural history. It regularly offers films, ranger talks, and hikes led by rangers.

Many hiking and strolling trails emanate from Skyline Drive, some less than a mile (1.6 km) long and others taking more than a day to traverse. Many of the trails provide scenic overlooks. From some are visible old homesteads. Fox Hollow Nature Trail passes the remnants of the Fox homesite; Snead Farm Trail leads to a homesite and barn. Both trails are in the park's northern district. Along Compton Peak Trail visitors see geological features of the area, including columnar jointing resulting from igneous intrusion that cooled with a crystalline structure. Along Fort Windham Rocks Trail are excellent examples of Catoctin lava formations.

The Limberlost is a trail slightly more than a mile (1.6 km) long that goes through one of the park's few regions with virgin forest. The Story of the Forest Nature Trail illustrates how Shenandoah reverted from meadowlands to forest. Deadening Nature Trail brings a variety of ecosystems into view, including meadows and fern rock gardens. The trail takes you to the summit of Loft Mountain, which offers beautiful vistas. The 4-mile (6.4-km) Neighbor Mountain Trail passes paper birch trees, rare in this part of the country.

Reforestation has brought back many animals that were driven from the area when their habitat was destroyed. Among species that visitors might view are white-tailed deer, black bear, snakes, wild turkey, ruffled grouse, and turkey vultures. The mountain streams are among the last preserves of the eastern brook trout.

The forests of the region generally consist of broad-leaved trees—hickory, dogwood, sugar maple, and white oak—and some conifers. Fall brings beautiful color changes. Wildflowers, including stiff gentian, southern harebell, autumn sneezeweed, asters, bloodroot, and violets, abound throughout the spring, summer, and fall.

Address: Shenandoah National Park; Front Royal Entrance: Skyline Drive & Stonewall Jackson Highway, Front Royal, VA; Thornton Gap Entrance: Skyline Drive & US Highway 211 East, Luray, VA; Swift Run Entrance: 22591 Spotswood Trail, Elkton, VA; Rockfish Gap Entrance: Skyline Drive & Blue Ridge Parkway, Afton, VA.
Phone: (540) 999-3500.
Days and hours: Open daily except in bad weather. Some sections are closed at night during deer-hunting season. Most visitor centers close from late Nov.–Mar.
Fee: Yes.
Tours: Self-guided; some ranger-guided walks are offered.
Public transportation: None.

Handicapped accessible: Mostly. Limberlost Trail is fully accessible.
Food served on premises: Restaurants and cafés.
Website: www.nps.gov/shen/index.htm

NEWPORT NEWS

Mariners' Museum and Park

This international maritime museum is in Virginia's Tidewater region, along one of the fingers of the coastal plain cut by the James and York rivers. Newport News takes its name from Captain Christopher Newport, who, whenever he arrived from England, would deliver the latest news to the colonists gathered there. The city, on one of the world's largest harbors, became a major port for large ships and home of Newport News Shipbuilding and Drydock Company, where many of the ships were built.

Scholar and philanthropist Archer M. Huntington founded the Mariners' Museum in 1930. He was the son of rail magnate Collis P. Huntington, founder of the dry dock company. (See California State Railroad Museum, Sacramento, California.) This museum presents about 3000 years of sea technology: human conquest of the sea; how the sea has influenced civilization; harvesting food from the sea; and use of the sea for transportation, war, and recreation. The facility, divided into a series of galleries, displays only 3% of its 35 000 artifacts at any one time. Congress designated this site as America's National Maritime Museum in 1998.

The Age of Exploration Gallery emphasizes the scientific and technical developments of the 15th through 17th centuries, and the effect they had on world exploration. Ship models, rare books, illustrations, maps, navigation instruments, and other materials

Mariner's Museum.
Courtesy of the Mariner's Museum.

related to shipbuilding, ocean navigation, and cartography of this period tell the story. The oldest navigation instrument in the collection is the nocturnal and quadrant, made and signed by Ionnes Baptista Iusti in 1565. Mariners used the nocturnal to tell time at night until the chronometer was invented in the 18th century. The quadrant side helps travelers calculate latitude.

The Miniature Ships of August F. and Winnifred Crabtree collection includes 16 tiny, handcrafted historic vessels displayed, replicating their full-size counterparts. This exhibit follows the evolution of boat construction from early dugout canoes and rafts, to vessels made by the Egyptians, Romans, and Normans, and then on to mid-19th-century sea craft.

Defending the Seas presents the US Navy's role, including theory of sea power, in which a country's international powers are bound up with maritime control. Five areas focus on five periods of sea power in the history of the US Navy. The Antique Boats Gallery, decorated as a 1930s dealer's showroom, contains several antique Chris-Crafts. Carvings shows figureheads from all over the world. Maritime Paintings and Decorative Arts examines maritime history from an artistic perspective. The Ship Models Gallery explores the many ways people have used ship models. The International Small Craft Center contains nearly 150 vessels from around the world, including a Venetian gondola and a Chinese sampan.

The US *Monitor* Center is the National Oceanic and Atmospheric Administration's (NOAA) official repository for preservation of the warship USS *Monitor*, the famous iron-clad warship from the Civil War, which sank at the end of December 1862, and was discovered in 1973. Also on exhibit are iron artifacts including the screw propeller and anchor. Conservation techniques are explained. The revolving gun-turret and 20-ton (18-tonne) steam engine are housed in the Batten Conservation Complex. Through large windows visitors view the artifacts, immersed in special solutions in the Wet Lab. Treatments include immersion in alkaline solution, electrolytic reduction to halt oxidative corrosion, stabilization of the material, and then salt-removal. Iron is treated via electrolytic sodium hydroxide and direct current is fed through the artifact, drawing chlorides out and loosening encrustation by living creatures. There is an additional Dry Lab, not on view to the public.

In the front lobby is a 10-foot (3-m) tall first-order Fresnel lens system weighing 1 ton (0.9 tonne), constructed in Paris, and used in the Cape Charles Lighthouse from the summer of 1895 through 1963, at the mouth of the Chesapeake Bay. (See Whaling Museum, Nantucket, Massachusetts; and Chesapeake Bay Maritime Museum, St. Michaels, Maryland.)

The museum is located in the largest privately owned free park in the United States, surrounding a lovely lake suitable for fishing and boating. The 5-mile (8-km) Noland trail around artificial Lake Maury provides an opportunity to investigate the local ecology. The Mariner's Museum also runs a library holding the largest maritime-history collection (over 2 million items ranging from photographs to manuscripts, charts, and maps) in the Western Hemisphere.

Address: Mariners' Museum and Park, 100 Museum Drive, Newport News, VA 23606.
Phone: (757) 596-2222.
Days and hours: Mon.–Sat. 9 a.m.–5 p.m., Sun. 11 a.m.–5 p.m. Closed Thanksgiving and 12/25.
Fee: Yes.

Tours: Self-guided.

Public transportation: HRT buses stop about 0.7 mi (1.1 km) from the museum, then walk to the site.

Food served on premises: Lunch café.

Handicapped accessible: Yes; wheelchairs available upon request.

Website: www.marinersmuseum.org/

VIRGINIA BEACH

Virginia Aquarium & Marine Science Center

Virginia Beach is the only city that borders both the Atlantic Ocean and the fragile Chesapeake Bay. This land, some 29 miles (47 km) of coastline along the along Virginia's Eastern Shore, was once a desolate strip of sand, known as a graveyard for ships lost at sea along the Atlantic Coast. Today it is the home of the Virginia Aquarium & Marine Science Center. Although the Virginia Aquarium & Marine Science Center is involved in research, it is primarily an educational institution.

Two pavilions that sit on this land, the Bay and Ocean Pavilion and the Marsh Pavilion, are separated by a nature path 0.3 miles (0.5 km) long. The larger of the two structures is the Bay and Ocean Pavilion. The main exhibit here follows the movement of water as it proceeds across Virginia. Along this path sits the very large Norfolk Canyon Aquarium, which includes three types of sharks and a number of varieties of rays, as well as other sea creatures, all found in Virginia.

The Chesapeake Light Tower Aquarium, with a capacity of 70 000 gallons (260 000 L), showcases sea turtles (an endangered species), one of the objects of research of this institution, along with various fish (such as Atlantic spadefish, cobia, permits, and lookdown fish) that live about 15 miles (24 km) from the Virginia coastline.

An Indonesian volcanic island stands in for Virginia's volcanic geography a billion years ago, though modern Komodo dragons, Indonesian finches, and more populate the exhibit. For the Silurian period (430 million years ago), when seas and swamps dried out, leaving salt deposits, the Coastal Sahara Desert exhibit displays seahorses, cuttlefish, cobras, scorpions, and hedgehogs. The Malaysian Peat Swamp acts as the model for the Carboniferous Period (over 300 million years ago) in Virginia, with animals from the Malaysian ecosystem including snakehead fish, Southeast Asian turtles, the Malayan leaf frog, and the tomistoma crocodile. Finally, for the Triassic period (over 200 million years ago), in which continents split apart, the Red Sea exhibit recalls Virginia's Piedmont Plateau, revealing live coral, eagle rays, and a zebra shark that swim among other fish from that ecosystem.

Whales are included in the science center in the exhibit Whales: Voices in the Sea. The numerous interactive displays provide the calls of many species of whales, interviews with marine scientists, and the issue of marine conservation.

Because of its location—on the last salt marsh in Virginia Beach with direct access to the ocean—several exhibits focus on this ecosystem. The Salt Marsh Room has dioramas and interactive games examining this biome and answering such questions as: How much salt is in a salt marsh?

To reach the outdoor salt marsh, visitors walk through an aviary filled with indigenous salt-marsh birds. While strolling along the marsh's boardwalk, one-fifth of a mile (0.3 km) long, look closely to see fiddler crabs, wild birds, and other animals.

One of the museum's major research projects involves sea turtles. Of the seven existing species of sea turtles, six are endangered or threatened. Four of these sea turtles visit Virginia waters: loggerhead, leatherback, Kemp's ridley, and green. In cooperation with the Virginia Institute of Marine Science, the Columbus Zoo (see Columbus Zoo and Aquarium, Columbus, Ohio), and the US Fish and Wildlife Service, the museum is studying nesting and hatching possibilities to give the fragile species a better chance at survival. The museum is also part of the Sea Turtle Stranding and Salvage Network of the National Marine Fisheries Service, a group that seeks to preserve the sea turtle population. (See Miami Seaquarium, Miami, Florida.) A special exhibit has aquariums with adult and juvenile loggerhead turtles.

The largest seawater aquarium in the museum is the 50 000-gallon (190 000-L) Chesapeake Bay Tank, which re-creates the bay from the Chesapeake Bay Bridge to the north and the first island of the Chesapeake Bay Bridge-Tunnel to the south. The floor of this hall slopes downward to simulate walking under the waves to deeper waters. In "deeper" areas, the visitor comes face to face with red drums and other marine creatures. In a touch tank representing shallow areas of the bay, there are horseshoe crabs and other marine life for the visitor to handle.

Behind the Scenes tours of the site teach, in more detail, about the center's sea turtles and harbor seals, and how aquariums operate. There is a 65-foot (20 m) catamaran that offers seasonal trips for whale- and dolphin-watching.

Address: Virginia Aquarium & Marine Science Center, 717 General Booth Boulevard, Virginia Beach, VA 23451.
Phone: (757) 385-3474.
Days and hours: Daily 9 a.m.–5 p.m. Closed Thanksgiving and 12/25.
Fee: Yes.
Tours: Self-guided.
Public transportation: Local HRT buses stop about one block from the site, then walk. During the summer, trolleys stop at the aquarium.
Handicapped accessible: Yes.
Food served on premises: Café and snack stands.
Special note: National Geographic 3D Theater at the Light Tower Aquarium, boat trips, and Behind the Scenes tours cost an extra fee, and require advance reservations.
Website: www.virginiaaquarium.com/

WILLIAMSBURG

Colonial Williamsburg

A visit to this site is a step back in time to the eve of the American Revolution. Williamsburg was the capital city of colonial Virginia, the oldest, largest, richest, and most populous of England's colonies. (See Jamestown Settlement, Williamsburg, Virginia.) The differentiation between town and country was not clear in the complex community of 220 acres (89 hectares) and almost 2000 inhabitants.

After the Revolution, Williamsburg remained the capital of Virginia for another 80 years, until the capital moved to Richmond. Williamsburg continued to be a prosperous city and its 18th-century buildings were used until the 20th century, when the historic section was restored to its original form to serve as a living-history museum (See Old Sturbridge Village, Sturbridge, Massachusetts).

The Reverend W. A. R. Goodwin, rector of Bruton Parish, initiated the Colonial Williamsburg project in the 1920s by contacting John D. Rockefeller, Jr., and interesting him in efforts to restore the city. Restoration began later in the decade based on in-depth research and information revealed in archaeological excavations. Today, the 301-acre (122 hectare) Historic Area has public buildings, shops, houses, taverns, and gardens that were preserved or re-created in or adjacent to the site. Eighty-eight original buildings and many others are reconstructed on original sites, and the restoration process continues. The historic area is part of the modern city of Williamsburg. Employees of Colonial Williamsburg use many of the restored and reconstructed houses as their own homes.

Several archaeological displays, including exhibits at the Anderson's Blacksmith Shop & Public Armoury, explain how the town was restored. While wandering through the town, the visitor can watch trades- and craftspeople using tools and methods of the 18th century. Among them are blacksmiths, silversmiths, gunsmiths, carpenters, weavers, and coopers. Several forges are in operation. James Anderson, who created and repaired arms and armor, employed blacksmiths to operate his forges, which have been reconstructed behind his house. The weavers use typical colonial technology, including period looms and spinning wheels. Occasionally demonstrations of fabric-dyeing using vegetable dyestuffs of the era are held.

Shoemaking, printing, bookbinding, and wig-making are demonstrated. Clementina Rind, who operated a public printing business, published for the House of Burgesses and from 1773–1774 was editor of the *Virginia Gazette*. She was the first woman in the city to operate this kind of establishment. At the printing office, post office, and bookbindery, once the communications center of the town, the visitor can watch newspapers, handbills, and books being printed on an 18th-century printing press and can see books being bound. The town also has a functioning sawmill, brickyard, and carpenter's yard. Stores replicating those of the 18th century sell typical products of the day.

At the Octagonal Magazine, erected in 1715 to store firearms and military equipment, period firearms are displayed and costumed interpreters demonstrate their use. A replica of a 1750 Newsham patent fire engine, which stands near the guardhouse, is taken through the streets and demonstrated during warm weather.

The Pasteur & Galt Apothecary Shop displays blue and white Delft jars containing traditional remedies. The shelves hold surgical instruments, books on the latest medical theories, splints, and a fracture box for stabilizing broken limbs. Also visit the Herb Garden, behind the apothecary shop, where medicinal plants grow.

Several of the homes, such as the Governor's Palace, display the period's household technology. On view are the palace's kitchen, gardens, and stable areas. On the grounds, a wheelwright demonstrates his craft. The Colonial Garden is a reconstruction of a typical wealthy homeowner's garden, with vegetables and herbs typical of the period. In the winter, colonial-era bell jars and oiled-paper tents are used to protect many of the plants. A Kitchen shows the culinary technology of the 18th century, including ceramic jars and wooden barrels for food-storage, and wrought-iron cooking implements.

Virginia Governor Fauquier, a member of the Royal Society of London, established a hospital for the mentally ill at Williamsburg in 1773, the earliest facility of its type in North America. It was destroyed by fire in 1885 but has been reconstructed. Inside the Public Hospital, compare a patient room from the late 18th century with one from the mid-19th century to see how treatments for the mentally ill evolved over the years. Displays show equipment and explain theories and treatment of the mentally ill from 1773 to 1885. (See Glore Psychiatric Museum, St. Joseph, Missouri.)

In the DeWitt Wallace Decorative Arts Gallery there is an extraordinary collection of American antiques and a display of 18th-century timekeeping, surveying, and measuring equipment.

A number of pastures located throughout the town are stocked with livestock selected from rare breeds of cattle (American Milking Red Devon), horses (American Cream Draft, Canadian), sheep (Leicester Longwool), and poultry (Dominiques, Red Dorkings, and more), to show animals similar to those that would have been found in colonial America.

The country's second-oldest institution of higher education, the College of William and Mary, sits at the edge of the historic area. The 1695 Wren Building is the oldest academic building still used in America.

Address: Colonial Williamsburg, 101 Visitor Center Drive, Williamsburg, VA 23185.
Phone: (888) 965-7254.
Days and hours: Daily 9 a.m.–5 p.m. There are often additional evening programs.
Fee: No fees to walk around the town; fees to enter exhibits.
Tours: Self-guided; guided tours are available at extra charge. In the Visitor Center, get daily schedules to determine what demonstrations and events are happening; not all houses are open every day.
Public transportation: Amtrak trains stop some blocks away; walk or take a taxi to the site. Shuttle-buses run regularly around the perimeter of the grounds.
Handicapped accessible: Some houses are not accessible.
Food served on premises: Taverns, restaurants, cafés. Plan ahead, for eating establishments are often crowded.
Special note: Lodging is available on the historic premises. Lodging and restaurants require early reservations.
Website: http://www.colonialwilliamsburg.com/

Jamestown Settlement and Historic Jamestowne

Two modern facilities, very near each other, examine the history of Jamestown: Jamestown Settlement and Historic Jamestowne. Jamestown Settlement, a re-creation of the original settlement, is operated by the Jamestown-Yorktown Foundation, an agency of the Commonwealth of Virginia.

Historic Jamestowne, next door to Jamestown Settlement, reveals through active archaeology, along with a trove of archeological finds installed in a museum, the original layout of the fort. It is located in Colonial National Historic Park on the site of the original settlement.

Jamestown, originally known as James Fort, was the first permanent European settlement in Virginia, in the first English colony in the New World. The Virginia Company of London sponsored James Fort in 1607 as a profit-making enterprise in the country of Tsenacommacah, ruled by the Powhatan Confederacy. This area soon became the English colony of Virginia. The name of the settlement was changed to Jamestown and became the capital of Virginia in 1616.

The fort was founded on an island in the James River because the surrounding water was deep enough to anchor ships and the location seemed a good, defendable choice. But the land turned out to be a swampy, disease-ridden place with brackish water, and not at all easy to defend. Mortality was high, and 80% of the population perished in 1609. Of the 214 settlers, only 60 survived. This period was known as "Starving Time."

The Virginia Company sent eight men from Poland, Germany, and Slovakia in 1608. These newcomers produced glass, then they cultured silkworms, and in 1613 they finally found their economic base in tobacco. They planted this crop in all open spaces; tobacco even became their currency.

Jamestown served as Virginia's capital until 1699, when the statehouse burned for the fourth time and the Virginians decided to move their capital to higher ground 5 miles (8 km) away. They chose Middle Plantation, between the York and James rivers, and changed the town's name to Williamsburg. (See Plimoth Plantation, Plymouth, Massachusetts; Colonial Williamsburg, Williamsburg, Virginia.) Jamestown survived as a town for another 100 years before it was fully abandoned.

Jamestown Settlement

Jamestown Settlement has indoor exhibits and an outdoor living-history display. Indoors, a docudrama film tells the story of the original Jamestown, from its beginnings in 16th-century England through its first quarter-century of existence. The museum galleries present dioramas, audio effects, graphics, and artifacts such as tools, weapons, furnishings, household utensils, navigation instruments, and maps presenting a complete and detailed picture of the town.

The Jamestown settlers came at a time when great innovations in cartography, ship design, and navigational equipment made the long, arduous trip across the Atlantic possible. There are indications that *The Tempest* by William Shakespeare was based on a shipwreck that happened to a group of settlers on their way to Jamestown.

The local indigenous Powhatan people, who arrived about 10 000 years ago, had their own technology. How they lived on their land, the evolution of their technology, and their lifestyles before the English came are presented. Exhibits follow the Virginia colony's progress from a humble outpost at Jamestown struggling for survival to a successful permanent colony, and note the government's move to Williamsburg in 1699.

Three acres (1.2 hectares) of outdoor living-history exhibits are staffed with costumed interpreters who encourage visitors to participate as they carry out the tasks of people who lived in the settlement. The Powhatan Indian Village was designed from eyewitness drawings by Englishman John White, who tried unsuccessfully to colonize the area some 20 years before Jamestown was founded. This village has several lodges or longhouses containing domestic furnishings, a garden, and a ceremonial dance circle. There are demonstrations of how the Powhatans prepared food, tanned animal hides, fashioned tools from bone and antlers, and made flint projectile points and pottery.

Moored alongside a pier in the James River harbor are full-sized re-creations of the three ships used by the British who colonized Jamestown in 1607. The largest ship, the *Susan Constant*, has interpreters who discuss shipboard life during the four and a half-month voyage from Europe and demonstrate the crafts and skills the travelers needed to survive. James Fort is a re-creation of the small, triangular fort built in 1607. It contains buildings of wattle-and-daub construction and thatched roofs. Wattle, or woven sticks between the beams, is filled with daub, a mixture of mud, dung, and straw. Around these buildings, costumed interpreters perform animal husbandry, agriculture, blacksmithing, meal preparation, and military musters in 17th-century style.

Address: Jamestown Settlement, 2110 Jamestown Road, Williamsburg, VA.

Phone: (757) 253-4838.

Days and hours: Daily 9 a.m.–5 p.m. (until 6 p.m. mid-Jun.–mid-Aug.). Closed 12/25 and 1/1.

Fee: Yes.

Tours: Self-guided.

Public transportation: Amtrak station in Williamsburg, taxi to Jamestown.

Handicapped accessible: Yes in most areas, but limited accessibility on ships. Scooters are available.

Food served on premises: Café.

Website: historyisfun.org/Jamestown-Settlement.htm

Historic Jamestowne

The original Jamestown site is adjacent to Jamestown Settlement. Upon arrival, enter the Visitor Center for an introductory film and exhibits before following a path to the ruins of the town, James Cittie. At Dale House, one of the re-created buildings, potters demonstrate how they perform their craft using 17th- and 18th-century designs. Notice that the foundations of the homes in the New Town, built after the profitable tobacco trade was established, are considerably larger than those in the Old Town, indicating that as the economy improved, house construction became more sophisticated. Excavations of the foundations reveal that some structures were covered with whitened brick, suggesting that the inhabitants had the money and time to decorate their houses.

A highlight of Jamestowne is the reconstructed glasshouse, where costumed craftspeople demonstrate glassblowing as it was practiced here in the 17th century. Glassblowing was one of the first industries in Virginia. The ruins of the original glass furnaces, built in 1608, are nearby. (See Blenko Glass Company, Milton, West Virginia.)

Visitors can tour the ruins by themselves, or take a guided tour with a park ranger. Along the way, paintings and audio messages provide information about the excavated foundations. Two driving loops, one 3 miles (5 km) and the other 5 miles (8 km), afford the visitor the chance to see the island's wilderness much as the colonists knew it.

The nearby Voorhees Archaearium is the archaeological museum housing the thousands of artifacts uncovered from James Fort. Among the artifacts on display include numerous iron implements, implements, weapons, armor, Native America beads and pottery, skeletons of several inhabitants, coins, glass, and even a Nuremburg pocket sundial from ca. 1610 (which would not have worked properly at the more southern latitude of Jamestown). An extensive display discusses forensic analysis of the first known victim of colonization cannibalism in North America, "Jane," a 14-year-old English girl, who was apparently eaten during the "Starving Time" of 1609–1610.

Address: Historic Jamestowne, Colonial National Historical Park, 1368 Colonial Parkway, Jamestown, Virginia 23081.

Phone: (757) 856-1200.

Days and hours: Daily 9 a.m.–5 p.m.

Fee: Yes.

Tours: Self-guided or guided.

Public transportation: Amtrak to Williamsburg, then taxi.

Handicapped accessible: Yes. Scooters are available.
Food served on premises: Café.
Website: www.nps.gov/jame/index.htm

West Virginia

Milton

Blenko Glass Company

Glassmaking, a major industry in West Virginia, is illustrated in detail at this factory. The state is an ideal place for making glass because it has the necessary ingredients: sand; soda ash, also known as technical-grade sodium carbonate; calcium carbonate or limestone; and a metal oxide, such as lead oxide or boric oxide. For clear glass the preferred sand is high in silica and low in iron oxide and alumina. Another important component is broken glass bits or cullet. Glass production requires a considerable amount of fuel because sand must be heated to about 2500–3000°F (1370–1650°C) before it will melt.

Early in our nation's history the hub of the glass industry was New England, where wood was plentiful. As the forests became depleted, however, glass production moved southward and settled near the coal and gas fields of Appalachia. New, improved glassmaking furnaces required these higher-energy-yielding fuels.

West Virginia became a popular location for glass manufacturers not only because of its large bituminous coal and natural gas supplies, but also because of its proximity to rail lines for sending the glass to market. In the third quarter of the 19th century, Collis P. Huntington, president of the Chesapeake & Ohio Railroad, built a railroad through West Virginia as part of the rail link between Maryland and Ohio. (See California State Railroad Museum, Sacramento, California.) He also built a new town and named it after himself. Huntington, West Virginia, became a busy industrial center and shipping point in a region where glass manufacturers flourished.

William Blenko, a glassmaker from London, shipped stained glass to the United States in the mid-1890s and then came over himself to find a place to produce glass for church windows and other uses. The business failed and he went back to England, but he later returned with his son. In 1922 they established the Blenko Glass Company in Milton, West Virginia, about 20 minutes from Huntington. Over the years this company expanded to make a variety of decorative glass items, and it is still a successful manufacturer. The Blenko Glass Company provided glass for Saint Patrick's Cathedral and the Cathedral Church of Saint John the Divine in New York City, the rose window in the National Cathedral in Washington, DC, the Harkness Library of Yale University in New Haven, and the Heinz Chapel in Pittsburgh. Both Eleanor Roosevelt and Mamie Eisenhower owned Blenko glassware.

The Visitor Center has exhibits tracing the history of glassmaking and displays on stained glass from nine leading American stained-glass studios. It also exhibits historical glassware, including a Country Music Award, gifts from US presidents to dignitaries, globes once used for lighting at the US Capitol, and original Williamsburg stemware. A small modern model of the first American glassmaking facility at Jamestown, Virginia is on display. The original glass factory was built in 1608. (See Jamestown Settlement and Historic Jamestowne, Williamsburg, Virginia.) Set into the wall of the museum is a contemporary piece by Martha Farley, produced by Old Dominion Glass Company in Ashland, Virginia, entitled the Early American Glass Blower.

On a self-guided walking tour of the factory, visitors can see craftsmen producing hand-blown and stained-glass with contemporary designs using the same tools and methods that have been around for hundreds of years.

Address: Blenko Glass Company, 9 Bill Blenko Drive, Milton, WV 25541.
Phone: (304) 743-9081.
Days and hours: Factory, Mon.–Fri. 8 a.m.–noon, 12:30 p.m.–3:15 pm. Museum, Mon.–Fri. 8 a.m.–4 p.m., Sat. noon–4 p.m.
Fee: Free.
Tours: Self-guided.
Public transportation: None.
Food served on premises: None. A number of fast-food restaurants are on East Main Street (US Route 60).
Handicapped accessible: Yes.
Website: www.blenko.com/

4

Mississippi Valley

Arkansas

HOT SPRINGS

Hot Springs National Park

On the western slope of Hot Springs Mountain are 44 springs that discharge 850 000 gallons (3.2 million L) of hot water a day. These springs, 143°F (62°C) at the surface, are formed of rainwater that has seeped into the mountain's dense Bigfork chert and Arkansas novaculite for 4000 years. The mildly alkaline water is heated in the bowels of the earth, then rises along layers of rock and bubbles to the surface. When the hot water evaporates, it deposits dissolved minerals, forming calcareous tufa rock. This tan-colored rock, visible from the park's Grand Promenade, supports unique algae life. Because the water is sterile just beneath the surface, NASA used it to store rocks brought back from the Moon while they were being examined for signs of life. The chemical composition of the minerals in the water is primarily bicarbonate ions, silica, and calcium ions.

The springs contain 37 different algae, the only plant life able to grow in these hot waters. A rare species known in very few places in the world, *Phormidium treleasei*, was discovered here in 1932. Today, most of the springs are covered to prevent contamination. Park grounds are blanketed by dense forests of oak, hickory, and short-needle pine. Songbirds and small animals fill this forest. Spring, when the rosewood and dogwood are in bloom, is a beautiful time to visit.

Local Native Americans knew about the springs thousands of years ago. After the Louisiana Purchase in 1803, President Thomas Jefferson sent the Hunter-Dunbar expedition to explore the springs. In 1832, to protect the springs, the federal government set aside part of the land as Hot Springs Reservation, which makes this site the oldest federally protected land in the country. By the 1870s the area became a health spa, becoming known as the "American Spa." When it was designated the 18th National Park in 1921, Hot Springs had several bathhouses where visitors could relax and be treated for such infirmities as rheumatism, kidney disease, liver problems, and polio. In the 1960s the resort had declined and the bathhouses, many of which were elegantly constructed of marble and stained glass, began closing. Now only two original bathhouses, Buckstaff and Quapaw, remain to provide visitors with the traditional services. The city of Hot Springs National Park, adjacent to the park, has several spa facilities.

The National Park Service has restored the Fordyce Bathhouse to its former glory and now uses it as the park's Visitor Center. It was once Hot Springs' grandest bathhouse. Here visitors can explore 24 rooms that served the ailing. In the Ladies' Bath Hall, personal attendants helped women through regimens of tub baths, sitz baths, vapor cabinets (which encased everything but the head in steam), and high-pressure needle showers. The elegant Men's Bath Hall had a central fountain flowing with medicinal drinking water from the springs. In the Pack Room bathers lounged on porcelain cots while being

Thermal Water Cascade at Arlington Lawn, Hot Springs National Park.
Courtesy of Hot Springs National Park.

wrapped in steaming towels and then took needle showers before moving to the Cooling Room, where they rested and cooled off. Later, perhaps, the bathers had a massage or other treatments upstairs.

The Hydrotherapy Room displays fascinating equipment for treating patients. Here there are sun-ray cabinets in which light bulbs created a hot, dry atmosphere; sprays; douches; sitz baths; and electrotherapy baths infused with a mild current. In the Chiropody Room visitors learn about what was considered "the latest" in foot-care. The Electro-Mechano Therapy Room has exercise equipment designed by Gustav Zander, a 19th-century Swedish physical therapist whose treatments and devices were used internationally. The Men's Massage Rooms have electric massage devices and an electric shock massager that was used to make the muscles contract and relax. A nine-minute video in the Exhibits Room explains traditional bathing procedures. The room also has displays about bathing cures.

On the third floor is the Hubbard Tub Room, where a licensed physiotherapist used the large tile-covered tub, installed in 1939, to treat patients. Hubbard tubs are still used for physical therapy. Both men and women took advantage of the third floor's large, attractive gymnasium, which contains exercise equipment, punching bags, and other gear still popular today. Jack Dempsey, Joe Lewis, and Babe Ruth are among early 20th-century sports greats who came here.

The park, the second-smallest in the National Park system, encompasses more than 5500 acres (2200 hectares) and 47 hot springs. Although most visitors spend their time seeing the city and bathhouses, there are a number of mountain hiking trails and scenic drives to explore. Park rangers offer walks through the park, some of which lead to hot springs, as well as mountain hikes and amphitheater programs describing the area's human and natural history. A half-mile-long (0.8-km) Grand Promenade, just behind the bathhouses, meanders between several covered springs and affords a lovely view of the park. Visitors can climb a 216-foot (65.8 m) observation tower at the summit of Hot Springs Mountain to see a grand vista of the Ouachita River Valley and environs.

Address: Fordyce Bathhouse (Hot Springs National Park Visitor Center), 369 Central Avenue, Hot Springs, AR 71901. GPS: Latitude: N 34° 30' 43.6391" Longitude: W 93° 3' 13.6398".
Phone: (501) 620-6715.
Days and hours: 9 a.m.–5 p.m. daily. The Fordyce Bathhouse Visitor Center is closed Thanksgiving, 12/25, and 1/1. Other park facilities have varying hours; check website in advance.
Fee: Free.
Tours: Self-guided in the bathhouse and park. Regularly scheduled ranger walks.
Public transportation: Greyhound Buses stop in Hot Springs.
Handicapped accessible: Yes.
Food served on premises: None. There are nearby restaurants in town, mostly on Central Avenue (Rte. 7).
Website: www.nps.gov/hosp/index.htm

Kentucky

LEXINGTON

Kentucky Horse Park and American Saddlebred Museum

Kentucky Horse Park
Patrick Henry, governor of Virginia, owned this property in the Kentucky Territory; in 1777 he granted 9000 acres (3600 hectares) of the land to William Christian for Christian's service in the French and Indian War. Part of this property became the Kentucky Horse Park, which opened in 1978.

Owned by the Commonwealth of Kentucky, Kentucky Horse Park is the only place in the world dedicated to the long history of horses and their relationship to mankind. The park's 1000-acre (400-hectare) expanse of rolling hills carpeted by native Kentucky bluegrass is in the heart of the world's most concentrated collection of horse farms. More than 40 breeds of horses are housed here in order to inform the public about the relationship between humans and the horse. Visitors can tour the park on foot or take a tram or carriage tour.

Breeding of horses began at this site by the 1850s. The 1866 house formerly owned by S. J. Salyers, a Thoroughbred horse breeder, now contains the park's offices. The 1879 Big Barn, one of the largest barns ever built (476 ft. [145 m] long and 75 ft. [23 m] wide), is still in use, as is the 1897 track.

At the entrance to the park is a bronze statue of the great racehorse Man O' War, who was born at Faraway Farm near Lexington, and who set world records and American records for speed. At the Visitor Information Center visitors can get information about the

Kentucky Horse Park.
Courtesy of the Kentucky Horse Park.

park's many activities, all related to horses. A film that discusses the relationship between human beings and horses runs throughout the day.

As visitors tour the park, they will see the day-to-day workings of a Kentucky horse farm. In the Breed's Barn one can have a close-up look at horse breeds in their stalls. The Big Barn houses draft breeds of the Kentucky Horse Park. Visitors may stop at the harness maker's shop and the farrier's shop, where leather goods and tack are created. The staff members and craftsmen are happy to talk about their skills.

During the 30-minute Parade of Breeds, performed twice a day, different breeds and their riders appear in authentic costumes representing the region of the world where the horse originated. The riders demonstrate the horses' gaits while an announcer discusses their unique traits. A variety of breeds are displayed. There is no scientific definition of a breed, but more colloquially, a breed is considered as possessing characteristics that breed true over the course of generations. Hundreds of breeds exist, generally divided into three sorts of temperament. For endurance and speed, there are "hot bloods"; for heavy and long work there are "cold bloods" (i.e., certain ponies and draft horses); and for specific riding purposes there are crosses of the two, "warmbloods."

On the grounds is the International Museum of the Horse, a state-of-the-art facility founded in 1978 with interactive exhibits on the history of how people and horses have coexisted. Considered the largest and most comprehensive in the world, the museum explores the history of the horse and how this animal has affected human civilization. Exhibits trace equine history from the *Eohippus*, a small horse-like mammal that lived 58 million years ago, to the present. A special wing added in 2010 features the Arabian horse, a breed found in pedigrees of nearly every other breed. Visitors can see bits, stirrups, spurs, saddles, horse-drawn vehicles, and racing memorabilia from different eras. Especially interesting are the 17th-century German cage stirrups.

Also included in the museum is the International Library of the Horse, holding over 8000 books, open to scholars by appointment only.

Address: Kentucky Horse Park, 4089 Iron Works Pike, Lexington, KY 40511.
Phone: (606) 233-4303.
Days and hours: 9 a.m.–5 p.m. Closed Mon. and Tue. from Nov. to Mar. Closed Thanksgiving, 12/25, 12/31, and 1/1.
Fee: Yes.
Tours: Self-guided; guided tours are run twice a day (call ahead for times).
Public transportation: None.
Handicapped accessible: Yes.
Food served on premises: Restaurant and snacks. Campground store offers food for overnight campers.
Special note: The horse park has a campground. See website for details.
Website: www.kyhorsepark.com

American Saddlebred Museum
Although the American Saddlebred Museum is on the park grounds, it is not part of the Kentucky Horse Park. It focuses on the oldest breed of horse registered in America. The American Saddlebred horse resulted from crossbreeding Thoroughbreds with American

American Saddlebred Museum.
Courtesy of the American Saddlebred Museum.

horses and Narragansetts, the mainstay of American pioneers on their westward trek. Used by both Union and Confederate forces during the Civil War, American Saddlebred horses proved superior on the battlefield. After the war they became very popular as show horses, and remain so today.

The Museum maintains a library of 3000 volumes used for equine genealogical research.

Address: American Saddlebred Museum, 4083 Iron Works Parkway, Lexington, KY 40511.
Phone: (859) 259-2746.
Days and hours: Daily 9 a.m.–5 p.m. Closed Mon. and Tue. from Nov. to Mar. Closed day before Thanksgiving, Thanksgiving Day, 12/24, 12/25, 12/31, and 1/1.
Fee: Yes.
Tours: Self-guided.
Public transportation: None.
Handicapped accessible: Yes.
Food served on premises: None. The Kentucky Horse Park has venues.
Website: www.asbmuseum.org

Monroe Moosnick Medical and Science Museum and Hopemont

Monroe Moosnick Medical and Science Museum

Transylvania University, in the center of historic Lexington, is the oldest college west of the Allegheny Mountains. Nineteen years after its 1780 founding, it opened a medical college, which set about collecting scientific equipment and books to better educate its students (see College of Physicians of Philadelphia and the Mütter Museum, Philadelphia, Pennsylvania). These later became part of the collection in the Monroe Moosnick Medical and Science Museum, which visitors can enter by appointment only.

The museum is named for Dr. Monroe Moosnick, longtime chemistry professor at the university and curator of the collection. Administrative offices are housed in Old Morrison, an elegant Greek Revival structure with massive white Doric columns and pediment, built in 1833. The campus is on the north end of the antebellum Gratz Park. Items in the museum displays range from skeletons and taxidermy to wax models of organs and limbs, to ancient physics, chemistry, meteorology, and medical equipment. Artifacts from the collection are displayed also on a rotating basis in the Brown Science Center's lobby.

The museum's first major acquisitions were made in Paris between 1780 and 1821. Many were already old and venerated for their historic importance; others, especially those needed for classroom demonstration, were the latest of the time. Among pieces purchased from Antoine-Hippolyte Pixii, the Parisian instrument maker, are Atwood's Machine, designed to illustrate the laws of falling bodies; a Saussure hygrometer that uses human hair to measure changes in atmospheric moisture; and a vacuum pump. When a new medical building was constructed in 1839, a second European shopping expedition yielded a Wollaston's goniometer for measuring the angle of crystal faces and a Wedgwood pyrometer for measuring temperatures between 1000°F and 3000°F (540°C and 1650°C) in crucibles and furnaces. Also acquired at this time were a wax model arm for the anatomy classes; 40 canvasses of medical plants for the pharmaceutical classes, painted at the Jardin des Plantes in Paris by medical illustrator Dr. A. Chazal; and a thunder house made by Watkins & Hill to show that a pointed rod attracts lightning better than a rounded rod. When the rod is struck by a flash of static electricity, gunpowder (inside the model of a house) explodes. (See Harvard Museums of Science and Culture and Warren Anatomical Museum, Cambridge and Boston, Massachusetts.)

In 1865 the Kentucky Female Eclectic Institute merged with Transylvania University and added its equipment, including a 1791 planetarium; a lunarium and tellurion showing how the Earth's movement creates day, night, and seasons; and continuous-movement pumps made by Charles Chevalier.

Address: Monroe Moosnick Medical and Science Museum, 300 North Broadway Road, Lexington, KY 40508.
Phone: (606) 233-8213.
Days and hours: By appointment only.
Fee: Free.
Tours: Self-guided.
Public transportation: Lextran buses regularly stop at the Transylvania University campus.
Handicapped accessible: Yes.
Food served on premises: Campus cafeteria, during the academic semester.
Website: libguides.transy.edu/SpecialCollections/Moosnick

Hopemont

On the south side of Gratz Park is Hopemont, also known as the Hunt-Morgan House, which recognizes the contributions of John Wesley Hunt and family members who succeeded him. Hunt, the first millionaire west of the Allegheny Mountains, build this Federal-style town house in 1814. The most famous of his descendants are Confederate General John Hunt Morgan and Dr. Thomas Hunt Morgan. The latter's work in genetics, confirming the theories of Gregor Mendel, won him the 1933 Nobel Prize for Medicine and Physiology. He documented his study of fruit flies' chromosomes in his book The Theory of the Gene.

The house has a beautiful Palladian window, a three-story cantilevered staircase, a charming garden, a walled courtyard, a collection of historic furniture, early 19th-century portraits, and fine porcelains. On the tour, docents relate the family's history.

Address: Hopemont, 201 North Mill Street, Lexington, KY 40507.
Phone: (859) 233-3290.
Days and hours: Apr.–Oct.: Wed.–Sun. 1 p.m.–4 p.m., Sat. 10 a.m.–3 p.m. Special tours with advanced booking only, Nov.–Mar.
Fee: Yes.
Tours: Guided only.
Public transportation: Lextran buses stop at the site.
Handicapped accessible: No.
Food: None. A variety of restaurants are found a few blocks southwest of the site.
Website: www.bluegrasstrust.org

LOUISVILLE

Louisville Zoological Gardens

A visit to the sprawling Louisville Zoological Gardens, known as "the Zoo," seems to transport us to a far-off land. Some 215 species of animals live on 134 acres (54 hectares) of naturalistic exhibits, representing a number of zoogeographic areas: Africa, Arctic climates, Australia, tropic zones, oceanic islands, and South America. Be prepared to do some walking, or ride the tram for a tour.

This zoo emphasizes the importance of respecting the animals and plants with which humans share the planet. Widely known for its pioneering techniques in conservation and animal breeding, the zoo participates in Species Survival Plans® to save 14 different animals from extinction. This facility was the first site to transfer an embryo from an exotic member of the equine family (a Grant's zebra) to a domestic equine (a quarter horse). Along with three other zoos in the country, it is trying to preserve the black-footed ferret, the most endangered mammal in North America. The last 15 ferrets in the wild were captured in 1986 and placed in a breeding program, with the hope of returning them to their native Wyoming habitat. The zoo has plans to exhibit the ferrets in the future. (See Cheyenne Mountain Zoo, Colorado Springs, Colorado; and Phoenix Zoo, Phoenix, Arizona.)

Of particular renown are the zoo's woolly monkeys, one of the largest colonies of these animals in the world. In another popular exhibit, visitors can watch polar bears frolic on land, and see them through a 30-foot (9.1-m) panoramic glass window as they swim underwater.

The Africa area, with its beautiful vistas, is inhabited by springbok, lions, giraffes, a variety of cranes, and other animals. Elephant Encounter provides a view of how the zoo staff members enrich the lives of the elephants. A 4-acre (1.6-hectare) Gorilla Forest displays a troop of about 10 gorillas in a naturalistic habitat that also houses pygmy hippos.

While strolling through the Australian Walkabout, note the hidden barriers that safely separate onlookers from the wallabies, wallaroos, emus, and other animals. Adjacent to the Australian exhibit is Lorikeet Landing, an aviary where guests can meet lorikeets, colorful birds that populate Indonesia, Papua New Guinea, and Australia.

Designed to replicate an Alaskan-style mining town, Glacier Run showcases polar bears, grizzly bears, sea lions, and Steller's sea-eagles which inhabit a 50-foot (15 m) aviary. Visitors view the polar bears through a variety of windows above and below the waterline.

The HerpAquarium is a combination of aquarium and herpetarium housing 110 species of reptiles, amphibians, fish, mammals, birds, and invertebrates. Among the biomes represented here are the diurnal (daytime) and nocturnal (nighttime) desert, the diverse worlds of water, and a neotropical rain forest with its 85°F (29°C) temperature and 85% relative humidity. More than 100 species live in the rain forest, including a Cuban crocodile. During the periodic rainstorms, note how the animals change their behavior.

Visitors can view displays of particular animals and the stress they encounter by living in these island environments. The Indonesian Village area, for example, has Sumatran tigers, Malayan tapirs, and orangutans. Penguin Cove is the residence for little penguins that live in Australia and Tasmania.

The South America exhibit has a realistic arrangement of tall, rocky cliffs, a waterfall, a stream, and foliage in which lynx and pumas live as they would in the wild. It includes a spacious display of bald eagles, the majestic and endangered symbol of the United States.

In keeping with the philosophy that the zoo is a living classroom, it contains the MetaZoo Education Center. This facility encourages on-site study of animals and specialized educational programs in naturalistic settings.

Address: Louisville Zoological Gardens, 1100 Trevilian Way, Louisville, KY 40213.
Phone: (502) 459-2181.
Days and hours: Summer 10 a.m.–5 p.m. daily. Hours may change in other seasons. Closed Thanksgiving, 12/25, and 1/1.
Fee: Yes.
Tours: Self-guided.
Public transportation: Local buses go from downtown directly to the zoo.
Handicapped accessible: Yes.
Food served on premises: Cafés.
Website: louisvillezoo.org

Museum of the American Printing House for the Blind

The American Printing House for the Blind is the oldest national company in the United States serving people who are blind. It is also the world's largest publishing house of materials for the visually impaired. Although it was formed in 1858, its first tactile book did not appear until 1866, when operations restarted at the end of the Civil War. Today the facility produces recorded materials and books in braille and large type.

In 1879 Congress passed the Act to Promote the Education of the Blind, which allocates funds to produce books and other instructional materials. The law specifies that these materials be given free to public institutions in the United States that teach blind students younger than college age. The annual appropriation from Congress has grown from $10 000 in 1879 to many millions of dollars today. In addition to these moneys, the Printing House is financed by sales of materials it produces, public fundraising, and contract work.

The museum at the Printing House "explores the history of the education of people who are blind." Exhibits include the development of tactile methods for printing and writing, the work of Louis Braille, the history of the Printing House, and modern methods for educational instruction in various fields such as music, mathematics, and science. A notable collection at the museum includes a variety of mechanical braille writers for embossing purposes, ranging from a Hall Braille Writer (ca. 1892) manufactured in Chicago, styled like a contemporaneous typewriter, to an IBM Model D Braille Electric Typewriter. When visitors come to the museum, they should also take the tour of the Printing House.

The Printing House uses braille for its tactile materials. This writing and printing system for the blind consists of 63 possible combinations of dots, arranged in six-dot configurations called cells. On a tour through the facility, visitors can see braille books being produced and proofread. Words on each page are embossed on both sides from one metal plate, which is made by a high-speed, computer-controlled plate embosser.

Three methods are used to convert words into braille. Optical character recognition is a technique in which computers translate each letter into a braille symbol. When this is not possible, the key disk operator does a letter-by-letter transcription for the computer, which then translates the material into braille. Another choice is the braille transcription editor, a computer that produces the braille dot by dot.

Diagrams are made on a PEARL machine (plate-embossing apparatus for raised lines) that is similar to a sewing machine. The written material is then proofread by a blind person who reads the text aloud to a sighted person.

For books, special paper that is slightly dampened is used for printing braille; this ensures proper embossing. Magazines, however, can be printed on dry paper because they are used short-term and do not need to hold up over time. The pages are then collated, assembled on metal rings, and riveted into hard covers. Machinery processes the magazines so the raised dots are not flattened.

Publishing large-type materials for people with impaired vision often requires redrawing diagrams to enlarge them. The type size is 18 point (.25 in. or 0.64 cm high) or larger. When only a few copies are needed, standard printed materials are enlarged to about 150% of the original size and reprinted.

In the Talking Book studio, formed in 1936, the contents of books, magazines—such as *Reader's Digest* and *Newsweek*—and catalogs are recorded as they are read aloud. The readers are carefully selected for their diction and expressiveness. During a tour, the visitor can listen to the recording process unless the material has been deemed "not for public preview." The first such talking book produced here was a recording of *Gulliver's Travels*.

The facility also produces materials for teachers of reading, social studies, mathematics, and science designed to enhance students' sensory, motor, and conceptual development. The Department of Educational Research, formed in 1952, conducts research to improve the education of people who are blind, have multiple disabilities, or are about to enter school or a vocational program.

Address: Museum of the American Printing House for the Blind, 1839 Frankfort Avenue, Louisville, KY 40206.
Phone: (502) 895-2405.
Days and hours: Mon.–Fri. 10 a.m. and 2 p.m.
Fee: Free.
Tours: Groups of 10 or more may make advance reservations for a guided tour; individuals and smaller groups may take guided tours at 10 a.m. or 2 p.m.
Public transportation: Local buses go from downtown Louisville to the site.
Handicapped accessible: Yes.
Food served on premises: None. A few restaurants are on Frankfort Avenue, a few blocks east and west of the site.
Special Note: Regular events and lectures occur on-site; check the website for details.
Website: www.aph.org

WaterWorks Museum

By the side of the Ohio River stands a grand temple and tower, completed in 1860, which is a monument to the Victorian Age and its industrial might, known as the Louisville Water Works pumping station. Inside the pumping station is the Louisville WaterWorks Museum. This museum tells the story of the structure and tower, and how it brought enough water to supply the entire city.

Louisville, like many cities of the early 19th century, suffered from typhoid and cholera. In 1819, Dr. Henry McMurtrie viewed the local well water as "extremely bad" and called Louisville "the graveyard of the west." An Englishman, Dr. John Snow, was able to link contaminated water with typhoid and cholera in 1854. After several attempts by the city council to fund a company to make a reliable water supply, the Louisville Water Company became committed to the project also in 1854. Construction of a pumping station started in 1857.

The pumping station was situated upriver from the city's slaughterhouses, to avoid the water contamination from this industry. Designed by Theodore R. Scowden, an engineer, and his assistant, Charles Hermany, the pumping house is in Classical Revival style. Scowden said that the complex should be "regarded as the most elegant and commodious for water works purposes in the country." The engine room was housed in the center of the structure, boiler rooms flanked the central portico, and a water tower was constructed in the front. Modern water towers are generally large tanks sitting on legs or a column, but this tower was a standpipe, 169 feet (52 m) tall, to equalize the intermittent pulses of pressurization as the engines operated. The complex was a park-like setting, deliberately designed to draw the entire family away from the polluted city atmosphere and into greenery and bucolic landscapes.

The waterworks pumped water into a nearby hilltop reservoir (now the site of a Veterans Administration hospital) with an elevation of 90 feet (27 m) where the particulate debris could settle before delivery to customers. After a new reservoir was built elsewhere in 1879, the original reservoir was converted into a swimming and recreation area.

Pumping to the reservoir was achieved by two Cornish-beam engines, originally developed in England, and fabricated by Union Foundry in Louisville. Each engine had a cylinder 70 inches (180 cm) in diameter, the beams each weighed 22 tons (20 tonnes), and the engines worked alternately, with a capacity of six million gallons (23 000 m^3) each of water per day. Three boilers powered each engine; the system operated for almost half a century and was removed in 1911. The engines are gone from the engine room, and the huge area is now a hall for parties and weddings. Museum staff members show visitors photographs of the machinery as originally situated. The pump room went 45 feet (14 m) below the main floor, and the beams were supported by four massive cast-iron columns.

The wooden water tower was felled in 1890 by a tornado, leaving the brick base. In 1899 a new iron tower was completed. Decorating the base of the tower are 10 statues of Greek gods and the seasons, and a Native American hunter with his dog. Inside, the brackets that once supported a spiral stairway around the standpipe, and the remains in the floor of four smaller standby pipes around the central pipe, are on display. These standbys were removed as scrap metal to support the war effort in World War II.

The museum was created in 2014 inside the pumping station. The house and tower were designated a National Historic Landmark in 1971.

This museum displays original artifacts related to the WaterWorks, including an old mud wagon, used to pumped out water from the area around a broken main, an original section of cast-iron pipe that was in service for 150 years, and a section of wooden pipe that led to Camp Zachary Taylor, a military camp set up during World War I because the water quality in the area was considered high. An exhibit describes an award-winning technique, Riverbank Filtration, for water-collection in Louisville, which included constructing a gravity tunnel one-and-a-half miles (2.4 km) long, 150 feet (46 m) below ground and a system of wells to collect groundwater. Several short videos describe aspects of water-system repair after flooding.

This museum tells the history of improved water quality. Originally the mud and dirt merely settled to the bottom of the reservoir, which did nothing to remove microorganisms or toxic compounds. This exhibit, using charts, shows how the death rate dropped immediately after a sand-filtration scheme came on-line in 1909. By 1914 Louisville implemented chlorine as a disinfectant, and the death rate in the city fell even more.

Displays and videos describe several flooding events during the last century and a half, how the pumping station was temporarily shut down during the floods, and the massive efforts undertaken to restore water service in a matter of hours or days.

A walk of the grounds around the building and tower provide a lovely view of the Ohio River.

Address: WaterWorks Museum, 3005 River Road, Louisville, KY 40207.
Phone: (502) 897-1481.
Days and Hours: Wed.–Sat. 10 a.m.–5 p.m., Sun. noon–5 p.m. Closed Mon. and Tue., and on major holidays.
Fee: Yes.
Tours: The museum is self-guided. Guided tours enter the water tower. Guided group tours of at least 10 people are available with advance arrangements.
Public transportation: Buses run to Country Club Road and Zorn Avenue, then walk 0.8 miles (1.3 km) to site.
Handicapped accessible: Mostly. Entering the water tower requires climbing several stairs.
Food served on premises: None. There are restaurants in Louisville.
Website: www.LouisvilleWaterTower.com

NICHOLASVILLE

Harry C. Miller Lock Collection

The Miller name is inextricably bound to Lockmasters Security Institute, a company that educates locksmiths in their craft. In the 1940s, Harry C. Miller was a master locksmith in Fairfax, Virginia, mostly consulting with US government agencies, including the Department of Defense, servicing and installing locking mechanisms. At that time, he also invented a number of locks for protecting classified information for the government, such as sliding deadbolts, electromechanical locks, changeable combination locks, and manipulation-proof combination locks. In 1949 he came to an agreement with the lock company Sargent & Greenleaf for mass-production of his combination locks. Within two years Miller took control of the company. Soon thereafter he formed a training institute within Sargent & Greenleaf, separating the training from the company, and forming Lockmasters, Inc.—though Miller retained control of Sargent & Greenleaf. By 1982 he sold the lock company to investors, and Lockmasters, Inc., to his son, J. Clayton "Clay" Miller.

At that time, Lockmasters conducted training classes in hotels around the United States. Clay Miller decided to situate the institute in a more central, fixed location, and he moved the training institute to Nicholasville, a town south of Lexington. Currently the Lockmasters Security Institute, comprising 22 000 square feet (2000 m^2) of space, is found on a road in an industrial park.

Locks and keys were known in antiquity, though the first metal locks may have been made in England in the late 800s. Precision metalworking in a reproducible way in the late 1700s allowed a variety of new types of locks to be invented. Harry C. Miller also amassed a vast collection of all kinds of locks, keys, safes, and locksmithing tools, numbering in the thousands. Only a portion of Miller's collection is displayed within cabinets in the Lockmasters Security Institute's corridors. Miller's collection includes a variety of safe locks, padlocks, time locks, and time-delay locks. Time locks refuse to open until a set time, and were popular in 19th-century banks; time-delay locks open at a set time, or earlier if the proper combination is entered.

While the collection includes examples of locks and accessories from around the world, the bulk of the collection is of American locks dating from the latter part of the 19th century through the first half of the 20th century. Labeling is minimal, mostly limited to the year and manufacturing company for a particular specimen, but many of the artifacts do have QR codes, which visitors can scan and link via their cell phones to video or audio explanations of particular artifacts.

The oldest object in the collection is a lock from about 1303 from Islamic lands, made of steel with brass inlay, inscribed with Arabic that reads, "Be beholden to your servant, for one day his position may be elevated and you will be beholden to him." Locks of all types are on display, from major companies in the United States, both early and modern. Of particular interest is the first padlock, made of brass, by Linus Yale, Sr., in 1840, from Newport, New York. (His son founded the Yale Lock Manufacturing Company.) Another Linus Yale specimen is an eight-lever key lock installed early in the 1860s during Abraham Lincoln's administration, and removed during remodeling in 1925. Other interesting items include an 1870 iron lock from the jail in Tombstone, Arizona, and 20th-century high-security padlocks used by the US government to protect various nuclear weaponry and materials.

Among the samples from around the world are locks from East Germany and the Soviet Union made during the Cold War; and a wooden-and-steel lock fabricated by the Bambara ethnic group of Mali about 1860 by using a can for the metal parts. Also on display are several locks from the late 1700s and early 1800s made by Joseph Bramah, of England, who created the unpickable "Challenge Lock." There are locksmith's tools and kits, an American Padlock 1045 Key Machine to create keys from blanks, and a Rocket cutting torch on view. Standing near these cases of locks are some old examples of small safes.

Address: Harry C. Miller Lock Collection, Lockmasters Security International, 2101 John C. Watts Drive, Nicholasville, KY 40356.
Phone: (866) 574-8274.
Days and hours: Mon.–Fri. 8 a.m.–5 p.m. Closed major holidays.
Fee: Free.
Tours: Self-guided.
Public transportation: None.
Handicapped accessible: Yes.
Food served on premises: None. There are fast-food restaurants on US Route 27.
Special Note: We recommend that visitors bring a device able to scan and link QR codes to play audio and video recordings about the specimens.
Website: www.lsieducation.com/museum/index.html

Louisiana

NEW ORLEANS

New Orleans Pharmacy Museum

Known locally as *La Pharmacie Française*, this charming gem in the Vieux Carré, or French Quarter, is in a beautiful 19th-century Creole-style townhouse typical of the area. The house was built in 1823 for Louis Joseph Dufilho, Jr., who in 1816 became America's first licensed pharmacist. In 1950 the site was converted to a museum.

Originally the first floor was a pharmacy, the second floor an entresol (storage space), and the upper two stories a living area. The first and second floors are open to the public as a museum. Although one can visit the museum without a guide, touring with a docent may provide a clearer understanding of 19th-century apothecary techniques, medicine, illness, and health care. (See Stabler-Leadbeater Apothecary Museum, Alexandria, Virginia.) Visitors can step outside into a walled courtyard behind the house, which contains a medicinal herb garden.

The museum offers exhibits that illuminate historical health issues and place these issues into perspective. The original Belgian stone floor still shines, and the walls are lined with beautiful, hand-carved rosewood cabinets made in Germany circa the 1850s. Shelves hold original, hand-blown apothecary jars, many filled with medicinal herbs and crude medications. There are jars containing rare patent medicines, some with original labels, packaging, and instructions. Also on display are leech jars and bloodletting

Prescription Area, New Orleans Pharmacy Museum.
Courtesy of the New Orleans Pharmacy Museum.

devices, as well as surgical instruments from before and after the Civil War. In the eclectic collection of medical artifacts are a seamen's medicine chest typically used on whaling boars, baby bottles and nipples from as long ago as 1820, and an 1840 stethoscope.

To ensure that no one tampered with his medications, Dufilho used a brass seal to secure each package with red wax. His symbol, almost identical to that of Portuguese apothecaries, consisted of a serpent (symbolizing the animal kingdom) wrapped around a palm tree (the tree of knowledge) and planted in soil (symbolizing minerals).

At the prescription counter, visitors can see cachet wrappers, pill tiles, suppository molds, and chemicals used to compound prescriptions, some 150 years old. Guests can view pharmacopoeias outlining how to prepare and use specific medications, and see prescriptions dating back to 1880. At one time cosmetics were the domain of pharmacies. An antique cosmetic counter with curved glass contains perfumes, face creams, rouges, and other cosmetics.

One section of the museum is devoted to Voodoo, a form of religion based on belief in witchcraft and magical rites. In the 19th century, *gris-gris* potions used as medicines by Voodoo practitioners had an important influence on medicine and pharmacy in New Orleans. Prescriptions could be obtained from Voodoo princesses (female Voodoo practitioners) or the apothecary shop. A collection of *gris-gris* bottles and recipes dating from 1900 is on view.

Soda fountains originated in pharmacies as a way to make medicines more palatable to patients. The museum displays a magnificent Italian black and rose marble fountain, circa 1855. Here crushed ice and rock salt cooled the mineral and seltzer waters dispensed through the brass faucets, to which fruit nectars and phosphate flavors were added. Because Coca-Cola was used long ago as a flavoring agent in medicines, Coca-Cola memorabilia are on display.

The second floor offers a display of Dr. J. William Rosenthal's extensive collection of all manner of spectacles. Dr. Rosenthal, who attended medical school at Tulane University, was an ophthalmologist in New Orleans during the latter half of the 20th century. He collected eyeglasses and ocular aids of all sorts, from bifocals to Eskimo glasses, optical fans from the late 1700s, and spyglasses, and was a member of the Board of Trustees of the museum. There is also a model physician's study and sick room. Temporary exhibits on health and pharmacy are showcased here as well.

While in the French Quarter and thinking about health and pharmacy, walk about eight blocks (0.6 mile, 1 km) northwest along Chartres Street to Denis House (600 Esplanade Avenue at Chartres Street). Mounted on the wall of Denis House is a commemorative plaque honoring Dr. Jokichi Takamine, who may have lived here. Takamine, a chemist, discovered the hormone adrenaline in 1900.

Address: New Orleans Pharmacy Museum, 514 Chartres Street, New Orleans, LA 70130.
Phone: (504) 565-8027.
Days and hours: Tue.–Sat. 10 a.m–4 p.m. Closed holidays.
Fee: Yes.
Tours: Self-guided; or docent-led Tue.–Fri. 1 p.m.
Public transportation: Streetcars and buses stop within five blocks of the museum, then walk.
Handicapped accessible: Limited.
Food served on premises: None. A variety of restaurants are nearby.
Website: www.pharmacymuseum.org

The Audubon Nature Institute

The Audubon Nature Institute administers a variety of natural-science sites and parks in New Orleans. For the purposes of this book, we concentrate on four of those sites.

Audubon Aquarium of the Americas

The Audubon Aquarium of the Americans is located in the 17-acre (6.9 hectare) Woldenberg Riverfront Park, among a beautiful array of trees and shrubs, including swamp red maple, Chinese parasol, sago palm, southern magnolia, confederate jasmine, myrtle, cherry oak, and others. Several distinct and beautifully designed environments within the aquarium display creatures from waters in North, Central, and South America, and the adjacent seas. At the Great Maya Reef, with its 132 000-gal (500 000-L) tank, a transparent walk-through tunnel made of 5-inch (13-cm) thick acrylic allows visitors to view the reef much as a scuba diver might. The tank is teeming with sea life— one could come face-to-face with many types of sea life, including a moray eel.

A multilevel Amazon Rain Forest has thick vegetation, with a fog machine creating jungle humidity, and birds and butterflies darting freely about. As visitors walk beneath a 20-foot (6-m) waterfall, the aroma of the jungle surrounds them. From a suspension bridge over the exhibit, a Native American hut is visible, a structure that suggests the major role human beings play in the ecology of this habitat.

The Mississippi River exhibit, a tribute to the vanishing Louisiana marshes, has underwater displays of endangered lake sturgeon, 100-pound (45-kg) flathead catfish, and primitive, shark-like paddlefish visible through a glass window. To round out the atmosphere, the exhibit has a pier and a Cajun trapper's cabin. The 400 000-gallon (1.5-million L) Gulf of Mexico tank includes an oil rig platform, providing information about this industry's effect on the gulf. The rig forms an artificial reef that houses lemon sharks, tiger sharks, tarpon, and amberjack.

There are many smaller exhibits, including a touch tank where visitors can pick up and examine small creatures, a Sea Otter gallery, seahorses, jellies, and a South African black-footed penguin habitat. The GEAUX FISH! exhibit explains the important role fishing has to Louisiana, including game fish and commercial fishing. The Water Quality Control Room has a viewing window through which behind-the-scenes operations needed to maintain the aquarium are visible.

Address: Audubon Aquarium of the Americas, 1 Canal Street, New Orleans, LA 70130.
Phone: (504) 565-3033.
Days and hours: Tue.–Sun. 10 a.m–5 p.m. Closed Mardi Gras and 12/25.
Fee: Yes. Admission charged until one hour before closing.
Tours: Self-guided.
Public transportation: RTA buses stop 1 block away, then walk across the street and around the building to the entrance. Riverfront and Canal Street streetcars stop here. Algiers and Gretna ferries also stop nearby.
Handicapped accessible: Yes. Wheelchairs are available (first-come, first-served basis).
Food served on premises: Café, grill, and snacks.
Special Note: Additional fees required for the animal encounters.
Website: audubonnatureinstitute.org/aquarium

Audubon Zoo

The zoo, which opened in 1938, is in Audubon Park in the uptown part of the city. Both the institute and park are named after John James Audubon, the naturalist and artist. To get to the park, located a short distance up the Mississippi River, visitors can either travel by land on the famous St. Charles streetcar or take the Zoo Cruise, which runs four to five times a day. The park's 300 acres (120 hectares) originally formed a plantation owned by the daughter of Étienne de Bore, who in 1795 invented the refining process for granulated cane sugar. Renowned landscape architect Frederick Law Olmsted designed the park, which is filled with more than 1000 varieties of ornamental plants from around the world. (See Frederick Law Olmsted National Historic Site, Brookline, Massachusetts.) Sculptures scattered throughout the grounds add a lovely touch.

Plants and animals in this zoo are grouped together by region to create natural habitats. Because some of the 1500 animals of 360 species that live here are rare and endangered, the zoo participates in several Species Survival Plans® to guard against their extinction. These animals' exhibits are clearly labeled. Visitors can stroll through the various exhibits or take a narrated train ride around the zoo.

Asia re-creates the habitats of several landmasses, large islands, and archipelagos, most of which are in tropical zones. Among animals living here are Malayan sun bears, axis deer, great yellow tigers, and sarus cranes. The World of Primates focuses on evolution, displaying primate families in order of ascending intelligence level. Audubon Aviary, built in the 1930s by the Works Projects Administration (WPA), was renovated into a state-of-the-art rain forest containing exotic and rare birds and butterflies. There are many smaller exhibits, including a touch tank where visitors can pick up and examine small creatures, and a South African black-footed penguin habitat.

Various grassland environments inhabited by reptiles, insects, plants, and hundreds of other life forms, all in ecological balance, occupy one side of the zoo. Located here are an African savanna and South American pampas. The zoo places great importance on these exhibits' message about understanding the food chain and the cycles of life. Nearby is Jaguar Jungle, a Mayan rainforest—with re-created ruins of a Mayan temple—showcasing jaguars, spider monkeys, macaws, ocelots, and giant anteaters, along with a Bat House revealing nocturnal creatures.

No other zoo in the country has an exhibit like the exciting Louisiana Swamp. Unlike a marsh, which is a wet lowland dominated by grass, a swamp is a wet forest. The exhibit has Louisiana irises, gum trees, hackberry trees, and other local flora, and such fauna as alligators, snakes, great blue herons, and black bears. In a small museum within the exhibit, visitors learn key details about the plants, the wildlife, and the people who live in the Louisiana swamps. At the Cypress Knee Café, designed to look like a trapper's cabin, people can buy Cajun "fast food" and hear Cajun music.

Monkey Hill, within the zoo, is the highest landform in the city, even though it is only 27.5 feet (8.4 m) high. This spot, famous with local residents, is not a habitat for primates, as the name implies; it is simply a hill built as a WPA project to show the children of New Orleans what a hill looks like.

Address: Audubon Zoo, 6500 Magazine Street, New Orleans, LA 70118.
Phone: (504) 861-2537.

Days and hours: Tue.–Fri. 10 a.m–4 p.m., Sat. and Sun. 10 a.m.–5 p.m. Closed Mardi Gras and 12/25.

Fee: Yes. Admission charged until one hour before closing. Swamp Train costs extra.

Tours: Self-guided and Swamp Train (narrated tour around the zoo).

Public transportation: RTA buses stop nearby, as does the St. Charles Avenue streetcar.

Handicapped accessible: Yes. Wheelchairs are available (first-come, first-served basis).

Food served on premises: Cafés and snack shops.

Website: audubonnatureinstitute.org/zoo

Audubon Louisiana Nature Center

The Audubon Louisiana Nature Center, in Joe Brown Park in eastern New Orleans, has a special location on 86 acres (35 hectares) of coastal wetlands abundant with life, rebuilt after Hurricane Katrina in 2005. The center, opened in 1980, examines the interdependence of people and the world of nature. Not far from the city's hustle and bustle, several self-guided trails lead through woodlands to view squirrels, rabbits, birds, and butterflies in their natural habitat. One of the trails is a raised boardwalk accessible to people in wheelchairs and children in strollers. Naturalists conduct talks at the amphitheater. Guided walks and bird-watching expeditions are scheduled regularly.

The Interpretive Center has interactive exhibits explaining the ecology and biology of the local flora and fauna. A small planetarium features shows that explore the universe. In the Wildlife Garden just outside the complex, visible through windows, are native plants that draw wildlife.

Address: Audubon Louisiana Nature Center, 11000 Lake Forest Boulevard, New Orleans, LA 70127.

Phone: (504) 861-2537.

Days and hours: Trails, daily 8 a.m.–4 p.m.; Interpretive Center, Wed.–Sun. 10 a.m.–4:30 p.m.

Fee: Free for trails. Fee for planetarium.

Tours: Self-guided, and ranger chats.

Public transportation: Local buses from downtown stop at the Center.

Handicapped accessible: Partly.

Food served on premises: None.

Website: audubonnatureinstitute.org/nature-center

Audubon Butterfly Garden and Insectarium

Housed in a 19th-century US Custom House in the heart of New Orleans, the Insectarium, opened in 2008, claims to be the largest museum in North American for insects and similar invertebrates. A Termite Gallery demonstrates the effect termites have on human habitations using a colony of Formosan subterranean termites. The Butterfly Garden allows visitors to walk through an enclosed area housing hundreds of butterflies. In the Metamorphosis Gallery, the life cycle of insects is on display with special glass cabinets for viewing pupae of exotic butterflies hatching. Louisiana Swamp Gallery recreates the sounds and sights of the local biome, including aquatic insects and alligators. Bug Appétit is an unusual exhibit where chefs use insects (rich in protein) as ingredients for foods, and visitors are encouraged to sample the delicacies.

Address: Audubon Butterfly Garden and Insectarium, 423 Canal St, New Orleans, LA 70130.
Phone: (504) 524-2847.
Days and hours: Tue.–Sun. 10 a.m.–4:30 p.m.
Fee: Yes. Insect adventures cost extra.
Tours: Self-guided.
Public transportation: Riverfront and Canal Street streetcars stop here. Algiers and Gretna ferries stop nearby.
Food served on premises: Snacks—and Bug Appétit.
Handicapped accessible: Yes. Wheelchairs are available (first-come, first-served basis).
Special note: The Custom House is a federal building; visitors must walk through security to enter the museum.
Website: audubonnatureinstitute.org/insectarium

Mississippi

JACKSON

Mississippi Agriculture and Forestry Museum and the National Agriculture Aviation Museum

This living-history museum examines life in rural Mississippi, transportation, its link with the soil, and the ways in which agriculture has shaped the state's history. The museum has several components: a Heritage Center, a farm, a small reconstructed town, a forest trail, and an arts and crafts center. Throughout the site visitors see costumed interpreters making repairs, tending animals, selling wares, or doing seasonal chores. They are more than willing to explain their activities and answer your questions.

The Heritage Center, an indoor facility, has dioramas, videos, explanatory signs, and historic pieces—some dating back to 15th-century Europe—that describe the historical development of agriculture and forestry by relating these industries to transportation. An exhibit about the era of water transportation focuses on early lumbering and sawmill industries, early construction methods, cotton production, plantation and farm life, and the development of farm implements. An interesting item displayed here is a *caralog*, a two-wheeled cart that moved logs from the woodlands to the riverbank.

An exhibit on the rail era reviews how lifestyles in the Mississippi region changed after the Civil War. Thanks to the complex system of railroad lines that developed, farmers were able to obtain more advanced implements such as cultivators, planters, and plows, and were able to sell their produce to more distant markets. Several model railroad layouts (HO and O gauges) portray Mississippi towns circa 1970.

In the 20th century—the road era—cars, trucks, and tractors came on the scene and once-remote rural areas became accessible. These new vehicles made it possible for farmers to change their crop from cotton to vegetables. With the advent of refrigerated trailers, these vegetables became available to the entire nation.

The National Agricultural Aviation Museum and Hall of Fame, situated inside the Heritage Center, looks at the interrelationship of aviation and Southern agriculture.

On view are aircraft used for dusting and spraying crops and instruments such as boll weevil catchers, hoppers for putting chemicals or seeds into spreaders, and spreaders used for dispensing chemicals and seeds. One display explains how boll weevils are eradicated.

One of the more interesting pieces of aircraft on display is the Stearman Model A75, the most popular biplane ever built. The Stearman company began producing biplanes in the 1920s and developed the Model A75 in 1934. Boeing Aircraft purchased the Stearman Company at the beginning of World War II and continued to make this model for the war effort. After the war, thousands of these planes became available on the surplus market and were converted into agricultural aircraft. Many still fly over rural Mississippi as crop dusters.

The Fortenberry Parkman Farm, which operated from 1860 to 1960, was moved to this spot from elsewhere in Mississippi and restored to its 1920 appearance. It consists of a farmhouse, barn, outbuildings, and farm garden. Interpreters perform daily chores such as plowing, planting, maintaining the buildings, and caring for the animals. Beyond the farm is a portable sawmill where one can see how tools and equipment shown in the Exhibit Center were actually used to prepare lumber.

Crossroads Town, alongside the farm, shows a typical rural commercial center in 1920s Mississippi. Although this was never a real town, it exemplifies towns that grew up alongside roads to provide supplies and service to nearby farm communities. Many of the buildings were moved here and are original to the period, while others are reconstructions.

This town includes a shop where blacksmiths made and repaired wagon wheels, fixed broken farm tools, and shod horses, and a 1920s filling station that not only serviced cars but also provided lubricants and fuel to nearby farms. In the doctor's office, drugstore, and veterinary infirmary, visitors learn how rural medicine worked. The general store sold products needed by townspeople from cradle to grave, such as food, needles, overalls, books, and shoes. It also served as the town social center—a place to share news and gossip. The Bisland Cotton Gin, built in 1892, was a working mill until 1954. Visitors can watch it in operation as it separates seeds and hulls from cotton fibers. A smithy, grist mill, sawmill (with diesel-powered circular blade), and sugarcane mill (powered by mules walking in a circular track) are all included in the town. At the edge of town are the Epiphany Episcopal Church, built in 1897, and the schoolhouse. Just beyond the town is the 300-yard (270-m) long Nature Trail, along which 55 species of native trees—mostly hardwood, plus pine and cypress—are identified in an area of about 10 acres (4 hectares).

A Mississippi-style Victory Garden showcases the foods individuals grew during the World Wars in Mississippi, including blueberries, muscadines, and sweet potatoes. Also of interest is a compost station and potato-tuber viewer.

At the Mississippi Heritage Arts and Crafts Center is a log structure near the main entrance where visitors can watch artisans at work and buy their wares. The Fitzgerald Collection, behind the Exhibit Center, includes more than 17 000 Native American artifacts such as clothing and arrowheads.

Address: Mississippi Agriculture and Forestry Museum, 1150 Lakeland Drive, Jackson, MS 39216.
Phone: (601) 432-4500.

Days and hours: Mon.–Sat. 9 a.m.–5 p.m. Closed Martin Luther King, Jr., Day, Presidents' Day, Confederate Memorial Day, Memorial Day, 7/4, Labor Day, Armistice Day, Thanksgiving, 12/24, 12/25, and 1/1.

Fee: Yes.

Tours: Self-guided.

Public transportation: Buses run from Union Station downtown to within 0.7 mile (1.1 km), then walk; or take a taxi.

Handicapped accessible: Yes.

Food served on premises: Café.

Website: www.msagmuseum.org

Tennessee

Memphis

Mississippi River Museum

In a 52-acre (21-hectare) park on Mud Island, in the middle of the Mississippi River, is a museum dedicated to the river, its history, and its people. To reach it, a pedestrian bridge or monorail is necessary. Inside the museum, 18 exhibit galleries cover the history of the

Mississippi River Museum.
Courtesy of the Mississippi River Museum at Mud Island, Memphis, Tennessee.

river from the days of the Native Americans, explorers, and early settlers up through the present. Spanish explorer Hernando de Soto "discovered" the river in 1541, calling it "the father of waters." (See Mississippi River Visitor Center, St. Paul, Minnesota.)

A short video tells the story of major Mississippi River floods and the havoc that they caused. A series of scale models shows typical river transport: a 1780 log raft, a flatboat from around 1800, an 1815 keelboat, a side-wheel steam packet from 1825–1831, and another side-wheel steam packet from 1866–1876. Visitors can board a full-size main deck and can view the cargo area, furnaces, boilers, and coal bin. One of the galleries explores the crucial transition from steam to diesel power.

Displays about the history of engineering along the Mississippi look at bridges spanning the river and erosion control along the riverbanks. Initially, willow trees were used to shore up the banks; later, concrete blocks took their place.

Five galleries cover the Civil War and the Mississippi River. In one of these halls, visitors find themselves in the midst of a battle, first with a view from the Confederate bluff, and then from the front third of a Union ironclad gunboat.

Displayed in the music galleries are folk instruments ranging from a five-stringed banjo to musical bones, a bottleneck slide, and a cigar-box fiddle. The Yellow Dog Café gallery is dedicated to the famous musician W. C. Handy and his style of music, yellow-dog blues. Other musical technology includes a variety of recording equipment and radios.

The River Room has a 4000-gallon (15 000-L) aquarium in the center containing fish typical of those found in the river. Dioramas display birds that use the river along their migration route. Several exhibits of steam engines line the room's walls.

A highlight of the facility is the Mississippi River Museum Walk, an outdoor pathway along a scale model of the river that takes up the equivalent of five city blocks, and even has flowing water. Pedestrians follow the course of the river from the lower Mississippi to the Gulf of Mexico.

Address: Mississippi River Museum, 125 North Front Street, Memphis, TN 38103.
Phone: (901) 576-7241.
Days and hours: 10 a.m.–5 p.m. daily.
Fee: Yes; admission includes monorail.
Tours: Self-guided.
Public transportation: Buses in town stop near the walkway to Mud Island.
Handicapped accessible: Yes.
Food served on premises: Café.
Website: www.mudisland.com/c-3-mississippi-river-museum.aspx

OAK RIDGE

American Museum of Science and Energy and Oak Ridge Site–
Manhattan Project National Historical Park

To understand the historical and scientific significance of the Oak Ridge site–Manhattan Project National Historical Park, visitors should start their tour at the American Museum of Science and Energy. The US Department of Energy initiated this museum in 1949 to educate the public about energy resources and how to use them. Displays are augmented

American Museum of Science and Energy. *Photo* by Matt Mullins.

by live demonstrations, interactive exhibits, and audiovisual materials. Topics range from the once-secret Manhattan Project and the development of the atomic bomb to the history of the government facilities of Oak Ridge.

Upon entering the museum, visitors start in the middle of the exhibit hall, at the display Introducing Oak Ridge, which explains its contribution to the production of enriched uranium for the atomic bomb up through current research. A short video and interactive map here explain about the local area. From this central display, visitors then branch off to examine five different galleries with themes about Oak Ridge: the Manhattan Project, describing the original "Secret City"; National Security, examining Oak Ridge's role in dismantling nuclear weapons; Big Science, which describes some of the researchers' work at Oak Ridge, including nanotechnology and supercomputing (note the 3D-printed full-size model of an Army vehicle, as well as a computer cabinet from one of the fastest computers, Summit); Energy Leadership, which describes Oak Ridge's contribution to everyday life; and Environmental Restoration, that is, how the city is a pioneer in remediation of earlier environmental problems. The opposite side of the central area describes what the future holds for the area.

In an auditorium regularly throughout the day, staff members present the "Atoms and Atom Smashers" program, which includes demonstration of a Van de Graaff generator, which audience members can touch for "hair-raising" effects, as the museum likes to say. There is also a gallery with temporary exhibits.

The US Department of Energy runs three-hour public bus tours several times a week around the Oak Ridge complex (except winter months), for adults and children at least 10 years old. Tour highlights include the Y-12 History Center, the Graphite Reactor, the East Tennessee Technology Park, and a drive past the Spallation Neutron Source.

A few miles from the museum is the Oak Ridge National Laboratory and Graphite Reactor. The laboratory is made up of several buildings on 92 square miles (240 km^2) of rural land the US government bought in 1942 for the purpose of constructing a secret weapons plant. The government created the Manhattan District Corps of Engineers to produce the necessary components and construct an atomic bomb. More than 75 000 people secretly moved here for the Clinton Engineer Works part of the Manhattan Project. (See Bradbury Science Museum, Los Alamos, New Mexico; and the National Museum of Nuclear Science & History, Albuquerque, New Mexico.)

In 1942, a month after Enrico Fermi and his team achieved the first nuclear chain reaction in the Chicago Pile, the E. I. du Pont de Nemours & Company of Wilmington, Delaware began to build the Graphite Reactor, originally known as the "Clinton Pile" after the nearby town of Clinton. (See Hagley Museum and Library, Wilmington, Delaware.) On November 4, 1943, after loading 30 tons (27 tonnes) of uranium, the reactor went critical. The uranium was contained in aluminum cylinders 4.1 inches (10 cm) long and 1 inch (2.5 cm) in diameter. After the reactor was properly loaded, the chain reaction was spontaneous. The Graphite Reactor operated from 1943 to 1963 and is now a National Historic Landmark. The reactor-loading face (through which cylinders containing uranium were fed) is visible to the bus-tour participants, as are the control room, an experimental facility where materials were tested, and an exhibit area with displays on Oak Ridge's current research. Visitors can examine original notebooks and gauges that recorded the start-up of the world's first functioning nuclear reactor.

Today, Oak Ridge National Laboratory is among the country's largest and most varied research-and-development facilities in the fields of applied energy and engineering development. It studies such diverse areas as fusion, fission, conservation, fossil fuels, and physical science. It also operates sites that researchers at universities and private industries can use. Although the federal government owns the plant and equipment, it is operated under contract by UT–Battelle.

The Y-12 National Security Complex, a plant built in 1943, was responsible for enriching uranium using an electromagnetic separation process. An exhibit at the Museum of Science and Energy explains more details about Y-12, and a stop on the bus tour is at the Y-12 History Center. Natural uranium contains mostly uranium 238, but only about 0.7% is uranium 235, which is the active fissile material in a fission reaction. The process to enrich uranium to contain necessary concentrations of U-235 was an important part of the Manhattan Project.

Later, another plant, the K-25 Gaseous Diffusion Plant, came on line to use the gaseous diffusion method for enriching bomb material. The K-25 Building, the world's largest building in 1945, was closed in 1964, and the nation has been recycling highly enriched uranium ever since. Visitors can stop at an overlook on Highway 58 to view the footprint of the K-25 plant site. (The major structures at the K-25 plant have been removed as part of an environmental-remediation and reindustrialization project.) The K-25 History Center here explains the operations of this 2900-acre (1200-hectare) research laboratory.

Address: American Museum of Science and Energy, 300 South Tulane Avenue, Oak Ridge, TN 37830.

Phone: (865) 576-3200.

Days and hours: Mon.–Sat. 9 a.m.–5 p.m., Sun. 1 p.m.–5 p.m. Closed 1/1, Thanksgiving, 12/24, and 12/25.

Fee: Yes. Additional charge for the bus tour.

Tours: Self-guided. Guided bus tours of the area run several times a week.

Public transportation: None.

Handicapped accessible: Yes.

Food served on premises: None. A variety of fast-food and casual restaurants line South Illinois Avenue.

Special Note: Register on-line for the public bus tour, and check on-line for its current schedule. For adults, bring photo identification.

Website: amse.org

5

Midwest

Illinois

BATAVIA

Fermi National Accelerator Laboratory

This laboratory, colloquially known as Fermilab, is situated 30 miles (48 km) west of Chicago. Owned by the U.S. Department of Energy, it is managed by the Fermi Research Alliance, which includes the University of Chicago and the Universities Research Alliance. Fermilab, opened in 1967, explores the basic structure of matter by researching high-energy particle physics. It was named after Enrico Fermi, an expert in theoretical and experimental physics who won a Nobel Prize in 1939 for his work leading to the first nuclear chain reaction (see American Museum of Science and Energy and Oak Ridge Site–Manhattan Project National Historical Park, Oak Ridge, Tennessee). Enrico Fermi died in 1954.

Current research at Fermilab focuses on "big" questions about the universe, such as: Why do the laws of nature operate the way they do? How did the universe begin? And even, why is there anything at all? To do this, researchers tackle experiments dealing the Standard Model of physics, which includes 12 subatomic particles, four forces, and the Higgs boson to provide matter with mass. About 600 local researchers plus around 4000 visiting scientists from other institutions work here.

The center of visitor activities is Robert Rathbun Wilson Hall, where exhibits explain what Fermilab does. Dr. Wilson, the laboratory's first director, believed that this building should be both an attractive cultural center and an effective research facility. He designed the basic structure, inspired by the Saint-Pierre Cathedral in Beauvais, France. It has 16-story twin towers joined by one of the world's largest enclosed atriums, with many internal crossovers. Under Wilson's direction, several sculptures were placed on the grounds, including *Broken Symmetry*, which Wilson designed and fabricated in the Fermilab machine shop. Sculptures by Wilson and other artists are displayed inside the facility as well.

Most of the exhibits in Wilson Hall are on the 15th floor. A scale model of the laboratory provides an overview of the facility. A short video introduces the complex and the work done here; a slide show presents both historic and contemporary laboratory themes, including close-ups of sections not open to the public.

A large window facing north gives a fine view of the grounds, including the outline of the mounds from the buried split-beam lines, which stretch from the main accelerator to several experimental halls. The Electron Beam Display shows the path that charged particles follow when they are bent by a magnetic field to keep the particles in a circular path within the accelerator. The main accelerator's purpose is the study of subatomic particles.

Fermilab. *Photo* by Reider Hahn.
Courtesy of Fermilab.

From the east window, visitors can discern the 4-mile (6.4-km) circumference of the Tevatron, once the world's most powerful particle-collider, and decommissioned in 2011. The ponds around parts of the perimeter once provided cooling for the copper-coil magnets. The ponds around parts of the perimeter once provided cooling for the copper-coil magnets. The 600-acre (240-hectare) area in the center of the accelerator ring contains the Prairie Restoration Project, where seeds have been planted from species that grew in the original prairie of the Chicago region. A herd of bison grazes from the grounds, and a stand of trees provides cover and food for deer. Other areas of the complex, whose total land encompasses 6800 acres (2800 hectares), are also devoted to saving the prairie.

A full-size model of a half-section of the now-defunct main accelerator tunnel is on view in the Wilson Building, because the actual tunnels of the accelerator are off-limits to visitors. A mock-up of the equipment buried in the 4-mile (6.4-km) accelerator ring is shown in the model.

From the southwest window, there is a view of the building that houses the first three stages of acceleration: the Linac, or Linear Accelerator, the Booster, and the Main Injector, Fermilab's flagship accelerator. When voltage is applied at the proper radio frequency, particles in the Linear Accelerator move faster—sometimes called "riding the surf," to an energy of 400 MeV. In the Booster Accelerator, protons are accelerated up to 8 gigavolts and are injected into the Main Injector for further acceleration. When a proton collides with a target, a spray of particles results, including muons and muon neutrinos, which scientists study here.

One of the important experiments to be undertaken at Fermilab is Deep Underground Neutrino Experiment (DUNE), to study the properties of neutrinos, which are subatomic particles that only weakly interact with other matter. A powerful beam of neutrinos (and antimatter neutrinos) is generated at Fermilab, with a near detector (at

Fermilab) and a far detector 800 miles (1290 km) away at Sanford Lab Homestake in South Dakota. (See Sanford Lab Homestake Visitor Center, Lead, South Dakota.) Neutrinos oscillate among different flavors; do the neutrinos and antimatter neutrinos oscillate in the same way? The near detector studies the beam leaving Fermilab, and the far detector studies the beam after traveling the long distance (during which the neutrinos can oscillate). Another question this experiment may help resolve is, why is the universe made of matter and not antimatter? A third question that DUNE may help answer is, do protons decay?

Another project at Fermilab is the Muon g–2 (pronounced "gee minus two") experiment, to study properties of the muon, another subatomic particle—much heavier than a neutrino—that can be imagined as spinning and precessing like a gyroscope. The gyromagnetic ratio, or g, differs very slightly from 2, so the experiment is looking at the difference between g and 2 (i.e., $g–2$). The detector for Muon g–2 is a series of magnets weighing 17 tons (15 tonnes) and 50 feet (15 m) in diameter, built at Brookhaven National Laboratories in Upton, New York, and shipped to Fermilab, where a powerful beam of muons exists. The goal is to compare the predicted muon g from the Standard Model of physics with the experimental value of g. If there is a significant discrepancy, there may be new physics to replace the Standard Model.

Other major projects include Mu2e (muon-to-electron), which looks at a rare decay mode of muons; Dark Energy Survey, studying a repulsive force causing the universe to expand at an increasing rate; experiments on dark matter, which is a mysterious form of matter that composes most of the matter in the universe; and quantum effects for computing and other purposes.

A self-guided driving tour through the ground follows the tour pamphlet available at the reception desk in Wilson Hall. This tour passes the Test Beam Facility where particle-detectors are tested. The building has a unique, multicolored roof. Another set of buildings houses the silicon-detector facility.

On the Fermilab grounds is the home of the Leon M. Lederman Science Education Center, primarily designed for children and families to learn about physics.

Address: Fermi National Accelerator Laboratory, Pine Street and Kirk Road, Batavia, IL 60510.
Phone: (630) 840-3000.
Days and hours: Daily 8 a.m.–8 p.m. (till 6 p.m. in the winter months). See the website for details of building hours.
Fee: Free.
Tours: Self-guided. Downloadable audio guides are available. Guided tours are offered once a week. Minimum age of 14 for tours. Check the website for other tours of specific colliders and instrumentation. Private tours for groups are available for a fee. See the website or call about public talks.
Public transportation: Limousines from Chicago, as well as a call-n-ride bus service for the Batavia area.
Handicapped accessible: Yes.
Food served on premises: Cafeteria (Mon.–Fri.).
Website: www.fnal.gov

BROOKFIELD

Brookfield Zoo

Although the zoo opened in 1934, its history dates from 1919 when Edith Rockefeller McCormick gave almost 85 acres (34 hectares) of land to the Forest Preserve District of Cook County to establish a zoological park. The design concept, a daring one for American zoos, was laid out in a 1921 report recommending that the zoo be modeled after the Tierpark Hagenbeck in Hamburg, Germany. This facility would be organized as a place for families of wild animals to live in "surroundings reproducing as far as artificial and other means will permit the natural habitat of each."

Today the zoo covers 216 acres (87 hectares). It has formal plantings, natural woodlands, and over 4000 animals of more than 500 species in nearly 20 major exhibits. Amidst this beauty, the zoo teaches appreciations and understanding of wildlife. The Brookfield Zoo is heavily involved in Species Survival Plans® and believes that research on animal habits and habitats benefits species in the wild as well as those in the zoo.

One of the most impressive exhibits, Tropic World, is in an enormous structure about the size of one and a half football fields—among the largest zoo exhibit buildings in the world. Three tropical rain forests represent South America, Asia, and Africa. Each has naturalistic habitats without bars, in which a variety of species live together. There are several waterfalls, a stream, and simulated thunderstorms that ring out periodically in each section. The zoo staff believes these natural environments prompt animals to act as they do in the wild, which is important for the animals' well-being.

Sulcata Tortoise at the Brookfield Zoo.
Courtesy of Brookfield Zoo.

A cave at the entrance of the building leads into the South American area, complete with waterfalls and thick vegetation. Among the colorful wildlife in this habitat are a two-toed sloth, a giant anteater, and several species of monkeys, including golden lion tamarins.

The Asian section has a 22-foot (6.7-m) high peninsula, covered with banyan trees sitting in a mangrove swamp. Orangutans, white-cheeked gibbons, and small clawed otters scamper about; and red-vented bulbul, a magpie robin, and a black-throated laughing thrush fly overhead.

A guard post recalls the wildlife poaching that goes on in the African rain forests, a practice that depletes the ranks of creatures already facing survival problems. On view are western lowland gorillas, Angolan colobus monkeys, Schmidt's red-tailed guerons, Allen's swamp monkeys, and many free-flying birds.

A Lion House, built in 1934, has been transformed to accommodate two naturalistic, mixed-species exhibits that illustrate how living things depend on the environment for survival. The interior depicts the African desert and the Asian rain forest. Outside is an exhibit that examines how animals adapt to their environment in order to hunt prey.

African animals that live above ground, as well as those such as the naked mole rat that live below ground in caves, are housed in The Desert's Edge exhibit, also known as the African section. This exhibit gives helpful explanations of how plants and animals survive in habitats where temperatures are extreme and food and water are scarce.

The Clouded Leopard Rain Forest features a warm, moist forest native to Southeastern Asia. Among the rare animals housed there are the fishing cat, binturong, and clouded leopard from Borneo and eastern Nepal. Outdoor habitats along the south side of the building are home to several species of big cats. Visitors have a close view of these predators, which are particularly vulnerable to interruptions in the food chain because they are at the very top.

Completed in 1960, the original Seven Seas Panorama was the first inland exhibit of marine dolphins in the world. Rebuilt on the opposite side of the park in the late 1980s, it has an indoor section, suggestive of a Caribbean coast, where dolphins swim and display typical behaviors during daily performances. An outside seascape that looks like the shores of the Pacific Northwest contains pinnipeds—aquatic animals with finlike flippers, such as gray seals. This exhibit includes facilities for marine mammal acoustic and behavioral research.

The Feathers and Scales building has two habitats: a desert and an open aviary replicating a rain forest, home to a variety of tropical birds. In the center of the building are many reptiles and amphibians, such as snakes native to Illinois, as well as various species of frogs and lizards.

Address: Brookfield Zoo, First Avenue and 31st Street, Brookfield, IL 60513.
Phone: (708) 688-8000.
Days and hours: Daily 10 a.m.–5 p.m. Extended hours from Memorial Day to Labor Day.
Fee: Yes. Extra fees for in-park attractions and the Motor Safari.
Tours: Self-guided. Motor Safari trams offer a narrated tour while visitors hop on and off.
Public transportation: Metra trains stop at the Zoo Stop, then walk four blocks. Pace buses stop at the South Gate.
Handicapped accessible: Yes. Wheelchairs and scooters can be rented for a fee.
Food served on premises: Cafés and snack bars.
Website: www.czs.org/

CHICAGO

Adler Planetarium

A philanthropist and executive of Sears, Roebuck & Company, Max Adler believed that the public's understanding of the universe was "too meager, the planets and stars too far removed from general knowledge." After seeing the newly invented Zeiss Planetarium projector in Germany in the 1920s, Adler knew he wanted to enlighten the public. He gave this planetarium, which opened in 1930, as a gift to the people of Chicago.

The Adler Planetarium, in a park on a peninsula that extends a half mile (0.8 km) into Lake Michigan, is flanked by the Shedd Aquarium and the Field Museum of Natural History. The peninsula was once an island, but was connected to the mainland with landfill. Next to the planetarium is the Copernicus Monument, built in 1973 to commemorate the 500th anniversary of the birth of Mikołaj Kopernik, or Copernicus, on February 19, 1473, in Toruń, Poland. From the planetarium grounds there is a wonderful view of the Chicago skyline.

The building is constructed of rainbow granite in the shape of a regular dodecagon. The dome is made of lead-covered copper, and each of the building's 12 corners bears a different sign of the zodiac. The entrance plaza has a fully functional, 12-foot (3.7-m) high equatorial sundial created by the famous sculptor Henry Moore.

The exhibits inside, which occupy three floors, illustrate the solar system, space exploration, and the nature of stars, galaxies, and the universe as a whole. Displays cover navigation, telescopes, and early scientific instruments related to these fields.

Our Solar System examines several space expeditions and offers exhibits about the Earth, its moon, planets, other moons in the solar system, and Halley's comet. The Planet Wall uses three-dimensional scale models illustrating the Sun, planets, and the largest moons of the solar system. The Universe: A Walk Through Space and Time is an introduction to cosmology over 13.7 billion years.

The Adler has restored an Atwood Celestial Sphere, with 692 holes drilled in the positions of the stars in a metal surface 17 feet (5.2 m) in diameter. Designed by Wallace Atwood, a board member of the Chicago Academy of Sciences, and built by the Aermotor Windmill Company, it was used by the US Navy for navigational training. Regular guided tours of the Sphere are given.

Telescopes: Through the Looking Glass displays instruments dating from the 17th century to the present, including one used by Sir William Herschel and his sister Caroline, who together discovered the planet Uranus in 1781. The hands-on exhibit, Using Telescopes, teaches about telescopes and encourages visitors to try some out.

The planetarium has a fine collection of early astronomical instruments and surveying equipment, including pieces from as far back as the 12th century. Upon seeing some of the handcrafted instruments explorers used to find their way around the globe, visitors realize that navigation and astronomy go hand in hand. Among these pieces is the 20-foot (6.1-m) Dearborn telescope, the largest telescope in the world when it was built in the 19th century.

Astronomy in Culture compares astronomy of the Middle Ages (1200–1500 C.E.) to modern astronomy. Visitors are transported back in time to re-created scientific settings in Europe and the Islamic world to learn how sundials, astrolabes, armillary spheres, and other astronomical tools were used.

Adler Planetarium's most important offering is its sky show, designed as a "doorway to the universe." This one-hour visual experience is presented in two parts in two theaters. The first part, in the Universe Theater, is a multimedia extravaganza that teaches about the universe. To reach the second part, in the Grainger Sky Theater, visitors take an escalator ride called Stairway to the Stars. In the Sky Theater a digital projection system with 20 projectors (190° viewing angle, 8000 × 8000 pixels) re-creates outer space. After each Friday evening sky show, a 20-inch (51-cm) computerized Cassegrain reflector telescope in the Doane Observatory provides instant live pictures of space, which are projected on the Sky Theater dome. Doane Observatory, part of the planetarium complex, is the only observatory in the nation that is entirely for public use. When weather permits, safe Sun-viewing happens during midday hours.

Address: Adler Planetarium, Museum Campus, 1300 South Lake Shore Drive, Chicago, IL 60605.
Phone: (312) 922-7827.
Days and hours: Daily 9 a.m.–5 p.m., open until 9 p.m. Fri. Closed Thanksgiving and 12/25.
Fee: Yes. Tour of the Atwood Sphere costs extra.
Tours: Self-guided.
Public transportation: Buses stop on the driveway in front of the planetarium.
Handicapped accessible: Yes.
Food served on premises: Café.
Special Note: Adler After Dark programs, for adults only, are on the third Thursday evening of the month.
Website: www.adlerplanetarium.org

Field Museum of Natural History

One figure behind the launch of this museum was Frederick W. Putnam, an anthropology professor and curator of the Peabody Museum at Harvard University in the late 19th century. (See Harvard Museums of Science and Culture and Warren Anatomical Museum, Cambridge and Boston, Massachusetts.) He recommended that a permanent natural history museum be formed to house the great collections from the 1893 Chicago World's Columbian Exposition, an international fair highlighting many aspects of participating countries, including their natural history. Marshall Field I, the American merchant and philanthropist, donated $1 million to establish such a museum, and in 1894 the Field Columbian Museum opened in the Palace of Fine Arts building in Jackson Park.

In 1905 the museum changed its name to the Field Museum of Natural History, and in 1921 it moved to a location in Chicago's South Loop. Within a decade, two other fine science-related institutions, the Shedd Aquarium and the Adler Planetarium, were added nearby. (See John G. Shedd Aquarium and Adler Planetarium, Chicago, Illinois.) The original museum building in Jackson Park is now the home of the Museum of Science and Industry. (See Museum of Science and Industry, Chicago, Illinois.)

The Field Museum research for its collections encompasses biology, anthropology, zoology, botany, and geology (including paleontology). The nearly 40 million specimens in the museum make up one of the most important collections of natural history in the world. Fewer than 1% of these artifacts are on display at any given time, for the Field Museum primarily uses its collections for scientific research.

The exhibits focus on human cultures of the world and the physical environment's diversity. Exhibits include temporary exhibitions based on specific themes and permanent displays housing taxidermy, anthropological artifacts, models of plants, meteorites, and more. These exhibits generally are related to the active research that scientists at the museum do around the world.

One of the great names in taxidermy, Carl Akeley, was chief taxidermist at the Field Museum between 1896 and 1909, when modern taxidermy was in its infancy. His works still form the core of the priceless collection of animal mounts here. Akeley created the idea of the diorama, in which animal groupings are displayed in their natural habitats. (See Milwaukee Public Museum, Milwaukee, Wisconsin; and American Museum of Natural History, New York, New York.)

Thanks to the museum's scientific collecting expeditions at the beginning of the 20th century and the work of Akeley and his staff, the Field Museum has one of the most complete collections of zoological materials in the world. The two remarkable bull elephants locked in fierce combat in Stanley Field Hall are not just stuffed animals—they are accurate works of science and art created by Akeley. Among this craftsman's earliest works is the Four Seasons, a grouping of white-tailed deer in their seasonal habitats along the Michigan-Wisconsin border. Akeley's wife, Delia, made much of the foliage in the exhibits.

The Akeleys, as a couple, are also responsible for preparing the two African elephants that greet visitors in the Stanley Field Hall, alongside a 122-foot (37-m) long touchable cast of *Patagotitan mayorum*, a titanosaur from Argentina and the largest dinosaur species discovered to date.

A comprehensive exhibit, Inside Ancient Egypt, features a three-story, walk-through reconstruction of an Old Kingdom mastaba (shaped like a truncated square pyramid) tomb. The exhibit places rare artifacts from 5000 B.C.E. to 300 C.E. in their cultural context. For example, a re-created Nile River Valley marsh contains plantings and a *shaduf*, a water-lifting device.

In the Regenstein Halls of the Pacific exhibit, visitors are familiarized with the Pacific region's environment, population, animals, and plants by walking through a coral atoll; witnessing a Hawaiian lava flow; and examining a Marshall Islands outrigger canoe, a New Guinea village, a Maori meeting house, and a Tahitian market. Part of this exhibit contains more than 600 artifacts from New Guinea, New Britain, New Ireland, Vanuatu, and the Polynesian Islands to illustrate not only the natural world and technology of the Pacific Islanders, but also their rituals and their spiritual and artistic life. An exhibit explores the technology of Melanesian weaponry.

The Griffin Halls of Evolving Planet features the Hall of Dinosaurs, and contains the skeletons of an *Apatosaurus* and several other dinosaurs, all reassembled according to the latest theories about these ancient animals. Here on display is SUE, the largest, most complete, and best-preserved *Tyrannosaurus rex* found, at 40 feet (12.3 m) high. SUE was discovered by paleontologist Sue Hendrickson in 1990 in the Black Hills of South Dakota.

Several laboratories offer visitors the chance to see scientists work on the conservation of artifacts, preparation of fossils, and analysis of DNA. A variety of natural-habitat dioramas and displays explain reptiles, birds, mammals, and amphibians, especially from Africa and Asia. Gems and jade objects, the life of Northwest and Arctic peoples, human cultures of the Americas, and scores of other topics are also covered at the museum. A 3D theater provides regular showings.

On the third floor is the Field Museum Library, which is open to researchers and the public. Among the Library's holdings are the four-volume set of John James Audubon's *The Birds of America*; a collection of movies taken during scientific explorations from 1914 through the 1960s; and Carl Akeley's photographs from his expeditions in 1895 and 1906, which he used to guide his taxidermy.

Address: Field Museum of Natural History, 1400 South Lake Shore Dr, Chicago, IL 60605.
Phone: (312) 922-7827.
Days and hours: Museum: Daily 9 a.m.–5 p.m., last admission at 4 p.m. Closed 12/25. Library: Tue.–Fri. 1–4 p.m., with 48-hour advance notice and a request for materials.
Fee: Yes. Special exhibits and 3D movie are extra.
Tours: Self-guided.
Public transportation: Buses stop at the Museum Campus.
Handicapped accessible: Yes. Wheelchairs are available on a first-come, first-served basis. Vision-impaired tours are available on advance request.
Food served on premises: Restaurant and café.
Website: www.fieldmuseum.org

International Museum of Surgical Science

A group of surgeons gathered in Geneva, Switzerland, in 1935 and formed the International College of Surgeons to "promote the science and art of surgery ... for the benefit of the public, students, and the medical profession," as its bylaws state. This organization, with 63 member nations, is now based in Chicago. It operates the International Museum of Surgical Science from the historic 1917 Eleanor Robinson Countiss House that adjoins its headquarters. The group's international medical collection, one of the most complete in the world, traces surgery and related sciences from Neolithic brain surgery to present-day medicine.

Historically at the museum, many exhibits formerly were organized according to the country and civilization that made contributions to modern surgery. Currently the museum emphasizes the development of medical specialties such as orthopedics, radiology, and obstetrics and gynecology, though specific exhibits relating to Japan and Taiwan still remain.

In its review of surgical history, the museum covers trepanning, used in Europe from Neolithic times until the advent of modern medicine, and in South America by the Incas for 4000 years. The term, derived from the Greek *trypanon* or "borer, auger," refers to making an opening in the skull to let demons escape. Trepanning was effective in relieving pressure on the brain created by cranial swelling. Some archeologists believe that up to 66% of patients who had this surgery survived. Visitors can see a collection of 2000-year-old skulls treated by trepanning and tools used to perform this procedure, which were recovered from ancient Peruvian tombs.

Bloodletting, a medical treatment widely used for more than 2000 years—well into the 19th century—was based on the hypothesis that the humors (blood, phlegm, black bile, and yellow bile) needed to be kept in balance. To achieve this equilibrium, doctors released "bad" blood from patients' bodies using an array of specialized equipment, some of which is on view. A beautiful bleeding plate dating from the 15th century was probably used by Joseph Lister in the 19th century.

Optical room, International Museum of Surgical Science.
Courtesy of the International Museum of Surgical Science.

The taboo in European culture against human dissection greatly inhibited the development of surgery. After Flemish anatomist Andreas Vesalius broke the taboo against human dissection in the 16th century, surgery quickly became a science. The Anatomical Illustration Gallery has a scale model of the anatomical theater built in 1594 at the University of Padua that Vesalius used for public lecture and anatomical observations. (The Padua theater is now a museum.) The room also contains 18th-century anatomist Bernhard Siegfried Albinus's medical textbook, *Tabulae sceleti et musculorum corporis humani*, open to its most famous engraving: a skeleton in front of a rhinoceros named Clara, which was at the time completing a 16-year tour of Europe.

The museum covers the long history of caesarian sections to save the life of an unborn child when natural childbirth was impossible. Performed as far back as 300 BCE in Egypt, the caesarean was a risky procedure for the mother until recent times. The

Obstetrics and Gynecology exhibit displays a painting commemorating the first successful caesarian section performed in South America, in 1844.

The Windows to the World: The Science of Sight & The Ophthalmic Art exhibition has an illustration of Suśruta, a surgeon practicing circa 7th century B.C.E. Suśruta stressed cleanliness in operating rooms some 25 centuries before Hungarian physician Ignaz Semmelweis reported its importance. Suśruta's writings emphasized short hair, clean hands, and "sweet smelling dress" for surgeons, and an operating room "fumigated with sweet vapours."

An exhibit on French contributions to medicine explains the impetus behind René Théophile Laënnec's invention of the stethoscope in 1816: to maintain the modesty of a female patient while allowing the physician to listen to her chest. Dr. Laënnec carried his stethoscope inside a top hat.

The display of x-ray equipment traces the evolution of this process from its introduction in the late 19th century by Wilhelm Conrad Roentgen, a series of x-ray tubes, and Emil Grubbé's original x-rays taken about 1910. A display on the 19th century includes an apothecary, the predecessor of the American drugstore, and a period dentist's office.

In the Library, and open only to scholars with an appointment, is the Thorek Manuscripts and Rare Books Collection, housing over a thousand medical works spanning 1200 years.

Address: International Museum of Surgical Science, 1524 North Lake Shore Drive, Chicago, IL 60610.

Phone: (312) 642-6502.

Days and hours: Mon.–Fri. 9:30 a.m.–5 p.m., Sat.–Sun. 10 a.m.–5 p.m. Closed on Memorial Day, Easter Sunday, 7/4, Labor Day, Thanksgiving, 12/24, 12/25, 12/31, and 1/1.

Fee: Yes.

Tours: Self-guided. Guided tours are available by prior arrangement.

Public transportation: Buses stop a half-block away, then walk to the site.

Handicapped accessible: Yes.

Food served on premises: None. There are some restaurants about 6 blocks south on Division Street and 7 blocks west on Clark Street.

Website: imss.org

John G. Shedd Aquarium

Opened in 1930 as part of the Grant Park Museum Complex along the banks of Lake Michigan, this aquarium is one of the largest in the world. The original aquarium building, an octagonal structure, is of classic Greek style to blend with the nearby Field Museum of Natural History. Aquatic motifs repeat throughout the building: in the marble tile of the floor and walls, the bronze entrance doors, and the light fixtures suspended by metal octopuses. This building contains 32 000 birds, fish, amphibians, reptiles, mammals, and invertebrates (not including every coral polyp) representing 600 species found in oceans, rivers, and lakes around the world. Three galleries are devoted to saltwater aquatic life, and three to freshwater exhibits. The aquarium was donated by John Graves Shedd, president and chairman of Marshall Field & Company, a large Chicago-based department store.

The aquarium's goal is to collect, care for, study, and exhibit fish, other aquatic animals, and plant life. When it was first built, a million gallons (4 million L) of salt water were brought from Key West, Florida, using 160 railroad cars. Today six different systems prepare salt water and fresh water, and distribute them to various displays through 75 miles (120 km) of piping.

The first gallery, with saltwater exhibits, features beautiful, brilliantly colored Caribbean reef fish and contains mangrove and turtle-grass habitats. A second gallery transports visitors to another part of the world, the shallow seas of the Indo-Pacific, populated with coral reef species. Gallery three has animals from northern tide pools and cold oceans around the world.

Among the permanent aquarium exhibits is Marine Jewels, a display of small invertebrates that illustrates the unusual ways these creatures adapt to their environment. A re-created version of the Fox River sandstone bluffs, complete with waterfall, pools, and rock work, houses the playful river otters. The Sea Anemones exhibit has more than 70 types of anemones, their relatives, and symbiotic associates. The anemone, or ocean flower, is closely related to corals and jellyfish.

The Tributaries exhibit spotlights small streams that feed larger tropical rivers, and the wildlife that live in them. Many of these fish have long been favorites in home tanks and pools. The Asian motif reminds visitors of the art of fishkeeping practiced for thousands of years in Asia.

This aquarium is one of a small number of aquatic facilities involved in the Species Survival Plan® to save the land snail, genus *Partula*, one of the 12 most endangered animal species in the world.

The centerpiece of the building, in the rotunda, is the 90 000-gallon (340 000-L) Coral Reef Exhibit, which contains 350 tropical Caribbean reef fish. This naturalistic habitat simulates day-to-night cycles so visitors can see how the fish behave during daylight and nighttime conditions. Divers enter the tank daily to feed the fish and to talk to visitors about these creatures.

Chicago's Oceanarium doubles the original facility's space and expands its scope to include more species of aquatic mammals and birds. The main exhibit area in the four-level structure re-creates the rocky coastline of the Pacific Northwest and southeastern Alaska. Five cold seawater pools that contain Pacific white-sided dolphins, false killer and beluga whales, and seals are located here. Other creatures shown here include crabs, flounder, sea stars, and sea cucumbers. Multilevel exhibits emphasize the natural history of these animals, their adaptation to the sea, their native abilities, and the need for their conservation.

Public walking trails wind through the pool areas. The North Trail continues beyond the pools to a tidal pool inhabited by starfish, sea anemones, crabs, and mussels. The South Trail leads to Seal Cove, the 400 000-gallon (1.5 million-L) beluga whale exhibit, and the 2 million-gallon (7.6 million-L) tanks housing dolphins and Pacific black whales. A lower level features an underwater viewing gallery and a penguin habitat that re-creates a rock formation of the Falkland Islands. Gentoo penguins, identified by a white patch over each eye, reside here. Their unusual habitat is a series of 12 rocky islands near Antarctica, one of which is the Falklands.

The Oceanarium includes an amphitheater with dolphin shows. An added treat at the Oceanarium is a magnificent view of the Chicago skyline.

Address: John G. Shedd Aquarium, Museum Campus, 1200 South Lake Shore Drive, Chicago, IL 60605.
Phone: (312) 939-2426.

Days and hours: Mar.–Oct.: daily 9 a.m.–6 p.m.; Nov.–Feb.: daily 10 a.m.–5 p.m. Closed 1/1 and 12/25.
Fee: Yes.
Tours: Self-guided.
Public transportation: Buses stop within 0.4 mile (0.6 km) of the site.
Handicapped accessible: Yes. Wheelchairs available on a first-come first-served basis. Tactile maps and audio descriptions available for the sight-impaired. Sign-language interpreters are on hand for the hearing-impaired.
Food served on premises: Cafes, restaurant, snack stands.
Website: www.sheddaquarium.org

Museum of Science and Industry

When Julius Rosenwald, chairman of Sears, Roebuck & Co., returned from a trip to Munich, Germany, that had included a visit to the Deutsches Museum, he decided that Chicago needed a participatory science and technology museum, with enough space for all the wonderful displays it would house. The Palace of Fine Arts was originally built for the 1893 World's Columbian Exposition. After the exposition this structure was used by the Field Museum of Natural History. Because the Field Museum moved to the Grant Park Museum Complex—where it is still located—the Palace of Fine arts was empty, and the Museum of Science and Industry moved there in 1933. Today this museum is the largest science museum in the Western Hemisphere, comprising 400 000 square feet (37 000 m^2) of display area, and is situated between Lake Michigan and the University of Chicago.

Because of the large size of this facility, the museum has put in place some very large and unusual exhibits that would be difficult to place in many other sites. There are more than 2000 exhibits in 75 major halls on three floors. In keeping with the museum's goal of being a "giant informal classroom," it provides education and entertainment while introducing nature to visitors and showing them how all branches of science are connected.

Coal Mine, the first exhibit installed in the museum, and dating from 1933, is a reproduction of an actual southern-Illinois bituminous coal mine with two elevator shafts and a 50-step staircase. Visitors descend in an authentic hoist to see the mine's workings. When inside, they ride on a work train to examine the coal walls from a genuine Illinois No. 6 seam and see mining equipment and machinery.

A 3500-square-foot (330-m^2) model train display, The Great Train Story, highlights modern railroad operations and the important role the railroad industry played in shaping the American economy. The current layout, with 1400 feet (430 m) of track, was unveiled in 2002, replacing the 1941 version. The museum grounds feature historically important trains, such as the Pioneer Zephyr, the world's first streamlined passenger train. In 1934 it broke speed records by averaging 77.6 mph (125 kph) over the 1015 miles (1630 km) of its trip from Denver to Chicago. Steam locomotive 2903 is a 304-ton (276-tonne) piece of equipment that operated on the Santa Fe Railroad from 1943 to 1955, hauling passengers at 90 mph (145 kph) and freight at 60 mph (97 kph). Locomotive 999 broke all existing speed records when it raced over the rails at 112.5 mph (181 kph) in 1893. The museum has other vintage vehicles on display, including automobiles, airplanes, as well as a German U-505 submarine captured in 1944. (See Steamtown National Historic Site, Scranton, Pennsylvania; B&O Railroad Museum, Baltimore, Maryland; National Railroad Museum, Ashwaubenon, Wisconsin; and

California State Railroad Museum, Sacramento, California.) Two airplanes from World War II, donated by the British government, reside here, as well as a model of the first Wright Brothers plane.

The Henry Crown Space Center, a wing built in 1986, is dedicated to humanity's relationship with outer space. The collection of historic artifacts includes an original Apollo 8 spacecraft and a lunar excursion module (LEM). There are flight simulators for visitors to try their hand at piloting historic aircraft, experiencing liftoff in the space shuttle, and more. The space exploration section covers electronics, aviation, industrial medicine, and home life in space.

One of the early exhibits (from 1938) and still a favorite is the Whispering Gallery, where participants stand on the brass footprints and speak quietly into the parabolic plastic dishes at either end of the ellipsoid-shaped gallery. The sound is focused to the opposite end of the gallery, where listeners can easily hear the whispers.

Many other galleries explore scientific and technological topics such as agricultural technology, human health, three-dimensional printing, numbers, and more. The museum also sponsors temporary exhibits. (See The Franklin Institute, Philadelphia, Pennsylvania.)

There are also exhibits on biology; for example, Genetics: Decoding Life explores the DNA molecule and genetics. In this area, a chicken hatchery examines how chicks develop from eggs, and genetically modified frogs as well as plants resistant to droughts are displayed.

Given the size of this facility and the number of exhibits available, we can only provide a small sample of the vast material covered in this museum space.

Address: Museum of Science and Industry, Chicago, 5700 South Lake Shore Drive, Chicago, IL 60637.
Phone: (773) 684-1414.
Days and hours: Mon.–Fri. 9:30 a.m.–4 p.m., Sat.–Sun. 9:30 a.m.–5:30 p.m. Summer hours 9:30 a.m.–5:30 p.m. daily. Closed 12/25.
Fee: Yes. Free on special days for Illinois residents. Certain exhibits charge extra fees.
Tours: Self-guided; some exhibits have guides as part of the exhibit visit.
Public transportation: Trains stop at Millennium Station, then walk 0.4 mile (0.6 km); buses stop at S. Hyde Park and 56th St., then walk 0.2 mile (0.3 km).
Handicapped accessible: Yes. Wheelchairs are available.
Food served on premises: Café, large food court, and snack machines.
Special Note: OMNIMAX Theater.
Website: www.msichicago.org

Indiana

INDIANAPOLIS

Indiana Medical History Museum

In the latter part of the 19th century, neurologist Edward Spitka noted that mental hospital physicians were "expert in everything but the diagnosis, pathology, and treatment of

Clinical Laboratory at the Indiana Medical History Museum.
Courtesy of IMHM/Thomas Mueller Photography.

insanity." Other neurologists pointed out that physicians at mental hospitals did not receive adequate training in psychiatry, neurology, and scientific inquiry. It was out of this concern that the administrator of Central Indiana Hospital for the Insane (now Central State Hospital) from 1893 to 1923, Dr. George F. Edenharter, decided to build a 19-room research and clinical laboratory on the hospital grounds where physicians and medical students could study diseases of the mind and nervous system. The site was completed in 1896. Over the years it served as a teaching facility, research center, and laboratory, and in 1969 it became a medical history museum.

The Old Pathology Building where the museum is housed is the oldest surviving pathology laboratory in the nation, and is virtually unchanged from when it was built. Because of its history and age, the two-story brick building with its amphitheater, dissection and autopsy rooms, library, photography room, and laboratories is on the Historic American Building Survey and the National Register of Historic Places.

The museum's purpose is to help the public understand the importance of medicine to society and the developments that have led to America's advanced medical treatments and healthcare. The most important artifact here is the building itself, a state-of-the-art teaching facility in its time. It was lit by three sources: gaslight, electric lighting, and natural light from strategically placed skylights. All the original cabinets, built-in oak microscope cabinets, and file cabinets are still here. Since the turn of the 20th century, no woodwork has been painted over and no new desks have been installed. The oak-paneled, acoustically perfect teaching amphitheater is particularly impressive. It has eight semicircular wooden tiers of seats: around 115 of the original straight-back, cane-bottom chairs are still in place.

Some of the important work accomplished at this facility included systematic autopsies. The autopsy suite displays an original cast-iron autopsy table and a wooden refrigerator from around 1920 that has two body trays. A speaking tube goes from the autopsy room to the records room on the second floor. Here the secretary, who was a physician, transcribed the notes immediately, without being present during the autopsy. Next to the Pathology Building is an ice house, or dead house, where cadavers were preserved. The ice house now displays an exhibit showing what a typical doctor's office in rural areas looked like from the mid-20th century. The idea is to illustrate how such doctors navigated the transition from medical practices of the late 1800s to the modern world. Artifacts on display were originally from the practice of Dr. Marion Scheetz of Lewisville, Indiana. Another building contained a mortuary where free funerals were performed—an offering that helped induce families to permit autopsies.

From the autopsy room, tissue samples went to the histology laboratory, one of several original laboratories in the building, where slides were prepared. Other laboratories were for bacteriology, histology, and clinical chemistry. The bacteriology laboratory was often used to test the safety of food served at the hospital.

By the 1920s and 1930s, neurosyphilis in its tertiary stage had become a major cause of institutionalization for mental disorders. Much research into this disease was carried on here. In the 1920s, Austrian physician Julius Wagner-Jauregg noted that syphilis progresses at a slower rate in patients with malaria, and suggested that syphilis patients be infected with live malaria parasite. He received the 1927 Nobel Prize for this medical breakthrough, which remained the accepted medical treatment for syphilis until penicillin was introduced in the 1940s. Dr. Walter Bruetsch, who introduced the Wagner-Jauregg treatment to this hospital, became famous for his syphilis research. The museum library contains his syphilis research notes.

In the photography studio, the hospital photographed patients being admitted and developed the photographs in the darkroom. Enlargements of microscope slides were prepared here as well, using a photomicrography box camera on display that was invented in 1895.

Of special interest among the museum's 15 000 medical and healthcare artifacts from the 19th and 20th centuries are the diagnostic instruments, surgical equipment, pharmaceutical bottles, dental equipment, and "quack" devices. The museum also has archives containing all postmortem records from patients treated at this hospital, including autopsy records, tissue slides, lantern slides, and pathological specimens, and it mounts rotating displays from its collection of portraits and medical artwork.

Just south of the Old Pathology Building is a Medicinal Plant Garden, in which over 120 different types of trees, shrubs, and plants with medicinal uses are grown. Each plant is labeled as to origin and use, as well as scientific and common names. The garden was installed in 2003.

The museum often offers special lectures and presentations.

Address: Indiana Medical History Museum, 3045 West Vermont Street, Indianapolis, IN 46222.
Phone: (317) 635-7329.
Days and hours: Thu.–Sat. 10 a.m.–4 p.m. Wed. is available for pre-scheduled group tours only. Closed 7/4, Thanksgiving Day and the following Fri., 12/25, and 1/1.
Fee: Yes.
Tours: One-hour guided tours. Parties of seven or more should call ahead at least three weeks in advance.

Public transportation: Buses stop at Washington and Hancock Streets, then walk 0.6 mile (1 km) to the museum.
Handicapped accessible: Limited. The second floor is not accessible.
Food served on premises: None.
Website: www.imhm.org

MUNCIE

National Model Aviation Museum

With the rise of true airplanes following the Wright Brothers' invention came the increase in serious model aviation as well. (See the Carillon Historical Park, Dayton, Ohio.) Here in Muncie is the home of the Academy of Model Aeronautics, founded in 1936, the official organization in the United States overseeing model aeronautics competitions, and the Academy's museum, the National Model Aviation Museum, on 1100 acres (450 hectares) of land, including fields for competition. In fact, should visitors arrive during a competition, they may well hear the whine of model aircraft engines from a distance.

The National Model Aviation Museum, "dedicated to the preservation of aero modeling history," holds over 11 000 objects in its collection, ranging from historical and current aero models, to engines, documentation including diagrams and plans, magazines and other literature, as well as posters, awards, and a variety of ephemera associated with model aviation. Upon entering the main hall of the museum, visitors may be

National Model Aviation Museum.
Courtesy of the National Model Aviation Museum.

overwhelmed with nearly a visual encyclopedia of model aircraft in wall displays and hanging from the ceiling.

The introductory hall holds several exhibitions recounting the earliest history of model aviation, from the late 1800s. These areas are divided into a number of display spaces beginning with The Early Years, exhibiting models and advertisements from the first decades of the 20th century. Initially, model aviators flew "free flight" aircraft: these airplanes were not controlled from the ground. They may have been powered by "gummi-bands" (rubber bands), pull-strings, solid chemical rockets, or launched from throwing devices. One unusual model aircraft displayed is the Autoplan, powered by a small tank of compressed air, built by the Bing Toy Company in 1914. By the 1930s, internal-combustion engines were the primary source of power for model aircraft. After Lindbergh's famous transatlantic flight in 1927, commercial kits produced by toy companies to build one's own airplanes became popular.

Another section of the main hall describes "control line" aviation, developed in 1937 by Oba St. Clair in Oregon. He used wires to control his model, *Miss Shirley*, from the ground because he wanted to keep control of his aircraft where he lived in the forest. Control line aviation, using wires 20 to 70 feet (7 to 20 m) long, is still used today, including some competitions where the aircraft have a crepe-paper tail, and the dueling pilots try to slice off each other's tail. It is a challenge for multi-pilot competitions not to get wires entangled.

Simultaneous with the rise of control line aviation was radio-controlled flight, which became possible in the late 1930s. This type of model aviation is also exhibited in the main hall. Bill and Walter Good built the *Big Guff* in 1937, the first successful radio-controlled airplane, which had only rudder control. A tube-transmitter on the ground transmitted impulses to a two-tube receiver in the aircraft. A typical early transmitter-receiver pair is shown here. For these early radio-controlled aircraft, not only flight-control was a problem; frequency-control was not easy as well. Nearby model aviators sometimes caused interference on a particular model aircraft's flight, and the result could be a pile of splinters instead of a smooth landing. Not until the 1970s did reliable transistor-powered radio control become available, with complete control of ailerons and rudder. With the advent of digital electronics, such receivers became very small and lightweight. On display are a variety of transmitters and receivers.

One section of the main hall shows a variety of model engines from the United States and around the world. A few model aviators have become interested in building miniature jet engines, which are also on display here.

Another gallery relates some of the uses that model aircraft have at work. NASA often builds models to test before attempting to construct full-size aircraft, and several are on display here. An 11-ounce (310 g) craft with a large photoelectric array hanging down from its belly is an example of the first aircraft solely powered by a beam of light: its five-watt engine was powered in a darkened area using a spotlight. The museum also displays several US military reconnaissance and target drones.

The Adventure Hangar is a gallery of interactive exhibits designed to teach the visitor about the basic principles of flight, and to demonstrate how model aviation works using flight simulators. One corner of the museum is a window onto the Restoration Shop, where volunteers recreate or occasionally restore model aircraft in a machine shop. Visitors can ask questions of the volunteers while they work. The Lee Renaud Memorial Library stores literature on model aviation, where visitors may walk in to a view over 5000 books and 850 magazines. At the in-house archives, researchers can request in advance both two- and three-dimensional artifacts, such as model planes, parts, awards, engineering drawings, and instruction manuals held in long-term storage.

Address: National Model Aviation Museum, 5151 East Memorial Drive, Muncie, IN, 47302.
Phone: (765) 289-4236.
Days and hours: Open daily 10 a.m.–4 p.m. during the summer; Tue.–Sat. 10 a.m.–4 p.m. other times of the year. See the website for seasonal and holiday hours.
Fee: Yes.
Tours: Self-guided. Guided tours upon advance request for an extra fee.
Public transportation: None.
Handicapped accessible: Yes.
Food on premises: None. There are restaurants in Muncie.
Website: www.modelaircraft.org/museum

Iowa

WATERLOO

John Deere Company

John Deere, born in Rutland, Vermont, in 1804, apprenticed as a blacksmith and later earned his living in that trade. His high-quality workmanship earned him an excellent reputation and created a great demand for his services, particularly to repair and create farm tools. Like many natives of Vermont who went west when the local economy deteriorated, Deere moved to Grand Detour, Illinois, in 1836, seeking adventure and opportunity.

Deere quickly established a forge and began repairing the equipment of local farmers, just as he had done in Vermont. The cast-iron plows the pioneers brought from the East were made for the light, sandy New England soil. Midwestern soil clung to the plows and brought them to a crawl. After studying the problem, Deere designed a plow with a highly polished metal blade and a moldboard that would turn the soil with a clean slice, making a neat furrow. In 1837 he made a prototype out of a broken saw blade and tested it in a nearby field.

Deere began manufacturing these new plows before he had any orders, contrary to the general practice at the time of making "bespoke" items (on specific order). This new approach to business had been brought to New England by Samuel Slater, who revolutionized manufacturing in America. (See Old Slater Mill National Historic Landmark, Pawtucket, Rhode Island.)

Deere faced many problems. Few banks were on the frontier to provide capital, transportation was poor, and steel was limited. His first few plows were made from whatever steel was at hand. In 1843 Deere imported a shipment of steel from England. The long, tedious route to get the steel to Grand Detour proved to be costly, but sales went up to 400 units that year. Deere moved his business to Moline, Illinois, along the Mississippi River, where he had access to waterpower and transportation. By 1846 he persuaded the Jones & Quiggs Steel Works of Pittsburgh to produce the type of rolled steel he needed for production, the first slab of steel ever rolled in the United States. Just 10 years after he made his first plow, Deere manufactured 1000 plows in one year.

The Deere Company continued to experiment and improve the plow. In 1868 it was incorporated as Deere & Company. John's son Charles joined the leadership of the business, which continued to prosper and increase its product lines. By 1907, the year of John Deere's death, the company made plows, cultivators, corn and cotton planters, and many other implements.

In 1918, when agriculture was moving from animal to mechanical power, the company bought the Waterloo Gasoline Engine Company of Waterloo, Iowa, and added tractors to its manufacturing line. By the end of World War II, the John Deere Company was one of the 100 largest manufacturing businesses in the nation. Today it is a world leader in the production of farm, forestry, construction, and lawn-care equipment for home and commercial use.

Some of this history is presented during the 1.5-hour tour at the John Deere production plant on the edge of the twin cities of Waterloo and Cedar Falls. Each of the other John Deere plants in the Midwest conduct similar tours.

The tour starts with a 20-minute film that gives a corporate overview. Because guests ride through the factory on an open cart, each visitor is provided with safety glasses. The tour guides, who are retired John Deere employees, share their vast knowledge of factory production.

One plant in Waterloo assembles only tractors; another builds diesel-engine parts; a third manufactures parts for drivetrains. Visitors see the sheet metal being shaped, the parts being assembled, the painting, customizing, and testing. Although the plants are highly automated, certain functions still require the human hand to ensure accuracy and precision. About the only major component that is not produced by any Deere plant is tires. On the tours visitors also learn about the unique tractor designs required by various countries around the world. Other facilities run by John Deere include a factory in East Moline, Illinois, which manufactures combines and front-end equipment; the Seeding Group in Moline builds planters; a Des Moines (Iowa) plant constructs sprayers, harvesters and tillage machinery, and grain drills; and a factory in Ottumwa manufactures various types of balers and windrowers. The Gator Works in Horicon, Wisconsin, builds utility vehicles. All offer factory tours at various times of the day; see the website listed in the following for details.

The John Deere Historic Site in Grand Detour, Illinois, has John Deere's home, the foundation of his original workshop—uncovered during an archaeological dig—and a replica of that workshop. The Visitor Center at the historic site contains period artifacts. Around it, a two-acre (0.8-hectare) natural prairie with wildflowers and tall grass yields insights into problems with the land that early pioneers faced.

Address: John Deere Tractor Cab Assembly, 3500 East Donald Street, Waterloo, IA 50703.
Phone: (319) 292-7668. Reservations are required 48 hours in advance.
Days and hours: Mon.–Fri. 8 a.m., 10 a.m., 1 p.m.
Fee: Yes.
Tours: By appointment in advance only. Minimum age is 13 for the tour, and all participants must wear closed-toed shoes.
Public transportation: None.
Handicapped accessible: Yes.
Food served on premises: Café.
Website: www.deere.com/en/connect-with-john-deere/visit-john-deere/factory-tours
Address: John Deere Engine Works, 3801 West Ridgeway Avenue, Waterloo, IA 50704.

Phone: (319) 292-5347.
Days and hours: Mon.–Fri. 9 a.m. and 1 p.m.
Fee: Yes.
Tours: Mon.–Fri. 9 a.m. and 1 p.m. by appointment only. Reservations are required 48 hours in advance. Minimum age is 13, and participants must wear closed-toed shoes.

Other information is the same as the Tractor Cab Assembly.

Address: John Deere Drivetrain Operations, 400 Westfield Avenue, Waterloo, IA 50701.
Phone: (319) 292-7668.
Days and hours: Mon.–Fri. 1 p.m.
Fee: Yes.
Tours: Mon.–Fri. 1 p.m. by appointment only. Reservations are required 48 hours in advance. Minimum age is 13, and participants must wear closed-toed shoes.

Other information is the same as the Tractor Cab Assembly.

Michigan

BLOOMFIELD HILLS

Cranbrook Institute of Science

George and Ellen Scripps Booth, newspaper magnates, purchased an old farm on which they founded the Cranbrook Educational Community in the quiet suburb of Bloomfield Hills northwest of Detroit in 1904. Ellen's father was James Scripps, who published the Detroit Evening News. The Booths named the property Cranbrook, after Cranbrook, England, where George Booth's father was born. Initially the Booths used the property as a summer retreat, and hired architect Albert Kahn to design a home for them. By the 1920s, the family joined forces with other town residents to create a local school, then hired architect Eliel Saarinen to design an entire campus of school buildings and museums. Often called just "Cranbrook" by the locals, the 319-acre (129-hectare) complex now includes private elementary and high schools, a graduate-level academy of art, gardens, art museum, and historic homes. The portion of Cranbrook of interest for purposes of this book is the Cranbrook Institute of Science.

The Institute of Science, first opened to the public in 1930, is housed in a structure originally designed by the architect Eliel Saarinen and unveiled in 1936, built in an Art Deco style. In 1998 an additional section designed by Steven Hall expanded the galleries. The focus of the Institute's collection of over 200 000 artifacts is natural history (especially of Michigan) and cultural anthropology.

Just beyond the main lobby is the Our Dynamic Earth gallery, which describes natural life and diversity of the past and present. A *Tyrannosaurus rex*, nicknamed "Black Beauty," so called because of the color of the fossil remains, looms over the exhibits, which tell how life evolves. The Ice Ages Come and Go area reveals how the Earth's climate has changed over the past million years, and includes both mastodon fossils and a reconstructed full-size model of a mastodon, comparing it to modern African and Asian elephants as well as mammoths. After the Ice Ages, a small exhibit of Michigan mammals weighing over 100 pounds (45 kg) that have survived to this day is also on view.

The second major section of the museum is called The History of Us, an interactive exhibit which compares and contrasts artifacts from many cultures around the globe, showing the commonality of their uses. A variety of tools, clothing, and ritual objects are presented.

Every Rock Has a Story is an exhibit on the fundamentals of geology and geological processes. One special example displayed is a large section of Watersmeet gneiss, found in Michigan's Upper Peninsula, dating to 3.65 billion years ago, the oldest rock found in the United States. There are examples of Petoskey stone, which is a fossilized coral *Hexagonaria percarinata*, peculiar to Michigan, and is the official state rock. Petoskey stone, when polished, shows the regular hexagonal formations of the coral.

The Institute is known for its generous mineral collection, some of which is housed in the Mineral Study Gallery. Besides the cases of minerals and gems found in Michigan, there are displays organized by chemical structure (oxides, sulfide, molybdates, etc.) and descriptions of crystal structure. All the minerals are labeled with their original provenance. The collection began as George Booth's personal samples, which were stored in the Booths' attic. The weight of the rocks began to warp the attic floor, and Ellen insisted they be removed from the home. Of the 11 000 mineral samples now held by the Institute, about 15% are on display.

The Institute of Science also has a planetarium presenting star shows regularly, using a Digistar projector. An observatory is open in evenings—weather permitting—with a 20-inch (51-cm) CDK research-grade system operated by remote control. There is also a 6-inch (15-cm) Takahashi refractor and a Lunt solar telescope for live observation of the Sun.

A courtyard within the Institute's structure is the home of the Erb Family Science Garden, portraying various Michigan biomes from meadow to northern woodland using appropriate trees, flowers, shrubs, and grasses.

A small auxiliary building behind the Institute is the ExploreLab, a hands-on area with changing installations.

Also of interest is the Art Museum, built in 1942, housing contemporary art, as well as the Cranbrook House with its formal gardens.

Address: Cranbrook Institute of Science, 39221 Woodward Avenue, Bloomfield Hills, MI 48304.

Phone: (248) 645-3200.

Days and hours: Sep.–May: Tue.–Thu. 10 a.m.–5 p.m., Fri.–Sat. 10 a.m.–10 p.m., Sun. noon–4 p.m.; Jun.–Aug.: Mon.–Thu. 10 a.m.–5 p.m., Fri.–Sat. 10 a.m.–10 p.m., Sun. noon–4 p.m.

Fee: Yes. Planetarium shows and ExloreLab incur an extra fee.

Tours: Self-guided.

Public transportation: None.

Handicapped accessible: Yes.

Food served on premises: Café.

Website: science.cranbrook.edu

DEARBORN

The Henry Ford

This complex of museums, archives, plus a factory tour, standing on 93 acres (38 hectares) of land, is a world-class facility. Together the combined organizations explore 300 years of industrial development in America, and tell how rural, agrarian America grew into an urban, industrial society.

Henry Ford Museum of American Innovation

Henry Ford, an avid collector of industrial and historic artifacts, conceived the idea in 1919 of a museum and village in which to display his collection. This complex, the realization of his vision, opened in 1929 with the theme "to create a better future we must understand the past." According to the museum's literature, the purpose of this revolutionary type of museum and village was and still is to "collect, present, and interpret to a broad public audience, the American experience with special emphasis on the relationship between technical change and American history."

The entrance to the Henry Ford Museum is a scaled-down replica of Independence Hall, a reminder of the collection's American emphasis. The building itself is designed to resemble a factory of the 1920s. It has 12 acres (4.9 hectares) of exhibits under one roof, which explore such diverse topics as transportation, communications, agriculture, industry, decorative arts, photography, leisure, and entertainment. As noted in the museum's literature, it is "a three-dimensional encyclopedia of American experience."

The museum places artifacts in context with their function whenever possible and explains their place in the lives of people who made and used them. Exhibits include audiovisual presentations, hands-on activities, and operating machinery.

Made in America is a multimedia exhibit the size of a football field that explores the history of American industry and production technologies, and takes a look at the people involved. It also examines the effects of historical choices on productivity, product quality, environmental responsibility, and quality of life. Some artifacts come from the museum's vast collection of power machinery—the most extensive in the world—including steam engines, boilers, generators, early motors, transformers, electrical transmission equipment, hot-air engines, and shop machinery.

The exhibit is divided into three sections. The first part, Making Things, is a tour of a historical plant to see people at work and the machines and systems used in production. Visitors learn about state-of-the-art manufacturing, planning, management, mass-production methods, advertising, mechanization of the trades, and craftsmanship. Among the artifacts on display are a 1992 Unibilt inverted monorail conveyor that moves materials on a production line, a 1991 GM France P-150 six-axis articulated robot arm, and a 1929 Corning Ribbon Machine #3, used to produce up to one million lightbulb blanks per day.

Making Power, the second section, is an in-depth look at the history and technology of various power systems. It traces the evolution of wind, water, and steam power and examines the trade-offs needed to develop power systems. On display is a 1903 turbine-generator, decommissioned in 1990 from a Spokane River power station, with a cutaway to expose the mechanism. In acknowledgment of American industrial roots in England, the exhibit has the world's oldest intact steam power equipment, a circa-1750 Newcomen engine from Lancashire. Among the 19th-century pieces is an 1891 triple-expansion engine-driven generator.

The last section, Making Choices, illustrates how industrial decisions are made. An original film with clips that date from 1904 shows this decision process. In addition, touch screens present issues and historical perspectives on industrial decision-making.

The three-dimensional, multimedia Driving America exhibit takes up one-third of the museum. Visitors learn of the tremendous impact cars have had on the way 20th-century Americans worked, played, and conducted their daily lives. The exhibit includes more than 100 automobiles as well as motel rooms, a 1940s service station, a diner, recreational vehicles, and even roadside advertising.

Another exhibit, showing agricultural implements, ranging from wooden plows and early hand tools to large tractors and 20th-century combines, is the most comprehensive of its type in the world. It presents the history of agricultural technology, including dairying, soil preparation, planting, cultivating, and harvesting. Two interesting devices are the lactometer for determining the richness of milk and the Babcock tester for finding milk's buttermilk content in 10 minutes.

Domestic Life follows the changing tastes and lifestyles of Americans from 1600 through the 1900s. A complete 1930s kitchen and a collection of fireplaces, furniture, iron stoves, textiles, and washing and drying machines illustrate these developments.

The museum displays telegraph equipment, telephones, televisions, and other communications devices, an extensive array of printing and typesetting machinery, and an excellent collection of lighting devices and fixtures. If all that is not enough, the museum exhibits such eclectic items as Admiral Richard Byrd's 1929 Ford Trimotor airplane, Abraham Lincoln's chair from Ford's Theater, Buckminster Fuller's Dymaxion House (a futuristic aluminum structure from 1948), and a 600-ton Allegheny locomotive, one of the largest ever built.

Greenfield Village

The outdoor Greenfield Village, made up of more than 80 buildings from various times and places and arranged in seven districts, has the look of a New England town. Each structure is the site of an important event that helped shape the United States.

The man Henry Ford admired most was Thomas Alva Edison. (See Thomas Edison National Historical Park, West Orange, New Jersey; and Edison and Ford Winter Estates, Fort Myers, Florida.) Ford gathered what could be found of Edison's Menlo Park laboratory complex, moved these pieces from New Jersey, and painstakingly reconstructed them in the village. The Menlo Park laboratory complex is generally considered to be the first industrial research and development park. Ford chose October 21, 1929, to open Greenfield Village because it was the golden anniversary of the invention of the practical incandescent light. He commemorated this event with a re-enactment in which Edison participated.

Henry Ford moved the boyhood home of his good friend Harvey Firestone, founder of the Firestone Tire and Rubber Company, to the village. It is now a working, 7-acre (3-hectare) farm that represents rural life of more than 100 years ago. On the farm,

19th-century corn and other antique seed varieties are grown, and merino sheep, speckled dark Brahma chickens, shorthorn cattle, and Poland China pigs are raised. Herbs and vegetables grow in the small, 19th-century kitchen garden.

Other historic buildings here are an English Cotswold cottage and forge, which acknowledge the English roots of the United States; the 1800s Heinz house, where horseradish was produced before ketchup became the company's major product; a three-quarter replica of the Edison Illuminating Company 1886 Station A, the first Detroit electricity plant; Orville and Wilbur Wright's home and cycle shop; and Luther Burbank's garden office from Santa Rose, California. (See Wright Brothers National Memorial, Kill Devil Hills, North Carolina; National Air and Space Museum, Washington, DC; Carillon Historical Park, Dayton, Ohio; and Luther Burbank Home & Gardens, Santa Rosa, California.)

This beautiful park also contains an operating machine shop, a 19th-century doctor's office, a glassmaking exhibit, and demonstrations of printing and pottery-making. An exhibit area memorializes George Washington Carver, who spent his career developing hundreds of useful products from peanuts and sweet potatoes. (See Tuskegee Institute National Historic Site, Tuskegee Institute, Alabama; and George Washington Carver National Monument, Diamond, Missouri.)

The upper-middle-class farmhouse where Henry Ford was raised has been moved to the grounds. It reveals much about Ford. As a boy, he was so inquisitive he would dismantle every clock that came into the house. To keep one clock in working order—and to keep Henry out—it was built into the wall and covered by protective glass.

The Benson Ford Research Center, devoted to American heritage, contains more than a million three-dimensional objects and 25 million letters, documents, photographs, posters, painting, records, rare books, catalogs, manuscripts, newspapers, sheet music, and maps. The Edsel Ford Design History Center collects and interprets materials related to the history of product design in America.

Ford Rouge Factory Tour

To see a modern industrial factory, take the Ford Rouge Factory Tour. Visitors board buses at the Henry Ford Museum and drive to the factory. There are two theater presentations, one in the Legacy Theater, and a second in the Manufacturing Innovation Theater. The tour continues through a truck factory.

Address: The Henry Ford, 20900 Oakwood Boulevard, Dearborn, MI 48121.
Phone: (313) 982-6001.
Days and hours: Daily 9:30 a.m.–5 p.m. Closed Thanksgiving and 12/25.
Fee: Yes.
Tours: Self-guided.
Public transportation: Taxis.
Handicapped accessible: Yes, except for historic village buildings with steps and narrow doorways.
Food served on premises: Café, diner, snacks.
Special note: Allow at least one day to see each facility.
Website: www.thehenryford.org

Negaunee

Michigan Iron Industry Museum

The first iron forge in the Lake Superior region, now part of the Michigan Iron Industry Museum, is hidden deep in the forested ravines along the Carp River in Michigan's Upper Peninsula. Because the mission of this museum, near the Carp River Forge, is the examination of the history of Michigan's iron forges and their impact on the economy and culture of the region and the nation, iron mining is a crucial part of Michigan's story. (See Lake Vermilion–Soudan Underground Mine State Park, Minnesota Discovery Center, and the Hull-Rust Mahoning Mine, Soudan, Chisholm, and Hibbing, Minnesota; and Cornwall Iron Furnace, Cornwall, Pennsylvania.)

Michigan contains three of the six principal iron ranges in the United States. The Marquette range, where this site is located, lies solely in Michigan and is the heart of one of the world's richest sources of iron ore. The other two iron ranges in Michigan, the Menominee and Gogebic, lie partly in Wisconsin. The Marquette range is approximately 40 miles (64 km) long and varies from 3–10 miles (5–16 km) wide.

Douglass Houghton, Michigan's first state geologist, conducted a systematic exploration of the Upper Peninsula in the 1840s and found iron and copper in the region, but did not recognize the quality of his find. A subsequent survey by William A. Burt, deputy surveyor for the United States, identified a deposit of unusual magnetic disturbances in the ground. Burt is renowned as the first person to use the solar compass in land-surveying, a technique that provided a true north-south direction

Michigan Iron Industry Museum.
Courtesy of the Michigan History Center.

and thus a truer survey. Burt used his patented variation compass to survey this region. He failed, however, to mention his discovery in his official report. A year later, in 1845, the search for iron ore began in earnest and a major discovery was made in the area.

By 1870, 20% of all iron produced in the country came from the Marquette range. Before iron was discovered here, copper ore was often sought. The Jackson Mining Company, founded in 1845, set out to mine copper but changed course when company officials realized the significance of the iron ore find. They eventually altered their corporate name to the Jackson Iron Company.

The Carp River (or Jackson) forge near the museum operated from 1848 to 1855, manufacturing wrought iron in small quantities. Mismanagement, inefficient production techniques, isolation from markets, inclement weather, and poor transportation to Lake Superior caused the venture to fail. Even though the forge was built near a river, rapids made it unusable for transportation. The costs of overland shipping were too high, particularly in the wet spring months when the roads became too rutted for use.

Another problem for local iron producers was that the Upper Peninsula's forests did not have enough wood to sustain the voracious appetite of the forge. The ironworkers had to go farther and farther afield to find enough wood to keep the forge operating. Although other iron forges opened in the Upper Peninsula, they learned the same lesson as the Jackson Forge: the area was exceptional for mining but unsuitable for processing. Despite its failure, the forge served as the seed of the successful Michigan iron-mining industry. The raw ore was transported to Lake Superior and then on to iron-smelting facilities on the lower Great Lakes.

Archaeologists have studied the site of the Carp River Forge. Trails with interpretive signs lead to an overlook of the site. On the museum's grounds is a rare vertical-boiler steam locomotive, the *Yankee*. It was made in Scotland and used from 1868 to 1891 to pull ore cars out of the Jackson iron mine in Negaunee.

Inside the museum, visitors learn about the history of the Upper Peninsula, the people of the region who worked the mines, and the iron industry's effect on the inhabitants of the Upper Peninsula and the nation. A 23-minute documentary, *Iron Spirits*, is shown regularly in the auditorium. The museum has hands-on exhibits, and displays of artifacts such as mining tools from about 1844 to the 1960s, many used locally. Larger artifacts on view include a Ford Model T, Model A, and a World War II jeep.

The site has re-created a full-size, walk-through, underground iron mine to help visitors understand the process of getting the iron out of the ground.

Address: Michigan Iron Industry Museum, 73 Forge Road, Negaunee, MI 49866.
Phone: (906) 475-7857.
Days and hours: May–Oct.: daily 9:30 a.m.–4:30 p.m.; Nov.–Apr.: Wed.–Fri. and first Sat. of the month 9:30 a.m.–4 p.m. Closed on state holidays, but open Memorial Day, 7/4, and Labor Day.
Fee: Free.
Tours: Self-guided.
Public transportation: None.
Handicapped accessible: Yes, but limited accessibility on the paths. A wheelchair is available.
Food served on premises: Snacks only; picnic area.
Website: www.michigan.gov/ironindustrymuseum

Minnesota

GRAND RAPIDS

Forest History Center

Forests are one of Minnesota's greatest resources. The Forest History Center interprets the history of Minnesota's forests from the last geologic glacial period and explores the interrelationship between forests and human beings. This center contains an indoor museum, an outdoor living-history logging camp, and nature trails.

Exhibits in the indoor museum review the natural history of forests and the culture of people who live in them. Computer-simulated logging machinery allows visitors to try a hand at operation. A small theater plays a video recounting the dangers of forest fires. There are video interviews of people who live in and manage the forest. One display highlights products manufactured from local Minnesota wood, from tiny (toothpicks) to large (hockey sticks). The museum discusses the role of forest conservation and society's efforts to preserve Minnesota forests. A multimedia presentation reviews geologic history, starting with the Ice Age up to the present.

An outdoor living-history exhibit is a reconstruction of an eight-building logging camp, designed to house 70 people. It reflects life in a typical camp at the turn of the 20th century. The year 1900 was selected for representation because it was the peak year for harvesting pine trees in Minnesota and occurred just before mechanization began to alter the lives of lumberjacks. Typical logging camps of the period were rough-cut and designed to last only one or two seasons. After the trees were played out, the camps were abandoned and the loggers moved on.

Costumed guides explain the significance of the buildings, tools, and artifacts, and describe the inhabitants' lifestyles and the role of the Mississippi River in logging and log drives. The surroundings convey a sense of the tough experience these men had.

Horses used to haul logs and equipment through the forest were housed in horse barns like the one here, which has room for 24 horses. A barn boss cared for and fed the horses that the teamsters harnessed and drove. An office store or *wanigan* supplied the necessities of life to the loggers, including patent medicines and tobacco. The foreman and the scaler, who determined the amount of lumber to cut, had offices in the wanigan.

Communal living was the norm. Loggers slept in a bunkhouse where privacy was almost impossible. Men slept two to a bunk and hung their sweaty clothes in the bunkhouse to dry each night. Body lice were a common affliction. No camp—this one included—was complete without an outhouse.

The severe northern winters required that food be stored properly. Visitors can see the root house where food stocks were kept at an even, cool temperature and the cook's shanty where mass meals were prepared for the hearty appetites of the lumberjacks.

A filer's shack shows where huge crosscut saws were sharpened, an endless task. A sharp saw was the key to an efficient lumberjack. The sawyers who sharpened saws often worked late into the night so the loggers could be out working early the next day. Blacksmiths were also critical at such camps because they kept equipment operational and the horses well shod. In winter the blacksmith used caulked shoes on the horse to improve traction on the ice-covered ground.

A logging sleigh, used to haul logs, and a water wagon, used to spread water on the winter roads, are on view. After water spread on the roads had frozen to a smooth glaze, a rutter cut even ruts for the logging sleighs to follow. Roads were regularly re-iced throughout the winter. The send-up men loaded the logging sleighs while the top loader made sure the load was balanced. A typical load consisted of 20 logs, which produced 5000–6000 board feet (12–14 m³) of pine.

After a series of disastrous forest fires in 1911 that cost many lives, the Minnesota Forest Service was formed to protect northern Minnesota against forest fires. Reconstructed on the museum grounds is a 1934 Forest Service patrolman's complex consisting of a fire-tower cabin, tool cache, and garage. The exhibit clearly illustrates the isolated life of a ranger.

Five miles (8 km) of nature trails along the Mississippi River cut through second-growth forest with large stands of birch, maple, and pine. Most of the virgin forest was harvested by 1900. Along the three trails—River Trail, Swamp Trail, and Forest of Today—visitors learn about modern forest management practices as well as common trees of northern Minnesota and how they are used.

Address: Forest History Center, 2609 County Road 76, Grand Rapids, MN 55744.
Phone: (218) 327-4482.
Days and hours: After Labor Day–mid Jun.: Sat. 10 a.m.–4 p.m.; mid-Jun.–Labor Day: Tue.–Sat. 10 a.m.–5 p.m. Closed for all winter holidays.
Fee: Yes.
Tours: Self-guided and guided (during the summer).
Public transportation: None.
Handicapped accessible: Yes. Wheelchairs are available.
Food served on premises: Snacks only; picnic area.
Website: sites.mnhs.org/historic-sites/forest-history-center

SOUDAN, CHISHOLM, AND HIBBING

Lake Vermilion–Soudan Underground Mine State Park, Minnesota Discovery Center, and the Hull-Rust Mahoning Mine

At Soudan Hill near Lake Vermilion, one of the richest deposits of iron ore in the world was found in 1866. This spot in the Vermilion mountain range was the birthplace of Minnesota's mining industry.

Northern Minnesota has three major iron ranges: the Vermilion, the Cuyuna, and the Mesabi. In the language of the Native American Ojibwe, *mesabi* means "sleeping giant." Vermilion does obliquely refer to "red," as the name suggests, but this is a very free French translation of the Ojibwe *onamuni*, meaning "lake of the sunset glow." The iron ore from these mountains is pure enough that, when scratched, it appears red, and gave rise to another term for the ore, "blood stone."

Lake Vermilion–Soudan Underground Mine State Park
The Soudan Mine began operations in 1882. It was the first mine in Minnesota, and seven decades later, in 1962, it was one of the last underground mines to close. Because early efforts with open-pit mining showed it to be inefficient and unsafe, underground mining became fully operational here by 1892. At one time, many such mines dotted the Minnesota iron ranges, but the other mines have disappeared. The US Steel Corporation

donated this land to the state of Minnesota, which made it a state park. It is one of the few underground iron-ore mines in the world open to the public for tours.

The Visitor Center was originally the dry house, where miners changed and dried their clothes. Now the Center has a small museum with introductory displays about the mine. Visitors can see mining equipment, miners' artifacts, photographs from the mining era, and models of the mine's design. A mineral collection includes core samples drilled out of the mine, pyrite crystals, rose quartz, and more. Before descending into the mine, visitors are treated to rare film footage taken for US Steel captured during the last few years of mining at this site.

Each visitor receives a hard hat to wear during the underground tour. A three-minute ride in the "man-cage" (similar to an elevator system) once used by the miners goes down the 2341-foot (714-m) shaft to level 27, where iron mining last took place. The one-and-a-half-hour tour begins here.

First, each tourist climbs aboard an electric train for a three-quarter-mile (1.2 km) ride through the drift to the Montana "stope" (a raised excavation area), the deepest ore-producing cavity that was mined. The continuous natural airflow keeps the air fresh and at 51°F (11°C). Very little water flows in the mine, there are no dangerous gases, and there is no shoring or timbering because the rock body is extremely hard and stable.

As visitors walk through the mine and climb steps to the stope, they see drilling and ore-moving equipment, and get a sense of the working conditions in what was considered the "Cadillac of mines." The tour guides are knowledgeable about the operations and are happy to answer questions.

After leaving the mine, visitors should be sure to take the self-guided tour of the surface. The Engine House contains the 600-horsepower (450 kW) electric hoist system built in 1924, which still lifts the 6-ton (5.4 tonne) cages that now carry tourists. The Drill Shop has equipment used by the surface workers to keep the drill bits sharp enough to cut through the hard Soudan rock. Visitors can also see several open pits from the late 1800s. The first miners in the region risked their lives in these dangerous pits to remove iron ore from the ground.

The three-quarter-mile (1.2-km) Walking Drift tour focuses on geology of the underground mine. Participants learn how geologists discovered iron ore and search for geologic clues on a walk following the route that the miners took to and from their work. Among other facts, those on the tour learn about the tools used to find the ore.

A separate tour of science research at the mine takes visitors half a mile (0.8 km) underground to see where University of Minnesota physicists operated the MINOS experiment to search for neutrino oscillations (see Fermi National Accelerator Laboratory, Batavia, Illinois; and Sanford Lab Homestake Visitor Center, Lead, South Dakota), as well as the CMDS II experiment to search for WIMPS (weakly interacting massive particles). These experiments were shut down in 2016. Currently the mine hosts other research, mainly dealing with local bacteria and their uses.

A colony of bats, perhaps a couple of thousand, hibernate in the underground mine during the winter months. The park, which includes over 4000 acres (1600 hectares), is home to six species of bats: little brown bat, the big brown bat, the northern myotis, the tricolored bat, the silver haired bat, and the hoary bat. Visitors may occasionally encounter underground-dwelling bats while touring the mine.

Address: Lake Vermilion–Soudan Underground Mine State Park, 1302 McKinley Park Road (during the summer) or 1379 Stuntz Bay Road (winter entrance), Soudan, MN 55782.

Phone: (218) 300-7000.
Days and hours: Visitor Center: 9:30 a.m.–6 p.m.
Tours: Daily from Memorial Day to end of Sep. 10 a.m.–4 p.m., weekends in Oct. Call for off-season tour hours and specific tour times.
Fee: Yes.
Tours: Guided tours only. Call (866) 857-2757 for reservations.
Public transportation: None.
Handicapped accessible: Yes, except for Walking Drift tour. The mine floor is uneven in areas.
Food served on premises: None. There are restaurants in the nearby town of Tower.
Special note: Wear jackets and sturdy shoes. At the website are audio guides for touring the surface facilities. People with claustrophobia may find riding in the cage difficult.
Website: www.mndnr.gov\vermilionsoudan

Minnesota Discovery Center
The second stop on our mining tour is the Minnesota Discovery Center in Chisholm, about 45 miles (72 km) southwest of the Soudan mine. Inside the museum building complex is a variety of equipment from mining days on the Mesabi as well as local artifacts. Chisholm's reconstructed Main Street inside the museum, complete with shops, schoolroom, school bus, and print shop containing an early linotype machine, preserves the ambience of an early 20th-century mining town.

One exhibit contains a scale model of the entire taconite mining process from open-pit mine to transportation, shipping, and unloading on the docks of Duluth. There is an exhibit on lumbering, another industry of great economic importance to Minnesota. On display is a collection of local rocks and minerals, topographic maps of Minnesota, household appliances from the early 1900s, a 1908 horse-drawn school bus, and a 1916 bus from the Mesabi Transportation Company, which started in this area to transport miners from town to work and back. It is now known as Greyhound Lines, Inc.

Visitors can take a self-guided tour of the museum grounds, which include a variety of exhibits on early Minnesotan life and mining. There is a 150-foot (46-m) long replica of an underground mine drift from the early 1900s with miner's tools, drills, steam water pumps, jackhammers, and the electric mule that hauled miners and minerals in and out of the mine. Visitors must wear a hard hat on this tour. They can inspect large mining and transportation equipment such as tuggers (small, portable hoists), skips (buckets to lift ore or men from the mines), loaders, ore cars, and a 110-ton (100 tonne) Atlantic shovel made in 1910, one of only six produced. It was the fastest steam shovel ever made and is the only one of its type now in existence. The Ely Greenstone, weighing 24 tons (22 tonnes) and 5 feet 6 inches (1.7 m) in diameter, is the largest drilled core in the world.

On site is the Iron Range Research Center, which holds a large collection of historical and geological materials related to the area. Narrated trolley tours of the early 20th-century mining village of Glen Location are run regularly.

Address: Minnesota Discovery Center, 1005 Discovery Drive, Chisholm, MN 55719.
Phone: (218) 254-7959.
Days and hours: Memorial Day–Labor Day: Tue., Wed., Fri., Sat. 10 a.m.–5 p.m., Thu. 10 a.m.–8 p.m., Sun. noon–8 p.m. Winter hours are Tue., Wed., Fri., Sat. 10 a.m.–5 p.m., Thu. 10 a.m.–8 p.m.

Fee: Yes. Trolley tours are an extra fee.
Tours: Self-guided.
Public transportation: None.
Handicapped accessible: Yes.
Food served on premises: Picnic area. Food Court is open for special events.
Website: www.mndiscoverycenter.com

Hull-Rust Mahoning Mine

The last stop on this three-part mining tour is the Hull-Rust Mahoning Mine, the largest open-pit mine in the world. Visitors take a two-hour bus tour from the Minnesota Discovery Center in Chisholm, about 6 miles (10 km) away. Although the first iron-ore mines began as surface mines and then moved underground, this one did the opposite. The Mahoning Mine began around the turn of the 20th century as a small underground mine, and eventually grew into about 50 small surface mines. It was the first strip mine, or open-pit mine, on the Mesabi range, and it led to a new era as other mines followed suit and moved to open-pit operations. The wood-fired steam shovel made this kind of mining economically possible.

The 50-odd surface mines eventually merged, and mining continued until a vast man-made canyon was formed. It is 3.25 miles (5.2 km) wide, about 1 mile (1.6 km) long, and about 535 feet (163 m) deep.

Deposits of iron and silica were left behind after a sea retreated that had covered this part of Minnesota about a billion years ago. A combination of heat and the pressure from deposits covering the glacial drift compressed some of the iron and silica into the hard, flinty taconite being mined here today. The smaller deposits of softer iron ore have long since played out. When miners first reached the hard taconite, from which it was not economically feasible to retrieve iron ore, the original mining operations ceased. Late in the 20th century the Hibbing Taconite Company developed a cost-effective method of crushing the ore and extracting the iron.

The bus tour takes visitors to this working mine, where the three-story-tall dump trucks are loaded, ore is converted to taconite pellets, and finally loaded on the trains.

The road leading in and out of the observation area is of historic importance. Because Hibbing was built too close to the mine, right on top of the rich iron deposits, the town was moved in 1918. House by house, it was relocated 2 miles (3 km) south, where it is now, so the mining could continue and expand. Just beyond the northwest corner of the mining operations is a triple watershed from which water flows south via the Mississippi River, north to Hudson Bay, or east via the Bigfork and Red rivers to the Atlantic Ocean. (See Cornwall Iron Furnace, Cornwall, Pennsylvania.)

Address: Hull-Rust Mahoning Mine, Hibbing, MN 55746. Tours leave from Minnesota Discovery Center (see previous listing).
Phone: (218) 254-7959.
Days and hours: Mid-Jun.–mid-Aug. Thursdays only.
Fee: Yes.
Tours: Guided only, for ages 10 and over.
Public transportation: None.
Handicapped accessible: Yes.

Food served on premises: None.

Special note: Long pants and closed shoes.

Website: www.mndiscoverycenter.com/visit/hibbing-taconite-tours-minnesota-discovery-center/

St. Paul

Science Center of Minnesota and Mississippi River Visitor Center

Science Center of Minnesota

This natural history, anthropology, science, and technology center is housed in a large structure with 370 000 square feet (34 000 m^2) of indoor space and 10 acres (4 hectares) of outdoor space in downtown St. Paul along the Mississippi River, dating from 1999. The Science Center envisions that we live in "a world in which all people have the power to use science to make our lives better" and that "science is an essential literacy." To these ends the museum offers a large array of interesting exhibits on several spacious floors.

In 1971 paleontologists working in Wyoming found the remains of a *Diplodocus*, a very large, herbivorous dinosaur. They spent 23 years unearthing 3000 fragments and transporting them to the museum. Then they assembled the two-story-high creature according to the latest information on how it walked: standing on its toes with its tail held high. Other dinosaurs on exhibit in the Dinosaurs and Fossils area include an *Allosaurus* fossil and two *Camptosaurus* specimens, one a cast and the other a fossil unearthed at the same site as the *Diplodocus*. The fossil Camptosaur is one of the largest *Camptosaurus* specimens ever found. The museum also boasts the largest complete real *Triceratops* fossil in the world.

The Human Body gallery explores the internal workings of the human body, from the system level down to individual organs and tissues. In the Cell Lab area, visitors can extract their own DNA from cells and examine specimens under a microscope. Here too is an Egyptian mummy, donated to the museum in 1926. The Perception Theater plays a 20-minute show highlighting various optical illusions. When the Museum of Questionable Medical Devices shuttered its collection in 2002, much was donated to the museum, and some of this collection is also on view in the Weighing the Evidence exhibit. For example, there is the foot-operated breast enlarger from 1976; the Cosmos Bag—a coarse-cotton bag holding low-grade radioactive material—applied to arthritis joints, dating from 1928; and the ACU-DOT Magnetic Analgesic Patches (from 1979) to relieve minor aches and pains; all proving that there is no limit to wishful thinking for relief from perceived or real ailments.

The Experiment Gallery offers visitors the chance to try their own experiments, whether it be creating a tornado out of an air vortex, studying wave-formation in a wave tank, and more. Anthropology and the biological basis (or non-basis) for human races are explored in the RACE: Are We So Different? Gallery, in which genetic variation is explained, as well as the cultural basis and history of racism. A Native American exhibition showcases the Dakota and Ojibwe peoples, their methods for procuring food, clothing, and shelter and how they pass on the technologies they learn. Highlighted here are types of canoes used to carry people and goods down rivers, the uses of bison, and materials used for clothing.

Mississippi River Gallery examines the environment and history of North America's longest river—which runs right past the museum. Among the highlights of this gallery are Collector's Corner, where people can bring in natural items they have found, learn about them, and trade them for specimens others have collected. (See Mississippi River Visitor Center run by the National Park Service; and Mississippi River Museum, Memphis, Tennessee.)

A 90-foot (27-m) dome-screen Omnitheater offers regular shows. Throughout the museum, demonstrations are performed regularly.

Address: Science Museum of Minnesota, 120 West Kellogg Boulevard, St. Paul, MN 55102.
Phone: (612) 221-9444.
Days and hours: Tue., Wed., Sun. 9:30 a.m.–5 p.m.; Thu.–Sat. 9:30 a.m.–9 p.m. Closed Mon.
Fee: Yes. Omnitheater shows cost extra.
Tours: Self-guided.
Public transportation: Buses stop within several blocks of the site.
Handicapped accessible: Yes. Wheelchairs are available at the Box Office.
Food served on premises: Snack bar and café.
Website: www.smm.org

Mississippi River Visitor Center

The Mississippi River is such an overwhelming presence to the midwestern part of the United States that the National Park Service has set up a Visitor Center in the lobby of the Science Center of Minnesota. Here at the Center, which "serves as a gateway to adventure along the mighty Mississippi," visitors can see a variety of videos about the history of the Mississippi River and its ecosystem. There are interactive exhibits, regular programs, and more. (See Mississippi River Museum, Memphis, Tennessee.)

Address: Mississippi River Visitor Center, 120 West Kellogg Boulevard, St. Paul, MN 55102.
Phone: (651) 293-0200.
Days and hours: Tue., Wed., Sun. 9:30 a.m.–5 p.m.; Thu.–Sat. 9:30 a.m.–9 p.m. Closed Mon., Thanksgiving, and 12/25.
Fee: Free.
Tours: Self-guided.
Public transportation: Buses stop within several blocks of the site.
Handicapped accessible: Yes. Wheelchairs are available at the Box Office.
Food served on premises: None. There is food served in the Science Museum of Minnesota.
Website: www.nps.gov/miss/planyourvisit/mrvcabou.htm

Ohio

CLEVELAND

Dittrick Medical History Center

The Dittrick Medical History Center is on the third floor of the Allen Memorial Medical Library, the library of the Cleveland Medical Library Association. Exhibits show how medicine and medical instrumentation have advanced according to human needs, and explore the social and cultural climate at the time of specific advances, particularly in Cleveland.

The museum formed when Dr. Dudley Peter Allen and another surgeon, Gustav Weber, began collecting artifacts in the 1890s and donated their collection to the Cleveland Medical Library Association, a group that promotes the art and science of medicine. Initially, physicians from Case Western Reserve donated their personal items, such as the nearly complete armamentarium of Dr. Weber. These materials serve as a direct, tangible link with the past and a reminder that medicine is a long-developing science.

Dr. Allen's widow, Elizabeth Severance Allen, funded a building in memory of her husband, after his death in 1915. Construction of the Allen Memorial Medical Library was completed in 1926, and Dr. Howard Dittrick was named curator. Under his guidance, the museum acquired such a wide range of artifacts from Roman times to the present that today this facility is considered as "one of the most comprehensive collections in the USA." By 1998 the museum was renamed the Dittrick Medical History Center.

More than 75 000 artifacts held by the center are primary-source materials, considered by some historians as the equivalent of manuscripts documenting the history of medicine. A wide collection of 19th-century instruments includes microscopes, sphygmomanometers, and stethoscopes, including the first cylindrical models used in the early 1800s for amplifying the sounds of the heart and lungs. (See International Museum of Surgical Science, Chicago, Illinois; and National Museum of Health and Medicine, Silver Spring, Maryland.) On view are tongue depressors in many sizes, made out of the many materials used over the ages; electrotherapeutic devices once thought to cure illnesses; surgical instruments with their cases; and ophthalmoscopes, instruments with lenses and a mirror for examining the retina of the eye, first designed by Hermann von Helmholtz in 1850.

In the first gallery, which focuses on the transformation of medicine from a craft to a science, exhibits emphasize the contributions of Case Western Reserve staff members and graduates. Visitors see the 1875 office of Dr. John Dale, a local physician; a pharmacy from 1880; and a 1930s doctor's office.

Additional galleries examine the evolution of diagnostic instruments. Sphygmomanometers, or blood-pressure measuring devices, were invented by Samuel von Basch in 1880. Modern sphygmomanometers are based on an 1896 design by Scipione Riva-Rocci. The electrocardiograph evolved from the string galvanometer designed by the Dutch physiologist Willem Einthoven (1860–1927) to measure the impulse of the heart. When Wilhelm Conrad Roentgen discovered x-rays in 1895, their importance to medical diagnosis was immediately recognized. The golden age of radiology began in the 1920s when technical problems related to the use and recording of x-rays were resolved.

Other areas illustrated are hematology, first studied in 1852 by Karl von Vierordt; urinalysis, practiced for more than 2000 years as an aid to diagnosis; thermometers, first seen as medically important in the mid-19th century by Carl Wunderlich; percussion hammers, which became a key diagnostic tool after Leopold Auenbrugger learned how to determine the condition of internal organs from the sound made by a hammer tap to the area; the Skuy Collection of contraception, including ancient and modern methods; and otoscopes, noted by Wilhelm Fabricius von Hilden in the early 17th century as a useful tool for examining the tympanic membrane and external canal of the ear.

The library and archives include letters, notes, records, and other unpublished materials from individuals and institutions. The rare-book collection, one of the largest in the country, has European and American books from the 15th–19th centuries; herbals, materia medica, anatomies, and works on dermatology, epidemiology, psychiatry, and the natural sciences. Among the holdings are *Opera Quae Extant Omnia* by Adrian van der Spieghel from 1645, and *Inventum Novum ex Percussione Thoracis Humani*, written by Leopold Auenbrugger in 1761. It also has works by Charles Darwin and Sigmund Freud, as well as memorabilia and artifacts that are related to these two science giants.

Address: Dittrick Medical History Center, Allen Medical Library, 11000 Euclid Avenue, Cleveland, OH 44106.
Phone: (216) 368-3648.
Days and hours: Mon., Tue., Thu., Fri. 9 a.m.–4:30 p.m., Sat. 10 a.m.–2 p.m. Closed Sun., Martin Luther King, Jr., Day, Memorial Day, 7/4, Labor Day, Thanksgiving Day and the following Fri., 12/24, 12/25, 1/1.
Fee: Free.
Tours: Self-guided. Rare-books section by appointment only.
Public transportation: City buses stop at the Health Sciences Campus.
Handicapped accessible: Yes.
Food served on premises: None; there are many restaurants nearby, mostly on Euclid Avenue.
Website: artsci.case.edu/dittrick/

DAYTON

Carillon Historical Park

Edward Deeds, chairman of the board of National Cash Register Company, and his wife, Edith, saw a carillon in Belgium and wanted their hometown to have one of its own. The 151-foot (46-m) tall carillon for which this park is named was financed by the Deeds family and completed in 1942. The limestone tower holds 57 bells that are played on certain occasions each season.

The 65-acre (26 hectare) park contains historical displays related to the history of Dayton and the city's place in American technology. Some historic buildings were moved here and others were reconstructed to form a small midwestern village. Displays include original artifacts and devices invented in Dayton, largely in the Heritage Center of Dayton Manufacturing & Engineering.

The world's first airplane capable of sustained flight, the original 1905 Wright Flyer III, is housed in the Wright Brothers Aviation Center, designed by Orville Wright specifically to showcase this plane. Visitors view the plane from above to better understand how it operated. In 1908 it became the first passenger plane when the Wright brothers added modifications so the pilot and a second person could sit upright. The hall also contains Wright Brothers memorabilia and serves as a memorial to the duo. (See Wright Brothers National Memorial, Kill Devil Hills, North Carolina; and National Air and Space Museum, Washington, DC.)

Next to the Aviation Center is a replica of the Wright Cycle Shop, which contains the Spokes, Sprockets, and Spaceships exhibit, a collection of early cycles built in the

Wright Flyer III, Carillon Historical Park.
Courtesy of Dan Patterson.

Miami Valley. (See The Henry Ford, Dearborn, Michigan; and The Bicycle Museum of America, New Bremen, Ohio.) Other materials highlight the Wrights' earlier venture as job printers and publishers of small neighborhood newspapers. This museum claims to be the only one in the world to showcase the three professions of the Wright Brothers.

Dayton, once a major printing and papermaking center, still contains several of these businesses and related industries. (See Paper Discovery Center, Appleton, Wisconsin.) The 1930s urban Print Shop houses presses from various periods and more than 300 cases of type in 72 different fonts. It is one of the largest collections of letterpress type anywhere. Printing demonstrations are held regularly.

Of the approximately 200 automobile manufacturing companies formed between 1904 and 1908, 10 were in Dayton. Carillon Park has a collection of Dayton-made automobiles, including a 1908 four-cylinder Stoddard with room for five passengers. John and Charles Stoddard went into the automobile business in 1905 after they stopped making tractors. The Speedwell Company, which produced cars in 1907 and a few six-ton (5.4 tonne) trucks in 1911, was destroyed by floods and went out of business in 1911. The 1910 Speedwell on display is one of only a few cars built by the short-lived company. The Courier car shown was made by a subsidiary of the manufacturer of the Stoddard automobile. It has a 100-inch (254-cm) wheelbase and a "mother-in-law seat," or rumble seat, in the back.

Walter Chrysler developed a reputation for financial success in the automobile business when he saved the Willys-Overland Company from bankruptcy. He was then asked to save the Maxwell Company, and took over the reins in 1923. The 1923 Maxwell on display was part of the company's production that year. A few years later, Chrysler went into business for himself and put both Maxwell and Willys-Overland out of business.

Luzern Custer produced a three-wheeled, electrically powered chair for wounded veterans of World War I. This led to larger three-wheeled vehicles suitable for airports and golf courses. In 1939, 110 Custer motorized chairs were used by guests at the World's Fair. A 1921 Custer car is on view.

Charles Kettering's inventions, the self-starter and the electric ignition, caused a revolution in the automotive industry and led to the formation of the Dayton Engineering Laboratories Company, DELCO. Exhibits about the starter are housed in the replica Deeds Barn, where Kettering invented it with the help of Edward Deeds and several other people, known as the Barn Gang.

At DELCO, Kettering and Charles Midgley developed ethyl gasoline containing ethylene bromide and tetraethyl lead. The research group tried almost 5000 mixtures before settling on this mix as the best in 1921. Their chemical additions to gasoline significantly reduced engine knock, a serious problem for early automobile owners.

An early electrified cash register is part of the exhibit illustrating Kettering's work at the National Cash Register Company, or NCR. The No. 2, a 1200-horsepower (890-kW) model Corliss engine, is an example of a fuel-efficient stationary steam engine that produced direct power and generated electricity, thus making possible the industrialization of America. The Corliss engine was built by C & G Cooper Company in Mount Vernon, Ohio. It was originally installed at the NCR plant in 1902 and remained in operation until 1948.

An exhibit of railroad equipment at the Transportation Center includes pieces from the Barney and Smith Car Company, which pioneered wooden-bodied passenger railcars. In 1831 the Baltimore and Ohio Railroad (B&O) ran a competition for an improved railroad engine. Phineas Davis, a watchmaker form New York, won the contest with his Grasshopper locomotive, named for its shape and its up-and-down piston motion. The Grasshopper was the first successful steam railroad engine in the country. The one in the park is the oldest B&O engine in existence. (See B&O Railroad Museum, Baltimore, Maryland.)

Another important steam locomotive on exhibit at the Transportation Center is the John Quincy Adams, built in 1835. This piece of equipment—a "grasshopper" type known to be more efficient, using only one ton (0.9 tonne) of coal to travel 80 miles (130 km)—is the oldest American-built locomotive that is still in existence.

The Rubicon railroad engine from 1909 is another interesting piece. It was built along the principles of a thermos bottle and filled with 150 pounds (68 kg) of steam. This engine worked in the yard at the NCR plant in Dayton, running for 3–4 hours without producing any pollution. It was built in Lima, Ohio, at the Lima Locomotive Works, and was designed after earlier German steam engines. Among other exhibits in the park are a railroad bridge, a fire engine, a gristmill, a blacksmith shop, and the only operational brewery inside a museum in the United States.

Address: Carillon Historical Park, 1000 Carillon Boulevard, Dayton, OH 45409.
Phone: (937) 293-2841.
Days and hours: Mon.–Sat. 9:30 a.m.–5 p.m., Sun. noon–5 p.m. Closed Thanksgiving, 12/24, 12/25, 12/31, and 1/1.
Fee: Yes.
Tours: Self-guided.
Public transportation: Buses stop within 0.3 mile (500 m) of the site.
Handicapped accessible: Limited.

Food served on premises: Brewery and café.
Website: www.daytonhistory.org

New Bremen

The Bicycle Museum of America

The history of the bicycle spans little more than two centuries, beginning in Germany, during the "year without a summer" of 1816, after the eruption of Mount Tambora in the Dutch East Indies caused the global temperature to drop significantly. The chill in the Northern Hemisphere caused widespread crop failures, and some people were forced to eat their own work animals to survive. The lack of draft horses in Europe meant that surveyors were slowed down while measuring land areas. One of these surveyors, a civil servant of the German city of Karlsruhe, Karl Freiherr von Drais, who invented quite a few devices during his life, came up with a way to traverse the land faster while surveying, without requiring the use of now-scarce horses: a two-wheeled, steerable vehicle that became known as a *laufmaschine*, velocipede, or Draisine. The rider would straddle the device, and run. This was the earliest recognizable bicycle, although without pedals.

One of the largest collections of bicycles in the world is housed in Ohio, for five well-known manufacturers of bicycles once existed in Ohio—and the Wright Brothers got their start as bicycle manufacturers in this state. (See Carillon Historical Park, Dayton, Ohio; and The Henry Ford, Dearborn, Michigan.) The Bicycle Museum of America is a privately owned collection of about 900 bicycles and their accessories. The museum is located in a pair of side-by-side Victorian buildings on a main street in the small town of New Bremen.

This museum houses hundreds of bicycles, along with an archival collection of bicycle catalogues, memorabilia, and accessories. Displays include original artifacts, some restored, as well as reproductions based on careful research of diagrams and catalogues. Museum artifacts include one of seven known Draisines dating from 1816. One room

The Bicycle Museum of America.
Courtesy of The Bicycle Museum of America.

explores the early history of the bicycle in the 19th century. Kirkpatrick MacMillian added pedals to the front wheel in 1839, as shown in a reproduction on display—the last original one of these was destroyed in a fire. During the 1870s "high wheel era," bicycles had large front wheels and small back wheels for extra power during pedaling. In Britain these vehicles were known as "pennyfarthings," from the large diameter of the penny coin in relation to the small farthing. Drive chains were added in the 1880s, as shown by the Kangaroo in 1885.

Some less-famous alterations to bicycle construction began appearing during the late 19th century. The pennyfarthings were prone to causing riders to be thrown head-first in front of the bicycle, giving rise to the phrase "to take a header." In response, some manufacturers placed the small wheel in front—keeping the rider's weight toward the back—as seen in the American Star model from 1881. Other changes included treadles or up-and-down pedaling.

By the 1890s, the safety incorporated chain drive became standard, and the modern bicycle was complete. The Iver Johnson Rover is an example of this style. For women with the expansive skirts of the Victorian Age, bicycle-riding became possible with a lowered bar, along with skirt guards to keep the cloth from entangling in the mechanism. The museum displays a 1910 ladies' Dursley Pedersen.

In the museum is a collection of bicycle-repair tools and wrenches, a rim-drilling machine, and a rim-straightener. A variety of bicycle lamps and parts are displayed in a number of cabinets. One unusual example is a bicycle gun containing an ammonia solution, which the rider could shoot at "dogs, hoodlums, and burglars." The transition from 19th-century wooden bicycles to steel frames took place in the later 19th century, when precision engineering of the Industrial Age made it possible. By the end of the 19th century, aluminum became a cheap element, so the St. Louis Refrigerator & Wooden Gutter Company experimented with the Lu-Mi-Num in 1894, one of the earliest commercial aluminum-framed bicycles.

Children's bicycles are also displayed. Early tricycles often included a twisted seat for girls with skirts to sit "properly," and a drive-train operated by a rowing action (e.g., the Unzicker Tricycle from 1878) or with a movable handle that governesses could pull wagon-like when the child tired from bicycling.

Specialty and unusual bicycles are exhibited, such as the 1901 ice bicycle made by Wolff-American, which incorporated nails as a studded rear tire, and front sled- or skate-like glide for riding in snowy conditions. Marketed for teenagers was the 1955 Huffy bicycle with a three-tube AM radio in the chassis. Unfortunately this model was released just before transistor radios became commercially popular, and it had the tendency to run the battery-pack down rapidly. One of the strangest vehicles is the monocycle, in which the rider sits inside a large wheel that rides upon the ground: the museum has several in its collection, both old and new. There is a room of military bicycles. Early on, various militaries experimented with placing soldiers and weaponry on bicycles (see the 1901 Maxim Machine Gun Tandem Tricycle), which were movable and light compared with wagons, but by World War II, bicycles were mainly confined to use by message-runners.

The museum maintains a collection of bicycles made by the famous manufacturers Schwinn and Huffy. Huffy transferred all its bicycle manufacturing operations overseas by 1998. One historical piece is the 1896 Schwinn Tandem, a bicycle owned by the Schwinn family themselves, in which the parents, front and back, pedal, while a toddler sits between them on a small high-chair-like seat.

An introductory video shows bicyclists in period dress demonstrating cycling techniques on a variety of historic cycles, including mounting, dismounting, riding in formation, and more.

Among other collections of note in the museum are some rocks and minerals, Civil War memorabilia including a restored regimental flag, and a cabinet of electrical television-antenna rotators.

The museum archive is accessible to researchers with advance notice.

Address: The Bicycle Museum of America, 7 West Monroe Street, New Bremen, Ohio 45809.
Phone: (419) 629-9249.
Days and hours: Jun.–Aug.: Mon.–Fri. 9 a.m.–7 p.m.; Sep.–May: 9 a.m.–5 p.m., Sat. 10 a.m.–2 p.m.
Fee: Yes.
Tours: Self-guided.
Public transportation: None.
Handicapped accessible: Partly. The first floor is accessible; the upper level is not. Call in advance to arrange a virtual tour of the upper level using a portable device.
Food served on premises: None. There are several eateries in town.
Website: bicyclemuseum.com

TOLEDO

National Museum of the Great Lakes

Since its founding in 1944, the Great Lakes Historical Society has grown dramatically. Initially the Society opened a small museum in Vermilion, Ohio, in 1952, then added a research library a few years later, and the organization remained more or less the same for decades. But in the 1990s the Society implemented a strategic plan, broadened its activities, and took advantage of a newly built, but disused Toledo customs house on the Maumee River near where the river empties into Lake Erie. After installation of exhibits, the National Museum of the Great Lakes opened in 2014. Its mission is "to preserve and make known the important history of the Great Lakes."

The museum is the only one to portray and interpret all five Great Lakes—Erie, Huron, Michigan, Ontario, and Superior: the largest combined body of water by area on Earth—with 21% of all fresh water by volume on our planet and 84% of all the fresh water on North America. These lakes were formed at the end of the last Ice Age, roughly 14 000 years ago, during the retreat of glaciers and their resulting meltwater.

Visitors enter into the central circular exhibit area, The Great Lakes: A Powerful Force, for a short audiovisual presentation introducing them to the geography and importance of the Great Lakes. From the center, visitors can then branch out to examine five different major interactive exhibits about the lakes. Within these exhibits are numerous short videos describing particular topics in more detail.

Surrounding the central area is a ring of displays, Maritime Technology, portraying tools used by boats on the lakes. Half a dozen detailed scale models illustrate the change in shipping potential from early canoes to modern freighters. One exhibit shows how shipbuilders' tools have evolved from a chest of woodworking tools to a 1930s welding machine. There are models of steam-powered engines—some working—which display the motive forces that powered ships. A variety of instruments and dials used on ships are shown. Navigational tools and communications devices, from lanterns, bells, and megaphones to radio transmitters and radar, are exhibited, including a pelorus dating from 1900. A pelorus, named for Hannibal's pilot around 203 B.C.E., resembles a compass, but merely provides a relative direction based on prior sightings. A number of historical and modern life-jackets are also on view, as well as flare guns and emergency beacons.

Behind the ring of technology exhibits are four sectors portraying historical and economic aspects of the Great Lakes. The Exploration & Settlement sector tells the early pre-history and history of human transport across the Great Lakes, from canoes (and how to build them), to the British and French explorers and trappers of the 18th century. The Expansion and Industry sector discusses the rise of shipping companies needed to transport the growing amount of goods and raw materials (and passengers) during the 19th century. Safeguard & Support informs visitors about the dangers of maritime transport on the volatile weather conditions of the lakes, and means of protecting the ships. Shipwrecks & Safety displays aspects of historical shipwrecks, lifeboats, and other methods for rescuing passengers from shipwrecks. There are several Fresnel lenses from lighthouses on display, from a small fifth-order lens, to a large second-order lens dating to 1873, one of five ever used on the Great Lakes. (No first-order lenses were ever placed on lighthouses on the Great Lakes.) There is a Lyle Gun in the exhibit, invented by Captain David Lyle in 1877, mounted on a wagon. The wagon was pulled to the edge of the lake, and the gun, loaded with a long flexible rope having a range of up to 700 yards (640 m), was shot toward the shipwreck, allowing the unfortunate sailors to grasp the rope and be pulled to shore. The last major shipwreck on the Great Lakes, the Edmund Fitzgerald, in 1975 on Lake Superior, is described in detail, as well as some of the Historical Society's archaeological missions in the Great Lakes.

Along with the technological aspects of Great Lakes shipping, numerous examples of ephemera, such as china, cutlery, posters, and other materials, are on display.

A highlight of the museum is the *Col. James M. Schoonmaker*, the flagship freighter christened in 1911, and built in the Great Lakes Engineering Works in Ecorse, Michigan to transport iron ore, a raw material found in Michigan (see Michigan Iron Industry Museum, Negaunee, Michigan) and Minnesota (see Lake Vermilion–Sudan Underground Mine State Park, Minnesota Discovery Center, and the Hull-Rust Mahoning Mine, Soudan, Chisholm, and Hibbing, Minnesota). For several years thereafter, it was the largest ship that operated on the Great Lakes, weighing 23 600 tons (21 400 tonnes) when loaded. The ship is 617 feet (188 m) long, 64 feet (20 m) wide, 33 feet (10 m) deep, and towers over the museum itself, even dockside on the river next to the museum. Visitors can climb a rather steep steel gangplank to traverse the top of the ship, viewing various engineers' cabins, the mess and galley, and the preserved engine room inside, below deck. Toward the prow of the ship are guests' quarters, captain's quarters, and the pilot house— designed before radar was invented—with two wheels, nine stories above the water line. Two wheels were included only to impress the guests; only one was required to steer the ship. Note the huge deck winches that traveled along the length of the ship on steel rails.

Address: National Museum of the Great Lakes, 1701 Front Street, Toledo, OH 43605.
Phone: (419) 214-5000.
Days and hours: Mon.–Sat. 10 a.m.–5 p.m., Sun. noon–5 p.m. Closed Mon. from Labor Day to Memorial Day. Jan.–Mar.: closed Mon. and Tue. Last time to board the *Schoonmaker* is 4 p.m.

Fee: Yes. There is an additional charge to visit the *Schoonmaker*.

Tours: Self-guided.

Public transportation: Buses run to Front Street, and then walk 0.2 mile (0.3 km) to museum.

Handicapped accessible: Mostly: The museum is accessible; the *Schoonmaker* is not. People with fear of heights or difficulty walking may have problems boarding the ship and using the narrow, steep stairways. Wheelchairs are available for the museum itself; the museum recommends calling in advance to reserve.

Food served on premises: None.

Special Note: The *Schoonmaker* is closed from Nov. to Apr.

Website: inlandseas.org

WRIGHT-PATTERSON AIR FORCE BASE

National Museum of the United States Air Force

The world's oldest and largest military aviation museum is a national treasure house that pays tribute to the Air Force's role in US military heritage. This museum collects and preserves aircraft, educates and entertains the public, chronicles how military aviation and airpower developed, and explores aviation's effect on civilization. (See Steven F. Udvar-Hazy Center, Chantilly, Virginia.)

The museum is located on Wright Field at Wright-Patterson Air Force Base, near the spot where Orville and Wilbur Wright did much of their work developing the first powered aircraft. It displays over 350 aircraft and missiles in four buildings comprising more than nine acres (3.6 hectares).

The self-guided tour starts in the first building. The Early Years Gallery chronologically follows the development of aviation through the buildup to World War II. A prologue exhibit traces humanity's dream of flying, from the days of Greek mythology to the present. The exhibit includes copies of scale models by Leonardo da Vinci as well as Chinese kites. The gallery focuses, however, on the Wright Brothers era with a reproduction of the 1909 Wright Flyer, the first airplane the military purchased, and devices the Wrights invented, such as a wind-tunnel scale, an anemometer, and a 1916 wind tunnel. Memorabilia are also shown, such as a piece of fabric from the 1903 plane the Wrights flew at Kitty Hawk. (See Wright Brothers National Memorial, Kill Devil Hills, North Carolina.) The museum has a fully restored British Sopwith Camel, made famous in the "Peanuts" comic strip. Floating overhead is a rare Caquot Type R observation balloon of World War I vintage. Other artifacts on view are the only Curtiss P-6E Hawk, Martin B-10, and Douglas O-46A in existence. The largest plane in the Early Years Gallery is the Martin B-10, the first all-metal monoplane bomber produced in quantity, manufactured in the 1930s.

The World War II Gallery presents military aviation from the early 1940s. On display is a B-24 aircraft used in combat in North Africa during World War II, several World War II prisoner-of-war relics, and a trombone of the well-known band leader Glenn Miller, who was killed in 1944 on a military flight from London to Paris. Other important artifacts here include the B-29 Bockscar that dropped the first atomic bomb on Nagasaki, Japan; a B-25 Mitchell; and a Japanese Mitsubishi A6M, called a "Zero," because it was placed into service in the Imperial Year 2600.

Buildings 2 to 4 continue the path to modernity with exhibits reviewing the Korean War, Vietnam War, the Cold War, and current topics in global and space aircraft. Visitors can enter the fuselage of a B-29 aircraft. Topics covered include rescue amphibians and helicopters, civil engineering, and Operation Desert Storm. One of the two remaining Boeing YQM-94A Compass Cope B remotely piloted airplanes, used for photographic reconnaissance in the 1970s, is displayed. A section on space travel experimentation looks at the development of rocket engines. Another exhibit focuses on the Aerobee research rocket, in which two monkeys and two mice were used to test animal reactions to high acceleration and weightlessness in 1952.

A massive aircraft here is a B-36 bomber whose wingspan extends across the entire gallery. An XF-85 Goblin is dwarfed by its mother ship, the B-36. There is also a Russian MiG-15 that was flown by a defector from North Korea to South Korea for a $100 000 reward, and high-speed, high-altitude aircraft, such as an X-15 that tested at up to 4500 mph (7200 kph).

The Missile Gallery is contained in a 140-foot (43 m) tall structure shaped like a silo. It has a Boeing AGM-86B cruise missile and a Titan-IVB intercontinental ballistic missile. An early model on view is a Chrysler PGM-19 Jupiter intermediate-range ballistic missile, developed in 1956. This missile was deployed in Italy and Turkey under the North Atlantic Treaty Organization (NATO) until 1963, when it was replaced with more modern technology. (See Titan Missile Museum, Sahuarita, Arizona.) Modified versions of the Jupiter were used as a first stage to launch space satellites. To view the missiles from above, visitors can reach an elevated platform on the inside circumference of the silo. Visitors see the Stargazer gondola in which Captain Joseph Kittinger and William C. White soared to 82 200 feet (25 100 m) to gather data and test pressure suits in 1962.

Building 4 houses the most recent developments in military aviation, as well as presidential aircraft and many other flight vehicles. Exhibits in the Space Gallery area range from the Mark I Extravehicular and Lunar Surface Suit, the great-granddaddy of space suits, to an Apollo 15 capsule, a space-shuttle astronaut trainer, and a moon rock. A Space Shuttle Exhibit includes Crew Compartment Trainer CCT-1 that was employed to train astronauts. The gallery also covers newer technologies born out of the special requirements of space exploration, such as Velcro®, cordless drills, Mylar® food packaging, solar-powered calculators, compact discs, and the artificial hip.

A Research and Development Gallery holds the world's biggest group of test aircraft. Here is the world's only remaining XB-70A Valkyrie, which reached Mach 3 (over 2000 mph; 3200 kph), from the 1960s. Other artifacts showcase advances in materials science.

There are also a couple of flight simulators for visitors to test their flying skills. An outdoor park displaying aircraft and other aviation artifacts includes a replica of an 8th Air Force Control Tower, like those used by the Royal Air Force and US Army Air Forces in the United Kingdom, is here. A Nissen Hut, built of corrugated metal on a concrete slab base, stands nearby, showing what life was like for World War II personnel.

Address: National Museum of the United States Air Force, 1100 Spaatz Street, Wright-Patterson Air Force Base, OH 45433.

Phone: (937) 255-3286.

Days and hours: Daily 9 a.m.–5 p.m. Closed Thanksgiving, 12/25, and 1/1.

Fee: Free. Theater and interactive simulators incur a fee.

Tours: Self-guided; guided tours also offered daily.

Public transportation: Local buses stop on the opposite side of Harshman Avenue, and then walk 0.8 mile (1.3 km) to the museum.
Handicapped accessible: Yes. Wheelchairs are available on a first-come, first-served basis.
Food served on premises: Cafés.
Special Note: IMAX® theater.
Website: www.nationalmuseum.af.mil

Wisconsin

APPLETON

Paper Discovery Center

In the 1830s and 1840s, wheat was a principal crop in Wisconsin, and a series of grist mills sprouted up along the Fox River. But by mid-century, the frontier moved westward to Minnesota, and the mills saw a drop in flour production. The heavy, continuous flow of the Fox River, fed by the huge Lake Winnebago watershed, did not even halt during the bitter Wisconsin winter, for the river did not freeze. Thus mill owners and real-estate speculators looked for the "Next Big Thing" for Wisconsin riverfront property and water power, which was paper.

John A. Kimberly, Charles B. Clark, Havilah Babcock, and Franklyn C. Shattuck formed a partnership in the early 1870s, and built the Atlas paper mill in 1878. The brick structure, on the bank of the Fox River, measuring 120 feet by 240 feet (37 m by 73 m),

Paper Discovery Center.
Courtesy of the Paper Discovery Center.

was first used as a paper mill. As the structure became obsolete for factory production by the mid-20th century, it was transformed into a research and development center for the Kimberly-Clark company. Even repurposed as a laboratory, the building was outdated eventually, and the facility shut down in 1999, paving the way for the Paper Discovery Center, which opened in 2005. The center teaches the public about the manufacture of paper and its uses.

The major permanent exhibit is the manufacture of paper. Displays take the visitor from chipping of timber into fragments, pulping, bleaching, making into slurry, to forming. A variety of interactive displays tell the story, including flaps under which are several scents found in paper factories. Standing in the exhibit is the world's first triploid aspen tree (now dead) cloned by tissue culture in 1968 at the Institute for Paper Chemistry.

In the museum are several large-scale models of papermaking factories, including a model of a Fourdinier Paper Machine PM1 used in Mantova, Italy in the 1960s. Other equipment on display includes a conical brass Jordan Refiner built in 1939 for the Eastman-Kodak Company. Pulp is fed into the refiner, between an inner and outer cone fitted with blades that fibrillate (add hair-like tendrils to) the pulp, creating more surface area for the fibers to stick together. Also displayed is the water wheel from a 57-inch (145-cm) Morgan Smith turbine used in Kimberly-Clark's Badger Globe Mill.

A perennial favorite at the facility is the regular workshop held several times daily for visitors to try their hand at making paper. Participants start with sheets of paper, pulverize them in a blender, and form them between screens. The handmade sample of paper can be customized in color and pattern.

A series of paintings created in 1948 by Thomas Dietrich, the artist-in-residence at local Lawrence University, are also on display. These paintings depict the activities of paper-mill employees at various tasks in the papermaking process. There is a theater which shows video taken by drone of the Fox River and the various mill sites on its banks. A large kiosk houses the Paper Industry International Hall of Fame, which describes many important founders and builders of the paper industry. The site also has a gallery of changing hands-on exhibits.

Behind the factory on the bank of the river is a 1909 structure on the site where the first hydroelectric plant in the United States was constructed in 1882 by Thomas Edison. Edison built the plant for H. J. Rogers, a local paper manufacturer, to run his factory as well as his nearby home. This interested the city elders of Appleton, who wanted to illuminate the streets and power the newly invented electric streetcar. (See Seashore Trolley Museum, Kennebunkport, Maine.) The building is now occupied by a restaurant.

Address: Paper Discovery Center, 425 West Water Street, Appleton, WI 54911.
Phone: (920) 380-7491.
Days and hours: Mon.–Sat. 10 a.m.–4 p.m. Closed Sun. and holidays.
Fee: Yes.
Tours: Self-guided. Docents assist with the papermaking workshops.
Handicapped accessible: Yes.
Public transportation: Buses run from downtown Appleton to the site.
Food served on premises: None. Restaurants are nearby.
Website: paperdiscoverycenter.org

National Railroad Museum

With the rise of the automobile, modern highways, and air travel after World War II, railroad usage began to decline. Thus it was, in 1956, that a number of local railroad aficionados proposed a museum dedicated to preserving American railroad history and artifacts. Within two years, Congress designated the fledgling facility as the National Railroad Museum.

Encompassing 33 acres (13 hectares), a circular track of one mile (1.6 km) and over 70 pieces of rolling stock, the museum concerns itself with the entire American railroad story. Most of the artifacts on display date from the 20th century, from freight cars to locomotives. The museum restores about one piece of stock per year in an on-site restoration facility. Many of the items allow visitors to walk through and see the various areas for engineers, passengers, and other railroad workers. Numerous signs explain details about the specification, operation, or engineering behind the pieces.

The indoor Lenfestey Center is the climate-controlled portion of the museum. Here visitors find perhaps the most important cars and locomotives available to the public. The most famous of these is the Union Pacific Big Boy #4107. Built by the American Locomotive Company in 1941, this model was a freight locomotive carrying, after modifications, 32 tons (29 tonnes) of coal as fuel, along with 24 000 gallons (91 000 L) of water. The massive size of this locomotive in relation to the ability to negotiate curved track was made possible by the innovations of Anatole Mallet, a Swiss engineer of the late 19th century. He allowed his locomotives to be built on two separate portions, with the rear engine holding the boiler on an independent frame from the front portion. Twenty-five of this type of locomotive were created, of which eight survive. This model, #4107, plied the rails in Colorado, Wyoming, and Utah. Visitors can hear synthesized sound effects of the locomotive's engine, whistle, and more.

Another major train on display is the Dwight D. Eisenhower A4 locomotive with armored cars, used during World War II in England by General Eisenhower. The locomotive, with its distinctive curved, sweeping shape, was designed by Sir Herbert Nigel Greeley, who was inspired by the Bugatti racing car. To attain the speed required by passenger locomotives, this engine was equipped with not two, but three cylinders, and could attain an average speed of over 70 miles per hour (110 kph). Water consumption by the engine required four "water scoops" for refilling during travel from London to Newcastle, each being around 2000 feet (610 m) long. Attached to the locomotive are a couple of armored passenger cars with blacked-out windows for secret military use. It is estimated that 22 tons (20 tonnes) of steel armor were originally used to sheath the cars, but currently they are displayed with about one ton (0.9 tonne) of metal.

Other interesting pieces of stock include the Pennsylvania Railroad #4890, which was a class GG-1 electric locomotive running from Washington, DC, to New York City. This model could run at 100 mph (160 kph) with 20 passenger cars, drawing power from 11 000-volt catenary cables. It was used primarily in the late 1930s and 1940s. The Pershing #101 was a US Army locomotive built in 1917, and operated until 1961, built by the Baldwin Locomotive Works. Other cars of note include a 1929 dynamometer car used to measure the work an engine needed to pull railroad cars; a snow-plow locomotive, a mail car, a Pullman car, the Wisconsin 40 & 8 boxcar from the French Gratitude train sent to the United States after World War II; and cabooses, all housed in the climate-controlled building.

Outdoors are a variety of rolling stock, both under a roof in the McCormick Pavilion, and not, on uncovered display tracks. There is an Acrotrain from 1955 with its sweeping distinctive look, plus two attached passenger cars, all used in the Midwest and California, designed to combat the rise of the automobile. (A model is displayed indoors.) The oldest locomotive owned by the museum is here, too: a Lake Superior & Ishpeming 2-8-0 steam locomotive built in 1910 and used until 1962. The oldest cars include a Lake Superior & Ishpeming Coach #62, built in 1897.

One gallery in the museum features a rare collection of drumheads, illuminated logos on the back of the last car in a train, advertising the railroad company. These logos were generally round and approximately the size of a bass drum, hence the name. An interactive display connects particular drumheads to their travel routes across the United States.

The museum runs regular train rides daily on the circular track around the site, and offers repeated showings of a 25-minute documentary on freight locomotives. There are changing exhibits as well.

For scholars of American railroading, the museum maintains an archive of photographs, technical manuals, historic logbooks, and other items related to the history of railroading. Call in advance for access to the archive.

Address: National Railroad Museum, 2285 South Broadway, Green Bay, WI 54304.
Phone: (920) 437-7623.
Days and hours: Apr.–Dec.: Mon.–Sat. 9 a.m.–5 p.m., Sun. 11 a.m.–5 p.m.; Jan.–Mar.: Tue.–Sat. 9 a.m.–5 p.m., Sun. 11 a.m.–5 p.m.; 12/31 11 a.m.–2 p.m. Closed Mondays, Easter, Thanksgiving, 12/25, and 1/1.
Fee: Yes.
Tours: Self-guided.
Handicapped accessible: Mostly. Some rail cars are not accessible.
Public transportation: None.
Food served on premises: None. Restaurants are nearby.
Website: nationalrrmuseum.org/

MADISON

University of Wisconsin–Madison Geology Museum

In the middle of the University of Wisconsin campus in Madison, inside Weeks Hall, is where the Geoscience Department and the Geology Museum now reside. Begun by the University of Wisconsin Board of Regents in 1848—the same year that Wisconsin became a state—the university's Geology Museum is home to 120 000 specimens of rocks, minerals, and fossils. The founding of the museum itself can be dated to a letter sent by the Board of Regents to local communities on October 7, 1848, asking laypeople to "contribute specimens, and solicit their friends and neighbors to do likewise. Every neighborhood can do something." Today the Geology Museum encompasses 3000 square feet (280 m^2) of exhibit space and takes its mission of being the guardian of Wisconsin's geologic history seriously: the collection is for the people of Wisconsin. Often the director

The iconic Wisconsin mastodon skeleton at the UW Geology Museum. *Photo* credit UW Geology Museum.

can be seen talking with people bringing newfound samples for identification. Many of the samples on display were collected by students and professors at the university.

In the large vestibule, in front of the entrance to the museum, a rare Rand McNally relief globe (ca. 1980), 6 feet (1.8 m) high, rotates. Shining in earth-toned and blue hues on one wall of the vestibule is a tall stained-glass window by local artist Paul Dombrowski, depicting geological strata. There are also large slabs of fossil jellies stranded on an ancient beach on display.

One hall of the museum contains a variety of rocks and minerals in glass cases, sorted by chemical composition, such as elements, oxides, hydroxides, sulfides, etc. One display showcases meteorites, including a rock ejected from the planet Mars that eventually landed on Earth. There is a room illuminating fluorescent minerals with ultraviolet light to show their fluorescence. One important rock specimen is a sample of Australian Jack Hills metaconglomerate 3.0 billion years old, containing microscopic zircon crystals, the oldest known bits of our planet, dated to 4.4 billion years old. Another valuable rock is the oldest known fossil, a large fragment of Trendall Locality Stromatolite from Western Australia. Stromatolites are bacterial mats that form recognizable fossils. To illustrate what stromatolites are, near the fossil a large glass canister illuminated by a spotlight mounted in its top has a similar modern microbial colony growing inside.

Other important collections include a variety of Middle Cambrian fossils from British Columbia, the Burgess Shale fauna, including *Burgessia bella*, an arthropod, *Haplophrenitis*, with a cone-shaped shell, and *Vauxia gracilens*, a branched sponge. Also there is a fine collection of specimens in limestone from the Late Jurassic Period, originating in Solnhofen, in Bavaria, Germany. The museum is known for its collection of

Waukesha Lagerstätte fossils from the Early Silurian period, found in a Wisconsin deposit, which preserved many soft-bodied creatures.

A prize slab of fossil on display is the large colony of *Uintacrinus* crinoids. The historic mastodon found in Wisconsin, made up of parts of two similar-sized skeletons, one found in 1897 and one in 1898, looms over one hall. The museum also is caretaker of the first dinosaur to be displayed in Wisconsin, a hadrosaur found in South Dakota, and mounted in 1991. There are a variety of other fossils on view, from beetles to sharks to mammals.

A small display discusses the differences between *Brontosaurus* and *Apatosaurus*, and even contains one of the five larger-than-life models of a *Brontosaurus* egg hatching, fabricated for the Sinclair Oil Company's Dinoland display at the 1964–1965 New York World's Fair. The corporate symbol for Sinclair Oil was the *Brontosaurus*.

A window into the Fossil Preparation Lab lets visitors view paleontologists prepare specimens for conservation and display.

For all the museum's earnestness in making geology accessible to all, the facility has become known for a touch of whimsy in recent years. For example, in 2017 the museum sponsored a corn maze designed in the form of a giant cabinet of curiosities, featuring a 480-foot (150-m) long trilobite, as well as a "Fantastic Jurassic"–flavored ice cream.

If time permits, visitors might also visit the Discovery Building two blocks away. Constructed in 2010, the building is a consortium of research laboratories and public space on the ground floor. Check the schedule online (discovery.wisc.edu/) or call (608-316-4300) for regular science lectures and events. The floor of the Discovery Building's public area itself is made of limestone with many natural fossil inclusions.

Address: University of Wisconsin–Madison Geology Museum, 1215 West Dayton Street, Madison, WI 53706.
Phone: (608) 262-1412.
Days and hours: Mon.–Fri. 8:30 a.m.–4:30 p.m., Sat. 9 a.m.–1 p.m.
Fee: Free. Guided tours incur a fee.
Tours: Self-guided. Tour guides are available for downloading at the website. Guided tours can be scheduled in advance.
Handicapped accessible: Yes.
Public transportation: Buses run to campus.
Food served on premises: None. Nearby Discovery Building has cafés, and there are student eateries on campus.
Website: geologymuseum.org

MANITOWOC

Wisconsin Maritime Museum

Founded in 1969, the Wisconsin Maritime Museum now occupies three floors and 60 000 square feet (5600 m^2) in a large facility on the bank of Lake Michigan. Originally the site was created to commemorate the submarine-building industry, the site—designated "Wisconsin's Maritime Museum"—now examines all of Wisconsin's maritime history dating back to 1840.

An important exhibit is the Maritime History Gallery, which covers more than a century and a half of shipbuilding techniques. On display is a cross-section of a schooner, showing how such a ship was constructed in 1854. A variety of shipbuilder's wood-working tools are shown nearby, along with cutaway models. One case examines the contents of a sailor's "ditty bag" (personal bag of supplies), such as thread, beeswax, awls, and other implements. Other cases display navigation instruments including telescopes and compasses. Further along in the gallery are displays of various marine engines, from steam in the first half of the 19th century to diesel in the 1890s. Many of the examples shown were built by the local Kahlenberg company founded in 1895 by William and Otto Kahlenberg. By 1898 the firm began producing gasoline-powered nautical engines. One-cylinder and two-cylinder engines are on display, as are larger motors such as the Kahlenberg E-6, from the 1950s, which incorporated six cylinders and generated over 600 horsepower (450 kW). The transformation of ship construction from wood to rivets and metal plate in the mid-1800s is described, as well as the entry of pneumatic tools and welding. The importance of Wisconsin to submarine building is discussed, because some 28 submarines were built in this area. Visitors can peer through a periscope.

The 65-ton (59-tonne) triple-expansion steam engine from the *Chief Wawatam* car-ferry sits in an annex to the museum that was constructed around the steam engine in the early 2000s. This technological behemoth, built in 1911 and over two stories tall, can be controlled by a touch-panel display to show the moving parts.

A maritime museum must have a model ship gallery, and the Wisconsin Maritime Museum is no exception. A large number of scale models and half-hull models are on display. The Wisconsin-Built Boat Gallery offers a rotating selection of boats that were built in Wisconsin. There are also galleries with changing exhibits.

Behind the museum and partially submerged is the USS *Cobia*, a fully restored submarine that saw service during World War II. Visitors take guided tours that last about 45 minutes of the engine room, mess halls, bunks above the torpedoes, and other areas of the vessel. The submarine, nearly 312 feet (95 m) long, was built and launched in 1943, served in the Pacific Theater, and was decommissioned in 1946. She is now a National Historic Landmark. The radio set is still used by licensed ham-radio operators, and the radar system is considered to be the oldest working radar in the world.

The museum maintains an archive of important maritime collections comprising manuscripts, plans, blueprints, charts, ship's logs, periodicals, and maps. Scholars wishing to research the archive need to provide one week advance notice.

Address: Wisconsin Maritime Museum, 75 Maritime Drive, Manitowoc, WI 54220.
Phone: (920) 684-0218.
Days and hours: Jul.–Aug.: 9 a.m.–6 p.m.; mid-Mar.–Jun. and Sep.–Oct.: 9 a.m.–5 p.m.; Nov.–mid-Mar.: Mon., Thu., Fri. 10 a.m.–4 p.m., Sat.–Sun. 9 a.m.–4 p.m., closed Tue.–Wed. Closed Easter Sunday, Thanksgiving, 12/24, 12/25, and 1/1.
Fee: Yes.
Tours: Self-guided. Guided tours for the submarine, age 16 and older, with steep ladders and narrow spaces.
Public transportation: Taxis.
Handicapped accessible: Partly. The submarine is not accessible.
Food served on premises: None. There are restaurants nearby in town.
Special note: The close quarters in the USS *Cobia* may be a problem for those with claustrophobia.
Website: www.wisconsinmaritime.org/

MILWAUKEE

Discovery World

Milwaukee's first cargo pier was built at the base of what is now Clybourn Street, jutting out into Lake Michigan, during the winter of 1842–1843. Near this historic spot, occupying a prominent place on the shoreline, is Discovery World, a large modern structure devoted to hands-on learning for all ages about science and technology.

The earliest plans for Discovery World were laid in the early 1980s by Robert Powrie Harland, Sr., a local patent attorney who wanted to promote the idea of innovation. The current two-story location on the lake front was opened in 2006, and is divided into two broad zones: technology and water (a stand-in for the natural world). Most of the exhibits are corporate-sponsored.

Within the Technology Building are a variety of exhibits dealing with basic technology and the science behind the machinery. In the center of this area is a circular staircase enclosing a mechanical aluminum and steel model of DNA designed by Chuck Hoberman in 2006 that opens and closes at random times. The sculpture, 39 feet (12 m) tall and 12 feet (3.6 m) in diameter when extended, is reminiscent of Hoberman's isokinetic spheres that open and close via a scissoring action. Physics and You demonstrates simple ideas in physics, from air pressure to electrical circuits. Visitors can pump up and race pneumatic cars, create simple circuitry, charge up and launch metal rings via electromagnetic pulses, and sail air cars with wind. Most of these exhibits require the visitor to do something to see the effects.

One display of how motion is transmitted within machinery takes up several walls. Here visitors can see how dozens of types of pistons, gears, and levers all shunt motion from one place to another, via bevel gears, cams, screws, and other machinery parts. Rockwell International's Automation Everywhere exhibit demonstrates several different types of robots that can sort, place, and cut objects; play simple games; and open and close roofs. Clean Air Trek describes pollution in the air and how batteries and fuel cells are used to reduce emissions in hybrid cars. Energy and Ingenuity is a walk-in model of a nuclear power plant, along with a display of household items containing weak amounts of radioactivity. Other exhibits deal with household automation such as how thermostats work, welding pieces of metal, mining equipment, and a flying simulator. An energy exhibit showcases a van de Graaf generator in which a visitor places a chain-mail glove next to the generator to produce high-voltage sparks from the generator to the glove.

Discovery World.
Courtesy of Discovery World's Chris Winters.

There is a virtual-reality kiosk, where visitors can don special goggles with headsets, and experiment with games as well as virtually travel the globe.

Native Wisconsonian Les Paul's work with guitars and recording technology is explained ingeniously in the exhibit Les Paul's House of Sound. Among the variety of displays are an electric guitar side-by-side with its x-ray photograph showing the internal electrical pickups, "The Thing" (Paul's custom-built mixing board from the 1950s), interactive exhibits allowing the visitor to control the amount of special effects to enhance a recording, and demonstrations visualizing sound. An unusual device for visitors to try is a Reactable, a circular table-like horizontal display upon which visitors can place blocks which create sound patterns of rhythm and melody. The sound effects change when the blocks are moved or rotated. Multiple-head tape recorders and many electric guitars are displayed throughout the Les Paul exhibit.

Within the water zone, called the Aquatarium, are displays and exhibits related to the natural world and water usage. At the entrance is a large wave-making display, which demonstrates via user-controlled knobs how different types of ocean waves form. A sprawling scale model of the Great Lakes occupies a large area, which is accompanied by models of canal locks and small tanks holding native creatures. Nearby is a wall-mounted two-story Burke Dymaxion map of the Earth. The Dymaxion projection for mapping was invented by R. Buckminster Fuller, and involves dividing the Earth's surface into 20 triangular sectors, forming an unfolded, flat icosahedron, which gives less distortion than many other map projections. Also next to the Great Lakes model is a slab of fossilized Cambrian-period lakebed.

A popular area is the Reiman Aquarium, which holds many large and small tanks of 190 species of freshwater and saltwater fish and other fauna, from around the globe as well as local to Wisconsin, such as lake sturgeon, alligator gar, and paddlefish. Several touch-tanks allow visitors to gently feel what live horseshoe crabs, stingrays, and sturgeons are like.

On the second floor of the Aquatarium is a walk-on section of the reconstructed 1852 clipper schooner *Challenge* that plied Lake Michigan, showing the various major parts of a ship. Near the ship are hands-on examples of simple machines such as pulleys, wedges, and the Archimedean screw.

The City of Fresh Water and Liquid House demonstrate the importance of water, water treatment, and water conservation.

Seasonally the museum-owned sailing vessel *Denis Sullivan* takes visitors out on Lake Michigan for a two-hour journey. The *Denis Sullivan* is the world's only re-creation of a 19th-century, three-masted, Great Lakes schooner. The vessel teaches sailing techniques, weather-forecasting, and environmental science.

Address: Discovery World, 500 North Harbor Drive, Milwaukee, WI 53202.
Phone: (414) 765-9966.
Days and hours: Mon.–Fri. 9 a.m.–4 p.m., Sat.–Sun. 10 a.m.–5 p.m.; closed Mondays during the fall and winter. Closed Easter, Memorial Day, 7/4, Thanksgiving, and 12/25.
Fee: Yes. *Denis Sullivan* is extra.
Tours: Self-guided. Docents assist with various exhibits.
Handicapped accessible: Mostly. A few sections (such as the lower level of the Challenge) are not. Wheelchairs are available at the Ticket Counter.
Public transportation: None.
Food served on premises: Café.

Special note: Animal demonstrations occur on weekdays, and there are occasional science demonstrations. *Denis Sullivan* is open seasonally. Call or see the website for details.
Website: www.discoveryworld.org

Milwaukee Public Museum

In the 1850s, Peter Engelmann, a schoolmaster in Milwaukee, encouraged his students to collect ethnological and natural-history objects to study in his classroom at the German-English Academy. The collection soon grew so large that Engelmann needed a whole room to display it. These artifacts and specimens later passed through the hands of the Wisconsin Natural History Society of Milwaukee and in 1882 were offered to the city of Milwaukee, which launched a public museum that today ranks among the best in the nation for human and natural history. In 1992 the museum came under the aegis of the County of Milwaukee.

A young taxidermist at the museum, Carl Akeley, decided to create a new kind of exhibit in 1890 that would be more realistic and help viewers understand the interrelationship between animals and their natural surroundings. Instead of arranging the animals the traditional way—in neat rows in glass cases—he created a natural habitat and placed a group of animals in a wrap-around scene with the foreground blending into the painted background. Called Muskrat Group, this exhibit was the world's first diorama and is still on display.

Carl Akeley was trained at Ward's National Science Establishment, the major taxidermy house and chief museum-supplier of the day. Akeley was not happy with the traditional method of preserving skins by stuffing them with straw, wood shavings, or vegetable fibers. The results, he felt, were lifeless. As an artist and naturalist, he wanted a more realistic look. This led him to invent a new taxidermy method using preliminary drawings, life-size clay models, disposable plaster molds, and lightweight structures of wire mesh, burlap, and papier-mâché. His accurate, lifelike mounts set the standard for natural-history museums, and his basic process is still followed today. Akeley is often called "the father of modern taxidermy."

With Akeley at the helm, the museum became a leader in innovative exhibition techniques and continues in this role, as evidenced in small displays, in more dramatic, larger exhibits, and in dioramas. (See Field Museum of Natural History, Chicago, Illinois; and American Museum of Natural History, New York, New York.) Innovative designs at this institution include the Streets of Old Milwaukee, where visitors walk through a turn-of-the-twentieth-century Milwaukee Street, and the European Village, a reconstruction of houses from 33 cultures that formed Milwaukee's population base during the 19th century. Grand-scale exhibits like these help visitors understand many aspects of life, including technologies prevalent in the community.

On the ground floor the Hebior Mammoth is on display. Found in the early 1990s in Kenosha County and acquired in 2007, carbon-dating places the mammoth's age at 14 500 years old. It is a rare, nearly complete skeleton that has butchering marks on the bones, indicating a human presence at that time in Wisconsin. Also on view is an exhibit describing the "green" aspects of the museum's structure: The southern-facing wall has 234 photoelectric panels generating electricity for the museum, and there are over 1000 biotrays of plants on the tallest section's roof, cooling the building and producing oxygen.

A comprehensive Great Plains exhibit uses part of the museum's internationally known collection of Native American artifacts to illustrate facets of their life, crafts, and technology. The Bison Hunt features a realistic depiction of Native Americans in pursuit of their most important commodity, the bison.

Dioramas re-create habitats and cultural aspects of peoples of America, from the Northeast to the Southwest, including the woodlands of Wisconsin. In the biomes are broadleaf forests and Seminole and Cherokee Native American lands. A Hopi pueblo and ceremonial kiva illustrate Southwest culture; a section of the Grand Canyon and a cactus forest show some of the Southwest's natural wonders.

Many dioramas illustrate architecture from around the world, such as a Philippine stilt house and a Japanese home. Animals of India occupy a re-created Old Delhi bazaar, and a collection of seashells from the ocean floor is on display.

Two extensive exhibits in the African wing examine the flora, fauna, and other characteristics of the bamboo forest and the savanna waterhole that contribute to the culture and technology of the African people. The Third Planet—Earth exhibit draws upon plate tectonics and principles of volcanoes and earthquakes to illustrate how the continents were formed. Here is displayed a representation of the Silurian reef that covered the Milwaukee area some 400 million years ago. Examples of plants and animals inhabiting the Earth over the eons include trilobites, corals, echinoderms, brachiopods, and nautiloids.

A two-story walk-through diorama, the Rainforest, provides a close-up view of the both the canopy and the ground level. The exhibits have traditional and high-tech designs, videos, and interactive displays to help visitors understand the rain forest. These walk-through diorama designs are known as "the Milwaukee Style."

Other important collections include East African artifacts, the largest assemblage of typewriters in the world (more than 700), and guns made in Wisconsin and throughout the United States. The museum is involved in research including nondestructive techniques for studying mummies, the best way to collect data on dinosaurs without removing remains from the dig site, and approaches for saving endangered species.

A six-story-tall dome theater presents regular videos, and a planetarium has Digistar® 6 technology to offer sky shows. A two-story walk-in butterfly area houses butterflies in a tropical environment.

Address: Milwaukee Public Museum, 800 West Wells Street, Milwaukee, WI 53233.
Phone: (414) 278-2700.
Days and hours: Mon.–Fri. 10 a.m.–5 p.m., Sat. 9 a.m.–5 p.m., Sun. 11 a.m.–5 p.m. First Thu. of the month 9 a.m.–8 p.m. Closed 7/4, Thanksgiving, 12/25.
Fee: Yes. Extra fees for theater, planetarium, and guided tours.
Tours: Self-guided. Guided tours are offered on an occasional basis.
Public transportation: Buses stop within a couple of blocks of the museum.
Handicapped accessible: Yes. Wheelchairs are available.
Food served on premises: Café, coffee shop.
Website: www.mpm.edu

PRAIRIE DU CHIEN

Fort Crawford Museum

An American army expedition arrived in Prairie du Chien in the upper Mississippi River region shortly before the War of 1812 and built a fort to expel the British, who controlled the area's fur trade. Before long, the Americans lost the fort to the British and were forced to withdraw. The British renamed it Fort McKay and held it for several years, until the Treaty of Ghent was ratified, ending the war. The battle at Prairie du Chien is the only encounter of the War of 1812 fought in Wisconsin. Because flooding caused malaria to spread, a new fort was constructed on higher ground in 1816 and was named Fort Crawford after William Harris Crawford, who served as Secretary of War, US Secretary of State, and US Secretary of the Treasury. Soon, the United States regained control of the regional fur trade.

Fort Crawford is now a National Historic Landmark and home of the Fort Crawford Museum. The complex of three buildings includes the Fort Crawford Military Hospital, which is a reconstructed military hospital on the original foundations, and the Museum of Prairie du Chien, plus many exhibits. Here visitors step back to the age of frontier medicine, relive Wisconsin medical history, and learn about health problems and trends in the 19th century, as well as explore local culture of Prairie du Chien. (See National Museum of Medicine and Health, Silver Spring, Maryland; and U.S. Army Medical Department Museum, Fort Sam Houston, Texas.)

The Hospital is dedicated to the memory of Dr. William Beaumont. It recounts the fascinating story of a wounded French-Canadian trapper, Alexis St. Martin, who was treated by Beaumont in 1822 while the young surgeon was assigned to Fort Mackinac in Michigan's Upper Peninsula. St. Martin's injuries were serious: a hole torn in his abdomen penetrated his stomach, a piece of his lung was gone, and a rib was broken. Beaumont decided not to operate, fearing that St. Martin would die of shock and infection if he tried to close the wounds. St. Martin lived to tell the tale, but spent the rest of his life with a hole in his stomach. Knowledge of the digestion process was scanty until Beaumont used the opening in St. Martin's body to study it.

From 1825 to 1833 Beaumont served the army at a series of forts—Mackinac, Niagara, Crawford, and St. Louis—carrying out medical experiments on St. Martin at each stop. These experiments form the basis of our present-day understanding of digestion. While he was at Fort Crawford in the 1830s, during the Black Hawk Wars, the commandant was Colonel Zachary Taylor, who later became president.

The museum has a re-creation of Dr. Beaumont's office and a variety of 19th-century medical artifacts from Wisconsin, as well as a display of Native American herbal remedies and a refurbished 1890s pharmacy containing prescriptions and instruments used to prepare medications. Also on view are the re-created offices of a dentist and a physician from the 1890s.

A series of small dioramas depicting important milestones in 19th-century surgery were created for the 1933 World's Fair in Chicago. One shows Dr. Philip Syng Physick amputating a leg at the end of the 18th century, long before the days of antibiotics and anesthetics. Known as the "father of American surgery," Dr. Physick pioneered many medical practices, such as the excision of tonsils, for which he devised a special instrument. (See College of Physicians of Philadelphia and the Mütter Museum, Philadelphia, Pennsylvania; Pennsylvania Hospital, Philadelphia, Pennsylvania; and International Museum of Surgical Science, Chicago, Illinois.)

Another diorama re-creates the first ovariotomy, performed in 1802 by Ephraim McDowell. Surgeons of that period did not usually attempt to remove ovarian tumors because doing so was almost certain to bring death. McDowell succeeded, without the benefit of anesthesia, and his patient lived another 30 years.

Noted 19th-century surgeon John Collins Warren introduced many medical innovations, including the use of a needle and ligatures to close wounds. He was also one of the attending physicians during the historic surgery in which Dr. William Morton may have been to the first to administer ether to a patient. A diorama depicts the scene of this surgery. (See Crawford W. Long Museum, Jefferson, Georgia.)

Dr. James Marion Sims, a pioneer in gynecology renowned for his surgical feats, is the subject of the final diorama. He helped make the study and practice of women's reproductive medicine respectable. Sims is credited with developing new instruments and creating hospitals devoted to the special surgical and medical needs of women.

Also on the site is the Museum of Prairie du Chien, which tells the history of the town. There is a collection of fossils, Native American artifacts (especially spear points), and much about the local culture of the city from the 19th and early 20th centuries.

Address: Fort Crawford Museum, 717 South Beaumont Road, Prairie du Chien, WI 53821.
Phone: (608) 326-6960.
Days and hours: May 1–Oct. 31: daily 9 a.m.–4 p.m.
Fee: Yes.
Tours: Self-guided. Guided tours by appointment.
Public transportation: Taxi.
Handicapped accessible: Yes.
Food served on premises: Picnic area only.
Website: www.fortcrawfordmuseum.com

6

Great Plains

Kansas

BONNER SPRINGS

National Agricultural Center and Hall of Fame

Kansas City businessman Howard Cowden had a vision of a site to celebrate agriculture, on which he began to act in the 1950s. The legislation creating the center was passed with the help of such prestigious officials as Herbert Hoover, Harry S Truman, and Dwight D. Eisenhower. Bonner Springs was selected for the location of the center for several reasons: because it is near the major metropolitan city of Kansas City, it is close to the geographic center of the continental United States, and it is a major area for producing wheat, corn, milo, soybeans, cattle, hogs, and poultry.

In 1960, the US Congress created this facility by federal charter to educate the public about the food-production industry, to introduce the public to American agricultural innovators, and to honor farmers, who have made Americans so well fed. It is a living memorial on 270 acres (109 hectares) of rich, rolling countryside in the middle of the nation's farm distribution center and hard-wheat market. Exhibits emphasize agricultural innovators, their contributions, and how early farmers lived and worked.

The Hall of Fame exhibit, in the main museum building, honors such people as Squanto, the Native American who, before his death in 1622, taught the Pilgrims how to plant corn and fertilize their crops. George Washington is honored for his invention of farm equipment and improving soil conservation and crop-growing techniques. Thomas Jefferson is honored because he pioneered scientific farming practices. (See Monticello, Charlottesville, Virginia.) Other distinguished individuals listed are Louise Stanley, the first chief of the US Department of Agriculture's Bureau of Home Economics; and George Washington Carver, who developed hundreds of food and industrial products from the peanut and sweet potato. (See Tuskegee Institute National Historic Site, Tuskegee Institute, Alabama; and George Washington Carver National Monument, Diamond, Missouri.)

The main museum's gallery features temporary exhibits that vary from season to season. There is also an exhibit examining poultry farming.

Several pieces in the building, some original and some reproductions, are of important historic value. There is a 200-year-old plow designed and used by Native Americans, a 1903 truck, a comprehensive collection of barbed wire, and the walking plow President Harry S Truman used on the family farm before World War I. It was the museum's first acquisition, donated by Truman himself. Also on display is a sodbuster plow with a steel moldboard, illustrating the dramatic changes that have occurred in the agricultural business. The unique shape of this plow allowed farmers to cut through the thick grasses and heavy soil of the Great Plains as never before. An exhibit about chicken farming explains aspects of poultry and poultry-raising.

National Agricultural Center and Hall of Fame.
Courtesy of Mark Carver.

Ye Old Country Town focuses on the era before power tools replaced hand tools. Several fully outfitted shops typical of those in the 19th and early 20th centuries reveal the role of certain merchants, craftspeople, and artisans in building the most efficient food-production system in the world. There is a telephone office, a pre-1900s dentist's office, a veterinarian's office, a harness maker's shop, a blacksmith's shop, and a general store carrying appliances and patent medicines.

Behind the main building is the Museum of Farming, which has one of the largest and most varied collections of early farm machinery—about 10 times more pieces than in the Smithsonian's collection.

The outdoor portion of the museum, a living-history center called Farm Town U.S.A., helps the visitor understand farm life and its ties to town life. Exhibits include a replica of a circa-1900 farmhouse, shed, and barn, a one-room schoolhouse, and a replica of a blacksmith shop owned by three generations of a family from Independence, Kansas, containing more than 300 anvils as well as horseshoes and tools. Also on the grounds is the 1887 Morris Livestock Depot that was moved here from the edge of Kansas City. Visitors can ride a train that runs on narrow-gauge track encircling the town. A garden showcases various plants, including a section on pollination and bees and their hives.

Next to the town are a three-acre (1.2-hectare) lake and heavily wooded nature trail, along which are indigenous wildlife and descriptive signs identifying plants and trees. Self-guided trail brochures are available in the main building.

Address: National Agricultural Center and Hall of Fame, 630 North 126th Street, Bonner Springs, KS 66012.
Phone: (913) 721-1075.
Days and hours: Late Apr.–mid-Oct.: Wed.–Sat. 10 a.m.–4 p.m. Closed 7/4.

Fee: Yes.
Tours: Self-guided.
Handicapped accessible: Mostly. Wheelchairs are available on a first-come, first-served basis.
Public transportation: None.
Food served on premises: None
Website: www.aghalloffame.com

HUTCHINSON

Cosmosphere International Science Education Center & Space Museum

This extraordinary installation in America's heartland came into being because of the foresight and hard work of a local resident with a lifelong love of science, Patricia Carey. In 1962, she purchased a used star-projector, and opened a planetarium in the poultry building on the new state fairgrounds. Four years later the planetarium was incorporated into the new science building at Hutchinson Junior College, now Hutchinson Community College. During the 1970s the *Voyager* spacecraft flashed a signal from outer space, marking the start of construction on campus for a new building to provide science and space education to the public.

A highlight of the lobby displays is the SR-71A Blackbird spy plane of the 1960s and 1970s. Also on view are a space shuttle replica, the Northrop T-38 Talon supersonic jet

Cosmosphere.
Courtesy of the Cosmosphere.

trainer, an Apollo 11 Saturn V F-1 engine thrust chamber recovered from the ocean floor in 2013, and the 83-foot (25-m) tall Mercury Redstone rocket like the one that propelled Alan Shepard on his history-making space flight in 1961. Exhibits in the center chronicle space exploration from the first part of the 20th century. They begin with Robert Goddard's experiments with rockets—on view is a full-scale model of Goddard's 1926 rocket engine—and continue through current international space ventures. V-1 and V-2 rockets demonstrate early German efforts at rocket flight using liquid fuels. Very few V-2s are known to exist today. The "buzz bombs" (V-1s) that harassed the British during World War II did little to alter the course of the war, but they did much to advance the science of rocketry. The 57-foot (17-m) long V-2 was the first man-made object to travel beyond the Earth's atmosphere. Visitors can also see a Messerschmidt Me-163 Komet engine, the first practical production rocket ever to have a person on board.

Besides the German Gallery, there are a series of galleries explaining the history of early space flight after World War II, through the Cold War, which has American Vanguard, Redstone, Explorer, and Pioneer artifacts, as well as back-up Soviet Sputnik, Luna Sphere, Vostok, and Voskhod spacecraft ready for flight. The Cosmosphere has the largest collection of Russian equipment outside of Moscow and the largest collection of American pieces outside of the National Air and Space Museum.

Among the more than 1500 pieces of equipment on display are the only flown Mercury, Gemini, and Apollo spacecraft on view in a non-NASA facility, as well as a full-size lunar lander module. Many spacesuits are on display, including from all American manned missions and several from the former Soviet Union. (See Smithsonian Institution, Washington, DC; Steven F. Udvar-Hazy Center, Chantilly, Virginia; and Kennedy Space Center Visitor Complex, Titusville, Florida.)

In the Investigate Space: Astronaut Experience gallery, technical considerations of space flight are described, including key details often overlooked in space museums, such as where the toilet is on a spacecraft and how it functions, and specially prepared foods to deal with zero-gravity situations. This area also describes the Russian Mir space station, the International Space Station, and the future of space exploration, including private, commercial flights.

The Justice Planetarium uses a Spitz Sci-Dome XD digital projection system to present programs on space and astronomy. Nearby is Dr. Goddard's Lab area for live explosive demonstrations, run daily.

Museum staff members developed Space Works, a space-artifact restoration facility, out of necessity. In its early years, the museum could not compete with larger and better-financed organizations, and was forced to collect bits and pieces in need of repair, restoration, and installation. The staff members amassed manufacturer's drawings and plans, some of which were quite rare, in their effort to assemble a credible collection. Their restoration work preserved the retrieved Saturn V F-1 engines from the Apollo program, the Liberty Bell 7, and the Odyssey Apollo 13 Command Module. Today, Space Works' staff members perform work for both their own site and other museums, science centers, and media production companies throughout the world. The movie industry also calls on the staff members for help. Ron Howard, director of the movie *Apollo 13*, hired the center to develop accurate sets.

A two-story digital projection theater shows movies on a regular basis.

Address: Cosmosphere International Science Education Center & Space Museum, 1100 North Plum Street, Hutchinson, KS 67501.

Phone: (800) 397-0330.
Days and hours: Mon.–Thu. 9 a.m.–5 p.m., Fri.–Sat. 9 a.m.–7 p.m., Sun. noon–5 p.m. Closed on Thanksgiving, 12/25, and Easter.
Fee: Yes. There are extra charges for the various theaters and live demonstrations.
Tours: Self-guided. Guided docent-led tours are available for larger groups.
Public transportation: None.
Handicapped accessible: Yes.
Food served on premises: Food court.
Website: cosmo.org

LAWRENCE

KU Natural History Museum

The Natural History Museum on the University of Kansas campus is world-renowned. With nine million specimens of ancient and present-day animals and plants, it is the public face of the Biodiversity Institute. The facility is organized into two parts: an "inner" museum for research, and an "outer" museum for the public. The public museum has over 350 exhibits relating to Kansas, the Great Plains, and the world.

When the University of Kansas began in 1866, its faculty of three was elected by the Board of Regents. According to the book *Snow of Kansas* by Clyde K. Hyder, "The Methodists elected the first professor, the Baptists the second, while the Congregationalists and Presbyterians contended for the third." The Congregationalists won and named to the faculty Francis Huntington Snow, who remained its only professor of natural sciences for 16 years. One of Snow's tasks was to establish a cabinet of

Dyche Hall, home of the KU Natural History Museum and the Biodiversity Institute.
Courtesy of the KU Natural History Museum.

natural history, which grew to contain more than 12 000 specimens by 1877 and is the basis of the formidable collections the institution holds today.

Taxidermist Lewis Lindsay Dyche was hired to assist Snow 22 years after the university was formed. Dyche received his degree at the university and studied taxidermy under W. T. Hornaday, chief taxidermist at the Smithsonian. In 1893, 112 North American large mammals that Dyche mounted in a realistic manner attracted worldwide attention at the Chicago Columbian Exposition. A new Romanesque-style building opened in 1903 to house the mounts and the rest of the collection, which had outgrown its previous space. The 1903 building, later named after Dyche, is now on the National Register of Historic Places, and is the older half of the current museum.

Mounts from the 1893 Chicago exposition are in an exhibit called the Panorama of North American Plants and Animals, one of the longest uninterrupted dioramas in the world. This display shows animals and plants ranging from the high Arctic to Veracruz, Mexico.

Several stylized representations of groups of ancient and modern fish, amphibians, reptiles, birds, and mammals are arranged in loose interpretations of the evolutionary sequence. Fossils of plants, invertebrates, and vertebrates are displayed alongside the skeletons of their currently living descendants for comparison.

Kansas is well known for its fossils from the Pleistocene era. Many of these specimens are on exhibit, among which are fossils of elephant skulls and teeth. A variety of fossil fish, sharks, plesiosaurs, pterosaurs, mosasaurs, and turtles from the Cretaceous Kansas Chalk beds are on display. There is a simulation of the La Brea Tar Pits in California illustrating how fossils are excavated from tarry material. Several animal fossils removed from the pits are presented. (See La Brea Tar Pits, Los Angeles, California.) There is also an exhibit of rocks and minerals from Kansas, and a display of fluorescent minerals.

The ornithological collection Snow initiated now contains more than 124 000 specimens, among the largest university-related collections of its kind in the United States. On display are 260 mounts of songbirds, shorebirds, and birds of prey, many in dioramas.

The herpetology collection, known internationally for its size and diversity, emphasizes the Western Hemisphere. In one exhibit, live snakes in naturalistic settings and terraria of a number of species of anole lizards are on view.

In the Bee Tree, a beehive exhibit explores live bees working in the hive, and how they come and go from their hive through an opening in a window to the outdoors. Bugtown showcases the insect world, such as giant cave cockroaches and the blue death-feigning beetle.

Some major collections of the museum include entomology, invertebrate paleontology, and the McGregor Herbarium, known for its plants of the Great Plains. Nearby on campus is the Spencer Research Library, which has many rare books, manuscripts, and antiquarian maps from the 10th century to the present. The scientific collections include 275 herbals from 1471 to 1753; a major set of works by 18th-century Swedish botanist Carl Linnaeus, who devised a system of classifying plants; illustrated zoological works from the 16th–19th centuries; and more than 15 000 volumes on ornithology. The library has the world's largest collection of naturalist John Gould's hand-colored lithographs and more than 2000 of his original drawings and paintings. It also has architectural works from the 16th–20th centuries, including a Frank Lloyd Wright collection.

Address: KU Natural History Museum, 1345 Jayhawk Boulevard, Lawrence, KS 66045.
Phone: (785) 864-4450.

Days and hours: Tue.–Sat. 9 a.m.–5 p.m., Sun. noon–4 p.m. Closed on Mondays and major holidays.

Fee: Free. Suggested donation.

Tours: Self-guided. Printed guide-sheets are available at the front desk.

Public transportation: Lawrence Transit buses stop nearby.

Handicapped accessible: Yes.

Food served on premises: None. There are eateries in the nearby student union building as well as within a few blocks of campus.

Website: biodiversity.ku.edu

OVERLAND PARK

Museum at Prairiefire

In the small business district of Prairiefire in the municipality of Overland Park, Kansas, stands an interesting natural history museum, opened in 2014. This Museum at Prairiefire concerns itself primarily with the geology and paleontology of Kansas.

Designed by architect Jonathan Kharfen, the museum is constructed with Kansas limestone and artificial stone, punctuated with hundreds of different-shaped windows made with architectural dichroic glass from Goldray Industries in Calgary, Canada. "Dichroic" means that the material displays different colors when viewed at different angles. From the outside of the building, colors range from deep red to orange to yellow (depending upon the viewing angle), reminiscent of a prairie fire. From the inside, the colors show hues in the purples, blues, and greens, providing the appearance of being within a prairie fire. The building is one of just a few structures in North America to incorporate such glass. Modern dichroic glass was developed as an optical filter, to allow only certain wavelengths

The Museum at Prairiefire ignites the landscape in Overland Park, Kansas.
Courtesy of the Museum at Prairiefire.

of light to pass, while others are reflected. The glass in this building is "dichroic laminated glass," made by sandwiching a dichroic film between two panes of glass. Another special exterior material is dichroic metal panels that also shift slightly in color with viewing angle. The panels are stainless steel coated with a thin, grown layer of chromium dioxide whose color depends on the thickness of the oxide coating.

The Great Hall, inside the entrance to the museum, displays Mesozoic marine fossils, including *Xiphactinus audax*, a fish native to the Cretaceous seas that covered Kansas 75 million years ago, and the mosasaur *Platecarpus*, also native to Kansas from 85 million years ago. In addition, there is a cast of a rare *Tyrannosaurus rex* (not native to Kansas), discovered by Kansas native and paleontologist Barnum Brown (1873–1963), whose life and work is told on a nearby wall and alcove. In the drawers of a replica of Brown's desk are artifacts from his career. Brown, the advisor for the Disney movie *Fantasia* dinosaur sequence, was one of the most famous paleontologists of his day. Hanging above visitors are replica fossils of *Pteranodon* and *Nyctosaurus*. Note the gray interior ceiling panels as well as the gray floor tiling, which evoke smoke-filled skies and charred grass of the prairie, respectively.

The museum is a partner with the American Museum of Natural History (AMNH) (See American Museum of Natural History, New York, New York), displaying the AMNH's traveling shows in its Featured Exhibition Hall. These shows deal with all aspects of natural history on a changing basis, concerning subjects ranging from Charles Darwin to outer space.

Inside the museum are also two virtual-reality shows that demonstrate the compelling, life-like nature of virtual reality. Participants must wear special goggles and headphones. One of these shows is 360° Video VR Experience, which is a varying film-based exhibition showing documentaries in which the viewer can turn his or her head at any angle and see different perspectives of volcanoes, astronauts in low-earth orbit, or other films. The other show, in the Michael L. and Sasha Kahn Virtual Reality Theater, is a computer-based interactive rendering of real, famous structures like Stonehenge, which allows the viewer to "travel" to the site and learn without leaving Kansas. The computer-based VR system uses a powerful custom-built computer to generate images based on the position of the goggles with respect to detectors attached to the walls of the theater. Each of these virtual-reality exhibits lasts about 15 minutes per participant.

There are more changing natural-history exhibits in the Great Hall and Sprint Hall. The museum also hosts evening Science Happy Hours, with local expert speakers on scientific topics. Check the website for the current schedule.

On the second floor is a Discovery Center. While aimed primarily at children, some adults may appreciate the hands-on experience on subjects in various corners of the room from Astronomy to Zoology. Visitors can handle casts of dinosaur bones, fossils, and geological specimens, see some small live animals, and more. Out on the balcony there are fossils embedded in the Kansas limestone walls.

Behind the building is a small paved path next to wetlands, with occasional signage describing aspects of the wetlands ecology.

Address: Museum at Prairiefire, 5801 West 135th Street, Overland Park, KS 66223.
Phone: (913) 333-3500.
Days and hours: Mon.–Sat. 10 a.m.–5 p.m., Sun. noon–5 p.m.
Fee: Yes. Exhibition Hall, virtual-reality shows, and Discovery Room incur an extra fee.
Tours: Self-guided. Group tours are available by prior appointment.
Public transportation: Local buses run directly to the site.

Handicapped accessible: Yes.

Food served on premises: None. There are local restaurants within a block or two.

Special notes: Check the website for regular evening Science Happy Hours weekly (for adults only). Because of the popularity of the Kahn Virtual Reality Theater, the museum recommends that visitors book in advance for this show.

Website: visitthemap.org

WICHITA

Sedgwick County Zoo

To visit this zoo is to leave behind the outside world of Wichita. Hills built around the zoo help create this insular feeling, and earth-toned buildings blend into the natural setting. Along these hills are part of one of the largest regions of restored prairie in Kansas. More than 13 000 plants of over 470 different species, all labeled, make up a botanical collection, part of the zoo landscape.

The zoo is organized into geographical areas representing Asia, Africa, Australia, South America, and North America, as a home to nearly 3000 animals of almost 400 species. The areas are designed to simulate the natural habitats of the animals that live in here, with shrubbery, trees, and grasses similar to those in the native environments. Animals are housed in groupings with compatible animals. Overall, the zoo tries to encourage natural behavior while making people aware of environmental issues.

Staff members at the zoo are involved in the Species Survival Plan®, a national program that strives to ensure the survival of selected wildlife species. This zoo has won awards for successfully breeding threatened and endangered species.

The Amphibians and Reptiles building displays reptiles, amphibians, and fishes from around the world. Giant Aldabra tortoises are in an open exhibit that uses a cattle guard to prevent them from escaping. The king cobra lives in an almost barrier-free arboreal exhibit with a low, inconspicuous glass wall. Carrot-tail viper geckos tend to stay hidden beneath succulent plants.

The Tropics is a large, indoor, walk-through jungle exhibit with the humidity, sounds, smells, and colors of a tropical rain forest. More than 50 species of birds live in this natural habitat, alongside Indian fruit bats, spotfin archerfish, and broad-snout caiman. Visitors can see vampire bats in their bat cave, and fish swimming from the walk-through tunnel.

A naturalistic African section contains a large variety of animals in three separate areas. The Downing Gorilla Forest is home to lowland gorillas, black-and-white colobus monkeys, and okapi. Pride of the Plains re-creates an African savannah, complete with kopje rocks to shade the resident meerkats, lions, painted dogs, and Red river hogs. This zoo also claims the third-largest elephant habitat in the United States with the Elephants of the Zambezi River Valley area, covering five acres (2 hectares), an indoor home, and a massive elephant pool with a volume of 550 000 gallons (2.1 million L). Nearby is the Asian forest section, home to Malayan tapirs, ruddy shelducks, demoiselle cranes, more birds, and maned wolves. The Slawson Family Tiger Trek is home to Amur tigers from Asia, red panda, and Burmese brow-antlered deer.

Australia has over 50 mammals and nearly 200 birds, in a free-flight area that includes cockatoos, cassowary, kookaburra, eclectus parrots, and blue-crowned pigeons. The mammals include wallaroos and Tammar wallabies. Next door is the South America area, housing 150 birds, such as spoonbills, ibis, herons, and terns; yellow-footed tortoises; peccaries; and anteaters.

The indoor and outdoor Koch Orangutan and Chimpanzee Habitat affords the opportunity to spy rare chimpanzees and Sumatran orangutans in two breeding colonies. This exhibit was selected by the Jane Goodall Institute for Wildlife Research, Education, and Conservation as a research site for ChimpanZoo, a program studying the behavior of chimpanzees in captivity.

Visitors can follow an elevated boardwalk through a forest with wolves and through the barrier-free North American Prairie to see this land as it was several hundred years ago. Communities of bison, North American grizzly bears, prairie dogs, and other animals live among authentic plantings, re-creating a piece of local natural history and ecology. The river otter exhibit has an underwater viewing area.

Cessna Penguin Cove is the dwelling-place for Humboldt Penguins from Chile and Peru, where visitors can view them along a 52-foot (16-m) viewing area underwater. Cacti grow on the rocks designed to replicate South American shores; Inca terns nest here as well.

The unusual Children's Farms illustrates American, Asian, and African farms. These farms of the world have the largest collection of rare breeds of domestic animals in North America. Buffalo, karakul sheep, Asian fowl, Watusi cattle, Nigerian dwarf goats, and other animals are grouped according to their continents of origin in farms similar to those on which they are raised back home. Each farm grows regional crops. For example, the African farm grows teff, the smallest grain in the world, which is used to make Ethiopian bread. The Asian farm has a rice paddy, and the American farm grows assorted crops in a rainbow of colors.

Address: Sedgwick County Zoo, 5555 Zoo Boulevard, Wichita, KS 67212.
Phone: (316) 942-2212.
Days and hours: Daily, summer 8:30 a.m.–5 p.m.; winter 10 a.m.–5 p.m.
Fee: Yes. Boat tours cost extra.
Tours: Self-guided. A free narrated tram tour stops at various spots around the site.
Public transportation: Buses run from the Wichita Transit Center to within about 1.6 miles (2.6 km), then walk or take a taxi.
Handicapped accessible: Yes. Wheelchairs are available for rent.
Food served on premises: Restaurant and cafés.
Website: www.scz.org

Missouri

DIAMOND

George Washington Carver National Monument

According to George Washington Carver's own writings, he was born a slave shortly before the end of the Civil War in the mid-1860s on an isolated southwestern Missouri homestead, the Moses Carver farm now known as Diamond Grove, in Newton County, Missouri. Moses Carver was a slaveholder, and owned George's mother, Mary, since

1855. The property was declared a national monument and dedicated to this soft-spoken yet driven scientist, shortly after his death in 1943, by President Franklin Roosevelt. This national monument was the first dedicated not only to a non-president, but to an African-American, which testifies to his stature in the mid-20th century. The original property that covers 240 acres (97 hectares) of wooded areas and rolling hills is at the edge of the prairie.

Carver's mother was kidnapped during the turmoil at the end of the Civil War; her fate remains unknown. George was returned to his former owners, Moses and Susan Carver, who treated him well. He was a thoughtful boy, who wandered in the local woods and observed the local flora. Because of segregation in the local Diamond area, he moved to Neosho where a school for blacks was available. To improve his education, he moved again to Fort Scott, in Kansas, until he witnessed a lynching, which forced him to move

Boy Carver Statue by Robert Amendola.
Courtesy of George Washington Carver National Monument, National Park Service.

yet again to various other places in the Midwest. After high school, he was accepted into Highland College, but later was refused entry after the officials learned he was not white. Eventually he did matriculate at Simpson College in Iowa. There he showed a talent for both painting and science, then entered Iowa Agricultural College as the first African-American student, and received his bachelor's degree, then a master's degree at Iowa State University. After graduation, he was hired by Booker T. Washington to join the Tuskegee Institute's faculty.

At Tuskegee, he researched local crops and invented many new products. (See Tuskegee Institute National Historic Site, Tuskegee, Alabama.) Carver believed that religion and science were compatible. He was also a talented artist; this site in Diamond holds several of his paintings.

Among the amenities found in the monument are the Carver Trail—about .75 mile (1.2 km) long—which winds past the Boy Carver Statue, installed in 1960, representing a contemplative young Carver; the George Washington Carver Talking Bust (which plays an audio recording of Carver himself reciting the poem "Equipment" by Edgar Guest, extracted from a commencement address in Selma in 1942), and Carver's birth site, the bare outlines and few layers of a reconstructed 14-foot by 14-foot (4.3 m × 4.3 m) log cabin. Around 1881 a tornado swept through the property and destroyed most structures, including the cabin. After the tornado, the Carvers rebuilt, and the Moses Carver House still stands. Within the park are 140 acres (57 hectares) of tall-grass prairie restored to its original appearance.

At the Visitor Center, there is a classroom built to resemble those at Tuskegee Institute. Here demonstrations are regularly held. Exhibits about Carver's life and the research he led are on display. These displays include a classification of local specimens of rocks, fungi, and pollen. By using microscopes, as Carver did, visitors may examine such specimens.

Of particular interest to Carver was the amaryllis: The Visitor Center has rare color movies of Carver doing cross-pollination on an amaryllis in 1939. A display on Carver's analysis of weather variability pertaining to crops highlights his work, including crop-rotation to prevent depletion of nutrients, and growing cotton on different types of soil.

One of Carver's innovations was to bring the Tuskegee Institute's research to the local people, by having a "Farmer's College on Wheels" constructed. Financially backed by Morris Jesup, a four-wheeled, mule-drawn "Jesup Wagon" was built, which included plows, fertilizers, seeds, cream separator, charts, milk tester, and more. A reproduction of this wagon is displayed here. A few of the rare paintings by Carver that survived are on display, as well as local flora and fauna that inspired the young Carver. There is a 28-minute video about his life.

Address: George Washington Carver National Monument, 5646 Carver Road, Diamond, MO 64840.
Phone: (417) 325-4151.
Days and hours: Daily 9 a.m.–5 p.m. Closed Thanksgiving, 12/25, and 1/1.
Fee: Free.
Tours: Self-guided. Guided tours of the trail are run twice daily.
Public transportation: None.
Handicapped accessible: Mostly. The 1881 Moses Carver House can be reached only by stairs. Wheelchairs are available at the Visitor Center.
Food served on premises: Picnic area only.
Website: www.nps.gov/gwca

St. Joseph

Glore Psychiatric Museum

Established in 1874 near the town of St. Joseph, the State Lunatic Asylum No. 2 began with 25 patients under the mission of "the noble work of reviving hope in the human heart and dispelling the portentous clouds that penetrate the intellects of minds diseased." Under a variety of names, the mental hospital survived through the 1990s.

George Glore, a "ward staff" member, or psychiatric aide in today's terminology, began his employment at the Missouri Department of Mental Health in 1955. By 1967 he was working at St. Joseph State Hospital as an Occupational Therapist. In 1968 he organized an exhibition of early devices used to treat psychiatric patients in the 16th through 19th centuries. He and a number of patients constructed replicas, displaying them during Mental Health Awareness Week. These pieces became the heart of the Glore Psychiatric Museum's collection, which now numbers over 10 000 items, from Missouri institutions and places around the United States. Glore continued to shepherd the collection and its expansion through the years until his retirement in 1996.

When a new psychiatric facility was constructed across the street from the original center in 1997, the museum moved into the 1968 structure of medicine, surgery, and admissions of the original hospital complex. Today the museum is one of the largest sites in the country dedicated to the history of psychiatric treatment.

The museum spans three floors of the building. On the first floor is an introductory exhibit, Within These Walls, with a 10-minute introductory video explaining the site's history. Displayed on the second floor are Glore's original 1968 exhibits, which include various cages and apparatus designed to shock and surprise a patient back to reality, such

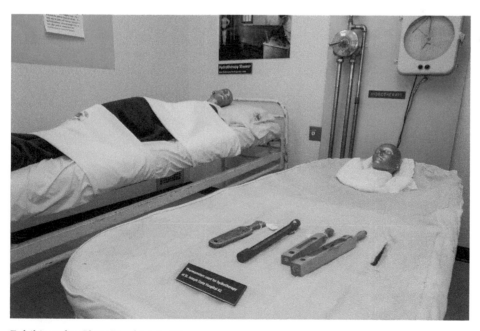

Exhibit at the Glore Psychiatric Museum.
Courtesy of the Glore Psychiatric Museum.

as the O'Halloran Swing, which rotated the patient to bring blood to his brain (nausea and vomiting were predictable side effects). Other apparatus from the 18th and 19th centuries include reproductions of a Hollow Wheel (a type of giant hamster wheel), Dousing Tub, Bath of Surprise, and Lunatic Box. A number of small rooms, each of which explains different types of treatment used at the site, such as EEG scans using an 8-channel Model 6 EEG Machine from Grass Instrument Company of Quincy, Massachusetts; tooth extraction (some psychiatrists believed infections were a source of mental illness) using a Ritter dental chair; lobotomies; cold and fever baths; and restraints. An exhibit space with changing displays is also here. One popular display is a mounted collection of over a thousand items extracted from one patient suffering from pica (a disorder in which patients ingest non-food objects) and who died in 1910.

The focus of the third floor is art created by the patients, ranging from sketches, paintings, sculptures, string art, to ceramics. Also displayed are church pews built and used by the patients, a room explaining music therapy (the hospital's rock-music band was well-known in the area), and an exhibit called Ward Quiet on psychiatric nursing.

In one hall are a series of rocking chairs, in which patients were required to sit and cause little trouble, especially during the early years of the institution. A library and archives are situated here; researchers may visit by advance appointment.

In the basement are displays explaining the on-campus farm that grew food for the institution, and showcasing a variety of farming tools from the late 19th and early 20th centuries. One room explains how this hospital, with hundreds of patients, was able to cook and feed them all, using industrial-sized mixing equipment (the 80-quart [76 liter] Hobart M-802 on display was purchased surplus from the US Navy for use at the Fulton, Missouri, psychiatric facility). A room used for occupational therapy contains various sorts of exercise and therapy equipment, ranging from a Full Body Immersion Hydro Massage HM 801 dating from the early 1950s, to a late 20th-century Lifestyler model 832 297530 treadmill.

A morgue and autopsy room are also in the basement. These two facilities were not just for the hospital, but were also used by the local community. Note the toothbrush holder hanging on the wall: it was a prime way that disease was spread throughout the hospital, for toothbrushes on the upper rows dripped onto those hanging below.

In the stairwell connecting the floors is a multistory mobile constructed by Sara Wilson, a former Interpretive Planner for the museum, consisting of pill bottles and containers from the institution; this is only a part of the extant collection of pill bottles. Be sure to visit the several restrooms on various floors of the museum. Each has a theme (for example, Freud, phobias, or optical illusions), with wall posters inside explaining these psychological ideas.

Across the breezeway from the first floor is a tunnel used by the hospital to transport equipment and patients. From 1988 to 1990, some patients were allowed to paint murals on these walls. Visitors may enter the tunnel and view the murals. Note that the tunnel is often afflicted with mold and dampness.

The museum suggests that young teens and children not enter the museum because some of the treatments can be disturbing. For those who do not wish to view the exhibits, there are small museums in another wing of the building about dolls, African-Americans, and Native Americans.

Address: Glore Psychiatric Museum, 3406 Frederick Avenue, Saint Joseph, MO 64506.
Phone: (816) 232-8471.

Days and hours: Mon.–Sat. 10 a.m.–5 p.m., Sun. 1 p.m.–5 p.m. Closed Memorial Day, Easter Sunday, 7/4, Labor Day, Thanksgiving, 12/24, 12/25, 12/31, and 1/1.
Fee: Yes.
Tours: Self-guided and guided. Group tours are available by prior arrangement.
Handicapped accessible: Yes.
Public transportation: Buses run near the site.
Food served on premises: None. There are eateries in St. Joseph.
Special Note: Some of the exhibits can be disturbing, and are not recommended for younger children.
Website: www.murr.missouri.edu

St. Louis

Missouri Botanical Garden

Established in 1859, the Missouri Botanical Garden is the oldest continually operated public garden in the United States, and is now a National Historic Landmark. It is a living tribute to its founder, businessman Henry Shaw, who made his fortune importing hardware from his native Sheffield, England. When he retired at age 39, he built two houses, one in town and one in the country. After a tour through Europe, Shaw decided to build a botanical garden on his country estate in southwest St. Louis. He sought advice from Sir William Jackson Hooker, director of the Royal Botanic Garden at Kew, England; German-born physician and botanist Dr. George Engelmann, who resided in St. Louis; and American botanist Dr. Asa Gray of Harvard University. Over the years, these three experts continued to advise Shaw.

Shaw designed his garden as a place of beauty, scientific study, and education. It contained a series of patterned gardens and a museum building with the same floor plan as Museum No. 2 at Kew Gardens. While on a trip to Germany in 1857, Engelmann purchased the extensive herbarium from the estate of scientist Johann Bernhardi, including items from Captain James Cook's first voyages to the South Pacific in 1768 and Charles Darwin's world expedition on the HMS *Beagle* in 1837. The museum is used by researchers and is occasionally open to the public.

This beautiful spot is now in the midst of the city that has grown around it. Shaw instructed in his will that the garden be maintained for the public for all time. Through horticultural displays, educational programs, and scientific research, the garden teaches about plants and their ecological role. Many of the garden's botanists are working to seek out, document, and classify unknown tropical plants before they become extinct, under the auspices of the Global Strategy for Conservation (GSPC). The garden runs the world's largest botanical database and the botanical information website, www.tropicos.org.

The contemporary Ridgway Center, a large, light, airy structure with a great vaulted ceiling, is the current entrance to the garden. With its modern design, it stands as a symbol of the management's forward thinking. Visitors can still see the original entrance, the Flora Gate, built by Shaw in 1858. Ridgway Center has a theater for performances and presentations, and an orientation theater that presents an introductory movie. Floral displays are mounted in the Orthwein Floral Display Hall several times a year. Monsanto Gallery is both an art gallery and a display area of botanical educational material. Rotating exhibits are shown in the Spink Gallery.

Missouri Botanical Garden's Central Axis.
Courtesy of Dan Brown.

The garden offers tram rides around its 79 acres (32 hectares). Passengers may leave and rejoin the tour at various stops. The ride is narrated by a garden staff member.

The Zimmerman Sensory Garden, especially appealing to people with visual or hearing impairments, contains plants with strong fragrances or textured leaves that visitors are encouraged to sniff and handle. Signs here are in braille; a large wind chime that appeals to the sense of hearing provides pleasant ambiance. (See Brooklyn Botanic Garden, New York, New York.).

An English Woodland garden grows unusual plantings of azaleas, dogwoods, and wildflowers. These grounds also include a 14-acre (5.7-hectare) Japanese garden, iris and day lily gardens, a bulb garden, a rock garden, and an azalea and rhododendron garden. More than 5000 roses fill two rose gardens with some of the finest specimens in the nation. The site is used by All-American Rose Selections, Inc., to select the top new varieties.

Tower Grove House, the country house of Henry Shaw, is open to the public. Inside is some of Shaw's original furniture. Behind the house is a 90-foot-by-90-foot (27 m × 27 m) Victorian-style maze of arborvitae and yew hedges that Shaw put in place in the 19th century.

The 1882 Linnaen House is the oldest continually operating display greenhouse west of the Mississippi River. For most of its existence it has featured camellias. This building was named after 18th-century Swedish botanist Carl Linnaeus, who developed the method of naming plants and animals still used today. On the façade are busts of Linnaeus and two American naturalists, Thomas Nuttall and Asa Gray.

A geodesic dome, the Climatron, inspired by designs of R. Buckminster Fuller, was built in 1960 to replace the old Palm House. It was the first geodesic dome used as a greenhouse, and the first to use plexiglass panels. During a later renovation, the plexiglass was replaced with film-covered, heat-strengthened glass. The advantage of this design is that it requires no interior supports and allows for more interior light. The dome covers half an acre (2000 m²) and houses more than 2800 plants of 1400 species.

The Climatron's display of tropical vegetation emphasizes the diversity and importance of rain forests. On display are tropical plants that are economically important, such as bananas, cacao, coffee, and rubber, and unusual plants such as the rare double coconut, which contains the largest seed of any plant. Here visitors also learn about the garden's world-renowned tropical botany research. Attached to the Climatron is the Temperate House, focusing on plants of the Mediterranean climate.

An 8-acre (3-hectare) Kemper Center for Home Gardening provides information about the use of plants in and about the home. Over 20 distinct gardens and displays provide practical and aesthetic ideas for residential gardens. On site are a plant doctor and a library for reference.

The Shaw Nature Preserve at Gray Summit, Missouri, about 30 miles (48 km) away, is part of the garden. It has 12 miles (19 km) of trails through a 2400-acre (970-hectare) Ozark landscape, tall-grass prairie, and plantings of conifers from around the world. Nineteen miles (31 km) due west of the botanical garden is the Sophia M. Sachs Butterfly House in Chesterfield, Missouri. This site has over 60 butterfly species indoors, many tropical plants, and an outdoor butterfly garden.

Address: Missouri Botanical Garden, 4344 Shaw Boulevard, St. Louis, MO 63110.
Phone: (314) 577-5100.
Days and hours: Daily 9 a.m.–5 p.m. Closed 12/25.
Fee: Yes. Free Wed. and Sat. 9 a.m.–noon for local residents. Extra charge for tram tours.
Tours: Self-guided and guided. Group guided tours are available by prior appointment.
Public transportation: Metro buses stop at the corner of Tower Grove Ave. at Shaw Blvd. and at the corner of Alfred Ave. at Shaw Blvd.
Handicapped accessible: Yes. Wheelchairs are available, first come, first served.
Food served on premises: Cafés.
Website: mobot.org

Saint Louis Science Center

This visitor-friendly attraction, one of the largest science centers in the nation, uses an interactive, hands-on approach to its displays. The museum consists of two buildings that are connected by a Skybridge, a unique bridge-tunnel that spans an interstate highway. The combination of bridge and tunnel serve as an exhibit gallery with a fascinating view of the highway, symbolizing the bridge between scientific and technical communities and the public.

The 700-foot (213-m) long Skybridge contains the Structures gallery, which focuses on architecture and engineering. Topics covered include arches, trusses, and domes. Topics explored include roads and why they are banked and curved. By using radar to clock speeding vehicles, the highway below becomes part of the display. The tunnel section examines a mine, a sewer, and a modern utility tunnel like those found beneath the city.

Visitors immerse themselves in science via the interactive exhibits in the main galleries. The lowest level and main floor are concerned primarily with the natural sciences. In the Dana Brown Fossil Prep Lab, scientists demonstrate how they prepare fossils for exhibiting, and are available for questions from the public. There is a simulated dig of a reconstructed Montana badlands. The Life Sciences Lab includes an aquarium of a Pacific coral reef, and microscopes under which visitors can examine their own cheek cells.

Experience Energy is an exhibit devoted to various methods we use to generate and use energy in our lives, including alternative and "green" fuels.

The Forest Park Building contains the James McDonnell Planetarium plus exhibits describing past and present space exploration and possible voyages of the future. Boeing Space Station covers two levels, and is concerned with research about the cosmos and future space travel. Galleries show how astronauts live, work, and research on the International Space Station, including clothing, environmental controls, food, and medicine. StarBridge explores how astronauts navigate, communicate, and use power to run the equipment onboard a spacecraft.

In the center of the Forest Park Building is a planetarium equipped with a Zeiss Universarium Model IX. Daily shows examine the night sky and other astronomical topics. In the lobby near the planetarium are Mercury and Gemini rockets, plus displays showing what astronauts' life may be like on Mars. In this building visitors may also try their hand at a variety of flight simulators, plus virtual reality equipment to simulate the experience on a space shuttle and space station.

Just outside, the GROW exhibit demonstrates food science. This area has greenhouses exploring botany, the water cycle, pollination, and fermentation, along with an F/A-18B Hornet airplane donated by Boeing. The combat jet can reach a supersonic speed of Mach 1.8 at an altitude of 40 000 feet (12.2 km). There is also an OMNIMAX theater which runs shows regularly. Daily science demonstrations are held in the CenterStage area, covering topics such as weather, cryogenics, combustion, and electricity.

Address: Saint Louis Science Center, 5050 Oakland Avenue, St. Louis, MO 63110.

Phone: (314) 289-4444.

Days and hours: Mon.–Sat. 9:30 a.m.–4:30 p.m., Sun. 11 a.m.–4:30 p.m. There are adult-oriented evening events and daily demonstrations: check the website in advance. Closed: Thanksgiving and 12/25.

Fee: Free. Fee for Planetarium and OMNIMAX theater. Some other areas and exhibits also have fees.

Tours: Self-guided.

Public transportation: Buses and Metro Forest Trolley (during spring and summer) stop here.

Handicapped accessible: Yes. Wheelchairs are available in the lobby.

Food served on premises: Cafés.

Website: www.slsc.org

Saint Louis Zoo

The idea of having a zoo inside the city of St. Louis was raised shortly after Forest Park opened in 1876, and citizens began donating animals. In 1891 Forest Park inherited a collection of exotic animals when the fairgrounds zoo closed. Expanding the facility became imperative in the 1890s when a herd of bison in danger of being exterminated was brought here and placed in natural surroundings.

But the impetus to make this an official zoo didn't occur until 1905, when the city bought the huge walk-through aviary the Smithsonian Institution had built for the World's Fair in St. Louis that year. This aviary's popularity made it clear that the people wanted a zoo, and so the Saint Louis Zoo opened officially on this site in 1913. Policies were set early, even before it was officially a zoo, and are still in place today. Its main purpose is to entertain and educate the public, and to be a survival center for endangered species, using natural habitats as often as possible.

The aviary from the World's Fair remains in use, in the area called Historic Hill. Animals residing there are different from those of the early 1900s; species from Illinois and Missouri, such as snowy egrets, yellow-crowned night herons, and roseate spoonbills are just a few residents of this open-air, walk-through cage. Opposite the aviary is the Bird House, which provides its inhabitants with natural habitats that encourage reproduction: lush vegetation, pools, sand seashores, and a simulated tropical rain forest. Nearly invisible wire separates visitors from the birds. The aviary holds the very rare Guam kingfisher and the horned guan, both part of Species Survival Plans® (SSP) for saving selected wildlife.

Bear grottos, built in 1922 and renovated in the mid-2010s, have no bars but instead use moats to keep the bears separated from viewers. This was one of the first zoos to

Amur Tiger at the Saint Louis Zoo.
Courtesy of the Saint Louis Zoo.

replace bars with open, moated enclosures. Grizzly Ridge includes outdoor habitat, a wading pool and waterfall, as well as grasses, mulch, and sand, all to stimulate the bears. Polar Bear Point houses polar bears and examines the lives of the Arctic peoples who rely on them.

Jungle of the Apes, an indoor rain forest exhibit, plus Fragile Forest, an outdoor habitat, is where groups of gorillas, chimpanzees, and orangutans live in an environment similar to their natural one. Keepers hide the food for these intelligent primates so they must forage for meals, and ropes are provided for them to climb and use for moving about. Young male gorillas live together in a successful bachelor group.

A 1929 Spanish-style Primate House provides naturalistic living space for social groups of primates. This exhibit area holds Coquerel's sifaka, Guereza colobus monkeys, cottontop tamarins, and other lemurs and monkeys.

Herpetology encompasses two types of cold-blooded vertebrates: reptiles and amphibians. In the Herpetarium, which has four biomes, are tortoises; meters-long pythons; the small, colorful, poison arrow frog; Chinese alligators; Aruba Island rattlesnakes; Madagascar tree boas; large aquatic salamanders called hellbenders; alligator snapping turtles; and Galapagos tortoises.

Other exhibits are the Waterfowl Lakes, Big Cat Country, Penguin and Puffin Coast, River's Edge, Sea Lion Sound, and Red Rocks (where hoofed mammals live).

The Monsanto Insectarium houses over 100 species of invertebrates (which make up as much as 99% of all animal species) in 20 different areas. A butterfly zone inside a geodesic dome is the dwelling place for butterflies, moths, and katydids.

The zoo offers a variety of behind-the-scenes tours, to show how zookeepers interact with animals, how the zoo cares for sick and injured animals, and some of the machinery used to conserve resources and water. There are tours uncovering recycling and composting, tours to the animal nutrition center, and tours bringing visitors up close with a variety of animals.

A relaxing, fun way to see the zoo is by riding the Zooline Railroad, which operates seasonally. The 1.5-mile (2.4-km) narrated tour stops at four stations along the way so passengers may get off and see the exhibits, then reboard and continue on their journey.

Address: Saint Louis Zoo, 1 Government Drive, St. Louis, MO 63110.

Phone: (314) 781-0900.

Days and hours: Spring, fall, and winter: daily 9 a.m.–5 p.m. Summer: Mon.–Thu. 8 a.m.–5 p.m., Fri.–Sun. 8 a.m.–7 p.m. Closed 12/25 and 1/1. Check the website for updates.

Fee: Free. Fee required for the Zooline Railroad, Children's Zoo, shows, and special tours.

Tours: Self-guided. Behind-the-scenes tours are also available regarding animals, animal-care, and ecological practices. For these tours, children must be at least age 8, and all participants must wear closed-toe shoes.

Public transportation: Spring through fall, the Metro Forest Park Trolley stops at the zoo. MetroLink and MetroBuses both stop within a short walking distance.

Handicapped accessible: Yes. Wheelchairs are available for rent on a first-come, first-served basis.

Food served on premises: Cafés, snacks, and bar.

Special Note: Check schedule for times of feedings and shows.

Website: www.stlzoo.org

Nebraska

HASTINGS

Hastings Museum

This multifaceted museum of natural science and pioneer history displays and protects collections of birds, animal life, Native American archaeology, and articles related to pioneer history. These collections were begun by Albert M. Brooking, a self-taught taxidermist who had a passion for collecting that began in his childhood. After reading of John J. Audubon's wish to paint every known American bird, Brooking decided to collect one of every bird in Nebraska, using taxidermy as a means to display his collection.

Part of Brooking's collection was shown for the first time in Hastings in 1911. By 1912 the collection had a home in the basement of the Hastings College Library. Other collectors, inspired by Brooking's work, offered pieces—such as A. T. Hill's Native American artifacts and Adam Breede's mounted wild animals from North and South America and South Africa. In 1937, after several moves, the collection settled in a new fireproof building of Hastings-made bricks, for years one of this city's major products. When it opened it was the largest museum between Chicago and Denver.

A special feature in this facility is the 25 accurate, detailed dioramas with scenes of North American mammals and birds. One of these scenes contains Dall sheep set in the mountains of Alaska, a type of sheep discovered in 1884 by Professor W. H. Dall. Other scenes show bison on the Great Plains, created by taxidermist Lewis Dyche (see KU Natural History Museum, Lawrence, Kansas), beavers at work near Hastings, ringneck pheasants, and large Merriam wild turkeys. The largest collection of whooping crane mounts in the United States in one scene occupies another display. Also on view are seashells, butterflies, insects, fish, and a display of rocks and minerals including fluorescent specimens.

A Transportation exhibit shows bicycles, horse-drawn and motor-powered vehicles, and motor vehicles used in Nebraska. The displays include a prairie schooner used by a pioneer family as a part of a wagon train that left from Ohio and settled near Geneva, Nebraska; a stagecoach known as a light running mud coach because of its speed; a surrey with a fringe on top used for courting; and a 1905 Cadillac automobile.

Several dioramas depict Native Americans who arrived in eastern Nebraska's wooded area from the Mississippi River Valley some 3000 years ago. These hunters and gatherers used the natural resources around them for food, fuel, and medicine. Exhibits show the chronology of life on the Plains, from the days of the historic Native Americans through the pioneer era. Visitors learn how portable tepees were used to create instant villages, and how—later—when the Native Americans became less nomadic, they constructed permanent houses. On view are models of a Pawnee earth lodge and a reconstructed pioneer sod house. Later the pioneers built frame houses. The museum contains the second frame house erected in Hastings, which is filled with household items and furniture, and a reconstructed general store.

On display are collections of tools from trades practiced by early settlers, and clothing worn by Native Americans with that of the pioneers, shown side by side. Among businesses established in Hastings over the years were fur trading, brick-making, and cigarmaking. One exhibit covers early medicines, healers and herbalists, and early medical and dental instruments.

An exhibit spotlights local inventor Edwin E. Perkins, who developed Fruit-Smack in 1927, a popular product that eventually became known as Kool Aid®.

The museum has one of the finest collections of Smith & Wesson® guns in the nation. A special vault room, open to the public, contains a large collection of coins and paper money. There are also several eclectic collections: an exhibit of World War II military hardware and shells; early radios; telephones; vacuum cleaners; and typewriters such as an 1874 Remington, a Caligraph #3 dated 1883, and a Franklin #7 dated 1891. A Cretaceous Sea gallery includes a sculpture of a 30-foot (9-m) *Tylosaurus*. A planetarium with a Digitarium® Epsilon Fixed Dome System runs regular sky shows.

Address: Hastings Museum, 1330 N Burlington Avenue, Hastings, NE 68901.
Phone: (402) 461-2399.
Days and hours: Tue.–Thu. 10 a.m.–4 p.m., Fri.–Sat. 10 a.m.–8 p.m., Sun. 1 p.m.–6 p.m. Closed Thanksgiving, 12/24, 12/25, and 1/1.
Fee: Yes. Additional fee for theater.
Tours: Self-guided.
Public transportation: None.
Handicapped accessible: Yes.
Food served on premises: None. There are a couple of fast-food restaurants within a few blocks of the museum.
Website: hastingsmuseum.org

Lincoln

University of Nebraska State Museum

On the same day that the University of Nebraska was established in 1871, the university's Board of Regents directed that a cabinet or museum of scientific specimens be instituted for teaching about the natural sciences. It occupied several different facilities until 1927, when Morrill Hall was built specifically as a museum. Because many fossil elephants were found in Nebraska, museum curator Erwin H. Barbour decided that the museum would have an Elephant Hall of grand proportions. A chronological procession of elephants that he created became world famous and is still on display.

The museum's function is to make the public aware of research and education in the natural sciences at the university, and bring about an appreciation of plant and animal life in Nebraska. Exhibits emphasize the importance of geologic time, the process of evolution, and the complexity and variety of nature.

Today, Elephant Hall contains one of the rarest collection of elephants in the world. Exhibits describe the evolution of 13 mounted skeletons of elephants and their extinct relatives. Among them are the Perfect Tusker, the only mammoth in North America; and the *Archidiskodon imperator maibeni*, one of the largest known mammoths. A collection of fossil elephant teeth is on view as well, and there are interactive exhibits to enhance learning.

The museum's vast entomology collection primarily has samples from the Great Plains and Neotropical regions. The facility has one of the largest collections of fossil mammals in the world and an invertebrate paleontology collection focusing on eastern Nebraska. It also has one of the largest parasitology collections in the world. The Herbarium, with

250 000 specimens, contains some of the first plant materials scientifically collected from the Great Plains.

Fossil Mammals Hall features extinct rhinoceroses, giant camels, ancient horses, and a giant hog, as well as other animals that once lived in Nebraska. The hall displays an extinct giant hornless rhinoceros, the only mounted skeleton of the largest species of American rhinoceros.

The Gallery of Ancient Life examines the origins of living things and the diversification of life in Nebraska during the Paleozoic era. It contains a diorama featuring the Pennsylvanian age and a terrarium and aquarium containing living descendants of life that once inhabited Earth.

The Mesozoic (Cretaceous) Gallery features a triceratops dig site, a walk-in fossil aquarium with a fully mounted mosasaur skeleton, and a *Xiphactinus* that is one of the largest fish skeletons ever collected from Niobrara Chalk. Embedded in the floor of the gallery is a 21-foot (6.4-m) plesiosaur neck and head. Also on display is a fully mounted *Chasmosaurus* skeleton, a Nebraska dinosaur. Multimedia exhibits use videos and computer simulations of eight North American Mesozoic excavation sites.

The Jurassic Dinosaurs Gallery focuses on the enormous creatures that walked the Earth between 225 million and 65 million years ago. It features mounted specimens such as a young and an adult *Allosaurus* and an adult *Stegosaurus* with its bony plates. Also on display is a *Plesiosaurus*, a 41-foot (12-m) long sea serpent. Although it is a marine reptile, not a dinosaur, it lived at the same time as these dinosaurs.

The Cherish Nebraska exhibit has carefully crafted dioramas illustrating wildlife in their natural habitats in specific parts of the state. In creating these displays, great attention was placed on putting animals and plants in proper relationship to each other. They show bobcats, whooping cranes now in danger of extinction, wolverines no longer found in Nebraska, and bison that today live only in preserves and large parks.

The interesting Migratory Waterfowl dioramas show many species of geese, ducks, and other birds. Nebraska is the Central Waterfowl Flyway—one of four routes across the United States used by birds as they migrate—and is the most important breeding ground for waterfowl in North America.

The Discovery Center is a large, hands-on facility where visitors can touch, smell, hear, taste, and study such objects as animal models, skins, bones, plants, pollens, fossils, geologic specimens, live animals, and written materials. The museum also has a planetarium that presents star shows. Other galleries highlight rocks and minerals, a variety of historical weapons, evolution, and the Native Americans of the plains.

The university operates several other public sites around the state, including Ashfall Fossil Beds State Historical Park, where visitors can watch the ongoing excavation and preservation of fossil rhinoceroses, three-toed horses, and other animals. Fort Robinson State Park Trailside Museum displays a variety of fossils, among which are mammoths, fossil rhinoceroses, and a giant tortoise. By prior arrangement, visitors can take field trips to regional fossil sites.

Address: University of Nebraska State Museum, Morrill Hall, University of Nebraska, Lincoln, NE 68588.

Phone: (402) 472-2637.

Days and hours: Mon.–Sat. 9:30 a.m.–4:30 p.m., Thu. open until 8 p.m., Sun. 12:30 p.m.–4:30 p.m. Closed Easter, 7/4, Thanksgiving, 12/24, 12/25, and 1/1.

Fee: Yes. Planetarium tickets cost an additional fee.

Tours: Self-guided. Guided group tours are available with prior arrangements.

Public transportation: Amtrak trains stop at Haymarket Station, then walk 0.9 mile (1.4 km) to the site.

Handicapped accessible: Yes.

Food served on premises: None. A variety of restaurants mostly are found a few blocks south of the museum, on Q and P Streets.

Website: museum.unl.edu

North Dakota

MEDORA

Theodore Roosevelt National Park

The North Dakota badlands, made up of ever-varying rock formations, were named by the pioneers based on the rugged terrain. Native Americans called this area *Mako Sica* and French trappers called it *Mauvaises Terres*, both of which translate to "bad lands." In 1864, General Alfred Sully said that this part of North Dakota reminded him of "hell with the fires put out."

Theodore Roosevelt National Park, in the heart of the badlands, preserves this unique landscape and commemorates the 26th president of the United States, who lived in this area off and on during the mid-1880s.

The parkland contains bands of lignite coal and petrified trees, indicating that the climate was wetter in ancient times than it is now. A local swamp once sustained a dense forest of conifers, including sequoia. Trees that fell into the water were eventually covered

North Unit badlands at the Theodore Roosevelt National Park.
Courtesy of National Park Service & Laura Thomas.

by volcanic ash, which, over time, with pressure from overlaying sediments, converted the trees and other decaying matter into lignite. Minerals filled other trees, replacing the wood to form petrified wood. Today petrified wood can be found throughout the park. (See Petrified Forest National Park, Grand Canyon, Arizona; and Ginkgo Petrified Forest, Vantage, Washington.)

The land uplifted and erosion carried the freed material down the eastern slopes of the Rocky Mountains, forming sedimentary layers in what is now North Dakota. During the glacial period the flow of the region's rivers changed, and these rivers, particularly the Little Missouri River, began eroding the landscape into the fantastic topography of today. The park's exposed hillsides reveal layered sediments of the alluvial plain formed by erosion.

The underbeds of lignite coal burned from time to time, baking the overlaying sediments into a natural red-brick material known as clinker or scoria. These hardened rocks are more erosion-resistant than the softer, unbaked materials. Bentonite clay, the soft gray or blue-gray layers of today's badlands, was formed from volcanic ash that drifted from distant volcanoes to the west. The result is the knobs, ridges, and buttes topped with red clinker caps that make up the unusual rock formations of the badlands.

Theodore Roosevelt first came to North Dakota in 1883 to hunt bison, and became so enamored of the land that he returned in 1884 to establish two ranches: the Maltese Cross and the Elkhorn. It was this region that may have inspired Roosevelt to seek the presidency, presumably because of the patriotic feelings its beauty evoked. He said once, "I never would have been president if it had not been for my experiences in North Dakota."

North Dakota also inspired Roosevelt to become a great conservationist. Though he was an avid hunter, he watched the quality of the land decline from overgrazing and witnessed the almost total destruction of big-game animal species. During his time as president he established the US Forest Service, national monuments, wildlife refuges, and national parks. He also signed into law the Antiquities Act, which enabled later presidents to declare national monuments—some of which are listed in this book.

The 110-square-mile (285-km^2) park has three sections: the North Unit near Watford City, the South Unit near Medora, and the Elkhorn Ranch Unit between the other two. The park's plateau grasslands have blue grama, western wheatgrass, and needle and thread grasses. Many prairie dog towns—clusters of burrows for prairie dogs—occupy the area. Cottonwood groves and sagebrush flats grow along the river bottomlands. The broken badlands terrain's north-facing slopes have Rocky Mountain juniper and green ash/chokecherry-dominated woody draws, while the dry, south-facing slopes have desert-type shrubs.

Elk, bighorn sheep, and bison that had been eradicated have been successfully returned to the park. The park also maintains a small herd of wild horses descended from animals that escaped from area ranches and a small herd of longhorn steers. Sharp-tailed grouse and wild turkeys live in the park, and golden eagles fly overhead.

The South Unit Visitor Center displays a collection of Roosevelt's personal memorabilia, ranching artifacts, and natural history items, plus an introductory video. Just behind the center is Roosevelt's restored Maltese Cross cabin, through which visitors can take a guided tour in summer or a self-guided tour the rest of the year. Painted Canyon Visitor Center, open only in summer, sits on the rim of the badlands and presents a magnificent view of the topography. The Scenic Loop Drive, 36 miles (58 km) long, has interpretive signs about the area's history and natural phenomena. Highlights of the Scenic

Loop Drive are the Coal Vein Trail, Boicourt Overlook, and Boicourt Trail, reached by a hiking trail. The Petrified Forest is reached from the park's western boundary.

The 70-mile (110-km) drive to the North Unit passes (from a great distance) the site of Roosevelt's principal home, the Elkhorn Ranch Unit. This unit is currently undeveloped and hard to reach because of poor-quality gravel roads. Elkhorn Ranch itself no longer exists, except for foundation stones. The Visitor Center in the northern unit has same video as the South Unit. A 28-mile (45-km) long scenic road has interpretive signs and places to part for good views of the scenery. There are also two self-guided walking trails, such as Caprock Coulee Nature Trail, which goes through a badlands coulee (deep ravine). (See Grand Coulee Dam, Grand Coulee, Washington.) This trail leads to the Buckhorn Trail, which crosses a prairie dog town. Of the seven varieties of prairie dog, only the black-tailed variety lives in the park.

Address: Theodore Roosevelt National Park, 315 Second Avenue, Medora, ND 58645.
Phone: (701) 623-4466.
Days and hours: All park units: daily. South Unit Visitor Center: 8 a.m.–6 p.m. in summer, 8 a.m.–4:30 p.m. in winter. Closed Thanksgiving, 12/25, and 1/1. Painted Canyon Visitor Center: 8:30 a.m.–4:30 p.m. Closed in winter. North Unit Visitor Center: Jun.–Sep.: daily 9 a.m.–5 p.m.; Oct.–May: Fri.–Mon. 9 a.m.–5 p.m. Closed Tue.–Thu.
Fee: Yes, per vehicle, and per person.
Tours: Self-guided. Guided tours of Maltese Cross Cabin and ranger-led programs and activities in the summer.
Public transportation: None.
Handicapped accessible: Yes. Accessibility is limited on trails.
Food served on premises: Picnic areas only. Restaurants are in nearby towns.
Special note: Roads may be closed in winter. The South Unit is on Mountain Time and the North Unit is on Central Time.
Website: www.nps.gov/thro/index.htm

Oklahoma

BARTLESVILLE

Woolaroc Museum & Wildlife Preserve

Woolaroc is a combination of three words—woods, lakes, and rocks—which represent features on the estate where Frank Phillips, cofounder of Phillips Petroleum Company, built a lodge as a country home in 1925. Today, the Woolaroc complex includes a museum of objects from the New World, with an emphasis on the Southwest; the lodge, open to the public; a nature and wildlife preserve; and the Indian Guide Center.

Phillips spent much time in the rugged Osage Hills southwest of Bartlesville, pursuing wildcat oil ventures. Because of his position with Phillips Petroleum, many rich and famous people throughout the world entertained him. Wanting to reciprocate in his own way, he built the lodge in his beloved Osage Hills for entertaining guests.

In 1929, Phillips sponsored Art Goebel in a nonstop air race from Oakland, California, to Honolulu, Hawai'i, in an airplane called *Woolaroc*. Goebel won. This airplane hangs 11 feet (3.4 m) from the floor in a sandstone hangar specifically built for this purpose, now one wing of the current museum. The plane was the museum's first display.

This eclectic collection includes many artifacts from a burial mound at Spiro, Oklahoma. On display are projectile points, shell beads, ear spools, shell gorgets, and shell necklaces made by Native Americans who clearly had a high degree of technical knowledge and skill. There is also a scale model of a Native American village re-created according to field-excavation data and written descriptions by 20 experts between 1540 and 1758.

Look for a Peruvian mummy that was well preserved by the chemical content of the soil and Peru's arid weather. Also displayed is an unusual collection of shrunken heads from the Jivaro Indians of Ecuador. These people first removed the skull bone and then applied heat to shrink the head.

An impressive collection of prehistoric Southwestern pottery includes examples of Anasazi, Patayan, Mogollon, and Hohokam work. A fine collection of Navajo blankets is also displayed.

Cultural artifacts from European settlers and cowboys are prominently presented. The collection of saddles includes one owned by Theodore Roosevelt, saddles used by movie stars, and Wild West showman Buffalo Bill's silver inlaid saddle. On exhibit is the Concord Express, an authentic stagecoach used from 1869–1912, which traveled about 625 000 miles (1.01 million km) back and forth between Fort Logan and Dorsey, Montana.

The firearms collection is one of the finest in the world. (See Springfield Armory National Historic Site, Springfield, Massachusetts.) Pieces on exhibit include all known

Woolaroc.
Courtesy of Woolaroc Museum & Wildlife Preserve.

varieties of Paterson Colt pistols and many Paterson long arms. A display of automatic pistols shows the evolution of these Colt firearms, including the first pieces from 1905 and the entire 1911 series. An interesting collection on view are arms from all known US manufacturers, many foreign copies, and experimental and pattern models. Here also are mass-produced Colt pistols and revolvers, current pieces, and specially-issued Colt commemorative weapons.

The Winchester collection has both carbines and rifles, from the 1866 brass-bound yellow boy to current production models. A Winchester rifle was the frontier weapon of choice.

Among the other displays are minerals classified and labeled by chemical composition, mounted birds, African game trophies, poison darts, and dinosaur eggs.

An outstanding collection of Western painting and sculpture provides an illustrated record of the people who settled this land and their history and technology.

Woolaroc's Indian Guide Center celebrates the culture and skills of Native Americans and provides demonstrations several times a year.

This entire complex sits on a 3700-acre (1500-hectare) wildlife preserve for more than 1000 animals, including American bison, elk, deer, Brahma cattle, Texas longhorn cattle, Scottish highlands cattle, zebras, yaks, and African pygmy goats. A 2-mile (3.2-km) drive passes through this scenic preserve, which includes lakes, clearings with tall grass, and sandstone ledges. This land used to be a place for Native Americans to gather and outlaws to hide.

Thunderbird Canyon Nature Trail begins near the Indian Guide Center. The mile-and-a-half (2.4-km) walk takes visitors through the Osage Hills for a firsthand look at the land that so attracted Phillips.

Bartlesville, the home of Phillips Petroleum, is 12 miles (19 km) from Woolaroc. A small exhibit hall in the main office building there describes the oil business, including its history, geologic principles key to the industry, present technology, and the industry's future.

Address: Woolaroc Museum & Wildlife Preserve, 1925 Woolaroc Ranch Road, Bartlesville, OK 74003.
Phone: (918) 336-0307.
Days and hours: Wed.–Sun. 10 a.m.–5 p.m. Closed Mon. and Tue., Thanksgiving, and 12/25. In summer, the site is also open Tuesdays.
Fee: Yes.
Tours: Self-guided.
Public transportation: None.
Handicapped accessible: Yes.
Food served on premises: Café, picnic area.
Website: www.woolaroc.org

OKLAHOMA CITY

Science Museum Oklahoma

This unique 8-acre (3-hectare) "mall" has exhibits, a botanical garden, and a planetarium under one roof. Just as in a shopping mall, visitors can select from a wide variety of stores, but instead of offering merchandise, they deal with history, art, culture, science, and technology. For one fee, visitors can see some or all of the offerings, as per the mission statement of this site, "We enrich people's lives by revealing the wonder and relevance of science." There is something for everybody, with over 350 000 square feet (33 000 m²) of space for hands-on science material and many interactive exhibits.

The center of the "mall" is the Science Floor, which houses a Segway® Park for visitors to try their hand at two-wheeled driving. Here also is the Resonant Pendulum, which allows visitors to move a 400-pound (180-kg) weight by piggybacking on the resonant oscillation of a pendulum.

Surrounding the Science Floor, on the first floor of the complex, are a variety of exhibit "stores." Tinkering Garage explores creativity and creation with problem-solving using everyday items. GadgetTrees illustrates basic physics of simple machines using wooden crafted objects, such as treehouses, slides, seesaws, and more. Light and electricity are examined in Light Minded, via colored shadows, strobe lights, a Tesla coil, and more. A small greenhouse and garden are packed with exhibits, from tropical displays to spices and fruits. The one-acre (0.4-hectare) plot has 10 different gardens: a children's garden, herb and vegetable gardens, native plants garden, water garden, Japanese garden, a waystation for Monarch butterflies, and more.

Also on the first floor is the Oklahoma Aviation and Space Hall of Fame, and an Aviation Wing. Memorabilia donated by famous pilots who came from Oklahoma are on display, as are a wide range of engines: a Curtiss OX5, a Wright R-975, a Cutaway Jacobs R-755, a Kinner R-370, and the Allison V-1710. The Aviation Wing houses vintage aircraft as early as 1910, such as the Bücker Jungmeister, the Wiley Post biplane, a Nieuport Bebe, a Fokker DR1, Lockheed F-104, Lockheed T-33, Stinson Voyager, Gulfstream Peregrine, the Travel Air 4000 flown by Louise Thaden in the 1929 National Women's

Science Museum Oklahoma.
Courtesy of Science Museum Oklahoma.

Air Derby, and the Ryan PT-22, as well as a number of drones. Here also is an Airborne Warning and Control Systems (AWACS) exhibit.

On the second floor is Destination Space, with a 1926 rocket designed by Robert Goddard; spacesuits from the Mercury, Gemini, and Apollo series of flights; an Apollo command-module simulator; and Mercury and Gemini space capsules.

Big Game Theory uses scaled-up examples of everyday board games to illustrate physics ideas such as catenary arches, light, and potential energy. Power Play uses sports to show reaction times and strength of the human body on rock walls, a tug-of-war, and more.

A variety of technological items are on exhibit throughout the museum. These include antique clocks, Navy artifacts, a 1929 Missouri-Pacific Railroad car built by the Pullman Company, and other objects. A statue of Sylvan N. Goldman shows him with his hands on his most important invention, the supermarket shopping cart.

Science Live is a series of daily science demonstrations in a dedicated auditorium. An area dedicated to the intersection of art and science is called smART Space, which houses changing exhibits of photography and other art related to scientific topics.

This facility is also home to the Kirkpatrick Planetarium with a 70-foot (21-m) dome.

Address: Science Museum Oklahoma, 2020 Remington Place, Oklahoma City, OK 73111.
Phone: (405) 602-6664.
Days and hours: Mon.–Fri. 9 a.m.–5 p.m., Sat. 9 a.m.–6 p.m., Sun. 11 a.m.–noon. Closed Thanksgiving, 12/24, and 12/25.
Fee: Yes.
Tours: Self-guided. Guided group tours by prior arrangement. Call or check the website for daily schedule of science demonstrations.
Public transportation: Buses run from downtown to the museum.
Handicapped accessible: Yes. Wheelchairs are available.
Food served on premises: Café, snack bar. Picnic area is near the gardens.
Website: www.sciencemuseumok.org

South Dakota

CUSTER

Jewel Cave National Monument

In 1900, while brothers Frank and Albert Michaud and their friend Charles Bush were prospecting, they heard wind rushing through a hole in the rocks of Hell Canyon. They enlarged the opening and entered to find a maze of cave passages, many lined with sparkling crystals of calcite. These prospectors filed a claim on what they called the Jewel Lode, but were disappointed when they found no valuable minerals. Next, they tried to make the site a tourist attraction, but that failed. President Theodore Roosevelt recognized the significance of the cave, by then well-known, and in 1908 he created Jewel Cave National Monument.

Jewel Cave is part of an underground cave system cut into the Paha Sapa limestone that encircles the Black Hills. This region, which has 68 of the world's 72 known calcite-crystal caves, has the second-largest concentration of cave systems in the world.

Although until 1958 less than 2 miles (3.2 km) of this cave had been explored, more than 90 miles (140 km) of passages are known today, and many more remain to be discovered. Jewel Cave is the third longest cave on the planet.

This cave was hollowed out when the seepage of acidic water about 30–50 million years ago caused limestone to dissolve, a process that continues today. The cave's vast display of crystal formations was created by mineral-laden water seeping through faults in the rock. Jewel Cave is particularly known for its dogtooth spar and nailhead spar, which give the cave its jewel-like quality. Some formations are particularly rare and unusual, such as the helictites: small knobby clusters of calcite.

Calcite crystals from Jewel Cave come in many colors. Some are translucent white, while others are yellow, red, or opaque white. The various colors are caused by iron oxide, manganese, and other minerals. An interesting formation discovered for the first time in this cave is scintillate, reddish chert rock coated with clear quartz crystals. Other unusual formations are hydromagnesite balloons—fragile, silvery bubbles—and moonmilk, a powdery substance that looks like cottage cheese.

Much of the labyrinth is closed to public because it has been set aside for scientific research. Even so, the different guided tours offered show the best parts of the cave. On each tour the guides explain geology, history, weather influences, and other relevant topics.

An 80-minute Scenic Tour follows a half-mile (0.8-km) loop that traverses five layers of the cave. Visitors reach the cave via an elevator that travels 230 feet (70 m) below ground. This tour, with 723 steps to climb, is considered moderately strenuous.

The Historic Tour, also half a mile (0.8 km) long, starts at the cave's natural entrance, some two miles (3.2 km) from the Visitor Center. Tour participants use candlelight lanterns and spend almost two hours exploring the cave, following some of the same calcite-coated passages that the early explorers used. This tour requires bending, stooping, and climbing numerous wooden staircases. Although it is more strenuous than the Scenic Tour, it does not pass through the most spectacular formations. No children under age 6 are permitted.

The Wild Caving Tour, two-thirds of a mile (1 km) long and taking four hours, is graded as very strenuous. Before starting, would-be spelunkers must prove their fitness by crawling through an 8.5-foot-by-24-inch (2.6 m × 61 cm) tunnel. Old clothes, boots, gloves, and appropriate gear are suggested. The park supplies hard hats and headlamps. The Wild Caving Tour provides visitors with a better understanding of this underground world because they see undeveloped portions of the cave in their natural state. The easiest tour to follow is the Discovery Tour, which is a brief guided tour, down 15 steps, into one room of the cave. It is wheelchair accessible.

Two square miles (5 km²) make up this park, a favorite for cross-country hikers. There are numerous hiking trails of various difficulties and lengths. Mule deer, elk, coyotes, marmots, and other animals inhabit the park's ponderosa pine forest; and visitors see many birds, including golden eagles and hawks.

The Visitor Center has exhibits explaining the geology and history of the cave and provide background information on spelunking. Naturalists give talks in the park about the plant and animal life.

Address: Jewel Cave National Monument, 11149 US Highway 16, Building B12, Custer, SD 57730.
Phone: (605) 673-8300.
Days and hours: Visitor Center: mid-May–mid-Jun. and mid-Aug.–mid-Sep.: 8 a.m.–4:30 p.m.; mid-Jun.–mid-Aug.: 8 a.m.–6 p.m.; winter: 9 a.m.–4 p.m.

Fee: Yes.

Tours: Check schedule for times of the various tours. There are no self-guided tours permitted. The Wild Caving Tour may be booked up to 28 days in advance. Historic Tours run only during summer months. Buy tickets (over 48 hours in advance) for the Scenic Tour using an online service: www.blackhillsvacations.com or (844) 245–6179.

Public transportation: None.

Handicapped accessible: Limited.

Food served on premises: None, picnic area.

Special note: Be prepared for the constant, cool temperature (49°F, 9°C). Wear low-heeled, securely fastened, comfortable shoes with rubber soles to avoid slipping on slick trails and stairs. Those with heart problems, claustrophobia, or other medical problems should not participate in the Spelunking Tour. For other tours, visitors should check with their physicians. Do not wear any clothes or shoes that were worn in a cave or mine outside of South Dakota, to prevent the spread of White Nose Syndrome (a bat disease). There are no self-guided tours. Some tours have restrictions on the age of children permitted.

Website: www.nps.gov/jeca/index.htm

HOT SPRINGS

Mammoth Site

The hot mineral waters that give this town its name have been known for millennia. The Native Americans called this place the Vale of *Minnekahta* (hot water), and the Sioux and Cheyenne tribes fought each other over the medicinal springs. European tourists came to drink and bathe at these curative waters during the 19th century. Paleontologists believe that, just like the Native Americans and Europeans, mammoths came to take advantage of the waters 26 000 years ago.

During the period of the great mammoths, this land was a steppe-tundra grassland. Pollen analysis reveals that there were few trees in the area but a good supply of shrubs, water-rooted plants, sedges, and green grasses. The region is about 3450 feet (1050 m) above sea level, and is capped with tilted beds of shale, limestone, and sandstone surrounding a metamorphic and igneous core. Underneath, caverns were caused by water dissolving the lower limestone rock. When the Permian-Triassic red spearfish shale above weakened, it collapsed and formed sinkholes. The collapsed rock or breccia helped to form a deep chimney with a funnel-shaped top. An underground stream filled it with warm water, 95°F (35°C), creating an enticing pond into which fell mammoths and other animals. They were unable to escape the slippery walls and died there. Animals continued to be trapped this way for millenia. Mammoth footprints left in the mud indicate that the water level may have varied at times. Eventually the sinkhole filled with sediments and was no longer a trap.

In 1974 this animal grave site was uncovered during excavations for a housing development. It contains the remains of more than 60 mammoths. Some skeletons are completely articulated; that is, the bones are in positions common to those of living animals. Because of the importance of this find, it became a museum, research institute, and study site. A building was placed over it to protect the site from the elements and to shelter visitors as they watch the excavations in progress. The bones are displayed exactly where they are found. This is the largest in situ exhibit of fossil mammals in the world.

Part of the Bonebed at The Mammoth Site.
Courtesy of The Mammoth Site.

Over 95% of the sinkhole is in the building, and only about 30% of the site has been excavated to date. So far, 58 Columbian mammoths and three woolly mammoths have been uncovered. An array of small animals such as peccary, camel, gray wolf, white-tailed prairie dog, fish, snails, and clams have also been found.

Most of the mammoths trapped were young adults. Researchers believe that their behavior was probably similar to that of modern elephants, and that they lacked the maturity and experience to judge the dangers. A few of the mammoths were older and, despite their wisdom, were likely drawn to the hole because of the ease of finding food and water.

The extinct, giant short-faced bear found here is rare. It is the first of this species discovered in the area of the Black Hills, and one of a dozen such skulls found in the United States. Scientists assume the bear was attracted by the dying mammoths and tried to feed on them, which brought about its own demise.

In the building, a short video provides an overview of the site and the work performed. Exhibits describe the geology of the area, how the sinkhole formed, and the evolution of the elephant. There is a large collection of Clovis and Folsom points on display. The site is open through the year, but paleontologists conduct research and excavation within the sinkhole only during the summer.

Address: Mammoth Site, 1800 US 18 Bypass, Hot Springs, SD 57747.
Phone: (605) 745-6017.
Days and hours: May 15–Aug. 15: daily 8 a.m.–8 p.m.; Sep.–Oct.: Mon.–Sat. 9 a.m. –5 p.m., Sun. 11 a.m.–5 p.m.; Nov.–Mar.: Mon.–Sat. 9 a.m.–3:30 p.m., Sun. 11 a.m.–3:30 p.m.; Apr.– May 14: Mon.–Sat. 9 a.m.–5 p.m., Sun. 11 a.m.–5 p.m. Closed Easter, Thanksgiving, 12/25, and 1/1.
Fee: Yes.
Tours: Self-guided tours of the museum; guided tours of the dig site.
Public transportation: None.
Handicapped accessible: Yes.
Food served on premises: None. There are a number of restaurants in Hot Springs, mostly along Route 18.
Website: mammothsite.com

LEAD

Black Hills Mining Museum and Sanford Lab Homestake Visitor Center

For an excellent introduction to gold-mining and related science, the Black Hills Mining Museum is the place to stop. In 1874, General George Custer discovered gold while on a military expedition in the Black Hills, or *Paha Sapa* as the local Native Americans, the Sioux, called them. Custer's discovery brought white men here in search of gold, silver, and precious gems. The Sioux fought against this invasion of their territory in a series of clashes, including the one that killed Custer and his troops at Little Big Horn in 1876. This battle did not, however, deter other fortune-hunters.

As prospectors poured in from all over North America, the town of Lead (pronounced *leed*) expanded quickly. For a time it was one of the largest cities in South Dakota. The town was named for the lead, or ledge, containing the gold that made this the largest gold-producing region in the Western Hemisphere. This city was a company town: without the Homestake Mining Company there would have been no town.

Black Hills Mining Museum
The Black Hills Mining Museum explains this gold-mining heritage of Lead and the Homestake Gold Mine, which operated for 125 years. The first floor of the museum exhibits various aspects of mining, artifacts used in mining, historic photographs, and a video about mining techniques. An exhibit about the miner's lifestyle consists of full-size dioramas showing such scenes as an assay office where the quality of ore is evaluated, a blacksmith shop, and a windlass shaft, a manually operated winch used to lower people and tools underground, and bring the ore to the surface. A relief map of the 36-square-mile (93-km^2) Lead and Deadwood region shows the mines in the area and when they came into being.

A model of the Homestake Mine gives visitors the opportunity to visualize the operations that took place in the ground nearby. Visitors have a chance to pan for gold, using the proper technique, in the museum. The facility's theater room presents a video about the Black Hills and the ways in which mining techniques and technology have developed.

The lower floor of the museum re-creates an underground level of the Homestake Mine. A guided tour goes past more than 20 full-size dioramas illustrating the history of hardrock mining. On view are an underground cage station, powder and cap magazines, mine locomotives, ore cars, and an underground ore dump. Chronological displays cover how drilling and blasting techniques changed over the years.

Address: Black Hills Mining Museum, 323 West Main Street, Lead, SD 57754.
Phone: (605) 584-1605.
Days and hours: Mid-May–early Oct.: daily 9 a.m.–5 p.m. Open the rest of the year by reservations only.
Fee: Yes.
Tours: Upper level is self-guided; lower level is by guided tour only.
Public transportation: None.
Handicapped accessible: Yes.
Food served on premises: None. Restaurants are within a few miles of the site, mostly along Route 85.
Special note: Additional fee required for gold-panning.
Website: blackhillsminingmuseum.com

Black Hills Mining Museum.
Courtesy of the Black Hills Mining Museum.

Sanford Lab Homestake Visitor Center

The Homestake Mining Company was organized within four years of Custer's discovery. Many men became rich on this gold, including George Hearst, father of William Randolph Hearst, who invested heavily in this company. What was once a solid mountain is now the remains of a spectacular, open-cut mine. Between 1876 and 2001, 40 million troy ounces (1.2 million kg) of gold, 9 million troy ounces (280 000 kg) of silver, and 6 million troy ounces (190 000 kg) of copper were removed. (There was a short hiatus during World War II when the mine was closed, and the Homestake Mining Company assisted with the war effort by manufacturing hand grenades.) By the end of 2001, however, because of economic reasons, this mine closed for good, with 370 miles (600 km) of tunnels dug reaching down 8000 feet (2400 m) below the surface.

Particle physicists took notice of the empty tunnels and depth of mining, found them useful for blocking cosmic rays that interfere with particle detection, and set up the Homestake Experiment in the late 1960s. The goal was to record solar neutrinos that the Sun emits during nuclear fusion. Theoretician John Bahcall calculated the rate of neutrino emission, and experimentalist Raymond Davis, Jr., set up the equipment, which consisted of a 100 000-gallon (380-m^3) tank of perchloroethylene placed 4850 feet (1478 m) below the surface. The experiment recorded a serious discrepancy between the theory and the experimental value: only one-third of the expected number of neutrinos were recorded. this became known as the "solar neutrino problem." The experiment ran until 1994. Refinements of the theory showed that neutrinos oscillated between different types, or "flavors."

A variety of experiments are running at Sanford Underground Research Facility (SURF) now, including a dark matter detector, geothermal energy, geophysical processes, particle physics, and more. The Deep Underground Neutrino Experiment, run in conjunction with Fermilab, will detect neutrinos produced at Fermilab, investigate neutrino oscillations, and more precisely determine their masses. (See Fermi National Accelerator Laboratory, Batavia, Illinois.)

The Visitor Center for Sanford Lab Homestake has a number of exhibits set up for the public to view about the experiments taking place deep underground at SURF. There are displays about the history of Homestake Mine and the local geology. In the Visitor Center there is an elevator cage from the mine, providing visitors with a simulated experience of descending 4850 feet (1478 m) underground. An awe-inspiring view from the deck of the Visitor Center allows visitors to see the rock strata and the open-cut mine reaching 1250 feet (381 m) deep. From the ceiling of the Visitor Center to the floor hangs a huge three-dimensional model of the entire Homestake Mine tunnel system at a scale of 1:1500.

The Visitor Center also offers one-hour tours of the area and a surface tour of SURF. Hoists built in 1939 are maintained and still operate for the scientists. Some details about Homestake Mine's process are discussed during the tour, as well as the waste-water cyanide treatment plant where cyanide solution is recycled. Gold was separated from the ore by cyanidation, because gold is soluble in cyanide solution. Lime was mixed with ore to neutralize the acid in the ore, then oxygen was added to facilitate the chemical process. Cyanide was percolated through the ore to remove the gold. The cyanide solution was passed over activated charcoal, which adsorbs the gold ions. Electrolysis on steel-wool cathodes then removed the gold ions.

Address: Sanford Lab Homestake Visitor Center, 160 West Main Street, Lead, SD 57754.

Phone: (605) 584-3110.

Days and hours: Daily 9 a.m.–5 p.m.

Fee: Free for Visitor Center's exhibit hall; guided tours cost an additional fee.

Tours: Self-guided for main hall; guided tours only for tour of area and research facility.

Public transportation: None.

Handicapped accessible: Yes.

Food served on premises: None. Restaurants are within a few miles of the site, primarily along Route 85.

Website: www.sanfordlabhomestake.com

7

Southwest

Arizona

Lowell Observatory

Lowell Observatory, high atop a mesa near downtown Flagstaff, is the largest privately operated, nonprofit, astronomical observatory in the world, and one of the oldest observatories in the southwestern United States. The site, which is open to visitors, was founded in 1894 by Percival Lowell, a businessman, author, scientist, and member of the illustrious Lowell family of Boston.

Dr. Lowell was intrigued by the notion that intelligent life might live in outer space. He was particularly interested in possible life on Mars, and set up this institution to explore that planet in detail. The site, selected because of its clear skies and good viewing conditions, became known as Mars Hill. Lowell's ideas about extraterrestrial life have probably contributed more to the popular and scientific interest in this subject than anything else.

In 1896, a specially ordered 24-inch (61-cm) Alvan Clark refracting telescope was put in place on the mesa. The research here quickly expanded in several directions. For example, Vesto M. Slipher, using the Clark instrument, made the first measurements of the red-shift recessional velocities of external galaxies. Today, after well over a century of service as a research tool, the Clark telescope (still a working instrument) is registered as a National Historic Landmark. Slipher's research with the Clark telescope was a partial basis for the work that led Edwin Hubble to conclude that the universe is expanding.

In 1902, Lowell predicted the existence of a ninth planet—called Planet X—beyond Neptune. The search for Planet X continued at the observatory until 1930, when Clyde Tombaugh discovered Pluto, the only planet first detected by an observatory in the United States. (Pluto is now classified as a dwarf planet by the International Astronomical Union.)

This institution has made many important astronomical contributions, including discovering the rings of Uranus and the hazy atmosphere on Pluto. The first discovery of atmospheric water in the atmosphere of a planet orbiting another star was made here. While working at the observatory, Dr. Henry Giclas discovered many white dwarf stars. Early measurements of the rotation of external galaxies gave him evidence that these galaxies are huge star systems much like our own Milky Way. Another discovery that was made here—the existence of transparent infrared optical windows—changed the way astronomical studies are conducted. Scientists now use these windows routinely for studying celestial bodies.

Today the complex has a number of telescopes, ranging from the historic Clark refractor to the state-of-the-art 4.3-m Discovery Channel Telescope, 40 miles (64 km) away in Happy Jack. Another telescope is at the Perth Observatory in Western Australia. Currently research is conducted on comets and asteroids, as well as exoplanetary

Lowell Observatory's 24-inch (61 cm) Clark Refractor.
Courtesy of Sarah Gilbert.

systems. Also under investigation are the evolution of massive stars, extragalactic star formation, and the brightness variability of stars.

Percival Lowell, in his 1906 book *Mars and Its Canals*, wrote: "To set forth science in a popular, that is, in a generally understandable form, is as obligatory as to present it in a more technical manner. If men are to benefit by it, it must be expressed to their comprehension.... The whole object of science is to synthesize, and so simplify...." In keeping with this philosophy, the observatory offers an extensive educational program for the public that is run from the Steele Visitor Center.

The Rotunda Museum, part of the Slipher Building and built in 1916, is where the observatory library once was. It displays historically important artifacts related to the Lowell Observatory such as Pluto-related instruments and devices V. M. Slipher used in his studies of recessional velocity.

Both self-guided and guided tours of the observatory begin with a brief show about Percival Lowell and the facility's history and current research. The guided tour then stops at three historic sites: the Clark dome with the 24-inch (61-cm) Clark refractor, the renovated Pluto telescope and dome, and the Rotunda Museum. In the evening—weather permitting—stellar observations are held using the Clark refractor.

Other interesting sites on the grounds include Lowell's mausoleum and a unique outdoor exhibit called the Pluto Walk, which illustrates the planets' sizes and distances in the solar system.

The Putnam Collection Center is an 8000-square-foot (740 m²) archives and research library, whose collection includes photographic plates, antique instruments, and historic papers. Access to the collection is for scholars only, but the public lobby displays highlights from the archives, including the V. M. Slipher spectrograph, Mars globes from 1894–1916, scientific books, Lowell's telescope from his teenage years, and Lowell's 1911 Stevens-Duryea Model "Y" car.

Address: Lowell Observatory, 1400 West Mars Hill Road, Flagstaff, AZ 86001.
Phone: (928) 774-3358.
Days and hours: Mon.–Sat. 10 a.m.–10 p.m., Sun. 10 a.m.–5 p.m.
Fee: Yes.
Tours: Self-guided and guided.
Public transportation: Taxi from downtown Flagstaff.
Handicapped accessible: Mostly. The Pluto Discovery Telescope is not.
Food served on premises: Snacks.
Special Note: Check the website or call for schedules of the variety of programs and talks offered daily.
Website: lowell.edu

Sunset Crater Volcano National Monument and Wupatki National Monument

On the dry, parched land northeast of Flagstaff are two national monuments with a common bond—the effects left of their landscape by a now inactive volcano. Sunset Crater Volcano National Monument has a 1000-foot (300-m) high cone of accumulated debris around a lava vent, as well as many volcanic formations. Wupatki National Monument is a group of ruins once inhabited by Native Americans. Each monument has parklike grounds for driving or hiking through, and a visitor center with exhibits on archaeology, anthropology, and the area's flora and fauna.

The world's largest ponderosa pine forest covers the area round Flagstaff, including parts of these two monuments. The ponderosa pine's twisted wood fibers give it the strength to withstand gusty winds, although others suggest that twisting offers resistance to heavy snow, or that it helps during competition for scarce water. It is even possible that only the dead, dried wood is twisted, i.e., the trees don't twist while living.

Also found here are the Apache plume, a member of the rose family; mullein, a velvet-leaved plant introduced by the European settlers; and rabbitbrush, a shrub often nibbled by browsing animals. Also of note are the white-barked aspens that stand out against the black lava flows. The piñon pine trees are slow growers: a tree with a trunk only 7–10 inches (18–25 cm) in diameter may be as much as 150 years old. Among wildlife inhabiting the region are Steller's jays, Abert's squirrels, eastern fence lizards, mule deer, bobcats, Clark's nutcrackers, and gopher snakes.

Sunset Crater Volcano National Monument

Sunset Crater is part of the San Francisco volcanic field that extends 2000 square miles (5200 km²) across the southwestern Colorado Plateau. This field of cones and lava flows was created during six million years of volcanic activity. The eruptions that caused Sunset Crater began around 1085 c.e. and lasted less than one year, for the lava flows present no evidence of the spring and fall seasonal winds, and that cinder-cone volcanoes are always short events. The Kana-a flow and the Bonito flow, which took place in 1150 c.e. and 1220 c.e., respectively, destroyed all life in their paths. Evidence of the Bonito flow consists of a jumble of jagged blocks and sharp, glassy rock called clinkers. This type of lava formation is known by the Hawaiian name 'a'ā (pronounced ah-ah) (See Hawai'i Volcanoes National Park, Hawai'i).

The volcano ejected rocks into the air and spewed them across the countryside. Cinders fell around the vent in the volcano, building a cone, and cinders covered 800 square miles (2000 km²). Light ash was more widely distributed. Iron oxide in the cinders gives the cone rim a permanent red glow.

John Wesley Powell, a Civil War soldier and later a geologist who became head of the US Geological Survey, explored the area in 1885. Powell was struck by the contrast of colors and the way, from a distance, that the red cinders on the cone seem on fire. Based on his description, the cone is called Sunset Peak.

At the base of the cone visitors can see the entrance to a 225-foot (68.6-m) long ice cave that is a lava tube, through which lava once flowed. Although the cave contains ice most of the year, it usually melts by August. Entry is prohibited to visitors because it began collapsing in the 1980s. Sunset Crater Volcano Visitor Center offers an active seismology station; the museum exhibits discuss the local geology. No hiking is permitted on Sunset Crater, but climbing is permitted on other volcanoes nearby.

The trails are open at night, so visitors can see the clear sky. The monument has certification as an International Dark Sky Park.

Address: Sunset Crater Volcano National Monument, 6082 Sunset Crater Road, Flagstaff AZ, 86004. GPS Coordinates: 35°22′09.0″N 111°32′36.6″W or 35.369162, -111.543507.
Phone: (928) 526-0502.
Days and hours: Trails are open 24 hours a day. Visitor Center is open daily 9 a.m.–5 p.m.; closed 12/25, and for inclement weather.
Fee: Yes.
Tours: Self-guided and guided. Check the website or call for schedules of the variety of programs and talks offered daily. Many ranger-led hikes are strenuous. Call or see the website for details.
Public transportation: None. Commercial tours visit from Flagstaff.
Handicapped accessible: Partly. Some hikes and tours are strenuous over rugged terrain.
Food served on premises: None. Restaurants are in Flagstaff, about 1 hour away.
Special note: GPS may be unreliable; the National Park Service recommends that you avoid unpaved unfamiliar roads. Visitors may bring telescopes to view the night sky.
Website: www.nps.gov/sucr/index.htm

Wupatki National Monument

North of Sunset Crater, within Wupatki National Monument, are more than 2600 prehistoric Anasazi and Sinagua sites, historic Navajo sites, and unique geologic features.

Because of volcanic eruptions, the Sinaguas—their name, given by the Spanish explorers, means "without water"—left the area for more than 400 years but eventually returned. The Anasazi's Kayenta and Cohonina tribes also migrated to the Wupatki area and found that the thin layer of ash left from volcanic activity extended and improved the growing season by absorbing moisture and conserving heat. There are indications that the climate changed and water became more plentiful. By 1225 the people of Wupatki left the area, and the pueblos were never again permanently inhabited.

A paved, 34-mile (55-km) loop road cuts through Wupatki National Monument's 56 square miles (145 km²). Self-guided walking trails pass ruins of five Native American communities, including the impressive Wupakti Pueblo, built in the 12th century. It has an amphitheater, a ball court, and 100 rooms that housed as many as 200 people. In some places the pueblo may have been two stories high. This sophisticated small city was likely on a major prehistoric trade route between areas to the north and east and as far away as present-day Mexico.

Signs along trails explain design and engineering techniques the Native Americans used to build this site and others in the park. Growth of trees that survived the 1064 eruption was stunted for several years. These trees, used in construction of the Wupakti Pueblo, enabled dendrochronologists to date the eruption accurately. The region's varied ecology incudes formation of Moenkopi sandstone, the prevalent red rock found throughout the monument, as well as gravel deposits, black lava flows, and basaltic lava flows. The near-vertical Doney Cliffs have a step-like fold known as a faulted monocline.

Address: Wupatki Visitor Center, 25137 North Wupatki Loop Road, Flagstaff AZ, 86004.
Phone: (928) 679-2365 or (928) 856-1705.
Days and hours: Scenic drive is open daily, 24 hours a day. Wukoki, Lomaki, and Citadel trails are open sunrise to sunset. Visitor Center is open daily 9 a.m.–5 p.m.; closed 12/25.
Fee: Yes.
Tours: Self-guided and guided. Check the website or call for schedules of the variety of programs and talks offered daily.
Public transportation: None. Commercial tours visit from Flagstaff.
Handicapped accessible: Partly. Some hikes and tours are strenuous over rugged terrain.
Food served on premises: Snacks and picnic tables only. Restaurants are in Flagstaff, about 1 hour away.
Special note: GPS may be unreliable; the National Park Service recommends that you avoid unpaved unfamiliar roads. Many ranger-led hikes are strenuous. Call or see the website for details.
Website: www.nps.gov/wupa/index.htm

GRAND CANYON

Petrified Forest National Park

Within the Petrified Forest National Park there are two districts: the Painted Desert at the north end and Rainbow Forest to the south. Petrified Forest National Park is located on 93 000 acres (38 000 hectares) of high, dry tableland that was once a huge floodplain.

Rainbow Forest Museum houses the park's geological and paleontological displays. Most of the area's exposed rock is sedimentary. Hills here are composed of clay and volcanic ash that were deposited on the floodplain between the ancient stream channels. Mesas above the clay hills are capped by layers of hard sandstone. This clay can absorb up to seven times its weight in water.

During the time of the floodplain, animals living here included crocodile-like reptiles, giant amphibians, and small dinosaurs. Tall, extinct trees—*Araucarioxylon*, *Schilderia*, and *Woodworthia*—as well as ferns and cycads provided foliage and cover for these animals. Trees growing in swamps and along streams fell into the waterways and were buried by sediments faster than they could decompose. The overlying sediment contained volcanic ash. Groundwater dissolved the silica from the volcanic ash, and the silica solution was transported by groundwater through the logs and replaced the wood cells, crystallizing as the mineral quartz.

Visitors can see samples of petrified logs at several places in the park. (See Gingko Petrified Forest, Vantage, Washington.) One of these is at Blue Mesa, where petrified logs rest on top of the sandstone caprock and erode out of the clay hills. The blue-gray color in the clay comes from manganese, and the pale red is caused by iron oxides. The most prominent pedestal log can be seen here, a petrified log supported by a thin neck of clay.

Semiprecious smoky quartz, clear quartz, and purple amethyst developed in the logs at Crystal Forest. Unfortunately most of the logs have been broken into pieces by profit-seekers from before the national park was established up through today.

The largest concentration of petrified wood and some of the longest petrified logs are at Long Logs Loop trail. All are of the extinct conifer *Araucarioxylon arizonium*, which lived in the area's Triassic Period swamps, slow streams, and low-lying hills.

The Giant Logs area contains massive petrified logs of many colors, which may have been created by deposits of iron and manganese oxide. The logs were broken by shifting earth as this land uplifted about 70 million years ago to form the Colorado Plateau. From the Jasper Forest Overlook visitors can see large quantities of petrified wood—with root systems still attached—that fell off the cliff and now rest at the base of the hill.

The beautiful Painted Desert landscape is awash with colors—red, orange, pink, and gray—as if from an artist's palette. The crusty soil, called Chinle Formation, consists of sandstone, conglomerate, and clay. Most of the sand and silt of these ancient formations was carried here by streams and rivers. As the clays eroded over the years, they revealed fossils and petrified logs buried beneath the surface.

Fossils of dinosaurs, reptiles, and plants of the late Triassic Period are on exhibit at the Painted Desert Visitor Center and the Rainbow Forest Museum. The fossil record at Petrified Forest National Park reveals so much information it has become an important world reference standard for the Late Triassic ecosystem. The species found in the fossil remains are typical of the Earth's tropic regions. Fossil ferns and scouring rushes, some growing as high as 30 feet (9 m) tall, are found throughout the Chinle Formation.

Among the ancient fish revealed in the formations is Chinlea, which is related to the coelacanth, a species of fish discovered off the coast of Africa in the late 1930s. Until the African discovery, the coelacanth was believed to be extinct. Another fish fossil unearthed in the Chinle Formation is the *Ceratodus*, a lungfish that may be related to modern lungfish found in the waters off the coast of Australia. The most plentiful fossil remains in the Petrified Forest belong to the giant reptile phytosaur. This 30-foot (9-m) long creature had numerous huge, pointed teeth set in a long snout.

There are indications that prehistoric people lived in the area for thousands of years, leaving a remarkable record. Visitors reach Agate House, an Ancestral Pueblo People dwelling, by following a trail through the petrified wood deposits. The Puebloans probably occupied the eight-room Agate House between 1100 and 1300 C.E. It is built entirely of colorful petrified wood and sealed with adobe. About 15 miles (24 km) from the Agate House is a 100-room pueblo with kivas (religious ceremonial rooms) built near the Puerco River before 1400 C.E. and open to the public. The prehistoric peoples also left petroglyphs, made by chipping through the dark mineral stain known as desert varnish to form a design on the lighter undersurface of the rock. Petroglyphs come into view in several spots in the park.

Most of the area's 9 inches (23 cm) of annual rainfall occur during summer, when heavy thunderstorms are common. Temperatures during the course of the year can vary greatly. Most of the vegetation is composed of grasses and brushes such as sagebrush and some cactus such as yucca. Among mammals that roam the park are the desert cottontail, jack rabbit, pronghorn, kangaroo rat, squirrel, coyote, prairie dog, badger, and porcupine. Resident birds include the red-tailed hawk, American kestrel (a small falcon), and golden eagle.

GPS coordinates: Painted Desert Visitor Center and Park Headquarters, Latitude 35.06543746738773; Longitude -109.78153824806213; Rainbow Forest Museum, Latitude 34.81517743163217, Longitude -109.86576497554779; Painted Desert Inn National Historic Landmark, Latitude 35.08343319608185, Longitude -109.78861391544342.
Phone: (928) 524-6228.
Days and hours: Daily. Park Road and Visitor Centers: 7 a.m.–6 p.m.; Painted Desert Inn National Historic Landmark: 9 a.m.–4 p.m.
Fee: Yes.
Tours: Self-guided and guided. Check the website or call for daily programs and ranger-led events.
Public transportation: None.
Handicapped accessible: Partial. Many tours are not handicapped accessible.
Food served on premises: Diner and snacks.
Website: www.nps.gov/pefo/index.htm

GRAND CANYON, ARIZONA; SPRINGDALE, UTAH; AND BRYCE, UTAH

Grand Canyon, Zion, and Bryce Canyon National Parks

These three parks on the Colorado Plateau, within 120 miles (190 km) of each other, have a geologic history inextricably intertwined. They are part of what is known as the Grand Staircase. Although the Grand Canyon National Park is in Arizona, Zion and Bryce Canyon National Parks are in southern Utah.

The Grand Staircase, a land formation created by geologic changes over the millenia, consists of a series of cliffs that retreat to the north, exposing rock layers formed by deposits of mud, silt, sand, and lava. These layers were uplifted more than a mile (1.6 km) above sea level, tilted sideways, and then eroded by wind, rain, and water. Each layer has its own distinct physical characteristics and color.

Grand Canyon National Park

The bottom step, the oldest in the Grand Staircase, is exposed in the Grand Canyon. It is the best known of the land formations because so many layers and such a broad expanse of rock are revealed along the canyon walls. The Colorado Plateau resulted from the upheaval of land about 65 million years ago. The Colorado River started carving this magnificent canyon about four to five million years ago. As time passed, it cut through layers of Kaibab limestone, sedimentary rock laid down as recently as 250 million years ago. The word *kaibab*, used to describe the Grand Canyon by the Native American Paiute people, means "mountain lying down." Eventually the river began cutting into the 1.7 billion-year-old Vishnu schist.

The Grand Canyon is up to 18 miles (29 km) wide in parts, and has an average depth of 1 mile (1.6 km). Although the canyon's average width is 10 miles (16 km), it takes five hours to drive from the north rim to the south rim because the highway skirts the rim for about 220 miles (350 km). The north rim, 1000 feet (300 m) higher than the south rim, has more annual precipitation and a cooler climate. The interior of the canyon, called the inner canyon, can be reached only by hiking on foot, taking a mule ride, or rafting down the Colorado River. In summer, the inner canyon's climate is desert-like and temperatures can exceed 100°F (38°C).

Of the seven climatic belts recognized in the world, six are represented in the Grand Canyon region, ranging from desert to arctic-alpine. The plant and animal life is equally varied. Because the north and south rims are separated by such distance, different life forms developed on each. For example, a tassel-eared squirrel known as the Abert's squirrel is found only on the south rim; another type of tassel-eared squirrel, the Kaibab squirrel, lives on the north rim.

This park's many museums and exhibits help explain the canyon's features and history. A museum at the Visitor Center at Grand Canyon Village has more than 200 000 objects representing such areas as archaeology, ethnology, biology, paleontology, and geology—only a small portion of which are on display. Yavapai Museum has geology exhibits. Tusayan

Bonito Lava flow at Sunset Crater Volcano.
Courtesy of National Park Service/ M. Ullman.

Museum and Ruins, in Desert View, has exhibits on the prehistoric ancestral Puebloan people and present-day tribes. Near the museum a self-guided trail leads to an 800-year-old ancestral Puebloan ruin. The Grand Canyon has had human inhabitants for at least 10 000 years; in the inner canyon the village of Supai is home to Havasupai Native Americans.

Addresses: North Rim Visitor Center, GPS coordinates 36°11'51"N 112°03'09"W. South Rim Visitor Center, 20 South Entrance Road, Grand Canyon, AZ; GPS coordinates 36°03'32"N 112°06'33"W.
Phone: (928) 638-7888.
Days and hours: South Rim is open daily, 24 hours a day. North Rim is open May 15–Oct. 15. Various visitor centers and museums in the park have more limited hours.
Fee: Yes.
Tours: Self-guided and guided tours. Check daily schedules for interpretive ranger talks and hikes.
Public transportation: Commercial bus tours run to Grand Canyon Village from Flagstaff and Las Vegas. The Trans-Canyon Shuttle runs between the North and South Rims. Grand Canyon Railway runs to the Grand Canyon Village from Williams. Within the Grand Canyon Village, visitors can ride the park's shuttle buses.
Handicapped accessible: Partly. Park shuttle buses and paved rim trail on the South Rim are wheelchair-accessible. Certain areas are not handicapped accessible; many sites were built before accessibility rules were created. Check the website for details.
Food served on premises: Restaurants and cafés at both South and North Rims.
Website: www.nps.gov/grca/index.htm

Zion National Park
Zion National Park, the middle step in the Grand Staircase, was originally called Mukuntuweap Canyon by the Native Americans, but the Native American name, unpopular with European settlers, was changed in 1919 to Zion, an abbreviation of "the Gates of Zion." This park is famous for its 2000-foot (600 m) vertical canyon walls of sandstone, which, in some areas, are separated by only 18 feet (5.5 m). These cliffs are among the tallest in the Grand Staircase.

Here the land formations reveal layers younger than those in Grand Canyon National Park. Rocks range from the Moenkopi formation of the Triassic period up to the Dakota sandstone formation of the Cretaceous period. Zion canyon, the largest and most visited canyon in the park, was formed by the endless cutting of the Virgin River. The oldest rock in the region, Kaibab limestone of the Permian era, is the youngest rock in the Grand Canyon. The various layers of sandstone, shale, limestone, and gypsum create beautiful colored bands.

Zion is the only one of the three Grand Staircase parks where most of the sightseeing, hiking, nature walks, and other activities take place on the canyon floor. At the Visitor Center are exhibits on the geology, animal life, and plant life found in the park.

Because of Zion's changes in elevation and multiple landforms, it too has climate zones that range from sun-drenched desert to alpine forest. The park contains a desert swamp, a petrified forest, a waterfall, more than 60 kinds of mammals, 270 kinds of birds, 13 kinds of lizards, 12 species of snakes, and 800 species of native plants. Among the park's wildlife are mule deer, roadrunners, golden eagles, mountain lions, and Gambel's quail.

Along the hiking trails, visitors are likely to see wildflowers such as columbines, scarlet monkeyflower, and maidenhair flower.

The Zion–Mt. Carmel highway, a major engineering feat of the 1930s, connects the lower canyon with the high plateau to the east and has two tunnels, one more than a mile (1.6 km) long, that were drilled and blasted through the cliffs.

Address: Zion National Park, 1 Zion Park Boulevard, State Route 9, Springdale, UT 84767.
Phone: (435) 772-3256.
Days and hours: Daily, 24 hours a day. Visitor Centers have more limited hours; call or see the website.
Fee: Yes.
Tours: Self-guided and guided tours. Check daily schedules for interpretive ranger talks and hikes.
Public transportation: None.
Handicapped accessible: Partly. Certain tours are not handicapped accessible; some trails have steep grades and rocky parts. Check the website for details.
Food served on premises: Restaurant and café.
Website: www.nps.gov/zion/index.htm

Bryce Canyon National Park
Bryce Canyon, the youngest and top step in the staircase, has at its base the Carmel Formation, into which the Paria River continues to cut. The top layer is the colorful Clarion Formation. The cliffs, once over 9000 feet (2700 m) high, have been eroded over

Cape Royal, Grand Canyon.
Courtesy of Grand Canyon National Park.

the millennia to their current 2000 feet (600 m). Despite its name, Bryce Canyon is not a canyon at all, but rather a series of amphitheaters carved by erosion.

The red rock spires, called hoodoos, are the most distinguishing land feature of Bryce Canyon. The spires are formed as rain and ice erode small pieces of the limestone cliffs, leaving tall, thin columns of rock. Their beautiful red and orange colors are from iron in the rock, and the purple and lavender are from manganese. The Paiute name for the area translates to "red rocks standing like men in a bowl-shaped canyon." "A hell of a place to lose a cow!" was the comment made by Mormon pioneer Ebenezer Bryce, who lived here and gave this place its name.

The park's forest areas contain three life zones: Upper Sonoran with piñon pine and juniper; the Transition Zone with ponderosa pine; and the Canadian Zone with dense fir, spruce, and aspen. Endangered peregrine falcons and threatened Utah prairie dogs live in the park, as do coyote, mule deer, mountain lions, Steller's jays, and ground squirrels. The Visitor Center has exhibits about the park's geology, animals, and plants.

Address: Bryce Canyon National Park, UT-63, Bryce, UT 84764. GPS coordinates for the Visitor Center: N37°38′24″ / W112°10′12″.
Phone: (435) 834-5322.
Days and hours: Daily, 24 hours a day. Visitor Center: May–Sep.: daily 8 a.m.–8 p.m.; Apr. and Oct.: daily 8 a.m.–6 p.m.; Nov.–Mar.: daily 8 a.m.–4:30 p.m.
Fee: Yes.
Tours: Self-guided and guided tours. Check daily schedules for interpretive ranger talks and hikes.
Public transportation: None.
Handicapped accessible: Partly. Certain tours are not handicapped accessible; some trails have steep grades and rocky parts. Check the website for details.
Food served on premises: Restaurant and café.
Website: www.nps.gov/brca/index.htm

HOOVER DAM

Hoover Dam

See Hoover Dam, Hoover Dam, Nevada.

ORACLE

University of Arizona Biosphere 2

Biosphere 2 might be considered one of the largest earth-science experiments ever attempted, and the engineering used to build this structure, covering 3.14 acres (1.27 hectares), deserves immense respect. The complex, built from 1987 to 1991 and enclosing 7.2 million cubic feet (204 000 m³), was originally planned to show how a closed ecological system could be replicated, and operate in an isolated environment such as outer space. The project's early goals called for the design of an enclosed facility with simulated ecosystems that could be occupied by human beings to demonstrate the interconnectedness of people and the environment. The name "Biosphere 2" is derived from being the second closed biosphere, because the Earth is the first "biosphere."

The structure is of steel tubing and 6500 windows of special glass with steel frames, designed with almost no atmospheric leakage (less than 10% annually), plus a 500-ton (450-tonne) stainless-steel floor liner to be sure of a closed ecosystem. With a leak-tight construction, alternating daytime solar heating and nighttime contraction meant a compensating mechanism was needed to avoid crushing variable pressures. Large steel-and-rubber diaphragms, nicknamed "lungs," were installed underneath the structure to allow variable volume inside.

Various crops were planted inside the structure, and fauna stocked included pygmy goats, chickens, dwarf pigs, and tilapia fish. Different biomes were installed: a fog desert area, a rainforest, an artificial ocean, and a mangrove swamp.

Two missions with crews of human beings were run, from 1991 to 1993, and for a few months in 1994. In terms of engineering, the experiment succeeded; as an analog to Earth's ecosystem, the experiment failed. A variety of problems ensued with each mission. Despite the crops being the most efficiently raised anywhere globally, the crews reported constant hunger from the low-calorie diet, and their metabolisms became more efficient. Trees were weak because of the lack of wind strengthening their limbs. Greater and more rapid cycles in the biomes occurred because of their small size. Animals tended to do poorly—except cockroaches and ants. Carbon dioxide built up, and a 0.25%-per-month decline in oxygen levels was measured, causing sleep apnea and fatigue. The extra carbon dioxide began reacting with concrete in the base of the complex, sequestering carbon and oxygen during formation of calcium carbonate. But perhaps most famously, interpersonal conflict caused serious factionalism among the crews.

After the two runs in the 1990s, Columbia University oversaw the campus, but it was eventually acquired by the University of Arizona in 2007, where various earth-science research projects still are underway. One major geological study is the Landscape Evolutionary Observatory, tracking how changes in the structure of soil influence the movement of water and plant nutrients. In addition, the study investigates how much mountain precipitation ends up downstream for human use, and how the quality of the water is influenced. A second, biological study is the Lunar Greenhouse, learning how to grow edible plants on Mars in a bioregenerative life-support system. Other projects also focus on sustainability, climate, and water. Large-scale ocean acidification was first demonstrated here.

Visitors can tour the complex and see the variety of habitats, both human and "natural." A marked boardwalk and trail crosses through much of the structure, both under glass, and underground, known as the Technosphere. Guides relate the history and show the living quarters, describe the engineering involved, and some of the pitfalls researchers encountered. Outside the giant building are wonderful desert vistas of the local sagebrush-covered hills in the Sonoran Desert.

Address: University of Arizona Biosphere 2, 32540 South Biosphere Road, Oracle, AZ 85623.
Phone: (520) 621-4800.
Days and hours: Daily 9 a.m.–4 p.m. Closed Thanksgiving and 12/25.
Fee: Yes.
Tours: Self-guided in Exhibit Center; guided tours only in complex.
Public transportation: None.
Handicapped accessible: Partly. Much of the complex has steep terrain or stairs.

Food served on premises: Café.
Special Note: Wear comfortable walking shoes.
Website: biosphere2.org

PHOENIX

Desert Botanical Garden

Desert Botanical Garden is part of a 1200-acre (490 hectare) city park called Papago Park. Conservation of the world's desert plants is a focal point of this 140-acre (57-hectare) garden museum, located deep within the Sonoran Desert. It displays more than 55 000 plants of nearly 4400 species on 55 developed acres (22 hectares) and is one of very few botanical gardens to meet the American Alliance of Museums' accreditation standards.

This garden certainly puts to rest the notion that a desert is a barren wasteland. On self-guided or guided tours, visitors can see a wide range of plants as well as exhibits explaining how plants, animals, and people adapt to the desert environment. The garden sometimes runs a popular self-guided Flashlight Tour that reveals night-blooming plants and small nocturnal animals.

The garden's Living Collection is made of 133 plant families. Succulent species are especially well represented in the collection, which contains 43% Cactaceae, or cacti, and 17% Agavaceae (fleshy-leaved tropical plants). Some of the nation's oldest plantings of cacti, agaves, and other desert plants line the Historical Garden Trails, which cover 20 acres (8 hectares).

The Desert Discovery Loop Trail takes about an hour to navigate. Radiating off the main trail are thematic walks that highlight examples of botanical taxa, growth forms,

Desert Botanical Garden.
Courtesy of Desert Botanical Garden.

and geographic origins. Along the trails are touch carts that contain plant materials to handle and examine. These carts are staffed by docents who explain the themes and significance of the plantings.

Plants & People of the Sonoran Desert Loop Trail is an interesting three-acre (1.2-hectare) display with a trail that covers one-third of a mile (.5 km). Visitors see five re-created Sonoran Desert habitats, along with re-created historic and prehistoric structures. The habitats include an upland chaparral, a saguaro forest, and a mesquite thicket. These displays examine ways that human beings adapted and contributed to the desert environment over the centuries and how they used native plants for food, fiber, medicine, and construction.

The Center for Desert Living is a shady area designed to give home gardeners ideas for creating landscapes and lush arrangements of plants in the desert. Desert plants that take little water to survive are promoted in the landscape exhibits. Information is available on how to use water-saving strategies and irrigation techniques. (See San Antonio Botanical Garden, San Antonio, Texas.) As visitors tour the backyard landscapes, patios, and raised vegetable garden beds, interpretive signs help them understand what they see.

The Sonoran Desert Nature Loop Trail exemplifies the basic ecology of the Sonoran Desert. Signs and sample plant-settings illustrate the desert's geography, wildlife habitats, and plant relationships. This trail offers a panoramic view of Phoenix, which surrounds the park.

Among other areas to explore is the Harriet K. Maxwell Desert Wildflower exhibit, especially beautiful in the spring, and the seven-acre (2.8-hectare) garden of Desert Plants. Because of the garden's natural desert habitat, many small animals live here, including jackrabbits, squirrels, desert tortoises, and native and migratory birds, from tiny hummingbirds to soaring desert hawks.

The botanical garden is also renowned for desert ecology conservation, with research done at the Hazel Hare Center for Plant Science. The Herbarium Collection, started in 1953, today has 83 000 specimens, primarily representing flora of the southwestern United States and Mexico. The Schilling Library is an archival collection of over 9000 books and 500 journals dealing with cacti, desert gardening, natural history, ethnobotany, botanical art, and wildlife. Special archives for botanists Edward F. Anderson and Lyman Benson are located here. For access to these special collections, advance notice is required.

Address: Desert Botanical Garden, 1201 North Galvin Parkway, Phoenix, AZ 85008.
Phone: (480) 941-1225.
Days and hours: Oct.–Apr.: 8 a.m.–8 p.m.; May–Sep.: 7 a.m.–8 p.m. Closed 7/4, Thanksgiving, and 12/25.
Fee: Yes.
Tours: Self-guided and guided tours. See the website for a calendar of events.
Public transportation: Buses run from downtown Phoenix to 0.2 miles (0.3 km) from the garden entrance.
Handicapped accessible: Yes. Wheelchairs and electric scooters can be rented on a first-come, first-served basis.
Food served on premises: Restaurant and seasonal café.
Special note: Bring hats, sunglasses, sunscreen, water, and comfortable shoes for hiking in the hot sun.
Website: www.dbg.org

Musical Instrument Museum

Robert Ulrich, the chairman of Target Corporation, visited the Musical Instrument Museum in Brussels, and decided to create a similar institution in the United States. Located on the northern outskirts of Phoenix, this massive modern structure faced with sandstone incorporates two floors and 200 000 square feet (1900 m^2) of exhibit space. The architecture is designed to evoke a canyon in Arizona. The museum, which opened in 2010, claims to be the largest museum of musical instruments in the world. It has nearly 14 000 musical instruments from practically every country across the planet and of many eras, of which about half are on display. Visitors use wireless headphones, which automatically activate videos demonstrating how each instrument is played as the viewer draws near. The idea that music is not just an art, but also a technology, is amply demonstrated in this vast facility.

There are a number of exhibits portraying how certain instruments such as the violin and piano are manufactured. On view is one of very few octobasses, so tall (11.4 ft.; 3.48 m) that the three wound metal strings (tuned to C_0, G_0, D_1) have to be played using a lever-operated system. The oldest musical instrument is a Chinese drum from the 5th millennium B.C.E.

The museum is organized largely according to geographic areas of musical instruments, divided into five regions: Africa and Middle East, Asia and Oceania, Europe, Latin America, and United States/Canada. In this last region, there are special exhibits concerning important American musical-instrument manufacturers such as Steinway & Sons and the Martin Guitar Company. Larger regions like Russia, China, and India are subdivided. There are areas devoted to East Asia, South Asia, Southeast Asia, Oceania, and Central Asia and the Caucasus. The United States/Canada section also is divided

Musical Instrument Museum.
Courtesy of Musical Instrument Museum

into various styles, such as Appalachian dulcimer music, New Orleans jazz, hip-hop, and college marching bands.

In the geographic galleries are displayed an entire Indonesian gamelan (traditional instrumental ensemble) and the largest playable sousaphone. Commonalities of instruments are clear: guitar-like instruments exist throughout the world. Even examples of electronic instruments such as the MiniMoog from the 1970s, a circa 1929 AR-1264 Theremin from RCA, and a Roland TR-808 rhythm machine from the early 1980s are on display.

Mechanical musical instruments, those that play themselves, have a gallery of their own, including player pianos, cylinder music boxes, and mechanical zithers. A unique instrument here is the 25-foot (7.6-m) wide "Apollonia" Orchestrion built in 1926 in Antwerp and rebuilt in 1950, which is demonstrated regularly throughout the day. An Experience Gallery allows visitors to try out certain instruments for themselves, such as drums, a gong, stringed instruments, and a theremin. (Wind instruments are not provided for hygienic reasons).

A Conservation Lab demonstrates how instruments in the museum's collection are preserved and restored. A STEM gallery explores connections between music and science, including physics of sound, how the human ear works, and hearing safety.

There are also changing exhibits, and a concert hall for seating 299 listeners. A separate gallery is devoted to important musicians such as Pablo Casals, John Lennon, and Elvis Presley, including memorabilia, instruments, outfits, and videos of concerts.

Address: Musical Instrument Museum, 4725 East Mayo Boulevard, Phoenix, AZ 85050.
Phone: (480) 478-6000.
Days and hours: Daily 9 a.m.–5 p.m.
Fee: Yes. Extra fee for behind-the-scenes tours.
Tours: Self-guided and guided.
Public transportation: Shuttle bus is available from the Mountain View Transit Center.
Handicapped accessible: Yes.
Food served on premises: Café.
Website: www.mim.org

Phoenix Zoo

When Robert E. Maytag, grandson of the founder of the appliance-producing Maytag Company, moved to Phoenix with his wife Nancy and their family, they found the city's cultural development lagging behind its population and commercial growth. The Maytags decided to do something about this situation and, because of their interest in wildlife, put their efforts into developing a zoo. In 1962 the Maytag Zoo opened, and changed its name the following year to the Phoenix Zoo.

Today this facility of 125 acres (51 hectares) in Papago Park houses more than 3000 animals, primarily in naturalistic outdoor exhibits. The animals selected for display must be able to adjust well to the warm, dry climate. Thirty species of all those on display are endangered or threatened.

This institution sees as its role the study and advancement of education, recreation, and science. Through its innovative exhibits, the zoo seeks to enhance visitors' appreciation of nature and teach people about the important role of animals in the earth's ecology.

The zoo's 6-acre (2.4-hectare) Arizona Trail is an immersion exhibit in which about 100 animals, including the bobcat and collared peccary, live in natural habitats. The trail has signs and graphics explaining the plant and animal life, local geography, native plant adaptations, and human effect on the environment. It emphasizes that the beauty of the desert is fragile and must be protected if we are to continue to enjoy it.

Throughout the zoo, signs denote animals on the endangered list and those participating in the Association of Zoos and Aquariums' (AZA) Species Survival Plan® (SSP) for preserving and breeding endangered species. Animal footprints are embedded in the walkways and illustrated on the exhibit signs throughout the zoo. Once-thriving Mexican wolves live in the Arizona Trail area, one of the very few places in the world where these animals are exhibited. The zoo has an on-site breeding program designed to help this species survive.

Africa Trail showcases animals from Africa, such as Grévy's zebras, cheetahs, and white rhinoceros. Within this area is an extensive exhibit created, in part, for the study of baboons, which are endangered animals. The exhibit houses troops of hamadryas baboons and mandrills, and resembles the natural habitat in which they would normally live. Only a thick plate-glass window separates the public from the baboons, so visitors can view them nose-to-nose.

Hamadryas baboons have a living range that extends from the middle of Africa into the Middle East. They inhabit rocky deserts that have some grass, and eat all kinds of grasses, seeds, roots, and bulbs. Mandrills, an animal on the endangered species list, are large, fierce baboons that roam the rain forests of west central Africa, and eat almost anything they find, from fruits and roots to insects and small mammals.

As part of its conservation efforts, the zoo has a trail called Desert Lives, which houses Arabian oryx—a highly endangered animal that is part of the AZA's SSP

Phoenix Zoo.
Courtesy of the Phoenix Zoo.

program. The last 11 in the world were brought to the zoo in the 1960s, where a breeding program has been underway since. Most Arabian oryx around the globe are descended from the Phoenix Zoo's original group. The other animal of note in Desert Lives is the Arizona bighorn sheep: its natural habitat of buttes is incorporated within the zoo's property.

Tropics Trail is a tour of the rainforest's animals. The inner part near the lake hosts an aviary of tropical birds, as well as Monkey Village, a walk-through home to squirrel monkeys. Lemurs and orangutans have separate exhibits within Tropics Trail. The outer part displays Asian Elephants, Galápagos tortoises, Sumatran tigers, and anteaters.

The zoo also breeds black-footed ferrets, believed to be extinct until a small group was found in Wyoming in the 1980s. (See Cheyenne Mountain Zoo, Colorado Springs, Colorado; and Louisville Zoological Garden, Louisville, Kentucky.)

For an intimate view of selected animals, take a tour of the Children's Trail. It is not just a petting zoo, but rather a scaled-down version of the larger zoo. Some of the animals here are on the endangered list.

Because the summer here is so hot and dry, zoo officials are concerned about public health. Various spots around the zoo house misters that cool visitors. Water fountains are everywhere, and zoo officials recommend that visitors stop to drink often.

Address: Phoenix Zoo, 455 North Galvin Parkway, Phoenix, AZ 85008.
Phone: (602) 286-3800.
Days and hours: Daily, Sep.–Oct.: 9 a.m.–5 p.m.; Nov.–Jan. 14: 9 a.m.–4 p.m.; Jan. 15–May: 9 a.m.– 5 p.m.; Jun.–Aug.: 7 a.m.–2 p.m. Closed 12/25.
Fee: Yes. Guided tours have extra fees.
Tours: Self-guided and guided tours. Call or check the website for daily events.
Public transportation: Valley Metro buses stop at the zoo.
Handicapped accessible: Yes.
Food served on premises: Cafés and snack bars.
Special note: The zoo advises that visitors bring sunscreen, water, and hats or umbrellas to shield against the Sun.
Website: www.phoenixzoo.org

SAHUARITA

Titan Missile Museum

South of Tucson is the only remnant of the United States' Titan II missile program that is accessible to the public. Originally operational in 1963, the Air Force Facility Missile Site 8, or Titan II ICBM Site 571-7, under the direction of the Davis-Monthan Air Force Base, was decommissioned in 1982 and opened for tours in 1986. It became a National Historic Landmark in 1994, and is now the Titan Missile Museum, one of only a trio of preserved missile launch sites. All tours are currently run by the Arizona Aerospace Foundation.

Above ground is a visitor center with an introductory video explaining the context of the Titan II missile site within the Cold War. From here, visitors take 55 steps or an elevator 35 feet (11 m) underground to tour the facilities, mostly painted in mid-20th-century industrial pale green, for about one hour.

The Blast Lock Area connected the world above ground to the missile complex. There are two sets of three-ton (2.7-tonne) steel blast doors and three-foot (0.9-m) concrete and steel walls to protect the crew from nuclear explosions, hazardous conditions, and even intruders. Two hundred feet (61 m) of steel-lined tunnels connected the Control Center, Blast Lock Area, and Missile Silo. Much equipment in the facilities was "hung" by massive shock absorbers, to protect the equipment and inhabitants from harm during the extreme vibrations caused by a nearby nuclear strike.

A three-story Control Center included bunk rooms for sleeping, a kitchen, and toilet. The launch control center had to be manned by two personnel, including one officer, at all times. From here the crew members monitored the status of the missile. In the launch control center, note the bright-red safe with several locks, holding the dual launch keys. When President Kennedy was assassinated in 1963, the Pentagon ordered these keys to be removed from the safe and placed near the consoles for a possible launch, which would have taken 58 seconds from turning the keys to liftoff. The bottom level of the control center included all the communications and power equipment, plus emergency rations and an escape hatch above ground.

The tour includes a view of the Titan II missile N-10 (the tenth missile built) itself, still poised for launch in the silo (its target is classified), but deactivated. Titan II missiles, used primarily from 1963–1987, were also used as launch vehicles in the Gemini space program, as well as the National Oceanic and Atmospheric Administration. In two stages, and at a length of 103 feet (31.4 m) and diameter of 10 feet (3.1 m), its presence looms over visitors to the silo. During the active period, the missile was kept with a hydrazine-based fuel on board, so it was ready to be launched at will, carrying a

Titan II ICBM on display in the launch duct at the Titan Missile Museum.
Courtesy of Chuck Penson.

9-megaton W-53 nuclear warhead. The lowest level of the silo is 140 feet (43 m) below the surface.

During the decommissioning process, each of the missile's fuel tanks and the re-entry vehicle's heat shield were perforated. The thermonuclear weapon was, of course, removed. The silo closure door at the surface was permanently set half-open, fixed with six blocks of concrete. In this way, Russian satellites can inspect from orbit that the missile is no longer useful.

Address: Titan Missile Museum, 1580 Duval Mine Road, Sahuarita, AZ 85629.
Phone: (520) 625-7736.
Days and hours: Nov.–Apr.: Sun.–Fri. 9:45 a.m.–5 p.m. (first tour at 10 a.m., last tour at 3:45 p.m.), no entry to the museum after 4:30 p.m., Sat. 8:45 a.m.–5 p.m. (first tour at 9 a.m., last tour at 3:45 p.m.), no entry to the museum after 4:30 p.m.; May–Oct.: Sun.–Fri. 9:45 a m.–4 p.m. (first tour at 10 a.m., last tour at 2:45 p.m.), no entry to the museum after 3:30 p.m., Sat. 8:45 a.m.–5 p.m. (first tour at 9 a.m., last tour at 3:45 p.m.), no entry to the museum after 4:30 p.m. Closed Thanksgiving and 12/25.
Fee: Yes.
Tours: Guided tours only.
Public transportation: None.
Handicapped accessible: Mostly. Certain tours are not handicapped accessible.
Food served on premises: Snacks and picnic tables only. There are restaurants in nearby Sahuarita and Green Valley.
Website: www.titanmissilemuseum.org

SUPERIOR

Boyce Thompson Arboretum

This 392-acre (159 hectare) living museum, at an altitude of 2431 feet (741 m) in the Sonoran Desert, focuses on 3900 species of plants from deserts around the world. On the 35 acres (14 hectares) open to the public are two miles (3.2 km) of nature trails visitors can follow to learn about desert vegetation. Guide materials are available at the Visitor Center, and interpretive signs are strategically placed along the trails. Besides showcasing plants that live in hot, dry climates, the arboretum researches the types of displays and the economically important plants that can survive in this environment.

The arboretum was created in the 1920s by Montana-born William Boyce Thompson, a mining engineer who made his fortune investing in mines. In 1917, while on a humanitarian expedition in Russia for the Red Cross, Thompson was impressed by the success that people living in arid parts of Siberia had with limited plant resources. Soon afterward, he launched the Boyce Thompson Institute for Plant Research at Cornell University, and then decided to create what he described as "the most beautiful and most useful desert garden of its kind in the world."

In the 1920s he selected an arboretum site near one of his mines, at the base of the 4400-foot (1300-m) high Picketpost Mountain. The mountain was named in 1870 when soldiers were assigned to the territory to watch for Apache Native American movement. The common term for this guard duty was to be "picketed." Later this outpost sent heliograph messages, using sunlight and mirrors, to military forts in the region. The

18-million-year-old lava flow at the top of the mountain is now called the Heliograph Formation.

Although many species in the arboretum exhibits are native, some have been introduced from elsewhere. The arboretum houses 805 kinds of cacti from 110 genera. Charles Russell Orcutt, hired in the 1920s as the site's first collector, planted its first cactus, *Echinocereus fasciculatus var. boyce-thompsonii*, a newly discovered variety. He also discovered *Echinocereus arizonicus* and *Mammilliaria viridiflora* about 11 miles (18 km) from the arboretum. The site's first plant propagator, Fritz Berger, went on to fame at the Huntington Botanical Garden. (See The Huntington Library, Art Collections, and Botanical Gardens, Pasadena, California.)

A special feature is the Chihuahuan Desert exhibit. The Chihuahua Desert is the largest of the four North American deserts and the primary desert of Mexico. Although more southern than the other three, its high elevation generally keeps it cooler than the Sonoran Desert. Freezing temperatures are common during the winter. Parts of the Chihuahuan Desert and the Sonoran Desert have comparable temperatures, and both deserts have similar precipitations. (The Mexican desert region has rain only in the summer, while the Arizona desert region has both summer and winter rainy seasons.) Many of the same plants and some closely related plants grow in both deserts.

Many plants commonly used for landscaping the southwestern United States, such as autumn sage and Texas ranger, originate in the Chihuahuan Desert. The arboretum hopes to expand the drought-resistant plants available for landscaping and to increase public interest in these plants. A 2.5-acre (1-hectare) demonstration garden helps home gardeners plan carefully to maximize shade, shelter, color, and privacy while conserving water. The garden explores erosion control, water harvesting, and irrigation techniques.

In 1925 a storage reservoir called Ayer Lake was constructed to irrigate the gardens. Water is pumped from nearby Queen Creek over a steep ridge to fill the reservoir, which has footpaths running alongside. A geological garden displays interesting local rocks from Silver King Wash and other nearby locations. The samples are sorted according to geological formations. At least 70 different free-ranging wild animal species inhabit the arboretum, some nocturnal and some active during the day. Nearly 150 species of birds are attracted to plants in the arboretum, including hummingbirds.

The historic Smith Building, which houses the Visitor Center, was built around 1925 of rhyolite, a native stone quarried nearby. The center contains exhibits about the arboretum, plants, the region, and other topics. In two public greenhouses attached to the center, visitors can see succulent plants and cacti from deserts around the world.

Address: Boyce Thompson Arboretum, 37615 East Arboretum Way, Superior, AZ 85173.
Phone: (602) 827-3000.
Days and hours: May–Sep.: daily 6 a.m.–3 p.m. (last admission 2 p.m.); Oct.–Apr.: daily 8 a.m.–5 p.m. (last admission 4 p.m.)
Fee: Yes.
Tours: Self-guided and guided.
Public transportation: None.
Handicapped accessible: Partly. Some trails can be steep.
Food served on premises: Picnic area. There are restaurants in town nearby.
Special note: Bring bottled water to stay hydrated.
Website: www.btarboretum.org/

TUCSON

Arizona-Sonora Desert Museum

The Sonoran Desert is considered by many to be the most visually appealing desert in the world. It has a greater variety of plant life than any other desert, and a wide range of animal life as well. Temperatures range from highs above 110°F (43°C) to lows dipping far below freezing. The region includes nearly all the world's biomes based on its diverse topography. Among the biological communities found in the Sonoran Desert are mountain woodland, grassland, tropical deciduous forest, and riparian habitats. Such topographical variety leads to the biological diversity on display at the Arizona-Sonora Desert Museum.

This outdoor museum celebrates the Sonoran Desert's diversity as a "fusion experience": part zoological park, part botanical garden, part natural history museum, part aquarium, and part art gallery. It depicts in microcosm all the desert's aspects while emphasizing the interdependence of the land, water, plants, wildlife, and people. Here, about 12 miles (19 km) west of urban Tucson, more than 240 species of animals and 1200 species of plants live on 98 acres (40 hectares), of which 21 acres (8.5 hectares) are interpreted. Interwoven trails explore different aspects of the desert. Docents are stationed throughout to answer questions.

As museum staff members point out, this desert does not suffer from lack of water; that is, if it had any more rain it would not be a desert. According to the site's literature, deserts are "places where the plants require a seasonal drought for survival." Just inside the museum's entrance is an orientation area that presents an overview of the Sonoran Desert, compares it with other deserts of the world, and introduces the museum's conservation efforts.

At the Reptile, Invertebrates & Amphibians Exhibit, visitors learn about centipedes, tarantulas, scorpions, the endangered San Esteban Island chuckwallas, gila monsters, snakes, and lizards. Endangered Mexican gray wolves are among animals the museum is trying to preserve. These wolves live in the Mountain Habitat, which represents a forested mountaintop between 4000 and 7000 feet (1200 and 2100 m) above sea level, typical of their natural home. The river otter, another endangered animal, lives in the Riparian Corridor, an exhibit that simulates the green ribbons of life around Sonoran Desert rivers.

A 60-foot (18-m) tunnel cast from a nearby rock formation is the setting for the Life Underground exhibit. This habitat illustrates the survival strategy of desert dwellers that retreat underground to escape the surface's environmental extremes. On the tunnel walls, visitors see examples of Sonoran Desert geology; tree roots are embedded and partially exposed, making the tunnel more realistic. Inhabitants of this exhibit include the kit fox, western tarantula, kangaroo rat, and tiger salamander.

The unique Hummingbird of the Sonoran Desert Region exhibit houses up to 10 species native to this area. The main enclosure is a walk-through aviary where birds live and fly freely. Just outside the main enclosure is the Hummingbird Garden, planted with vegetation that attracts these birds so people can observe them in the wild. More species of hummingbirds live in southeastern Arizona than anywhere else in the United States. Another walk-through aviary has roughly 20 other species of Sonoran Desert birds.

Unexpectedly, the museum has an aquarium, built in 2013, which includes an exhibition describing the freshwater areas and their animals within the Sonoran Desert, highlighting the Colorado River, and a gallery describing the Sea of Cortez with its variety of marine life.

Several gardens demonstrate the Sonoran Desert's great variety of plants. The Demonstration Garden offers the home gardener landscaping ideas. The Cactus and Succulent Garden contains more than 140 species of cactus and other Sonoran Desert plants. In the Convergent Evolution exhibit, plants from different parts of the world that evolved a variety of desert survival strategies are grown.

The Earth Sciences Overlook provides a good overview of the broad Avra Valley's geography and geology. The Earth Sciences Center has a replica of a living limestone cave in the process of forming. It contains an underground stream and exhibits on cave-dwellers, both bats and human beings. The Earth History Room, which reviews 4.6 billion years of geological history, includes a fine collection of minerals from the Sonoran Desert region.

Address: Arizona-Sonora Desert Museum, 2021 North Kinney Road, Tucson AZ 85743.
Phone: (520) 883-2702.
Days and hours: Oct.–Feb.: 8:30 a.m.–5 p.m.; Mar.–Sep.: 7:30 a.m. to 5 p.m.; Jun.–Aug.: Sun.–Fri. 7:30 a.m.–5 p.m., Sat. 7:30 a.m.–10 p.m.
Fee: Yes.
Tours: Self-guided.
Public transportation: None.
Handicapped accessible: Yes. Wheelchairs are available for rental on a first-come, first-served basis.
Food served on premises: Restaurants and snack shops.
Special note: The museum recommends avoiding midday visits during the summer months.
Website: www.desertmuseum.org

Kitt Peak National Observatory

The largest single collection of optical telescopes and instruments in the world is housed at this laboratory, part of the National Optical Astronomy Observatory (NOAO), which was formed in 1984 to advance American astronomy by providing major facilities for US astronomers. It consolidates three "mountaintops"—Kitt Peak National Observatory, Cerro Tololo Inter-American Observatory in Chile, and the Community Data and Science Center—and is run by the Association of Universities for Research in Astronomy. Besides doing individual research, astronomers at these three facilities develop, test, and evaluate new instruments and upgrade existing telescopes.

Kitt Peak, almost 7000 feet (2100 m) above sea level, is part of the Quinlan Mountains of the northern Sonoran Desert and is less than 60 miles (100 km) from Tucson. The site is on the Tohono O'odham Reservation. The Native Americans make this land available only as long as research is conducted here.

To introduce Kitt Peak, the Visitor Center has exhibits on astronomy, the observatory's history and research, and the local Tohono O'odham Native Americans. A welcoming lecture and film are presented several times a day, and there are two 16-inch (41-cm) telescopes plus a 20-inch (51-cm) telescope and a 24-inch (61-cm) telescope to use to search the heavens. From the center a self-guided walking tour passes several observatories and enters three facilities. Visitors can see several telescopes from the galleries or through special windows; other telescopes at the site are not open to the public. Small solar telescopes are also available for use in the SOLARIO observatory on the southern patio.

The first telescope visitors see is the McMath-Pierce Solar Telescope, among the largest solar telescopes in the world. The bulk of this instrument is below ground. Its light shaft is set to a slant of 32 degrees to keep the telescope constantly aligned with the celestial North Pole. Through a system of mirrors, the solar light beam reflects to instruments at the observatory's base. The above-ground portion is encased in copper and kept at a uniform temperature using a glycol/water solution. Each day it gives scientist a high-resolution image, 30 inches (76 cm) in diameter, of the Sun. Inside the McMath-Pierce facility is the Windows on the Universe Center, where visitors can attend planetarium shows, see a 6-foot (1.8-m) Science on a Sphere® globe, and learn about current research at Kitt Peak with interactive exhibits and data-visualization systems.

Next visitors see the middle-sized, 2.1-Meter Telescope, the fourth-largest telescope on Kitt Peak. It dates from 1964 and is an iconic symbol of the skyline.

The final site open to visitors is the Mayall 4-Meter Telescope, dedicated in 1973. The Mayall is one of the largest optical telescopes now operating in the contiguous United States. Although it weighs nearly 375 tons (340 tonnes), it is balanced so precisely that it can be moved by hand. The dome housing this instrument is 18 stories high. Inside are control instruments, darkrooms, laboratories, sleeping quarters, and two visitors' galleries. One has an exhibit describing the telescope and astronomical research for which it is used; the other provides a panoramic view of the observatory grounds, the Sonoran Desert, and the Quinlan Mountains.

The observatory offers nighttime stargazing programs; call or see the website for the schedule.

Kitt Peak sunrise.
Courtesy of Dean Salman.

Along the visitors' trail for is a display case with a petroglyph, or rock carving, estimated to be 500–800 years old. The petroglyph, found on a ranch near Kitt Peak, was probably made by ancestors of the Tohono O'odham.

Address: Kitt Peak National Observatory, Latitude 31.96, Longitude -111.598; 56 miles (90 km) west southwest of Tucson, AZ.
Phone: (520) 318-8600.
Days and hours: Daily 9 a.m.–3:45 p.m. Closed Thanksgiving, 12/25, and 1/1, and during inclement weather.
Fee: Free for the Visitor Center; guided tours incur a cost.
Tours: Self-guided in the Visitor Center; guided tours of the telescopes.
Public transportation: None.
Handicapped accessible: Partly. Both the 2.1-Meter Telescope and the Mayall Telescope are accessible only by staircases.
Food served on premises: Snacks only.
Special note: Visitors who are sensitive to elevation sickness should take care. The weather is significantly cooler than in urban Tucson. Non-US citizens must bring a passport or green card upon leaving the site to return to Tucson.
Website: www.noao.edu/kpno/

Pima Air & Space Museum

The third largest aviation museum in the United States, the Pima Air & Space Museum (PASM) was founded in 1976, and now displays 350 aircraft as well as over 125 000 artifacts related to aerospace. The idea for the museum started in the 1960s when, during the 25th anniversary of the US Air Force, commanders realized that World War II and later aircraft were being dismantled for their parts and their metal was being melted down. The first indoor gallery was opened in 1981, and several more galleries were constructed through the 1980s and 1990s. Currently the site contains more than 120 000 square feet (11 000 m²) of indoor displays and 80 acres (32 hectares) of outdoor display grounds.

Exhibits housed in the Main Hangar focus on early aviation, marine aircraft, aerial reconnaissance, and Women in Flight from the 18th century to now. One airplane of special note shown here is the last intact Martin PBM-5A Mariner. This type of craft, which were built largely in the 1940s, was the largest amphibious plane every built. A Lockheed SR-71A Blackbird also is displayed here. The Blackbird was the replacement for the famous U-2 spy plane. It was flown in the 1960s and 1970s, holding a variety of speed and altitude records, routinely exceeding a speed of Mach 3.

Hangars 3 and 4 house a variety of aircraft related to World War II. Here visitors can see the massive Consolidated B-24 Liberator, used during the fronts in Western Europe and the Pacific. A Boeing B-29 "Superfortress"—one of the most expensive projects from the mid-20th century, and flown during World War II and the Korean War—looms within the hangar.

Another structure, the 390th Memorial Museum, contains a B-17G "Flying Fortress," developed in the 1930s, which dropped more bombs than any other American airplane in World War II.

The Space Gallery describes machinery and events in the history of space exploration, from the X-Planes (including a mock-up of an X-15A), which were used in the late 1950s to study ballistic flight and re-entry via winged craft, and the early Space Race between the United States and the Soviet Union, to robotic landers on Mars. A Moon rock is on exhibit here as well.

A restoration building on this site is where artifacts are restored and maintained.

Outdoors are rows and rows of over 150 examples of aircraft, arranged by general types. There are helicopters, bombers, early fighter airplanes, training craft, NASA aircraft, tankers, commercial aircraft and civil aircraft, and more. Among the highlights here is the first Boeing 777-200 airplane, flown from 1994 to 2018, which was used as a testing airplane and then was owned by Cathy Pacific. Another commercial airplane here is the second Boeing 787 Dreamliner, built in 2009 with 80% carbon-fiber composites, which is a more fuel-efficient jet-liner. A tram tour is available for the outdoor exhibits.

Additionally, PASM runs exclusive air-conditioned bus tours several times each weekday of the 309th Aerospace Maintenance and Regeneration Group (AMARG) Facility—colloquially called the "Boneyard"—on Davis-Monthan Air Force Base nearby. Visitors on this one-hour tour see major highlights of the over 4400 aircraft stored in dozens upon dozens of rows in the dry desert climate. The area has a relative humidity below 20%, low rainfall, high altitude (2550 ft.; 777 m), and hard alkaline soil, all contributing to the choice of this area for outdoor storage after World War II. The Boneyard is the largest airplane storage facility in the world.

The craft belong to various US government divisions, including the Navy-Marine Corps, Army, Air Force, Coast Guard, and NASA. Many of the craft standing in the Boneyard are slated for recycling parts or restoration. Among some aircraft visitors may see are a Boeing YC-14, hundreds of A-10 Thunderbolt II "tank-busters," C-5s, and even a Boeing YAL-1 Airborne Laser (a high-powered laser mounted on a 747-400 airframe).

Address: Pima Air & Space Museum, 6000 East Valencia Road, Tucson, AZ 85756.
Phone: (520) 574-0462.
Days and hours: Daily 9 a.m.–5 p.m. Last admittance 3 p.m. Closed Thanksgiving and 12/25. Boneyard tours Mon.–Fri. only.
Fee: Yes. Tram tours and Boneyard tours cost extra.
Tours: Self-guided and guided. Tram tours of the grounds.
Public transportation: Limited. There are local buses, but they stop over 1.5 miles (2.4 km) from the site.
Handicapped accessible: Mostly. All of the buildings are accessible, but the outdoor grounds may be somewhat difficult in which to maneuver. Tours are accessible.
Food served on premises: Restaurant.
Special note: See the website for a schedule of lectures. Wear comfortable walking shoes, a hat, and bring a water bottle if you plan to wander in the outdoor areas. Reservations (see website) at least 10 days in advance are required for the Boneyard tours; security clearance and government photo identification are required—though non-US citizens are allowed.
Website: www.pimaair.org/

Nevada

HOOVER DAM

Hoover Dam

Hoover Dam, completed in 1935, is the highest dam in the Western Hemisphere: 726 feet (221 m) above bedrock. This arch-type gravity dam was built into the volcanic andesite breccia rock of Black Canyon primarily to control floods from the erratic Colorado River. Its base is 660 feet (200 m) thick, but its top is only 45 feet (14 m) thick. The dam forms Lake Mead, a massive reservoir that provides silt-free water to farms and homes, and is also a magnificent recreational spot.

Hoover Dam irrigates over 1 million acres (400 000 hectares). The Power Plant, completed in 1936, can generate more than 4 billion kilowatt-hours of electricity annually. The last additions to the dam system were made in 1961, when the 17th generator went into operation.

The dam authority controls water storage, water use, and electricity generation. The All-American Canal System distributes water for various irrigation projects. Revenues from the sale of power have paid back, with interest, the entire cost of the project.

Lava flowed in this area about 70 million years ago. The Little Colorado, a river that has since disappeared because of geologic changes, once flowed from the Rocky Mountains into Arizona. Eventually the Colorado River formed and began cutting the canyons we know today: Grand Canyon, Boulder Canyon, and Black Canyon. The river was named Colorado, which means "red" in Spanish, because of its thick, red silt. Today

Aerial view of Hoover Dam.
Courtesy of U.S. Bureau of Reclamation photographer Jamel Carry.

the Colorado River runs from the Rocky Mountains to the Gulf of California, covering some 1400 miles (2300 km).

Hoover Dam offers two guided walking tours: the longer Dam Tour and the shorter Powerplant Tour. Both tours include the Old Exhibit Building, where visitors learn the history of the dam and how it operates. On display is a model of one-twelfth of the continental United States—the area serviced by the dam. The Dam Tour includes visits to passageways in the dam itself, which the Powerplant tour does not cover.

The guided walking tours include a 10-minute film describing the dam's history. Visitors then visit a platform overlooking a 30-foot (9.1-m) diameter penstock pipe, and also explore one (the Nevada side) of the power plant's two similar wings. One wing, in Arizona, has nine generators; the other wing, in Nevada, has eight. Each generator is powered by an individual turbine located some 40 feet (12 m) below the generator floor. There are 17 hydroelectric generators with the capacity to generate 130 000 kilowatt-hours of power each. The generators are fed by four intake towers, two in each state. The towers rise out of the lake on the north side of the dam. Four pressure penstocks conduct the water to the turbines, which generate about 3.5 billion kilowatt-hours per year and provide electricity to cities as far away as Los Angeles. Electric step-up transformers convert the 16.5 kV created by the generators into 230 kV, ready for the transmission lines. Producing electricity this way saves an estimated 6 million barrels (10 billion liters) of oil annually.

Throughout the tour, guides describe how the dam was built and how it operates. Stops along the route feature diagrams and working models. Visitors also learn of the difficulties workers encountered as they risked their lives drilling the canyon walls and pouring concrete in weather as hot as 125°F (52°C).

The Alan Bible Visitor Center in Lake Mead National Recreation Area, a bit west of the dam, explores the interesting geology, flora, and fauna of the Lake Meade National Recreation Area. The botanical garden just outside the center grows a selection of desert trees, shrubs, and cacti.

Land elevations in the area around the dam range from 500 feet (150 m) to almost 7000 feet (2100 m) above sea level. Temperatures reach above 110°F (43°C), and rainfall is less than 6 inch (15 cm) annually. These conditions provide a wide range of climatic biomes. In springtime the desert regions near the dam explode with blossoms of brittlebush, lupine, sundrops, and other wild plants. The surrounding desert has giant Joshua trees, roadrunners, snakes, and lizards. Bighorn sheep often come down from the steep rocks to the lake for a drink.

Address: Hoover Dam, NV 89005. GPS coordinates: North 36.016222, West -114.737245.
Phone: (702) 494-2517.
Days and hours: Visitor Center: 9 a.m.–5 p.m. Last tickets for tours are 4:15 p.m. Closed Thanksgiving and 12/25.
Fee: Yes. Tours are an extra fee.
Tours: Self-guided in Visitor Center; guided tours for the dam.
Public transportation: Tour buses run from major nearby cities to the site.
Handicapped accessible: Yes. Wheelchairs are available for rental on a first-come, first-served basis.
Food served on premises: Restaurant.
Special note: Wear light clothing and a hat, use sunscreen and sunglasses, and carry water bottles on the tour during the summer. Arizona does not observe Daylight Saving Time, so

time on one side of the dam may differ from that on the other. The Dam Tour does not allow children under age 8.

Website: www.usbr.gov/lc/hooverdam/

RENO

University of Nevada, Reno, and Rancho San Rafael Regional Park

Nevada was founded primarily on industry and mining, and secondarily on ranching. At the northern end of Reno is the University of Nevada, Reno, and Rancho San Rafael Regional Park. The University of Nevada was founded in 1874 at the city of Elko, nearly 300 miles (480 km) east of Reno. But declining enrollment forced relocation of the university to Reno in 1885; mining engineering was an important department from early on. Rancho San Rafael Regional Park, 408 acres (165 hectares) in size, used for ranching, was purchased by Washoe County in 1979 as a public bond-issue, and plans were made for a botanical garden, children's amusement park, and other developments. We focus on several science- and technology-related sites within the university and park.

W. M. Keck Earth Science and Mineral Engineering Museum
The Mackay School of Mines Building, designed by architect Stanford White, known for buildings in the eastern United States, opened in 1908 on the campus of the University of Nevada, Reno. The building is set in a quadrangle deliberately patterned after that of the University of Virginia. The museum has been situated here ever since, and boasts original wooden cabinetry. (See Wagner Free Institute of Science, Philadelphia, Pennsylvania.)

W.M. Keck Earth Science and Mineral Engineering Museum.
Courtesy of University of Nevada, Reno.

Known for its wide-ranging collection of minerals, ores, mining equipment, and fossils, this facility has three levels.

The main floor houses hundreds of rocks and minerals, organized in various ways. Specimens are labeled with their name, chemical formula, and provenance. Along one wall are cabinets explaining the Dana system of mineral classification, such as elements, silicates, oxides, sulfides, etc. Other cabinets showcase specific elements and their minerals. Some display cases are organized by Nevada counties and their native minerals.

One case shows "thumbnail specimens," those that are about 1 inch (2.5 cm) in size, because these are small, easy to store, and often cheaper than large samples. There is a turquoise collection, and a case of rare minerals—those found in one place on Earth, or formed under peculiar conditions, or containing exceedingly rare elements. Other cabinets describe how precious stones are carved or cut, describe synthetic gemstones such as ruby and sapphire, and various minerals important for industry, such as salt, obsidian, talc, and garnet. One cabinet displays minerals that were discovered in Nevada, such as hübnerite and stetefeldtite. Hung on the walls are historic mining maps from the 19th and early 20th centuries.

The lower floor showcases the remaining portion of the Mackay Silver Dining Service, a set originally numbering around 1250 pieces. What is unusual is that this set's original silver provenance is known: the Comstock Mines in Nevada. On display nearby is a case of fluorescent minerals. The Schulich Document Research Lab, on this level, allows the public to view the Schulich Historic Certificate Collection, a set of stock certificates from the early years of the state of Nevada.

The mezzanine level of the museum was added in the 1920s. Along one side visitors can see mining paraphernalia; for example, a model of West End Mine near Tonopah, Nevada, was used in court litigation over mine ownership. A rare sample of blue mud is on display. Originally blue mud was discarded by miners because it clogged equipment used to recover gold. Later, blue mud was found to contain substantial amounts of silver—so the early miners were discarding about $3 of silver per dollar of gold recovered. A cabinet shows off samples of original ore from the Comstock Mine in Virginia City, Nevada. Many historic mining tools are on display, including drills, picks, balances, and crucibles. From the Comstock Lode, a mine cage, used to raise and lower miners out of the mines, stands in one corner.

On the other side of the mezzanine level, fossils are displayed. Cases show various fossils native to Nevada, describe techniques used to date fossils, and hold specimens of petrified wood. Perhaps the rarest fossil here is that of a giant sloth's tracks, formed 50 000 years ago in what is now Carson City. Paleontologists originally thought these tracks were made by giant human beings wearing sandals—until bones from actual ground sloths were unearthed at the La Brea Tar Pits in California, and the feet matched the tracks. (See La Brea Tar Pits, Los Angeles, California.)

Address: W. M. Keck Earth Science and Mineral Engineering Museum, Mackay School of Mines Building, 1664 North Virginia Street, Reno, NV, 89557.
Phone: (775) 784-4528.
Days and hours: Mon.–Fri. 9 a.m.–4 p.m., first Sat. of the month noon–4 p.m. Closed university holidays.
Fee: Free.
Tours: Self-guided. Group tours with advance notice.

Public transportation: Buses run from downtown Reno to North Virginia Street, then walk 0.3 mile (0.5 km) to the museum.

Handicapped accessible: Limited. Many display cases are too high for viewing from a wheelchair.

Food served on premises: None. The Student Union building has fast-food, and there are eateries within a few blocks of campus.

Website: www.unr.edu/keck

Fleischmann Planetarium and Science Center

At the northern end of the campus of the University of Nevada, Reno, sits an unusual structure, designed by local architect Raymond M. Hellman and built in 1963. With a roof in the shape of a hyperbolic paraboloid (i.e., a saddle), the facility was originally the Charles and Henriette Fleischmann Atmospherium-Planetarium. Dr. Wendell Mordy, an atmospheric scientist at the university's Desert Research Institute, planned the site for showing 360-degree panoramic time-lapse films of the atmosphere. Mordy sent out camera crews to capture such footage, along with Apollo rocket launches. Though the first of its kind, the Atmospherium gradually was overtaken by other planetariums with similar or more advanced multimedia technology. After two decades, the site was converted into a planetarium, and now has both astronomical shows and an Exhibit Hall.

Fleischmann Planetarium and Science Center on the campus of the University of Nevada, Reno from above.

Image Credit P. McFarlane.

The building has experimental features ahead of its time. One is a method of solar cooling and heating by means of 19 rotatable metal louvres, behind its front windows, that extend from floor to ceiling. Each louvre is white on one side and black on the other to reflect or absorb heat. Another method uses the roof's paraboloid form to avoid internal support columns and provide an uninterrupted internal volume. The guardrails in the lobby with hyperboloid wire designs were suggested by Dr. Mordy's daughter, Dr. Lee Anne Willson, then a high-school student (now a retired astrophysicist). The building is in the architectural mid-20th-century Populuxe style, and is listed on the National Register of Historic Places.

Inside the building is a globe-shaped Star Theater, with an internal dome 30 feet (9.1 m) in diameter, which presents astronomy and earth-science-related programs.

In the Exhibit Hall, a rare pair of extremely accurate 6-foot (1.8-m) Rand McNally relief globes of the Earth and Moon rotate. (See University of Wisconsin–Madison Geology Museum, Madison, Wisconsin.) The museum has the engineering prototype of the BIOCORE experiment on mice. In this experiment, Biological Cosmic Radiation Exposure, five mice were transported in their own little capsule to the Moon on the Apollo 17 mission, to see what effect cosmic rays would have on their biology. A large segment of the Quinn Canyon meteorite, found near Tonopah, Nevada, is on display. Discovered in 1908, the Smithsonian Institution sliced it in half, to show its internal crystalline structure. The other half of this meteorite is at the Field Museum. (See Field Museum of Natural History, Chicago, Illinois.) A time-lapse movie of the construction of the building itself is shown continuously.

Address: Fleischmann Planetarium and Science Center, 1664 North Virginia Street, Reno, NV 89557.
Phone: (775) 784-4812.
Days and hours: Sun.–Thu. 9:30 a.m.–4 p.m., Fri.–Sat. 9:30 a.m.–8 p.m.
Fee: Yes.
Tours: Self-guided.
Public transportation: Buses run from downtown Reno to the site.
Food served on premises: Vending-machine snacks.
Handicapped accessible: Yes.
Website: www.unr.edu/planetarium

Wilbur D. May Botanical Garden and Arboretum
University of Nevada, Reno, professor Edgar F. Kleiner, a friend of local resident Wilbur D. May, heir to the May Department Store chain, wanted to build a botanical garden and museum of May's artifacts within the newly purchased park. After May's death in 1982, the May Foundation endorsed Kleiner's project. Twenty-three acres (9.3 ha) of the park were earmarked for an arboretum, and the first two acres (0.8 ha) were opened in 1986. Development has proceeded rapidly since then, although—with much land undeveloped—this park is still young compared with most botanical gardens in the United States. Many trees and plants are labeled with Latin and common names, and placards are placed in some of the gardens describing the ecology represented.

Currently the facility has over two dozen special gardens, with nearly two miles (3 km) of trails. The arboretum is irrigated with a holding pond, Herman's Pond, plus city water and direct draw from the Truckee River. Among gardens of interest are the Kleiner Hardwood Grove and Oak Grove, in which deciduous hardwoods native to the eastern United States (maples and oaks) grow. Another is the Evans Creek Wetlands, in which cattails, cottonwood trees, and other riparian species exist in a wilder setting. The Fritz Went Native Garden and Rock Garden are home to Nevada native species including single-leaf pinyon, Utah juniper, curl-leaf mountain mahogany, and Mormon tea (*Ephedra viridis*).

For a spectacular view over the landscape and mountains, visit the Lear Garden's gazebo.

Next door to the arboretum is the Wilbur D. May Museum, which houses artifacts collected by Wilbur D. May on his world travels. Technological collections of interest include antique pistols and rifles, Japanese swords, Venetian glass, and stone slings.

On the university campus is a small Museum of Natural History on the third floor of the Fleischmann Agriculture Building, which focuses on the ecology of Nevada.

Address: Wilbur D. May Botanical Garden and Arboretum, Rancho San Rafael Regional Park, 1595 North Sierra Street, Reno, NV 89503.
Phone: (775) 785-4153.
Days and hours: Open daily. Spring and summer: 8 a.m.–9 p.m; autumn (after Labor Day to time-change: 8 a.m.–7 p.m.; winter (fall time-change to spring time-change): 8 a.m.–5 p.m.
Fee: Free.
Tours: Self-guided; guided tours with advance notice.
Public transportation: Buses run from downtown Reno to North Virginia Street, then walk 0.5 mile (0.8 km) to the site.
Handicapped accessible: Yes.
Food served on premises: None.
Website: www.maycenter.com

New Mexico

ALBUQUERQUE

The National Museum of Nuclear Science & History

This small, fascinating museum was created in 1969 to preserve and share with the public the history of atomic weapons and nuclear science. Originally sited on Kirtland Air Force Base, the facility was forced to move just east of Kirtland AFB because of security concerns after September 11, 2001. In Heritage Park, 9 acres (3.6 hectares) of outdoor exhibits behind the building, are large, historically interesting pieces, such as the B-52B bomber that dropped the last nuclear weapon tested in the stratosphere. It is armed with an M-61 rear-turret gun and can fly nearly 600 miles per hour (1000 kph). The 280 MM Atomic Cannon on display can propel a 600-pound (270-kg) nuclear shell more than 20 miles (32 km). This is one of 20 such cannons made for the military; only three remain. The Hound Dog AGM-28 A, deployed in 1959, was a long-range air-to-ground strategic missile developed as a "super-elusive miniature supersonic airplane" that could fly in unpredictable patterns at high and low altitudes. The short-range attack missile on display

National Museum of Nuclear Science & History.
Courtesy of National Museum of Nuclear Science & History.

replaced the Hound Dog in 1976. The Polaris missile is part of the arsenal of the Strategic Air Command and the Navy's submarine defense system.

From the lobby with a periodic table etched into the stone floor, exhibits inside the museum building trace the development of nuclear weapons from the early 1940s to the present in Los Alamos, New Mexico, Oak Ridge, Tennessee, and Hanford, Oregon. (See American Museum of Science and Energy and Oak Ridge Site–Manhattan Project National Historical Park, Oak Ridge, Tennessee; and Bradbury Science Museum, Los Alamos, New Mexico.)

The exhibit titled The Decision to Drop contains duplicate casings of Little Boy and Fat Man (ca. 1945), atomic bombs developed over three years by the Manhattan Project. Little Boy, a gun device with a 13-kiloton yield, was the first nuclear weapon used in warfare. It was detonated on August 6, 1945, over Hiroshima, Japan. The bomb was detonated by firing two masses of uranium-235 at each other inside the device; when the size, shape, and amount of uranium reached what is called a "critical mass," nuclear fission occurred. On display is a sample of Trinitite, a green glassy solid formed from fused sand during the tests at the Trinity bomb-testing site on July 16, 1945.

Fat Man, an implosion device with a 23-kiloton yield, was the second and last atomic device used in warfare. This bomb, detonated over Nagasaki, Japan, on August 9, 1945, worked by surrounding a critical mass of plutonium-239 by high explosives. When the inward pressure of the explosives squeezed the plutonium at the same time that neutrons were added to the plutonium, the device detonated.

Among artifacts on display in the museum are the Davy Crockett, a bazooka-type missile mounted on a jeep or hand-carried; Honest John, the first surface-to-surface rocket in the United States able to carry a nuclear weapon; and a Thor re-entry vehicle, all relics of the Mike Device tests at Eniwetok Atoll. Also on view is the Mark-6, the first mass-produced atomic bomb and part of the weapons arsenal from 1951 to 1962. It had a higher yield, reduced weight, and improved aerodynamics compared with earlier devices. The Navy's Hotpoint was the first atomic bomb designed for "laydown" applications: it employed a parachute so it would reach the ground intact before exploding.

Pioneers of the Atom tells interactively about the people who first uncovered the structure of the atom, from Marie Curie and Wilhelm Roentgen to Albert Einstein, Max Planck, Enrico Fermi, and Glenn Seaborg, using a timeline.

Radiation 101 teaches visitors about the basics of radiation: what it is, its sources, how much occurs naturally, and radiation's effects in our lives. The What's Up with U exhibit examines the element uranium, from mining to health effects from long-term radiation exposure and the radioactive decay products of uranium.

Uranium: Enriching Your Future shows how naturally occurring uranium ore is processed and enriched to enough U-238 for nuclear power in reactors. The Decision to Drop looks at the Manhattan Project and the Trinity bomb-testing site in New Mexico up through July 1945. There are videos, artifacts, and photographs to help lead the visitor. A photographic display compares Hiroshima and Nagasaki before and after the dropping of the two atomic bombs, from the city level down to individuals.

Cold War is an exhibit that examines the history of the rivalry between the United States and the Soviet Union from the 1950s through the 1990s. Artifacts include bombs, "broken arrow" accident casings, and missiles that were contained in stockpiles of the United States, the Soviet Union, and the United Kingdom. (A "broken arrow" is a nuclear-weapon accident without a risk of war.) Visitors also see pieces of nuclear weapons recovered from a 1966 crash in Palomares, Spain, of a B-28 nuclear weapon.

The question of what to do with nuclear waste is examined in the Nuclear Waste Transportation gallery. Here visitors see what happens during storage, handling, and transportation. On display is a TRUPACT-II stainless-steel canister 8 feet (2.4 m) in diameter and 10 feet (3 m) high, which can transport up to fourteen 55-gallon (210-L) drums of waste, and approved by the US Department of Energy.

The Nuclear Medicine exhibit traces the history of using radioactive materials and ionizing radiation to diagnose and treat human illnesses. Both early and current examples of nuclear medical technology are presented, from an early 1920s fluoroscope and a mock-up of a full-body gamma camera to a PRISM 2000XP gamma-ray camera.

Energy Encounter explores all sorts of civilian energy resources, but focuses on the one-sixth of our energy production that comes from nuclear power. The exhibit discusses the history and engineering of reactors, safety and accidents at nuclear reactors, and storage of waste. Supplementing the exhibit are scale models of the French Superphénix Fast breeder reactor, the Nuclear Generating Station in Palo Verde, Arizona, and the first nuclear-powered merchant ship, the NS *Savannah*, built in the 1950s.

Address: The National Museum of Nuclear Science & History, 601 Eubank Boulevard SE, Albuquerque, NM 87123.

Phone: (505) 245-2137.

Days and hours: Daily 9 a.m.–5 p.m.; open until 3 p.m. on 12/24 and 12/31. Closed Easter, Thanksgiving, and 12/25, and 1/1.

Fee: Yes. Guided tours are an extra fee.

Tours: Self-guided and guided.

Public transportation: Tour buses run from major nearby cities to the site.

Handicapped accessible: Yes.

Food served on premises: None. There are restaurants within a few blocks of the museum, mostly on Central Avenue SE.

Website: www.nuclearmuseum.org

New Mexico Museum of Natural Science and History

This is the place to learn about New Mexico's natural history from the Earth's creation nearly 5 billion years ago to the 21st century. The building, opened in 1986, was constructed specifically to accommodate the many innovative, interactive, educational, and entertaining exhibits. The theme "Timetracks—A Journey Through the Natural World" runs through and unifies the exhibits.

The displays are divided into several main zones: Emergence, Dawn of the Dinosaurs (Triassic), Jurassic Super Giants, New Mexico's Seacoast (Cretaceous), Age of Volcanoes (Tertiary), Evolving Grasslands, Cave Experience, and New Mexico's Ice Age (Pleistocene). Visitors can view them in any order, but to get the most from them it is best to follow the museum's planned route.

At the entrance to the museum is a bronze statue of a *Pentaceratops*, a large dinosaur with five horns on its face.

The museum holds the second-largest collection of Triassic vertebrate fossils in the nation. One of the more interesting exhibits is about what was originally called *Seismosaurus*, or Earth Shaker, of the Jurassic period—a dinosaur found in 1985 in the 150-million-year-old Morrison Formation near Albuquerque. This dinosaur was considered possibly longer and larger than any other known sauropod. On view are several vertebrae and part of the pelvis that give a sense of the animal's great size. Later it was found that the vertebrae were closer to the hips, and the estimate of the dinosaur's size shrank, though still considered to be over 100 feet (30 m) long. The fossil is now considered to be a form of *Diplodocus* instead. Nearby is a skeleton of a carnivorous dinosaur, *Saurophaganax*, a dinosaur similar to *Allosaurus*, from the late Jurassic Period.

To show how a living dinosaur may have acted, the museum has worked with the Kokoro Company in Japan, a firm that creates animatronic robots, to create an animated dinosaurian robot, the Bisti Beast. In the late 1990s, nearly complete fossilized remains of the Bisti Beast were found and collected from the New Mexican Bisti/De-na-zin Wilderness by Dr. Thomas Williamson, a paleontologist at this museum. The animal's zoological designation is *Bistahieversor sealeyi*, a 30-foot (9-m) long tyrannosaur that lived about 74 million years ago. Standing in the main atrium, the engineering marvel "breathes," shifts its eyes, and moves its head and tail, and is the only such robotic dinosaur in existence. The interior of the robot is constructed from metallic mechanical parts with a pneumatic pressure system, covered with a sculpted layer of urethane foam, and it is programmed to move based on best guesses from paleontologists.

A highlight of the museum is New Mexico's Seacoast, an area in which a variety of displays illustrate how the present land of New Mexico evolved. The Evolator Time Machine offers a 75-million-year ride back into New Mexico's geologic and evolutionary past, from the age of the dinosaurs, or Cretaceous Period, to the rise of the mammals, or Tertiary Period. The Evolator, a merging of "evolution" and "elevator," vibrates and rumbles to simulate a descent into the Earth. During the six-minute tour, layers of the Earth with fossils embedded pass the view ports and a video guide explains the sights.

In the Seacoast's arboretum, rare fossils—some from the Cretaceous age—are intermixed with living palms, ferns, and other modern vegetation similar to ancient species

Dinosaur and hot air balloon outside New Mexico Museum of Natural History & Science. *Photo* by Anne Green-Romig.
Courtesy NM Department of Cultural Affairs.

that once grew in the region. Its simulated coastal scene illustrates how a shallow sea with sandy beaches and marshy channels evolved into the modern, dry New Mexico. Displays compare and contrast the ancient with the modern: for example, a fossilized cypress stump is placed alongside a living cypress tree. A cast of the skull of a *Pentaceratops* is mounted on supports above the ground level. To appreciate the 320-degree angle of vision of the *Parasaurolophus*, visitors can look through a mask of this duck-billed dinosaur.

A television display in the Seacoast zone explains the role plate tectonics played in forming the Earth's surface in this part of the world. By turning a wheel, visitors see the Pacific and Continental plates collide to create mountains.

The walk-through exhibit Age of Volcanoes provides a look into a smoldering volcano. Visitors can see the lava, feel the walls tremble, smell sulfur, and experience the heat. An interactive Ice Age Cave has dripping stalactites and stalagmites, examples of animal life inhabiting this type of cave, and a special exhibit on bats.

FossilWorks is an area in the museum where specially educated volunteers demonstrate how fossils are prepared for display.

A 55-foot (17-m) domed planetarium with Sky-Skan Definiti® offers regular astronomical shows. Nearby hangs the Hall of the Stars, a huge tapestry with 1100 LEDs representing the night sky. The LEDs have the approximate color of the actual stars, and various constellations are marked. Messier objects are colored green. A large-screen theater offers popular films and natural-history documentaries. There are also temporary exhibits, rotating on a regular basis.

Address: New Mexico Museum of Natural History & Science, 1801 Mountain Road NW, Albuquerque, NM, 87104.
Phone: (505) 841-2800.
Days and hours: Daily 9 a.m.–5 p.m. Open 5:30 p.m.–9 p.m. on the first Friday of the month. Closed Thanksgiving, 12/25, and 1/1.
Fee: Yes. Planetarium and theater are extra.
Tours: Self-guided and guided. Guided tours last one to one-and-a-half hours.
Public transportation: Buses run from downtown to within about six blocks of the site, then walk.
Handicapped accessible: Yes. Wheelchairs are available upon request.
Food served on premises: Café.
Website: www.nmnaturalhistory.org

LOS ALAMOS

Bradbury Science Museum

Los Alamos National Laboratory is a key site for researching nuclear weapons, nuclear defense, energy, and other technologies, as well as chemistry and basic science. Their Bradbury Science Museum, a separate building in downtown Los Alamos, examines the laboratory's history and current research, with special emphasis on weapons development, alternative energy sources, and biomedical research. The museum is named for the second director of the laboratory, Norris E. Bradbury.

In 1943 the US government chose Los Alamos for Project Y, a code name for the program to design and build the world's first nuclear weapons, because of its isolation, good climate, and sparse population. The selected site already had housing, thanks to a boys' boarding school called the Ranch School, and was near government-owned land that the laboratory could obtain. Project Y was part of the US Army's secret World War II effort called the Manhattan Project. (See American Museum of Science and Energy and Oak Ridge Site–Manhattan Project National Historical Park, Oak Ridge, Tennessee; and The National Museum of Nuclear Science & History, Albuquerque, New Mexico.)

Radioactivity as a phenomenon of nature was known since 1896 when Henri Becquerel, working in France, conducted experiments on crystals of uranium salts. These crystals released light after exposure to sunlight. Becquerel measured the fluorescence of these crystals and came to believe that the rays he discovered were similar to the x-rays emitted by barium platinocyanide that Wilhelm Roentgen had detected earlier that year. This information set off a flurry of research. In 1938 Otto Hahn and Fritz Strassman, working in Berlin, replicated Enrico Fermi's work of four years earlier, when Fermi had bombarded uranium with neutrons and found that the atoms split into fragments. Hahn and Strassman learned that one of the fragments was barium and were deeply puzzled by the unexpected results. Nevertheless, they established that fissioning of nuclei was possible.

A group of scientists moved to Los Alamos in 1943 to try creating nuclear weapons based on the initial atomic pile built by Enrico Fermi in Chicago in December 1942. Physicist J. Robert Oppenheimer was recruited to lead one of the teams at this site. By 1945 the group had enough uranium-235 and plutonium to test their model, which would release the equivalent of 20 000 tons (18 000 tonnes) of TNT. The first bomb was tested over the desert of New Mexico; the next two were dropped on Japan.

After World War II the laboratory developed additional weapons and over the years grew in size and scope, until today it is a world-recognized research institution. Although it is operated by the Triad National Security, it is a national laboratory of the US Department of Energy.

The museum first opened in 1954. Inside the museum are over 60 exhibits describing the history of the laboratory and its past and present research on nuclear weapons and energy.

The History Gallery time line details the origin and history of the Los Alamos site. Background material printed on the wall describes the area's geology, Native American history, the life of early homesteaders in the region, and the Ranch School. Photographs, videos, original documents, and artifacts tell of the Manhattan Project. In this gallery visitors may watch a 15-minute film on the Manhattan Project and the laboratory's beginnings.

The Research Gallery explores the role scientists and engineers had in advancing basic scientific knowledge, increasing the energy supply of the nation, solving environmental problems, and developing advanced technology. Among the topics covered are the Neutron Science Center, research into space science, environmental monitoring and sustainability, nanotechnology, and biofuels. The laboratory developed the technology of flow cytology, which uses a laser to sort cells and determine the DNA, RNA, and proteins present.

The Defense Gallery focuses on the basics of atomic energy. On display are models of the only two nuclear weapons ever used in war. The first, Little Boy, was a gun-assembled uranium weapon detonated over Hiroshima, Japan, on August 6, 1945. The other, Fat Man, an implosion-type plutonium bomb, was tested over Alamogordo, New Mexico, on July 16, 1945, and was detonated over Nagasaki, Japan, on August 9, 1945. There is a discussion of the uses of supercomputers to model atomic and molecular processes.

The Weapons Exhibit examines the history of weapons development from design in testing. A display on the role of underground testing and verification includes an operating seismograph. Explosives and stockpile stewardship are two other topics covered in this gallery.

The theater presents a 15-minute film called *Stockpile Stewardship: Heritage of Science*, which tells about current work at Los Alamos National Laboratory. The TechLab has some interactive displays and changing exhibits.

Address: Bradbury Science Museum, 1350 Central Avenue, Los Alamos, NM 87544.

Phone: (505) 667-4444.

Days and hours: Tue.–Sat. 10 a.m.–5 p.m., Sun.–Mon. 1–5 pm. Closed Thanksgiving, 12/25, and 1/1.

Fee: Free.

Tours: Self-guided. Guided tours for large groups are available with advance notice.

Public transportation: Buses run from downtown to within about six blocks of the site, then walk.

Handicapped accessible: Yes. Wheelchairs are available upon request.

Food served on premises: Café.

Website: www.lanl.gov/museum/index.php

Texas

BEAUMONT

Edison Museum

Thomas Alva Edison's contributions to the electric industry are described at this small museum in the Old Travis Street Substation. The 1920s substation, which has been modernized and still operates, was one of the first facilities of this type to send electricity to downtown Beaumont and to power local rice farms and railroads, coming on line in 1929. Some of the distribution equipment originally used is on view. The collection was started by W. Donham Crawford, chairman of the board and chief executive officer of the Gulf States Utilities Company. In the early 1980s his widow left the collection to the museum.

The museum has the largest Edison collection west of the Mississippi River. Early models of many of Edison's inventions are on display, related to construction and electrical materials. Edison obtained 1093 patents during his lifetime for his inventions from the US government, more than any other person (until Shunpei Yamazaki in 2003), which visitors can peruse on a computer display in the museum. Some of these inventions are showcased here, including the Edison automobile spark plug, a 1918 nickel-alkaline battery, and attempts to extract rubber from the goldenrod plant.

In the museum, visitors can see some interesting historic pieces, including an early cylinder Edison Standard Phonograph from the mid-1890s, restored to original condition. This machine plays four-minute cylinder records with a "K" reproducer. Many historians believe that Edison liked his phonograph best of all his creations and thought every home should have one.

Also on view are examples of Edison's motion-picture camera, a variety of incandescent light bulbs in the Miracle of Light display case, a large 10 000 W incandescent bulb used for motion-picture illumination, and a spark plug. Marketing the Miracle of Light describes the efforts Edison made to promote his bulbs, including a segment of copper conductor retained from 1882 and a string of early Christmas-tree lights.

One interactive display plays short segments of early Edison motion pictures, such as "The Great Train Robbery" from 1903 and a film showing the devastation after the hurricane that destroyed Galveston in 1900; another display plays several examples of early

Edison Museum.
Courtesy of the Edison Museum.

recorded sound, including Edison's first recorded words. (See Thomas Edison National Historical Park, West Orange, New Jersey; and the Johnson Victrola Museum, Dover, Delaware.)

Invention Factories is a display describing the first research-and-development laboratories, including several items from his labs. One of the priceless pieces of personal memorabilia the museum holds is a scrapbook put together by Edison, Harvey Firestone, and John Burroughs to commemorate their camping trip together in 1916. It is one of six Edison scrapbooks in existence. (See Edison and Ford Winter Estates, Fort Myers, Florida.)

Address: Edison Museum, 350 Pine Street, Beaumont, TX 77701.
Phone: (409) 981-3089.
Days and hours: Mon.–Fri. 9 a.m.–2 p.m. Closed Sat., Sun., and major holidays.
Fee: Free.
Tours: Self-guided. Guided tours by prior arrangement.
Public transportation: Taxis.
Handicapped accessible: Yes.
Food served on premises: None. There are restaurants within several blocks in town.
Website: www.edisonmuseum.org

Spindletop Gladys City Boomtown and Texas Energy Museum

Spindletop Gladys City Boomtown, just on the south side of Beaumont, is a living-history museum re-creating the original oil workers' community for Spindletop Oil Field, where the age of liquid fuel began. The Texas Energy Museum in downtown Beaumont explains the history, science, and technology of the petroleum industry. Together, these two places offer a complete picture of the petroleum industry's birth, development, and significance.

Spindletop Oil Field was the beginning of financial success for the companies that later became the Texaco Corporation, Gulf Oil Corporation, Mobil Oil Corporation, and Sun

Oil Company. It all started when Pattillo Higgins, a Beaumont real-estate entrepreneur, became fascinated with the geology of the 15-foot (4.6-m) Big Hill near Beaumont. After studying a US government geology book, he concluded that oil was beneath the hill. Standard Oil Company, headed by the Rockefeller family, brushed aside one of Higgins's early requests for financial help as too speculative. In 1892 Higgins put together Gladys City Oil, Gas and Manufacturing Company, which he named after one of his Sunday-school students, and continued his crusade to find the resources to drill for oil. Several of his attempts at drilling failed. Captain Anthony Lucas, financed by the Mellon family of Pittsburgh, leased the land from Higgins's company and hired a drilling company with sophisticated equipment to penetrate more than 1000 feet (300 m). On January 10, 1901, Lucas struck oil.

The oil of the "Lucas Gusher" was under a salt dome, a hard, curved cap of rock formed when a thick layer of salt was thrust to the surface from a long-buried ancient ocean. The pressure of the oil was so great that 32 joints of drilling pipe were blasted out of the hole and flew 160 feet (49 m) into the air during the January 10 gusher. The thick, greenish-black oil spurted for 10 days before the drilling contractor could figure out how to cap the well. No source of oil of this magnitude had been seen in America before. The average well in Titusville, Pennsylvania, was producing about three to four barrels (400–600 L) a day, primarily for lighting. (See Drake Well Museum and Park, Titusville, Pennsylvania.) Drake and his team in Pennsylvania were the first to drill for oil anywhere in the country. Lucas and his group improved the drilling technology and, in doing so, produced crude oil in vast quantities—100 000 barrels (16 million L) a day.

The oil produced here virtually doubled the available crude overnight. Within one year there were more than 280 wells at the site pumping oil. The main production period lasted only from 1901 to 1906. Overproduction depleted the Spindletop oil field, and the town was virtually abandoned by 1913. Higgins's dream of a prosperous oil town went from boom to bust. Gladys City remained on the site until the 1950s, when it was removed to allow for sulfur mining.

Spindletop Gladys City Boomtown

Spindletop Gladys City Boomtown, the center of community life for workers at the oil field, has been re-created just a short distance from Big Hill. The town has 15 buildings, including an engineer's office, the Gladys City Oil Company office, a drugstore, a doctor's office, and a photography studio, each furnished with period artifacts and furniture. Among the oil equipment on display is a reproduction of a 64-foot (20-m) oil derrick. The site runs a two-minute re-enactment of the Lucas Gusher (using water instead of oil) regularly; see the website for details.

Address: Spindletop Gladys City Boomtown, Lamar University, 5550 Jimmy Simmons Boulevard, Beaumont, TX 77705.
Phone: (409) 880-1750.
Days and hours: Tue.–Sat., 10 a.m.–5 p.m., Sun., 1–5 p.m. Last admission time is 4:20 p.m. Closed Mondays, Easter, Memorial Day, Labor Day, Thanksgiving, 12/24, 12/25, 1/1.
Fee: Yes.
Tours: Self-guided. Guided group tours by prior arrangement.
Public transportation: Buses run from Beaumont to within about 0.2 mile (0.3 km) of the site, then walk.

Handicapped accessible: Mostly.
Food served on premises: None. There are restaurants within several blocks in town.
Special note: A schedule of Lucas Gusher re-enactments can be found on the site's website.
Website: www.lamar.edu/spindletop-gladys-city/index.html

Texas Energy Museum

The Texas Energy Museum, established in 1987, is the result of the integrated collections of two museums: the Western Company Museum of Fort Worth and Spindletop Museum (see preceding entry) of Lamar University in Beaumont. This state-of-the-art learning facility has exhibits that appeal to both the general public and those knowledgeable about the industry. An especially entertaining feature is the Cine-Robots with their "talking faces." These unique figures, with filmed images projected onto their faces so they resemble actual pioneers of the petroleum industry, talk about their contributions.

The first floor houses the Western Company Museum collection, considered by the Smithsonian Institution staff to be "the finest in the United States and Europe." Displays examine several areas of science and their relationship to energy use. The astronomy exhibit looks at the sun and its nuclear energy as the source of all energy, and the planet Earth as part of the solar system. Displays on oceanography and paleontology explain photosynthesis, organic sediments, fossils, and the origins of hydrocarbons. The geology exhibit focuses on tectonic, seismic, and volcanic forces, as well as the migration and accumulation of hydrocarbons and how they are affected by the porosity and permeability

Texas Energy Museum.
Courtesy of Texas Energy Museum.

of rock formations. It also reviews the structure of hydrocarbons, atomic combinations, and physical properties of crude petroleum. The technology exhibit illustrates the removal of petroleum from the ground, whether it is under the ocean or land, and explains how petroleum is transported and refined.

The Wall of Philosophy, History, and Economics explores the oil industry with photographs, memorabilia, and other materials. The information is set in chronological order and augmented with interpretive signs. A sand peel, or thin layer of sandstone, from the Red River serves as a beautiful wall mural while illustrating sediment buildup. A working seismograph reflects earth movement in Beaumont. On the wall is a seismogram of the Gulf of Mexico, showing what happens to the land when explosives are used for oil exploration. Among petroleum industry equipment is a model of a Western Triton III offshore-drilling rig, a section of pipe from the Alaska pipeline, and the Western Company symbol—a 1000-gal (3800 L) hydrochloric acid truck used between 1939 and 1948. The acid, injected into the tight rock, helped to release the petroleum.

On the second floor are exhibits about the petroleum history of Southwest Texas. Cine-Robots representing three generations of the fictional Willard family tell the story of their relationship with the oil industry, from the past to the present. Captain Lucas and Patillo Higgins, also Cine-Robots, relate the story of how Spindletop Gusher came to be and how it changed the petroleum industry forever. Other topics the exhibits cover range from the microscopic—hydrocarbon molecular structure—to the macroscopic—the birth of the universe. Also described are energy transformations; oil exploration, production, transportation, and refining; and contributions of the petroleum industry to American society, the economy, and modern technology.

The museum has an archive of photographs, maps, and more, available for scholarly research with advance notice.

Address: Texas Energy Museum, 600 Main Street, Beaumont, TX 77701.
Phone: (409) 833-5100.
Days and hours: Tue.–Sat., 9 a.m.–5 p.m., Sun., 1–5 p.m. Last admission time is 4 p.m. Closed Mon.
Fee: Yes.
Tours: Self-guided. Guided group tours by prior arrangement.
Public transportation: Buses run from Beaumont to within about 1 block of the site, then walk.
Handicapped accessible: Yes.
Food served on premises: None. There are restaurants within several blocks in town.
Website: www.texasenergymuseum.org

FORT SAM HOUSTON

U.S. Army Medical Department Museum

The role of this museum is to highlight medical field equipment and prominent medical officers, and to chronicle medical activities of the U.S. Army Medical Department from 1775 to the present. The Army Medical Department provides healthcare for soldiers and their dependents and aims to "conserve the fighting strength," as noted in the department's literature. Congress created the Army Medical Department in 1818, replacing the former Hospital Department, and designated the surgeon general its senior officer.

Originally this museum was part of a Medical Field Service School on the grounds of a former Native American school in Carlisle Barracks, Pennsylvania. The school and museum were transferred to Fort Sam Houston in 1946. The museum, now part of the Brooke Medical Center, became the Army Institute of Pathology, which, in turn, was renamed the U.S. Army Medical Department Museum.

Part of the museum's mission is to collect, preserve, and interpret objects relating to the development of the department and medical science since the 18th century. It serves both as a training aid and a pubic exhibit facility. The chronological exhibits of uniforms and equipment are augmented with graphics and text that explain the historical significance of the items on display.

Here visitors learn that one of the first assignments General George Washington gave the Hospital Department was to inoculate 20 000 men in the Continental Army against smallpox. This step, the first armywide treatment for smallpox, reduced the incidence of this illness considerably. Seeking a relationship between weather and disease, army surgeons collected and studied meteorological data until 1890. The Meteorological Registers published by the surgeon general's office led directly to the establishment of the U.S. Weather Bureau in 1890. Over the years the Army Medical Department has also kept detailed public health statistics.

When Army physicians were introduced to anesthesia in 1842, they rejected its use as too dangerous. (See Crawford W. Long Museum, Jefferson, Georgia.) Just a few years later, during the Civil War, both Union and Confederate doctors used anesthesia extensively for operations. It was during the Civil War that the Army Medical Museum, a sister institution to this one, was established to collect "specimens of morbid anatomy, surgical or medical, which may be regarded as valuable; together with projectiles and foreign bodies removed and such other matters as may prove of interest to military medicine and surgery." The Army Medical Museum collection is now housed at the National Museum of Health and Medicine. (See National Museum of Health and Medicine, Silver Spring, Maryland.)

Among items on display are early stethoscopes, Civil War surgical equipment, and ambulance vehicles, both restored originals and replicas. The collection of ambulances, from horse-drawn wagons to motor-driven vehicles, traces the history of medical evacuations and illustrates changes that occurred at various periods of military history. One interesting piece is a 1909 Studebaker Army ambulance.

An exhibit on Captain Thomas Hewlett, a medical officer captured by the Japanese during World War II, includes equipment he used to treat fellow prisoners. The Army Nurse Corps display features materials these women made while they were prisoners of war in the Philippines. Medical personnel who were POWs struggled to aid their ailing comrades and were forced to get by with makeshift substitutes for proper medical equipment. Non-American medical equipment that was confiscated from Germany, Japan, Russia, China, and Vietnam is also on display, as is equipment from several wars and POW memorabilia.

Address: U.S. Army Medical Department Museum, 3898 Stanley Road, Building 1046, Fort Sam Houston, TX 78234.
Phone: (210) 221-6358.
Days and hours: Mon.–Sat. 10 a.m.–4 p.m. Closed Sundays.

Fee: Free.

Tours: Self-guided.

Public transportation: Buses run from downtown San Antonio to within a couple of blocks of the museum.

Handicapped accessible: Yes.

Food served on premises: None.

Website: ameddmuseum.amedd.army.mil/index.html

FORT WORTH

Fort Worth Zoo

Just after the turn of the 20th century, three Fort Worth officials purchased a lion, two bears, an alligator, a coyote, a peacock, and several rabbits from a traveling carnival. The animals, placed in cages in the city park on the banks of the Clear Fork of Trinity River, became the first inhabitants of a newly formed zoo. During a major flood in 1911, all the animals were lost. But Fort Worth, not to be outdone by its rival city, Dallas, built a new zoo on 70 acres (28 hectares) of land designated for this purpose. In short order the zoo owned two panthers, a family of beavers, several cinnamon bears, and 21 prairie dogs. In the 1920s two American bison, a zebra, and an elephant were added. The zoo is now one of Fort Worth's oldest landmarks; it is the oldest site used continuously as a zoo in Texas.

The zoo has two roles: as a recreation center and as a conservation facility for wildlife, particularly for endangered species. The animal collection now consists of more than 7000 exotic and native animals of over 500 species. More than 100 of the zoo's species are in Species Survival Plans® for breeding and maintaining selected wildlife.

Because zoo officials believe an educated public will understand and support their conservation efforts, they have put great effort into designing exhibits that are both entertaining and enlightening. The staff is particularly proud of displays of unique animals such as bonobos and endangered black and greater one-horned rhinoceros. The bird collection includes endangered Storm's storks, great Argus pheasants, and endangered Harpy eagles. The zoo has one of the largest reptile and amphibian collections in the United States, with many fine specimens such as caiman lizards and gharial crocodiles.

Texas Wild! is an exhibit that leads visitors through the different regions of this state, because Texas has more species of animals than any other state in the United States. The overall message is that of "coexistence" with the land and its ecology. Here visitors can see white-tailed deer, burrowing owls, river otters, red wolves, and meet other domesticated and non-domesticated species in a petting ranch and touch tank.

Another special exhibit is the African Savannah, filled with reticulated giraffes, hippopotamuses, and African birds. The Museum of Living Art (MOLA), the herpetarium, is known for its naturalistic settings, extensive holdings, rare species, and excellent breeding programs. Its Fiji banded iguana is rare even in Fiji. The zoo is one of very few that exhibit the giant leaf-tailed gecko from Madagascar. The Australian Outback houses some Australian birds and a troop of red kangaroos, and the allied Great Barrier Reef is three saltwater tanks in which tropical fish, long-spined sea urchins, and blacktip reef sharks live. Asian Falls is set on a carefully landscaped hillside and has a 40-foot (12-m) waterfall. Nearby are Asian elephants and Malayan tigers.

The Fort Worth Zoo's World of Primates exhibit houses multiple ape species including the critically endangered western lowland gorilla.
Courtesy of the Fort Worth Zoo.

World of Primates is an indoor/outdoor facility covering 2.5 acres (1 hectare). Colobus monkeys and free-flying tropical birds are enclosed in a climate-controlled tropical rain forest with waterfalls and two enormous "living walls" covered with mosses and tropical plants. The connecting outdoor exhibits have lush islands surrounded by moats and wooded areas for such primates as gibbons, orangutans, bonobos, and mandrill baboons, and the critically endangered western lowland gorilla.

Address: Fort Worth Zoo, 1989 Colonial Parkway, Fort Worth, TX 76110
Phone: (817) 759-7555
Days and hours: Jul–Oct.: daily 10 a.m.–5 p.m.; Nov.–Feb.: daily 10 a.m.–4 p.m.; Mar.–Jun.: Mon.–Fri. 10 a.m.–5 p.m., Sat.–Sun., 10 a.m.–6 p.m.; Thanksgiving and 12/25 noon–4 p.m.; 1/1 10 a.m.–4 p.m.
Fee: Yes.
Tours: Self-guided.
Public transportation: Buses run from downtown Fort Worth to 0.7 mile (1.1 km) from the zoo, then walk.

Handicapped accessible: Yes. Wheelchairs are available on a first-come, first-served basis; motorized scooters can be rented.
Food served on premises: Cafés and snack bars.
Special note: Check the website for Wild Encounters and other animal shows.
Website: www.fortworthzoo.org

GALVESTON

Moody Gardens

W. L. Moody, Jr., and Libbie Shearn Moody, who were in the banking, insurance, and newspaper businesses in Galveston, set up the Moody Foundation in 1942 to improve the State of Texas. Among projects financed by the Moody Foundation is Moody Gardens®, "a public, non-profit educational destination utilizing nature in the advancement of re-habilitation, conservation, recreation, and research." Sited on 242 acres (97.9 hectares) of Galveston Island overlooking the Gulf of Mexico, the main attraction for purposes of this book is the three huge glass pyramids designed to educate the public about science. There is also a beach of white sand imported from Florida in 1988, a golf course, and other recreational attractions as well.

The first of the pyramids opened in 1993, the Rainforest Pyramid®, 10 stories tall. Here visitors encounter the biological diversity of rainforests found in Africa, Asia, and North and South America. Among the rare plants found in the pyramid is the Corpse Flower. In the Rainforests of the World exhibit, endangered animals include saki monkeys, ocelots, and Giant Amazon river otters (*Pteronura brasiliensis*) which grow to six feet (1.8 m) long. There are bat caves. In total, over 1700 species, including Chinese alligators, live here in the structure. After Hurricane Ike in 2008, refurbishment of the pyramid added a canopy walk, in order to see animals such as monkeys and sloths that live in the forest canopy.

The second pyramid is the 12-story pink-hued Discovery Pyramid®, which show-cases a variety of traveling exhibits related to science largely for children, and was opened in 1997.

The third pyramid is the blue Aquarium Pyramid®, which opened to the public in 1999. A total of 1.5 million gallons (5.7 million L) of water house about 10 000 living creatures from the North Pacific, South Pacific, South Atlantic, and Caribbean areas. Here are found sharks, stingrays, seals, and Humboldt penguins (which live in warmer climates), as well as five species of penguins from the colder waters of the South Atlantic. In the Gulf of Mexico Rig Exhibit, a two-story model of an oil-production platform links petroleum technology and marine life. There is a jellyfish gallery and a mangrove lagoon, and the Flower Garden Banks, a small tropical coral reef. The South Pacific Biome show-cases corals, fishes, and various invertebrates. The largest of the exhibits is the Caribbean area, which includes a walk-through tunnel, allowing visitors to see the creatures swim-ming overhead. Visitors can discuss topics with aquarium employees in the working Conservation Lab.

Address: Moody Gardens, One Hope Boulevard, Galveston, TX 77554.
Phone: (409) 744-4673.
Days and hours: Daily 10 a.m.–6 p.m.

Fee: Yes. Each pyramid has a separate fee. Animal encounters are also extra.
Tours: Self-guided.
Public transportation: Taxis.
Handicapped accessible: Yes. Wheelchairs are available, first come, first served.
Food served on premises: Café, pub, restaurant, food court.
Website: www.moodygardens.com

HOUSTON

Houston Museum of Natural Science

The top priority of this museum is to "enhance in individuals the knowledge and delight in natural science and related subjects." A dozen and a half galleries cover such diverse subjects as the petroleum industry, astronomy, space, Native Americans, paleontology, gems and minerals, malacology, and Texas wildlife. Other facilities in the museum complex are the Burke Baker Planetarium, the Cockrell Butterfly Center, and a large-screen theater. A satellite facility, the George Observatory, is located in Brazos Bend State Park, about an hour southwest of Houston.

Cullen Hall of Gems and Minerals is a highlight of the facility. More than 750 high-quality minerals and gemstones, lit only from the light in the display cases, give the hall an almost mine-like quality. The aesthetically appealing collection includes one of the world's best collections of tourmaline, a crystalline silicate, and one of the finest specimens of Brazilian imperial topaz ever found. Also noteworthy are the samples of Bohemian cassiterite, a tin ore; a sample of Japanese stibnite, a lead-gray mineral; a one million-carat amethyst geode; and the Ausrox Nugget (748 troy oz. or 23.3 kg) of gold, found in the Eastern Goldfield of Australia. Among the jewelry on display is a necklace with 13 perfectly matched amethysts.

The gallery has a reconstructed cave of fluorescent minerals has about 75 specimens from the all over the world with the unique characteristic of glowing under ultraviolet light. There is a particularly rich collection of willemite, franklinite, calcite, and esperite from northern New Jersey. (See The Sterling Hill Mining Museum and the Franklin Mineral Museum, Ogdensberg and Franklin, New Jersey.)

The malacology (seashell) collection has 200 000 specimens, primarily from the northwest Gulf of Mexico mollusk population. Many of the microscopic samples, however, are from the Persian Gulf, Alaska, and Mexican waters. Among the over 2500 samples on display in the Strake Hall of Malacology is a *Cypraea fultoni* that was taken alive. This cowrie—a mollusk with a glossy, brightly marked shell—is the rarest in the world and is usually found off southeast Africa. Visitors can also see a pair of *Pinna nobilis* shells, known for their thread holdfast that was sometimes woven into garments for warriors, as described in *The Golden Fleece*.

Displays in the Morian Hall of Paleontology include the skeleton of a huge, 140 million-year-old *Diplodocus* dinosaur with articulated and complete tailbones, a 45-foot (14-m) long *Tyrannosaurus rex* cast, a large Ankylosaurus model with plates on its back, an adult duckbilled *Edmontosaurus*, and an authentic dinosaur footprint. Visitors can examine prehistoric fossil skeletons of other animals, such as an early saber-toothed cat, a small horse, a breathing lobe-finned fish, a 330 million-year-old fossil of a coelacanth fish, and well-preserved 50 million-year-old garfish from Germany.

Petroleum, an integral part of Houston's history, is explored in the extensive Wiess Energy Hall exhibit, which contains fossils, models, and touchable displays. Video, interactive, and holographic displays tell the story of oil, its formation, its recovery, and its uses. On display are an oil rig, drilling equipment, and other tools used in the discovery of oil and gas.

The unusual 25 000-square-foot (2300-m^2) Cockrell Butterfly Center attached to the museum is a walk-through, cone-shaped, glass structure containing thousands of live, free-flying butterflies.

The John P. McGovern Hall of the Americas deals exclusively with Native American cultures of South and North America from prehistoric times to the present. Topics explored include native textiles, Plains beadwork, contemporary pottery, pre-Colombian gold artifacts, and basketry.

Unusual for science museums is the Welch Hall of Chemistry, with interactive displays allowing visitors to change the electronic arrangements of atoms in molecules and explore their chemical properties. Shining a Light on Molecules instructs visitors on photochemical reactions. There are exhibits on viral self-assembly and the formation of elements during the Big Bang. A wall-sized periodic table with samples of elements graces the exhibit hall, as does a giant ball-and-stick structure of DNA. (See Science History Institute, Philadelphia, Pennsylvania.)

The Farish Hall of Texas Wildlife has lifelike dioramas depicting various habitats in the state. Mounted animals on display include armadillos, Attwater's prairie chickens, and the collared peccary. A Foucault pendulum over 60 feet (18 m) long demonstrates the rotation of the Earth.

The Burke Baker Planetarium has a Digistar® 6 full-field projection system that provides images and special effects to enhance the regularly scheduled programs. In cooperation with the Johnson Space Center, the planetarium trains astronauts in star-field identification.

George Observatory, the museum's first satellite center, has three telescopes for public use, including a 36-inch (0.91-m) research telescope identical to an instrument at Kitt Peak National Observatory. (See Kitt Peak National Observatory, Tucson, Arizona.)

Address: Houston Museum of Natural Science, 5555 Hermann Park Drive, Houston, TX 77030.
Phone: (713) 639-4629.
Days and hours: Museum: Daily 9 a.m.–5 p.m. Planetarium: Daily 10 a.m.–4 p.m.
Fee: Yes. Planetarium, Butterfly Center, and special exhibits are extra. Guided tours incur an extra fee.
Tours: Self-guided. Guided tours are run daily.
Public transportation: Buses run from downtown Houston to within a block of the museum.
Handicapped accessible: Yes. Wheelchairs are available on a first-come, first-served basis.
Food served on premises: Café and fast food.
Special note: Check the website for After Dark late hours.
Website: www.hmns.org

Space Center Houston

The National Aeronautics and Space Administration (NASA) opened the Manned Spaceflight Center, the home base for US astronauts, in 1963. It was designed to resemble a university campus for space engineering and science, and to be a place where researchers could feel free to use their imaginations. Much of the land it was built upon was donated by Rice University.

The Manned Spaceflight Center, now called the Johnson Space Center, houses technical and scientific programs having to do with manned spacecraft and allied systems, a permanent inhabited space station, astronaut and mission-specialist training, and experiments related to space flights. (See Kennedy Space Center Visitor Complex, Titusville, Florida.)

This NASA complex has a visitor center, Space Center Houston, designed by Walt Disney Imagineering Corporation, that is neither a museum nor a theme park. Rather, it is an "experience center" that gives a hands-on sense of the adventure of space flight. At the 123-acre (49.8-hectare) complex, visitors can explore the past, present, and future of the space effort. An extensive area called Starship Gallery contains the big-screen Destiny Theater, which offers a 15-minute film on the history of space flight. In the gallery visitors learn about space exploration with Mercury, Gemini, and Apollo capsules; see the Skylab trainer and lunar-rover trainer; and watch re-creations of experiments performed in space and international cooperative efforts, such as the historic 1975 Apollo-Soyuz space rendezvous. Visitors can walk through a mock-up of the Space Shuttle's cargo bay and its manned maneuvering unit.

A Mission Briefing Center gives up-to-the-minute reports on the status of the Houston Mission Center, Kennedy Space Center launchpad, or other NASA flights. Here, mission briefing officers use live cameras and information reports to relay the latest news about the International Space Station.

The Space Shuttle mock-up *Independence*, which dominates Independence Plaza, re-creates the mid-deck and flight deck. Visitors can climb aboard to see the control panels, peer out the windows, and imagine what it would be like to fly the shuttle into space and back to Earth.

The Astronaut Gallery presents the opportunity to compare the design and function of original space suites dating from American's first space efforts. This is also the waiting area for the Space Center Theater, an ultra-high-resolution projection system coupled with a five-story screen. Films are shown regularly on astronautics and space research.

A shuttle replica mounted on the historic shuttle carrier aircraft NASA 905, in Independence Plaza at Space Center Houston.
Courtesy of Space Center Houston.

Behind-the-scenes guided tram tours of the Johnson Space Center are offered. Visitors can see Mission Control, various facilities around the complex, and Rocket Park, where a Saturn V rocket stands. The Saturn V was the type used to launch America's lunar missions. It is awe-inspiring to see the size of the rockets and imagine the power needed to lift such a mass from the ground. The first Manned Space Launch Vehicle, Mercury-Redstone, was only 83 feet (25 m) tall and weighed 62 000 pounds (28 000 kg); the Apollo-Saturn V rocket is 363 feet (111 m) tall and weighs 6.2 million pounds (2.8 million kg).

Address: Space Center Houston, 1601 NASA Parkway, Houston, TX 77058.
Phone: (281) 244-2100.
Days and hours: Daily 10 a.m.–5 p.m. Closed Thanksgiving and 12/25.
Fee: Yes. Guided tram tours are extra.
Tours: Self-guided; guided tram tours. See the website for times of guided tours.
Public transportation: Houston METRO Buses run from downtown Houston to the Visitor Center.
Handicapped accessible: Yes. Wheelchairs are available at the Guest Services Desk on a first-come, first-served basis.
Food served on premises: Diner.
Website: spacecenter.org

The Printing Museum

Two Houston printers, Vernon P. Hearn and Raoul Beasley, pooled their private collections and underwrote the construction of this small museum, created in 1982 to inform the public about the history of printing. This institution explores the effect that printing has had on the development of civilization, and encourages young people to put their literacy skills to maximum use. Collections added to the museum since it opened include historic printed pieces from the Printing Industry–Gulf Coast Collection and prints by Bohuslav Horák, an internationally known lithographer. Over 10 000 items are now held in the museum's collections.

The museum displays old printing presses, some of which are very rare. Among them are a working replica of Gutenberg's mid-15th-century press from Mainz, an 1830s Albion Press, and two stone lithograph presses from 1830. Acquisitions also include rare examples of printing, ancient woodcuts, stone lithographs, metal and wood engravings, and etchings. Precursors to printing, such as cylinder seals from Mesopotamia, fragments of ancient papyrus, and medieval illuminated manuscripts, are also exhibited. On display is the priceless Hyakumanto Dharani Scroll from 764 c.e., the oldest existing printed piece in the world. Created in Nara, Japan, the scroll bears a prayer to Buddha and a command from Japanese Empress Shotoku for its recitation.

Among other important documents here are a power-of-attorney form printed in 1563 by Juan Pablo, the first known printer of the Western Hemisphere, and two original pages from the Nuremberg Chronicle, printed in 1493 by Hartman Schedel. The museum also has a reproduction of a page from the Gutenberg Bible. This Bible, printed in 1452 in Mainz, Germany, was the first book published using movable type.

A piece signed by Eldridge Kingsley, America's first wood engraver, is on display. Other interesting examples of the technology of printing are a shipping receipt dated 1790 and printed in Boston; a leaf from the pamphlet *The Liberty of the Spirit of the Flesh* printed in 1750 by Benjamin Franklin; a very rare issue of the *Pennsylvania Gazette* from June 6, 1765, also printed by Franklin; and an issue of Isaiah Thomas's *The Royal American Magazine* from 1774. Thomas, who hired Paul Revere as his engraver, gained fame for signaling Revere from the Old North Church, "one, if by land, and two, if by sea."

The Hearst Newspaper Gallery includes a variety of newspaper printing paraphernalia, such as a 20th-century Linotype machine. More recent innovations are displayed in the 20th Century Gallery: the first Xerox® photocopier, four-color printing methods, offset printing, and an Apple® Macintosh® personal computer for home-based graphics.

A theater in the museum runs a 45-minute video on the history of communication and printing. The museum's special collections are available to scholars with advance notice.

Address: The Printing Museum, 1324 West Clay Street, Houston, TX 77019.
Phone: (713) 522-4652.
Days and hours: Fri.–Sat. 10 a.m.–4 p.m. Previously booked guided tours only Tue.–Thu. Closed Sun.–Mon., Martin Luther King, Jr., Day, Easter, Memorial Day, 7/4, Labor Day, Thanksgiving and the day after Thanksgiving, 12/25, and 1/1.
Fee: Yes. One-hour guided tours are extra.
Tours: Self-guided and guided (with prior reservations). See the website for times of guided tours.
Public transportation: Buses run from downtown Houston to within a block of the museum.
Handicapped accessible: Yes.
Food served on premises: None. There are restaurants within walking distance of the museum.
Website: printingmuseum.org

San Antonio

Natural Bridge Caverns

In the late 1800s, European settlers found this natural bridge, a 65-foot (20-m) wide rock formation spanning the sides of a sinkhole approximately 20 feet (6.1 m) deep. In 1960, as four young men on a spelunking expedition made their way through the sinkhole, they felt a cool updraft of air, which they followed for more than 60 feet (18 m) through an 11-inch (28-cm) wide crevice. Much to their amazement, they discovered a huge cavern. At a later date, several other caverns were found. Arrowheads and spear points found here, some dating to 5000 B.C.E., indicate that Native Americans used these caverns long ago.

These caverns are part of the Edwards Plateau geographic region, often called the Texas Hill Country. They are just 6 miles (10 km) from Balcones Fault, the escarpment of which separates the lowlands of the Gulf Coast region from the raised Great Plains of central North America. This fault is probably the result of the rapid falling and settling of the coastal regions of the vast inland seas that covered much of the continent's midwestern and western region in the Cretaceous period. The caverns are of limestone laid down by deposition of sediments of these seas. Two different strata make up this formation: Edwards limestone, the younger, and Glenrose limestone, the older.

The live caverns change constantly as times passes and new formations, or speleothems, develop. Water saturated with dissolved limestone seeps from the ground into

Sherwood Forest, in Natural Bridge Caverns.
Courtesy of Natural Bridge Caverns.

the caverns and drips down the walls. Stalactites, stalagmites, flowstones, and columns are found throughout the caverns. The calcium carbonate, calcite, formations, and walls range from orange to dark brown caused by the organic matter (mostly humic and fulvic acids) mixed with the calcite. Moisture keeps the colors brilliant.

The caverns were opened to the public shortly after they were discovered. An entrance was cut near the original air shaft and the natural bridge. Concrete walkways were put in and a Visitor Center was built by 1964. Today, more than one-half mile (0.8 km) of trail through the caverns is open to the public, and the expansion continues.

On regularly scheduled walking tours, guides take visitors past the natural bridge and into the caverns 185 feet (56 m) below ground, where temperatures remain a constant 70°F (21°C) under 99% humidity. The guides point out the geology and formations, such as rare, narrow stalactites called soda straws, short stalagmites called beehives, and multicolored flowstones called banded ribbons. Visitors who look carefully can see fossil seashells embedded in the cavern walls. Guides also discuss animal and plant life in the cavern. The Visitor Center has a geology exhibit explaining how the caves were formed, and there are displays of artifacts left in the caverns by the Native Americans.

Address: Natural Bridge Caverns, 26495 Natural Bridge Caverns Road, San Antonio, TX 78266.
Phone: (210) 651-6101.
Days and hours: Hours vary by date; see the website for schedule. Closed Easter, Thanksgiving, 12/25, and 1/1.
Fee: Yes.
Tours: Guided only.
Public transportation: None.

Handicapped accessible: Partly. Much of the underground portion is not recommended for wheelchairs.

Food served on premises: Snacks and picnic tables.

Special note: Wear comfortable shoes that can grip wet, slippery floors.

Website: naturalbridgecaverns.com

San Antonio Botanical Garden

This 38-acre (15-hectare) museum of living plant life opened in 1980 on land used as a waterworks and reservoir in the 1890s. Remnants of a limestone wall of the old reservoir now form a natural outdoor theater where performances and lectures are regularly held.

The site contains a series of formal and specialty gardens, each with its own theme. The Rose Garden grows familiar roses, antique blooms, plants mentioned in the Bible or cultivated during biblical times, such as oleander, myrtle, and fig, as well as plants that have significance to religions other than Christianity.

Cultivars of plants popular in past generations, such as marigolds and asters, grow alongside the flagstone walkways in the Old-Fashioned Garden. The Herb Garden has plants used by early Texas setters to season meals and prepare medicines. A Culinary Garden features an outdoor Teaching Kitchen. The Sensory Garden grows plants with interesting textures and scents, and has sculptures in ceramic and bronze that appeal to the sense of touch. Plants are identified with signs in Braille. The Kumamoto En Japanese Garden, a gift to San Antonio by its sister city, Kumamoto, was designed by landscape artists and craftsmen from Japan. It is a haven of rustic simplicity and peace amid stone walks, ponds, and bamboo fences.

The WaterSaver Garden and WaterSaver Lane are designed to illustrate to residents of south-central Texas the modern, high-tech irrigation and landscape-gardening methods that use a minimum of water. It is divided into three sections: one for plants that need

San Antonio Botanical Garden.
Courtesy of San Antonio Botanical Garden.

high levels of water, one for those using medium levels, and one for low-water-use plants. (See Desert Botanical Garden, Phoenix, Arizona.) Staff members are on hand to give local home gardeners additional information.

A two-acre (0.8-hectare) Lucille Halsell Conservatory, one of the largest conservatories in the Southwest, contains plants from around the world. This state-of-the-art structure, opened in 1988, has a series of tent-like glass pavilions placed partially below ground to shelter plants form the searing South Texas sun. Inside the pavilions, surrounding an inner courtyard, orangerie, and a pond, are 90 000 square feet (8400 m²) of greenhouses including five rooms: Exhibit Room, Desert Pavilion, Tropical Conservatory, Palm and Cyclad Pavilion, and Fern Grotto.

A main focus of the botanical gardens is native Texas flora and its relationship with Texas history. Eleven acres (4.5 hectares) in the Texas Native Trail are devoted to three distinctly different eco-regions regions of Texas—Hill Country, East Texas Pineywoods, and South Texas—each with its own temperature range, soil, and rainfall. In each section, human and botanical elements are combined to illustrate the interrelationship between the land and the history of the people. The Hill Country contains typical trees: Texas red oak, Uvalde big-tooth maple, live oak, and juniper. A German settler's house relocated to this habitat—the Schumacher Home, built in Fredericksburg, Texas, in the 1840s—is set on a meadow with native wildflowers as background: bluebonnets, Indian blankets, Mexican hats, and Indian paintbrush. The house was built with three separate construction styles, which illustrates the evolution of building techniques in the region.

A one-room cabin, built of hand-hewn logs in the 1850s, was moved to the East Texas Pineywoods display, where it overlooks a one-acre (0.4-hectare) lake surrounded by a forest of oak, sweet gum, cypress, and pine. Magnolia and dogwood also grow in the garden's modified soil, replicating East Texas sandy loam.

The South Texas area has scrub prairie with many chaparral species typical of the area, such as mesquite, cactus, sotol, and yucca. An adobe house, which illustrates construction techniques common in the Southwest, stands in stark contrast to the undergrowth.

A Bird Watch Structure allows visitors to see the variety of birds that the area hosts during bird migrations. The garden offers adult tours about an hour to an hour and a half in length. Each group can request to focus on a particular area within the facility.

Address: San Antonio Botanical Garden, 555 Funston Place, San Antonio, TX 78209.
Phone: (210) 536-1400.
Days and hours: Daily 9 a.m.–5 p.m. Closed Thanksgiving, 12/25, and 1/1.
Fee: Yes. Guided tours are an extra fee.
Tours: Self-guided and guided. For adult tours, register at least two weeks in advance.
Public transportation: VIVA bus stops next to the site.
Handicapped accessible: Yes. Wheelchairs are available on a first-come, first-served basis.
Food served on premises: Restaurant.
Special Note: Wear hats and sunscreen, and bring water bottles, especially during summer.
Website: www.sabot.org

San Antonio Zoo

This internationally known zoo is set along limestone cliffs from an abandoned quarry that rise about the headwaters of San Antonio River. The area is shaded by oak, pecan, and cypress trees. Tropical conditions exist along the riverbank, semi-arid weather prevails on the hillside near the cliffs, and desert-like conditions dominate the limestone outcroppings. Because of this unique locale, the facility is well suited for displaying animals from around the world.

The zoo, which began in 1914 with a few elk, buffalo, deer, and small carnivores, has grown into one of the largest and best in North America. The unique designs of the zoo's early habitats, very modern in their approach, were prototypes for many other zoos across the country. These habitats merge spring-fed canals and artificial rock with the natural limestone found in the park. More than 3500 animals of around 750 species live here in family groupings, alongside species with which they cohabit in their native environment. About 15 000 fish swim in the San Antonio River, which weaves its way through the zoo. The aim of this facility is for science and technology to coexist healthfully with the environment, improving life for animals, plants, and humans.

In the 1930s and 1940s the zoo started placing a priority on collection and breeding African antelopes, waterfowl, and other birds. Today the zoo is known for its large numbers of African antelope and for having the most varied collection of antelopes in the world. At this zoo, a leader in propagating animals, the first white rhinoceros was born in captivity in the United States.

The bird collections are also outstanding, and the captive breeding program is well known. This is the only zoo to exhibit the endangered whooping crane. It was among the first to hatch and rear flamingos, and now has four flocks. The zoo's wild, yellow-crowned night herons form one of the largest, most concentrated colonies of these birds in captivity, and the colony continues to grow.

Africa Live! is where a variety of African mammals that live around Lake Malawi and Lake Tanganyika are exhibited, including okapi, African wild dogs, elephants, zebras, rhinoceros, crocodiles, and giraffes. Through underwater windows, visitors can see cichlids and hippopotamuses. A termite mound is on display as well.

A walk-through, free-flight Australian Aviary is home to the buff-crested bustard, white-faced tree duck, lilac-breasted roller, and other birds. The zoo exhibits tree kangaroos and maintains one of the world's two breeding programs for these animals; the other is in Germany. San Antonio is one of very few cities to have a permanent exhibit of koalas, animals found almost exclusively in eastern Australia wildlife preserves. At the beginning of the 20th century, koalas were almost extinct. Zoo officials hope that the Queensland koalas here, on load from the San Diego Zoo, will breed. These small animals typically weight 6–10 pounds (3–5 kg) when fully grown and eat only eucalyptus leaves, which must be flown in fresh twice a week. Other animals from Australia include red kangaroos, wallabies, wallaroos, wombats, New Guinea singing dogs, and giant flightless emus.

The primate collection includes cotton-top tamarins, *Saguinus oedipus oedipus*, and emperor tamarins, *Saguinus imperator imperator*. The cotton-tops, from northern Columbia, are critically endangered as their habitat disappears. Deforestation of the rainforest is putting great stress on wild troops of these shy, secretive animals, which live deep in the dense vegetation of Brazil, Peru, and Bolivia. They are very rare both in the wild and in captivity.

The Fun Farm has not only great appeal for children but much to offer adults. Here visitors see small, intimate exhibits of animals from around the world.

Great care goes into the San Antonio Zoo's beautifully landscaped grounds. Because this organization is part of the plant rescue program, illegally imported plants confiscated by the US government are brought here and propagated.

Address: San Antonio Zoo, 3903 North St. Mary's Street, San Antonio, TX 78212.
Phone: (210) 734-7184.
Days and hours: Daily; hours vary, so see the website for a calendar.
Fee: Yes. Animal encounters may have an additional fee.
Tours: Self-guided.
Public transportation: Buses run from downtown to the zoo entrance.
Handicapped accessible: Yes. Wheelchairs are available for rental.
Food served on premises: Restaurants and café.
Special note: Wear hats and sunscreen, and bring water bottles, especially during summer.
Website: sazoo.org

Utah

BRYCE

Bryce Canyon National Park

See Grand Canyon, Zion, and Bryce Canyon National Parks, Grand Canyon, Arizona.

SALT LAKE CITY

Natural History Museum of Utah

This natural history museum, a place where curiosity is king, follows as its maxim this portion of *Little Gidding* from the *Fourth Quartets* by T. S. Eliot:

> We shall not cease from exploration
> And the end of all our exploring
> Will be to arrive where we started
> And know the place for the first time.

Increasing visitors' knowledge about the complex interrelationship of the land, oceans, atmosphere, plants, animals, humans, and bacteria is what this facility does best. To accomplish this task, it collects and preserves materials of the past, particularly those related to Utah, the Great Basin, the Great Plains, and the Colorado Plateau.

This museum, part of the University of Utah and located at the university's Research Park, opened in 1969 to serve students and scholars. In 1990, under legislative mandate, all the archaeological and vertebrate paleontological specimens recovered on state lands became the museum's curatorial responsibility. In 2011 the institution moved from the old library building into the specially designed Rio Tinto Center. The Rio Tinto Center sits on 17 acres (6.9 hectares) of terraced land in order to blend with the local environment. The structure is covered in copper sheathing that was mined in Utah's Bingham Canyon; the banded pattern represents rock strata. The museum sits on the Bonneville Shoreline Trail, overlooking the Salt Lake Valley.

The museum is most proud of its fine paleontological collection. Its dinosaur skeletons came from Dinosaur National Monument and the Cleveland-Lloyd Quarry, both in Utah. At the latter site, dug in 1939–1941 by a Princeton University team sponsored by Malcolm Lloyd, the museum runs an interpretive center. The Cleveland-Lloyd Quarry is rich in *Allosaurus,* a large carnivore, as well as the herbivores *Stegosaurus, Camarasaurus,* and *Camptosaurus.* Based on bones found at the site, two new dinosaurs were identified in the 1970s: *Stokesosaurus clevelandi* and *Marshosaurus bicentesimus.* The fossil mammals in the museum's vast collection come from the Ice Age and earlier. At one time, Utah periodically was covered by inland seas. The fossils from the period include shelled invertebrates, extinct fishes, and amphibians. Some of this collection is displayed in the Past Worlds exhibit, which includes the largest display anywhere of horned dinosaur skulls, and Triassic plants. Visitors can see specimens being prepared in the Paleontology Preparation Lab.

Mining's great importance to the economic well-being of Utah is explained in a geological exhibit called Gems and Minerals, geared toward people with a variety of backgrounds. The museum holds more than 6000 mineral and gemstone specimens, some unique to Utah, such as corkite, mixtite, and sulvanite. Within the Land gallery, visitors learn about the Wasatch Mountains, southern Utah deserts, and streams and wetlands of the region. Utah is unique in its collection of natural wonders, and the museum helps explain their characteristics.

Over 200 000 specimens make up the biology collection. They represent the major animal and plant groups of the Great Basin, Great Plains, and the Colorado Plateau. Almost 60 000 insects, thousands of amphibians, reptiles, and shells, and a huge collection of regional birds and mammals are held in the museum. Visitors to the facility learn about Utah and global biology in the exhibit Life, which examines "the web of life," starting with nanoscale of DNA molecules all the way up to entire ecosystems. One ecosystem examined in particular detail here is that of the Great Salt Lake.

The herbarium has more than 114 000 dried plant specimens from all over the world and a large collection of seeds. Some samples date from as far back as the 1870s. The museum also has a live collection of endangered western fish.

The natural world above the land is interpreted in the Sky gallery, weaving together astronomy, climate and weather, and the Sun. The Dark Sky exhibit explains why darkness, especially in a state like Utah with a low population density, is so important for the natural world. Included is a high-resolution image of the Milky Way galaxy, and controls for visitors to adjust, to see the effect that human-created lighting has on the nocturnal landscape.

The museum's important anthropology collection includes more than a half-million objects. In the First Peoples gallery, visitors can see evidence found in the Great Basin and Colorado Plateau of the ancient Desert Archaic, Anasazi, and Freemont cultures. On view are Anasazi pottery, baskets, tools, and textiles the museum collected just before the Glen Canyon Dam was built. There is a reconstruction of an actual dig excavated in the 1960s, and the Dry Caves Learning Lab explains how the dry climate of Utah preserves so much archaeological evidence.

A circular hall holds the exhibit Native Voices, which describes the eight tribes in Utah recognized by the US government: Northwestern Band of Shoshone Nation, White Mesa Community of the Ute Mountain Ute Tribe, Paiute Indian Tribe of Utah, Ute Indian Tribe, San Juan Southern Paiute Tribe, Navajo Nation, Skull Valley Band of Goshute, and the Confederated Tribes of Goshute. A variety of artifacts on display includes baskets, textiles, and ceremonial clothing.

Address: Natural History Museum of Utah, 301 Wakara Way, Salt Lake City, UT 84108.

Phone: (801) 581-6927.

Days and hours: Thu.–Tue. 10 a.m.–5 p.m., Wed. 10 a.m –9 p.m. Closed Thanksgiving and 12/25.

Fee: Yes. Guided tours have an extra fee.

Tours: Self-guided. Guided Highlights tours aimed at adults run several times a week; see the website for the schedule.

Public transportation: UTA Buses run to within a couple of blocks of the museum, then walk (uphill).

Handicapped accessible: Yes. Wheelchairs are available on a first-come, first-served basis.

Food served on premises: Café.

Website: nhmu.utah.edu

SPRINGDALE

Zion National Park

See Grand Canyon, Zion, and Bryce Canyon National Parks, Grand Canyon, Arizona.

8

Rocky Mountains

Colorado

National Center for Atmospheric Research Mesa Lab Visitor Center

The Rocky Mountains provide a beautiful, dramatic setting for the National Center for Atmospheric Research (NCAR) Mesa Lab, located in a 400-acre (162-hectare) nature preserve 600 feet (183 m) above the city of Boulder. Global weather, climate, and related topics are the focus of research at this center. NCAR also provides resources to the scientific community studying the atmosphere. The Mesa Lab is sponsored by the National Science Foundation, and managed by the University Corporation for Atmospheric Research. The latter group also manages other research sites for over 115 institutions of higher education.

The headquarters building, designed by the acclaimed architect I. M. Pei, consists of two six-story office towers on either side of a lower, center section that houses the library, meeting rooms, and exhibit areas. The block-shaped buildings are reminiscent of the ancient cliff dwellings in Colorado's Mesa Verde National Park, and their iron-red color blends in with the region's soil.

Throughout the year, visitors can take self-guided tours to see exhibit areas that cover research conducted at the center and show tools used to collect data. In summer, a guided tour is offered at noon. The four main research areas are: storms and weather phenomena of similar magnitude, climate, the chemistry of the atmosphere, and the Sun. Research is aimed at both understanding and, when, appropriate, managing complex atmospheric phenomena such as climate change, ozone depletion, severe storms, and solar magnetic fields.

NCAR Visitor Center shows a 10-minute introductory video in the NCAR theater. Galleries explaining about what NCAR studies include The Weather Gallery, Climate Discovery, Sun-Earth Connections, and Layers of the Atmosphere. There is also a gallery about the Mesa Lab building's architecture.

One exhibit illustrates the function of a microbarograph, which records barometric pressure on a drum and airspeed on a wind gauge. In the Rockies, down-slope winds have been known to exceed 130 mph (209 km/h). Such winds can be very destructive and so detailed records are vital to the health, safety, and economic well-being of the area's inhabitants.

Termites are an important subject of study because of the effect their vast number has on global weather. According to some estimates, the world has about a half-ton (0.45 tonne) of termites for each person. Termites consume wood and expel methane and other products into the atmosphere. As methane disproportionately traps heat, the resulting warming changes the atmospheric temperature, a factor in global warming. One display describes the worldwide distribution of termites and their effects on the climate. (See Audubon Butterfly Garden and Insectarium, New Orleans, Louisiana.)

Another exhibit area shows research aircraft used by the agency to study severe storms and support other research projects associated with microbursts, such as hail, thunderstorms, tornadoes, and violent winds. These dangerous weather-related events are known to kill many people and cause severe damage to property.

The aviation industry suffers from weather-related hazards such as wind shear that cause problems during takeoff and landing. Efforts to understand these dangerous phenomena, better predict them, and rapidly inform pilots of current and future dangers are underway here.

How clouds and lightning relate to each other and to electrical charges collected by clouds is under investigation here. An exhibit explains current research in this area and displays photographs of lightning.

Examples of early and current models of dropwindones, a type of research balloon carrying instruments to measure weather, are exhibited as well. NCAR staff members release them from high-flying aircraft to check the atmosphere over oceans and other hard-to-reach regions.

A cut-away model of the Sun, displaying the core to the solar atmosphere, explains the role that the Sun's energy plays in shaping our weather. Information about sunspots, solar flares, and similar phenomena is provided.

On their tour, visitors can see the NCAR computing facility and an exhibit with the shell of a CRAY-1A computer, the first supercomputer ever installed in a nongovernmental facility. An art gallery in the building showcases work exploring the relationships between art and science.

Outdoors, several hiking trails go through the nature preserve. Just outside the building is a topographical map of the region and signs explaining natural features, weather phenomena, and the various wildlife hikers might encounter while exploring the grounds.

Address: National Center for Atmospheric Research Visitor Center, 1850 Table Mesa Drive, Boulder, CO 80305.
Phone: (303) 497-1000.
Days and hours: Mon.–Fri. 8 a.m.–5 p.m.; Sat., Sun., and holidays 9 a.m.–4 p.m.
Fee: Free.
Tours: Self-guided. Free tablets with guided tours pre-loaded are available on site. Guided tours Mon., Wed., Fri. noon.
Public transportation: None.
Handicapped accessible: Yes. Wheelchairs are available for rental on a first-come, first-served basis.
Food served on premises: Cafeteria.
Special Note: Cafeteria only accepts cash. Check the schedule for when guided tours are offered.
Website: scied.ucar.edu/visit

COLORADO SPRINGS

Cheyenne Mountain Zoo

On Cheyenne Mountain overlooking Colorado Springs, almost 7000 feet (2100 m) above sea level, is the only mountain zoo in the United States. It was founded in the 1930s by Spencer Penrose, a well-known entrepreneur and philanthropist who needed a place to house his growing collection of exotic animals. He deeded the zoo to the people of Colorado Springs in 1938 as the Cheyenne Mountain Museum and Zoological Society.

To use this unique location to its best advantage, the zoo displays many animals from mountain regions around the world, along with animals from other geographic zones. The organization continues Penrose's commitment to the conservation of threatened and endangered animals by taking part in the Species Survival Plans® for 28 species. Because human preservation is closely linked to plant and animal survival, the exhibits examine the interdependence of life forms and their reliance on a shared environment.

Many species would not be alive today if it were not for zoos and their survival programs. For example, the zoo houses and breeds black-footed ferrets, once native to the region but now considered extinct in the wild. The zoo staff members hope eventually to reintroduce them into the wild. (See Louisville Zoological Gardens, Louisville, Kentucky; and Phoenix Zoo, Phoenix, Arizona.) The Conservation Center, which is not open to the public, houses the black-footed ferret recovery program, a bird-propagation unit, and a holding area and breeding facility for large primates and carnivores.

Many animals are in naturalistic habitats. Putting animals in homelike surroundings encourages more natural behavior, which makes them physically and psychologically healthier and thus better able to breed.

Southwestern-style architecture sets the ambiance at the Entry Plaza. One of the first exhibits beyond the plaza is the Rocky Cliffs, where Rocky Mountain goats and small regional animals such as yellow-bellied marmots live among rock outcroppings and vegetation that is native to the beautiful Colorado landscape. The exhibit is constructed with granite taken entirely from zoo property.

The other exhibit just beyond the entrance is African Rift Valley's giraffe area. Beside the elephants, hippopotamuses, meerkats, and black rhinoceroses, this facility has one of the largest herds of reticulated giraffes in captivity. This site is famous for having the greatest number of giraffe births of any zoo in the world.

Primate World has indoor and outdoor areas for viewing lion-tailed macaques, black and white colobus monkeys, guenon monkeys, golden-lion tamarins, orangutans, and gorillas. The three naturalistic habitats represent typical forest clearings. The Great Ape House, which is an indoor enclosure, has glass walls separating visitors from the animals. Videos on large screens describe the animals and their conservation.

At the far end of the zoo is the Australia Walkabout, home to emus, Matschie's Tree Kangaroos, Red-Necked Wallabies, and more.

Asian Highlands is situated along 1.5 acres (0.61 hectares) of mountainside. Near the entrance sits a replica of a yurt, a tent-like structure used by Asian nomads for shelter. A glass wall separates visitors from Amur tigers. Also on view are red pandas with their soft, chestnut-red coats and dark legs. They come from the temperate bamboo thicket forests of the Himalayan Mountains. Amur leopards from Manchuria, Russia, northern China, and North Korea are part of the exhibit as well. These animals are extremely rare in the wild. Nocturnal hunters, the leopards usually spend the day resting on tree branches or lounging on rock ledges. Visitors glean facts about plant and animal life form interactive displays and manipulate a model to learn how a cat's retractable claws work.

Wolf Woods, home to Mexican gray wolves, is a quarter-acre (one-tenth hectare) naturalistic mountain habitat. This zoo is one of several dozen facilities worldwide planning to breed this rare and endangered animal, with the aim of reintroducing it into the wild. Two displays explore wolf myths and facts.

Zoo admission includes access to a two-mile (3.2-km) scenic highway leading to the Will Rogers Shrine of the Sun. From the shrine is a spectacular view of the zoo and surrounding area. For those who enjoy heights, a Mountaineer Sky Ride—an open-air chair lift—provides a vista of the zoo grounds.

Address: Cheyenne Mountain Zoo, 4250 Cheyenne Mountain Zoo Road, Colorado Springs, CO 80906.
Phone: (719) 633-9925.
Days and hours: Daily 9 a.m.–5 p.m. Certain holidays have shortened hours; see the website for details.
Fee: Yes. Group tours, sky ride, and Animal Encounters incur an extra fee.
Tours: Self-guided.
Public transportation: None.
Handicapped accessible: Yes. Wheelchairs are available for rental on a first-come, first-served basis. There is a shuttle for transportation around the zoo.
Food served on premises: Cafés, snack shops.
Special note: Because of the high altitude and variable elevations in the zoo, we recommend that visitors pace themselves.
Website: www.cmzoo.org

Colorado Springs Pioneers Museum

By the end of the 19th century, Colorado Springs, nestled high in the Pikes Peak region, had become very prosperous because of a major gold discovery in nearby Cripple Creek. Beautiful brick and stone commercial buildings replaced the frame structures lining the streets. One of these, the El Paso County Courthouse, built in 1903 in Italian Renaissance Revival style, is home to the Colorado Springs Pioneers Museum. This lovely building, adorned with a clock tower, is now on the National Register of Historic Places. It is surrounded by a city park filled with flowers and grand walkways.

The museum is the result of efforts by a group of pioneers who formed an association in 1896 to preserve the memorabilia of the Pikes Peak region. By 1909 they started exhibiting their collection in part of the county courthouse. The exhibit moved to another building in 1937, but later outgrew it and returned in 1979 to the then-vacant courthouse, taking over the entire structure.

This museum celebrates the history and culture of the Pikes Peak area and offers permanent and temporary exhibits on topics of local and national interest, as well as exhibits with large sections on science or technology.

Because of its location high in the clear mountain air, the region became known as a favorite spot for tuberculosis patients. The beginning of the 20th century brought a proliferation of health spas and tuberculosis sanatoriums, the history of which is explored in the museum's extensive City of Sunshine exhibit. Roughly a third of all visitors to the area were seeking cures, and many of them stayed. The display includes photographs of tent camps where sanatoriums were established and patients were treated, and a web-based mapping tool to show visitors the history of such treatments. A reproduction of the medical office of Dr. Edwin Solly, physician and tuberculosis sufferer who emigrated here from England, shows how much he accomplished to bring health resorts to Colorado Springs. A renovated tuberculosis hut from the Modern Woodmen of America Sanatorium, portions of a historic pharmacy in Colorado Springs, a re-created

Colorado Springs Pioneers Museum.
Courtesy of the Colorado Springs Pioneers Museum.

sun-porch demonstrating the believed health benefits of the rest cure, plus an extensive section on tuberculosis treatments and fads are shown here.

Important pioneer work in medicine was accomplished in Colorado Springs. Local dentist Frederick S. McKay noted that a number of adolescent patients had permanently stained teeth. In 1916 he published a paper in which he assumed this problem was caused by unidentified contents in the local water. Subsequent work by others indicated he was right, and that the culprit was a high level of fluoride. McKay's work led others to study the relationship between fluorides and tooth decay. In 1946 a new word was coined: fluoridation. The museum has an extensive display on McKay's work and other research on fluorides, including some of his glass slides showing brown-staining, and a collection of dental instruments from the early 20th century. An interactive mapping display reveals where McKay took water samples for his research.

The Pikes Peak area, well known for pottery manufacture, has a gallery with 150 pieces devoted to the Van Briggle Pottery works, a well-known long-time pottery-maker still in operation. One of the founders of the works, Artus Van Briggle, was a victim of tuberculosis, and died in 1904, whereupon his wife, Anne, assumed the presidency of the company until 1912.

The museum encourages visitors to use its original, early 20th-century Otis birdcage elevator to travel between floors. Instructions are posted.

The Starsmore Center for Local History is the museum's archives, with more than 6000 cubic feet (170 m^3) of manuscripts, mostly in the era of 1870–1930. Public access to the archives is possible with advance notice. Holdings include letters, diaries, scrapbooks, oral histories, and 80 000 photographs.

Address: Colorado Springs Pioneers Museum, 215 South Tejon Street, Colorado Springs, CO 80903.
Phone: (719) 385-5990.
Days and hours: Tue.–Sat. 10 a.m.–5 p.m. Closed Sun. and Mon., Martin Luther King, Jr., Day, President's Day, Memorial Day, 7/4, Labor Day, Veteran's Day, Thanksgiving and the following Fri., 12/25, and 1/1.
Fee: Free.
Tours: Self-guided; guided tours led by docents.
Public transportation: Taxi. The site is only a couple of blocks from the Greyhound Bus Station.
Handicapped accessible: Yes.
Food served on premises: None. Picnicking is possible in the park surrounding the museum building. There are eateries a few blocks away from the museum.
Special Note: Visits to the archives need several weeks advance notice.
Website: scied.ucar.edu/visit

DENVER

Denver Botanic Gardens

These gardens focus on worldwide high plains and mountain horticulture, with an emphasis on the semiarid West. The 23-acre (9.3-hectare) gardens make the best possible use of the local environment, which at one mile (1.6 km) above sea level has a high-altitude climate. Plants from other regions are also part of the collection. The gardens are organized according to three themes: teaching gardens, display gardens, and environmental gardens.

The Denver Botanic Gardens seek to find and protect rare species native to the intermountain West. Seeds brought to Denver are propagated into a living collection used for research, education, and display. Some seeds are frozen and stored to keep them viable for decades. Staff members are committed to increasing the public's enjoyment and knowledge of plants and horticulture.

A one-acre (0.4-hectare) Rock Alpine Garden is known as one of the finest in the country. Some 2300 species from around the world that grow on high mountain summits and the Arctic tundra fill this lovely spot. Plants are divided into distinct zones according to soil type and are displayed in an English rock-garden setting with western American influence. Many of the region's "green" industries learn from the Rock Alpine Garden of new plants they can merchandise.

The Gates Montane Garden, the oldest garden in the facility, illustrates the transition zone from the plains to the alpine peaks of Colorado. Plantings are divided into three life zones: foothills, montane, and subalpine. Rich, diverse vegetation of Colorado makes up the Plains Garden. Grasses and other native prairie plants are brought together to re-create the seven plant systems of this region. The Roads Water-Smart Garden and Darlene Radichel Plant Select Garden have plants from the arid West and similar regions around the world to illustrate home-landscaping approaches that consume less water. Many of these plants can be used to create alternative landscapes not only in the West but also in gardens of the Eastern United States and Europe. The section includes rare and endangered xeric plants of the Rocky Mountain region. (See Desert Botanical Garden, Phoenix, Arizona.)

Summer in the Denver Botanic Gardens.

Photo © Denver Botanic Gardens; photo by Scott Dressel-Martin. Used by permission.

The award-winning Water Gardens, with their extensive waterway system, provide a sharp contrast to the semiarid gardens. Artistically arranged water lilies, aquatic plants, waterfalls, ponds, and rocks create a variety of natural sounds and visual effects.

The Japanese garden, *Shofu-en*, or Garden of Pine and Wind, is carefully landscaped as if done by nature, in the traditional Japanese perspective. Designed in 1979, it contains Colorado-native Ponderosa pines, water, Asian plantings, and more than 300 tons (270 tonnes) of crushed stone.

The Herb Garden, planted in a formal bowknot design, serves a utilitarian function by providing seasonings for vinegars, salts, and spice mixes that are sold to raise money for the gardens.

As a member of the Center for Plant Conservation, a national group seeking to preserve the rare and endangered plants of the country, the Denver Botanic Gardens has created a showcase called the Conservation Garden. Plants from the arid West that are threatened with extinction grow among shards of oil shale, as they do typically in western Colorado, eastern Utah, and southern Wyoming.

A lovely conservatory dominates the gardens. It features plants related to medicine, building materials, dyes, fibers, and foods. One wing grows collections of orchids and bromeliads. The conservatory's Lobby Court has changing exhibits, a library, and a herbarium.

A Science Pyramid explores steppe biomes using video, light, sound, and touchscreens. At the demonstration desk, docents and staff members often discuss topics from pollination to seeds.

The 700-acre (280-hectare) Chatfield Farms, part of the Denver Botanic Gardens, is located in the suburb of Littleton on the southern edge of Denver. Once only cottonwoods, willows, and other native plants grew here, but today it has much more, including several distinct High Plains habitats. There are two 19th-century farmhouses on the grounds: the Hildebrand, which, along with surrounding farmland, is being restored to preserve early Colorado farm life; and the Green Farm, which contains the Polly Steele Nature Center. A seasonal butterfly house shows off hundreds of butterflies, including mourning cloaks, painted ladies, and swallowtails.

Address: Denver Botanic Gardens, 1007 York Street, Denver, CO 80206. Chatfield Farms, 8500 West Deer Creek Canyon Road, Littleton, CO 80128.
Phone: (720) 865-3500.
Days and hours: Daily wintertime 9 a.m.–5 p.m.; summertime 9 a.m.–8 p.m.
Fee: Yes. Additional fee for Chatfield Farms.
Tours: Self-guided and guided. See the website for the tour topics and times.
Public transportation: RTD buses stop at York and Josephine Streets, and 12th Avenue.
Handicapped accessible: Mostly. A few areas are not accessible. Wheelchairs are available on a first-come, first-served basis.
Food served on premises: Two restaurants in the Denver location; none at Chatfield Farms.
Website: www.botanicgardens.org

Denver Museum of Nature & Science

With more than 400 000 specimens and artifacts related to anthropology, earth science, and zoology, this institution is the fifth-largest natural history museum in the country. The collection started in the 19th century when naturalist Edwin Carter collected and displayed 3300 mounted birds and mammals in his log-cabin museum in Breckenridge, Colorado. At his death in 1900, his collection was moved to Denver and became the Colorado Museum of Natural History. In 1908 a new building opened with a new name, the Colorado Museum of Natural History, and the collection grew. With increased funding through its host city, the name became Denver Museum of Natural History. In its centennial year, 2000, the museum rebranded itself as the Denver Museum of Nature & Science, with a newer, wider focus on science.

The 20-foot (6-m) tall *Tyrannosaurus rex* in the entrance hall is a replica of a rare and important specimen excavated in Montana in 1908, now displayed at the American Museum of Natural History. (See American Museum of Natural History, New York, New York.) Using rubber molds of the original, this dinosaur was assembled in 1987 in an upright position, with one leg up. A display on how dinosaur bones are assembled is part of the exhibit.

Complementing the dinosaur exhibits is the Fossil Preparation Laboratory or Family Laboratory of Earth Sciences. This highly sophisticated facility has a viewing window through which visitors can watch personnel prepare and study rock and mineral specimens and fossils of flora and fauna. In the large-specimen area, fossils are removed from rocks, and materials are readied for mounting. In the small-specimen preparation area, a video monitor wired to a special microscope shows scientists carrying out exacting work that the public rarely has an opportunity to see. The open storage area contains many fossil specimens that are part of the museum's reference collection.

Prehistoric Journey displays a diverse array of dinosaurs and other fossils, the fruits of this laboratory's work. The first dinosaur exhibit in the museum, created in 1936, resulted from a high school field trip on which a *Stegosaurus* was uncovered. Other specimens on exhibit, many from Colorado, are mastodons, mammoths, saber-toothed tigers, horses, and camels. The impressive, 12-foot (3.7-m) high imperial mammoth found in Nebraska is considered a young fossil at only 10 000 years old.

The Hall of Gems and Minerals has a crystal cave composed from pieces of two caverns in the silver mines of Chihuahua, Mexico. Gypsum crystals, some several feet (1 m) long, come from the Xochitl Cavern; crystals in the rest of the cave come from the El Potosí mine. The hall also displays minerals found in Colorado.

The museum has an extensive array of ecology dioramas and mounted wildlife, from polar bears on the frozen Arctic Ocean to the manatees of Florida's Crystal River. Among the scenes of Colorado ecology and wildlife is one that shows a coyote family near its den about 30 miles (50 km) from Denver. Another shows a moose, the largest member of the deer family, as it lives in the Talkeetna Mountains of Alaska. Its normal range is from the Alaska in the north, through Canada, and south into Colorado. A family of nesting golden eagles is part of a scene of the Pikes Peak region. A diorama containing pronghorns, often confused with antelope, highlights the only living members of a family of animals that survived from the last Ice Age. Once common on the prairie, pronghorns have been returned to Colorado thanks to a major recovery effort.

The hall entitled Botswana, Africa, has a unique life-sized diorama, Savuti Waterhole. It is populated by mounted specimens of chacma baboons, greater kudu, antelope, African elephants, warthogs, steenbok, and zebras.

Anthropology and archaeology are explored in the halls of Egyptian Mummies, North American Indian Cultures, South Pacific Islands, and Australia. Space Odyssey is an exhibit about the cosmos, including an area about Mars with a huge diorama of the bottom of an actual Martian canyon, and interactive displays on Martian crater formation and ancient water action.

Address: Denver Museum of Nature & Science, 2001 Colorado Boulevard, Denver, CO 80205.
Phone: (303) 370-6000.
Days and hours: Daily 9 a.m.–5 p.m. Closed 12/25.
Fee: Yes.
Tours: Self-guided.
Public transportation: RTD buses stop near the museum.
Handicapped accessible: Yes. Wheelchairs are available on a first-come, first-served basis.
Food served on premises: Café and snack bar.
Website: www.dmns.org

Denver Zoo

The mayor of Denver received an orphaned black bear cub as a gift in 1896 and tethered it near a haystack in City Park. From that humble beginning, the Denver Zoo has grown to 76 acres (31 hectares) containing more than 600 species. This organization is first and foremost a place to conserve wildlife. It also strives to enhance the public's appreciation of a world wildlife, a goal closer to being achieved as new habitats are constructed with increasingly naturalistic designs. These large, more sophisticated habitats are making it possible for the zoo to focus its collection by having fewer animals in larger enclosures and to include rarer species.

The zoo pioneered the use of naturalistic habitats with Bear Mountain, constructed in 1918. This exhibit includes castings from a rock formation outside Denver. The Denver

Zoo was the first zoo in the United States to build an exhibit this way, setting a standard for other zoos. Currently Bear Mountain still houses grizzly bears and coati.

Bird World includes an indoor, simulated rain forest and swamp, seashore, and treetop exhibits. The outdoor section, the Pheasantry, houses hundreds of species. Harp wire creates an almost invisible barrier between the animals and people.

Older exhibits that have been reworked with naturalistic living spaces include the Feline House, Sheep Mountain, and the Condor-Eagle Aviary. Northern Shores features an underwater viewing facility for observing polar bears and sea lions. This exhibit also contains Arctic foxes and wolves. The EDGE is a one-acre (0.4-hectare) enclosure for several Amdur tigers, with a pine forest mimicking their normal Asian habitat, and including two pools and connecting bridges over visitors' heads.

The zoo participates in Species Survival Plans® for saving and propagating selected animals and reintroducing them to the world. It serves as an urban refuge for many rare and endangered species, including a herd of Père David's deer, which have been extinct in the wild for 1200 years. It has a breeding herd of Mongolian Przewalski's horses, the only truly wild horses in the world. Both animals are being reintroduced into China.

This zoo is one of the few in the United States to exhibit the warthog from Africa, south of the Sahara, and the babirusa, an Indonesian island swine species endangered by rapid destruction of its habitat.

The highlight of the Denver Zoo is Tropical Discovery, a total-immersion exhibit that represents a tropical rain forest. It includes species from North and South America, Asia, and Australia, housed in two glass pyramids containing more than 6000 plants and 59 exhibits in 14 separate areas. This state-of-the-art facility makes extensive use of graphic panels, labels, and illustrations to help visitors identify and learn about the animals. Here is one of the largest Komodo dragon exhibits in the United States. Harmless lizards roam throughout the building. Upon entering Tropical Discovery, visitors arrive in the Jungle Clearing, under a canopy of lush vegetation near a large waterfall. A glass-fronted, 2000-gal (7600-L) pool contains freshwater fish, some familiar to home-aquarium owners and some exotic. The pool also houses South American turtles. Nearby are cameo exhibits of flamboyant flower beetles, Bell's horned frogs, Suriname toads, Tokay geckos, and others.

The Forest Clearing features a mixed grouping of mammals from Southeast Asia. Greater Malayan chevrotains or mouse deer, which are among the smallest hoofed animals in the world, graze the forest floor. The trees contain such interesting animals as Malayan tricolored squirrels and tree shrews.

Visitors can walk through a realistic-looking Mountain Cave exhibit to see vampire bats, neotropical fruit bats, and Mexican blind cave fish. In the Tropical Rain Forest are shy, endangered clouded leopards. The Temple Ruins, which acknowledges human beings as part of the environment, dominates the pyramid-shaped structure. In this part of the building are venomous snakes, such as the Asian monocled cobra, the type of snake often seen with Indian snake charmers.

A series of cameo exhibits in the Tropical Discovery focuses on tropical biodiversity. Special temperature-controlled environments shelter exotic creatures like poison dart frogs, whose poison natives apply to their darts, and Goliath bird spiders, which prey on small birds.

At River Pools a school of red-bellied piranha—South American cichlids whose fierceness has been overstated in movies—swim in search of their next meal. The Tropical Riverbank displays spectacular and bizarre fish of the South American Amazon drainage area. Its 11 000-gallon (42 000-liter) tank contains specimens of the world's largest fish, the arapaima or pirarucu, which can grow to 10 feet (3 m), and the anaconda, the New World's heaviest and largest snake, which can reach 33 feet (10 m) in length. Another unique animal in this exhibit is the semiaquatic capybara, the largest rodent in the world. Its weight sometimes reaches as much as 100 pounds (45 kg).

The Mangrove Swamp has animals that live on land, in brackish or fresh water, and in the sea. They all are influenced by the rise and fall of the tides. Among the saltwater fish are damsels, angelfish, wrasses, and the fascinating Southeast Asian archer fish, which spits droplets of water at insects to make them fall to the water below, and then eats them. Mudskippers spend time both in water and on land, where they use their special fins as limbs to walk.

The centerpiece of the Tropical Reefs area is a coral reef tank that holds more than 15 000 gallons (57 000 L). Species inhabiting the tank are primarily from the Caribbean Sea such as the colorful triggerfish, ominous green moray eels, and horned sharks. Visitors can watch the roughly 25 fish species through a bubble window.

In the Cypress Swamp, with its moisture-laden air and trees draped with Spanish moss, is a large diorama depicting the swamp at dusk. The alligator snapping turtle, a North American freshwater turtle, as well as largemouth bass and other fish, live in this exhibit.

Siamese crocodiles from Thailand inhabit the Tropical Wetlands. These crocodiles, believed to be extinct in the wild, live only in zoos and in commercial farms where they are raised for their valuable skin. The exhibit depicts an eroded riverbank, covered with plants and teeming with life.

The Discovery Center supplements and reinforces the conservation theme of Tropical Discovery and the zoo. It has a demonstration area displaying artifacts and live animals.

Address: Denver Zoo, 2900 East 23rd Avenue, Denver, CO 80205.

Phone: (720) 337-1400.

Days and hours: Daily 10 a.m.–4 p.m. Grounds close at 5 p.m. Closed Thanksgiving Day and 12/25.

Fee: Yes.

Tours: Self-guided. Animal Encounters incur an extra fee.

Public transportation: RTD buses stop at City Park, near the zoo entrance.

Handicapped accessible: Yes. Wheelchairs and scooters are available for rental on a first-come, first-served basis.

Food served on premises: Restaurants, café, snack shops.

Website: www.denverzoo.org

DENVER AND GEORGETOWN

History Colorado Center and Georgetown Loop Railroad® & Mining Park

Both of these sites are operated by the History Colorado, a group founded in 1879 that maintains 11 museums and historic sites to help the public appreciate the region's rich, colorful past.

History Colorado Center
History Colorado Center, in the heart of Denver, is the home base of the History Colorado organization since 2012. It explores the prehistory and history of the West,

with special focus on Colorado. Topics spotlighted include Native Americans of the region in prehistoric and more recent times, and the role of mining in Colorado's development. A major exhibit is an extensive timeline, accompanied by artifacts, which illustrates nearly 200 years of Colorado history.

Living West examines Colorado's people and their relationship to the land from Mesa Verde hundreds of years ago to the present, which includes a variety of Native American artifacts such as baskets, ceramics, and tools from the Pueblo peoples. The Dust Bowl of southeastern Colorado in the 1930s explains the scarcity and necessity of water. The current environment of the Rocky Mountains is also examined.

Written on the Land showcases the state's longest continuous residents, the Ute Nation. Among Ute technology displayed are items of beadwork, clothing, tools, wooden saddles, and baskets.

Destination Colorado is the re-created town of Keota, in the 1920s, with all the tools and tribulations of living in eastern Colorado from a century ago. Typewriters, farming implements, a General Store, and an early automobile are among the artifacts on display. Near this exhibit is the locally famous Denver Diorama, which depicts the city in the early 1860s. The diorama was built to 1/192 scale as a WPA project in the 1930s, and is now surrounded by SmartGlass, which becomes more opaque under brighter illumination, for protection from light.

Housed within the History Colorado Center's building is the Hart Research Library, which holds History Colorado's archives, including artifacts, newspapers, and photographs.

Address: History Colorado Center, 1200 Broadway, Denver, CO 80203.
Phone: (303) 447-8679.
Days and hours: Museum: Daily 10 a.m.–5 p.m. Closed Thanksgiving, 12/25, and 1/1. Hart Research Library: Wed.–Sat. 10 a.m.–3 p.m.
Fee: Yes.
Tours: Self-guided.
Public transportation: Buses stop near the site.
Handicapped accessible: Yes. Wheelchairs are available on a first-come, first-served basis.
Food served on premises: Café.
Website: www.historycolorado.org/history-colorado-center

Georgetown Loop Railroad® & Mining Park
In Georgetown, a 50-minute drive from Denver, a steam-powered train tours the Georgetown Loop, a National Historic Site in an area once rich with gold and silver ore. Visitors buy tickets at the Georgetown Devil's Gate terminal and either board the train there or start at the original depot in Silver Plume, which has maintenance and repair facilities open to the public. The train tour covers 6 miles (10 km) round trip (1 hour 15 minutes). It stops at several of the late-19th-century mines.

At the Lebanon Mine and Mill, visitors may take an optional tour (1 hour 20 minutes). Here visitors go 500 feet (150 m) into the mine to see the sampling and concentrating mill and walk the grounds to view the manager's office, change room, blacksmith shop, and toolshed. At the Lebanon Mine, manager Julius Pohle invented a pneumatic air pump with no moving parts to keep the mine from flooding. It was so effective that it became widely used in the mining industry to keep mines free of deep water. A second tour is the Lebanon Extension Mine Tour, 900 feet (270 m) into Leavenworth Mountain, following a section of the Hise ore body. Visitors see calcite dams, calcified hobnail bootprints, and more. The third tour visits the Everett Mine, with dim lighting similar to candles the miners used. Here visitors learn about the transition from hand- to mechanical drilling, tunnel construction, and blasting.

Along the train-tour route, passengers learn how two brothers, George and David Griffith, came to this area at the end of the 1850s as rumors of gold deposits in Pikes Peak country spread. George, who found gold while panning in Clear Creek, staked out the Griffith Lode, changing the history of the area forever. People began to move into Clear Creek Canyon and a camp formed that became known as George's Town, later Georgetown. As other gold-containing lodes were discovered, the Griffith mining district became a booming concern.

The gold boom lasted only about a decade, but soon silver was discovered and the area was revitalized. In the mid-1860s, Georgetown was the fastest-growing community in Colorado. Silver mining spread up the valley, and the town of Silver Plume was founded in 1870.

Over time, the quality of the silver ore declined, which meant more bulk ore had to be shipped. To improve the economic value of silver, the mining industry sought ways to reduce the amount of ore shipped out of the valley. Eventually new techniques were pioneered to isolate the silver from the ore and produce bullion. The first silver bullion from the Griffith Mining District was made by Lorenzo Bowman, a freed slave who worked at one time as a miner in Missouri's galena mines. One of the few miners who knew anything about smelting, he founded a smelter called Bowman and Company on Leavenworth Mountain.

When the increasing complexity of the silver ore required even more sophisticated techniques, a chemistry professor at Brown University in Providence, Rhode Island, provided assistance. Nathaniel Hill took samples of Colorado silver ore to Swansea, Wales, where the world's leading center for smelting copper was located. Hill found that the Welsh process could be adapted to Colorado silver ores and in 1868 opened the Boston and Colorado Smelting Company in Black Hawk.

A narrow-gauge railroad constructed to transport the ore, thus lowering freight costs and increasing profits, reached Black Hawk in 1872. A financial panic in 1873 halted railroad construction beyond Black Hawk, but eventually the Union Pacific Railroad decided to continue building the line. After the line reached Georgetown, the next two miles (3.2 km) of track to Silver Plume had to be built on difficult topography. This required the narrow-gauge railroad to rise 600 feet (180 m) while negotiating tight curves more efficiently than was typical in railroad construction of the day. Chief engineer Jacob Blickensderfer designed a high bridge at Devil's Gate, near George's Town, with a tack that looped back over itself in a helix. Devil's Gate Bridge was 300 feet (90 m) long and rose 100 feet (30 m) above the valley floor.

The Union Pacific never finished the line, ending it a few miles past Silver Plume. By the last quarter of the 19th century, the Georgetown Loop attracted large groups of tourists who clamored to ride on the engineering feat and see the beautiful scenery. In 1939 the bridge was dismantled and sold for scrap, and World War II took workers away from the mines, forcing their closure for good. Today, the tour train travels over a reconstructed Devil's Gate High Bridge built in 1984.

Address: Georgetown Loop Railroad® & Mining Park. Devil's Gate Depot, 646 Loop Drive, Georgetown, CO 80444. Silver Plume Depot, 825 Railroad Avenue, Silver Plume, CO 80476.
Phone: (800) 456-6777.
Days and hours: See website for exact schedule. Georgetown Devil's Gate Depot runs from late Apr. through Dec. Silver Plume Depot runs from late May through late Sep.
Fee: Yes.
Tours: Guided.
Public transportation: None.
Handicapped accessible: Partly; trains are accessible, but mine tours are not.
Food served on premises: Snacks.
Special note: Bring a heavy jacket because the mine's internal temperature is 44°F (7°C) year-round. No children under 5 in the mines. Closed-toed shoes are a must.
Website: www.georgetownlooprr.com

Estes Park

Rocky Mountain National Park

Indications are that human life came to these majestic mountains between 10 000 and 20 000 years ago, and that human habitation continued off and on over the centuries. Modern Ute and Arapaho tribes are the most recent Native American groups to live here. The first Europeans to travel these mountains and valleys were French fur traders. In 1859 Joel and Milton Estes, father and son, came to the valley that now bears their name. Few settlers followed, however, and the region remained sparsely populated. At the beginning of the 20th century, Enos A. Mills, a naturalist, writer, and conservationist, recognized the beauty and uniqueness of this land and launched a campaign to preserve its pristine glory. In 1915, it became Rocky Mountain National Park.

The Park has many special features, such as significant variations in the land, flora, and fauna, and changes in elevation. A full one-third of the park is above the tree line at 10 000–11 000 feet (3000–3400 m). It is alpine tundra, very unusual at this latitude and the major reason the area was made a national park.

To reach the uppermost region of the park, more than 12 000 feet (3700 m) above sea level, Trail Ridge Road was constructed. This east-west route, the highest continuously paved highway in the nation, goes through several ecosystems—the same climates and habitats travelers would encounter when going from Denver to the Arctic Circle. Changes are so dramatic that there can be 25°F (14°C) temperature drop and two to three times more precipitation at the peak than in the lower elevations. (See Hawai'i Volcanoes National Park, Hawaii National Park, Hawai'i.)

Rocks in the upper portions of the park are composed of gray, white, and black gneiss and schist. The base of the mountains is granite and pegmatite. When naturalists sought to form Rocky Mountain National Park, no gold or silver had been found within its proposed bounds—unlike in nearby areas—so no one was seriously opposed to the park.

The Continental Divide, sometimes referred to as the backbone of the continent, runs through the park from its northwest corner to the southeast. Very different weather patterns exist on each side, and rivers flow in different directions. On the eastern side water flows into the Atlantic Ocean, and on the western side it flows into the Pacific Ocean. The park contains the headwaters of the Colorado River, which from here begins its 1400-mile (2300-km) journey to the Pacific Ocean. (See Hoover Dam, Hoover Dam, Nevada.)

Those who drive along Trail Ridge Road find many wayside exhibits and self-guided trails that encourage them to explore the park's natural features. The park also has several visitor centers with explanatory displays and programs.

At the lower elevations, between 7000 and 9000 feet (2100–2700 m) above sea level, is the montane ecosystem, where elk, bighorn sheep, and other animals graze in grassy meadows. On the south ridge are ponderosa pine and juniper forests. The north side has Douglas fir, stands of lodgepole pine, and groves of aspen.

Climbing higher, the forests and animal life change. Subalpine forests are filled with Engelmann spruce and subalpine fir. The forest floor is covered with moss, orchids, and blueberry bushes. This is the land of the black bear. (See Lassen Volcanic National Park, Mineral, California.)

At the tree line is the *krummholz*, meaning "crooked wood" in German, which contains miniature forests of spruce and fir created by cold temperatures, scant water, and strong winds. Trees that are hundreds of years old may be no more than a foot or two high.

Beyond the krummholz is the alpine tundra. In summer the temperature rarely exceeds 60°F (16°C), and blizzards are common. (Park personnel advise visitors to bring appropriate clothes for the wide range of possible weather conditions.) Hardy plants with an interesting array of adaptive features have evolved because of the harsh conditions—winters lasting five months or more, long periods of below-freezing temperatures, and a short growing season. Many of the 200 species of tundra plants are no more than an inch or two (3–5 cm) high. More than a quarter of them are also found in the Arctic. Several contain anthocyanin, a chemical that acts like antifreeze, and have tiny hairs that protect the plants from losing heat and moisture.

Alaska has the only other expanse of tundra like this in the United States. Animals living in this part of the park include elk, deer, bighorn sheep, and the ptarmigan bird.

Old Fall River Road climbs along 9 miles (14 km) of forested land and ends in an open alpine tundra. Guide booklets are available at the beginning of the road and at the visitor centers. A walk along the 0.3-mile (0.5-km) Tundra Trail, which branches off this road at an altitude of more than 12 000 feet (3700 m), provides a close-up view of a world so fragile that visitors are asked not to wander off the path. Signs along the way help visitors understand what they are seeing.

Glaciers played a major role in shaping the parkland, creating canyons and leaving moraines, or deposits of debris, behind. The 0.25-mile (0.4-km) Moraine Park Nature Trail allows a good look at the five small glaciers still in the park. At the end of the glaciated valley the trail passes through ponderosa pine. Trail booklets are available at the Moraine Park Museum and the visitor centers.

Although many national parks have housing within their boundaries, this park has only camping facilities. Housing is available in nearby towns. Road access may be limited in winter because of snow and adverse weather. Because of its high altitude, Trail Ridge Road may be dangerous to those with heart conditions. All visitors should acclimate themselves to the altitude before exertion. Beaver Meadows Visitor Center and Kawuneeche Visitor Center offer a 20-minute introductory video.

Fall is a special time in the park. The leaves change to beautiful colors and elf bugling rings throughout the park. Years ago Enos Mills said to visitors of this park, "These are your fountains and gardens of life; kindly assist in keeping them." His words are just as important today.

Address: Rocky Mountain National Park, Estes Park, CO 80517.
Phone: (970) 586-1206.
Days and hours: Park: Daily 24 hours a day. Beaver Meadow Visitor Center: Daily 9 a.m.–4:30 p.m. Other visitor centers have more restrictive seasonal hours.
Fee: Free.
Tours: Self-guided. There are a variety of ranger-led programs.
Public transportation: None to the park. A shuttle runs regularly during the summer from Glacier Basin Campground to various locations within the park.
Handicapped accessible: Mostly. Some trails have limited accessibility.
Food served on premises: None; picnic areas only. A variety of eateries are in Estes Park.
Special Note: A wide range of possible weather conditions is possible in the park, including snow, so bring appropriate clothing. High altitudes may be dangerous to those with heart conditions.
Website: www.nps.gov/romo/index.htm

GOLDEN AND IDAHO SPRINGS, COLORADO

Mines Museum of Earth Science and Edgar Mine

These two facilities of the Colorado School of Mines offer an opportunity to see thousands of top-quality mineral samples and to take a half-mile (0.8-km) underground tour of a mine that once yielded high-grade silver, gold, lead, and copper. The mine is now a training facility.

Mines Museum of Earth Science

The Colorado School of Mines, founded in Golden in 1868, was transferred to the territorial legislature in 1874. Its purpose was to teach practical mining skills to students, who, after several classes, usually disappeared into the mountains in search of mineral riches. The Mines Museum of Earth Science began in 1882 as a teaching tool for the professors to illustrate mineral identification and blowpipe analysis.

Collections donated to the museum early on came from the Colorado Bureau of Mines; J. Alden Smith, state geologist during the end of the 19th century; and Frank C. Allison, a newspaper publisher and amateur mineralogist. Smith's personal collection is known for its tellurides, gold, silver, and fossils. Allison's collection, amassed over 35 years, has a large number of specimens from different gold-yielding localities in Colorado. Today the museum has more than 40 000 specimens from Colorado, the rest of the United States, and other countries. It displays 2500 of the best mineral samples, and 100 faceted gemstones.

In the foreign collection, the Brazilian pegmatites, igneous rocks sometimes rich in rare elements, are especially notable. The US display features the tristate region of Oklahoma, Kansas, and Missouri, which at one time produced more than half the nation's zinc. An exhibit from the eastern United States and Canada includes samples of rutile from Georgia, franklinite from New Jersey, and purple apatite from Maine. (See The Sterling Hill Mining Museum and the Franklin Mineral Museum, Ogdensburg and

Franklin, New Jersey.) There is also a systematic exhibit aimed at mineralogy students, in which the samples are organized by chemical composition, and a small collection of laboratory equipment that includes a late-19th-century binocular microscope.

The museum displays fossils and mining equipment. The Thomas Allen Mine Lamp Collection includes 40 lamps illustrating the development of mine lighting from the 17th century to the present. Clear Creek Cave is a walk-through diorama illustrating the formation of carbonate speleothems (cave deposits) in Precambrian gneiss from the Colorado Front Range.

Other items of special note are moon rocks from the Apollo 15 and 17 missions. From the 1939 Golden Gate International Exhibition are murals painted by Irwin Hoffman illustrating the history of mining. There is an introductory video about geology of the region.

Address: Mines Museum of Earth Science, 1310 Maple Street, Golden, CO 80401.
Phone: (303) 273-3815.
Days and hours: Mon.–Fri. 9 a.m.–4 p.m., Sat. 10 a.m.–2 p.m., Sun. 1–4 p.m. Closed 12/25 and 1/1. Call to confirm if the museum is open during university breaks.
Fee: Free. Fee for guided tours.
Tours: Self-guided and guided. Call at least a week in advance for tours.
Public transportation: Buses run from Denver to Golden, then walk 0.9 mi. (1.4 km).
Handicapped accessible: Yes.
Food served on premises: None. A variety of restaurants are several blocks away, mostly northeast from the museum.
Website: www.mines.edu/geology-museum/

Edgar Mine
Also known as Colorado School of Mines' Experimental Mine, the Edgar Mine is about 15 miles (24 km) away, near the town of Idaho Springs. It was named after the Edgar mineral vein discovered at the site in 1863 by Richard McNiel. The mine is cut into Front Range Precambrian gneiss and granite with Laramide porphyritic intrusive igneous rock. Among the rich yields from the mine were pyrite ores rich in gold, silver, and copper; galena-sphaleritic ores rich in lead and silver; composite ores; and telluride ores with free gold.

The school obtained the mine in 1921 for training mining engineers. Its objectives have expanded to include an extensive research program on mining-related topics such as tunnel detection, blasting, geophysical exploration, and drilling.

On a tour of the mine, visitors learn about mining technology and ongoing research at the site. Tours are conducted from the entrance to the Miami Tunnel, cut in 1870 to provide an easier way to haul ore from the mine's lower workings and to improve the mine's drainage. Although called a tunnel, which has two open ends, it is actually an "adit" because it has only one open end. The Miami Tunnel extends about 2000 feet (600 m) northwest and connects with a drift from the Edgar shaft.

The walking tour, led by a staff member or student, takes about 45 minutes and passes historic and modern mining equipment in situ, block-testing facilities, and numerous boreholes made for various tests. Although no atomic wastes are stored here, there is an atomic waste testing room. The rock in the tunnel consists of biotite gneiss, biotite schist, magmatized gneiss, pyrite, hornblende, feldspar crystals, quartz, and other minerals. Some years after the Miami Tunnel was first cut, it was connected to the Big Five Tunnel at the west end of Idaho Springs.

The tour also goes through the research and development center and classroom training center for mining engineers, both of which are located inside the mine. Facilities visitors see at the mine surface include an industrial shop, a dry house where miners change to dry clothes, a compressor house, storage facilities, and an underground powder magazine. A 2-ton (1.8-tonne) hoist, an air-ventilation system, and many mining tools are visible along the tour route.

Address: Edgar Mine, 365 8th Avenue, Idaho Springs, CO 80452.
Phone: (303) 567-2911.
Days and hours: Call for hours.
Fee: Yes.
Tours: Guided. Call in advance to book a tour.
Public transportation: Buses run from Denver to Idaho Springs, then walk 0.8 mi. (1.3 km).
Handicapped accessible: No. Unpaved paths and rough walkway.
Food served on premises: None.
Special note: Sturdy shoes suggested. Wear warm clothing, for the temperature in the mine is 52°F (11°C). Visitors must wear hard hats provided by the mine.
Website: tour.mines.edu/edgar-mine/

Idaho

POCATELLO

Idaho Museum of Natural History

This natural history museum on the campus of Idaho State University serves both the university and the public. It has more than a million specimens in its collection, only portions of which are on display. Established in 1934 as a historical museum, over the years its focus has changed to preserving and disseminating information about the natural history of Idaho and the intermountain Northwest.

The collections explore anthropology, paleontology, geology, botany, and zoology. Specimens mainly come from the extensive field-collection gathered by the university faculty and museum staff. The archaeology collections include documentation for more than 8000 sites, including field records, artifact catalogs, maps, and photographs. Many of the museum's exhibits are interactive. Especially interesting research at the museum explores the interdisciplinary relationships of geology, flora, fauna, and humanity in the intermountain West during the Quaternary period.

Marine fossils give clues to the nature of the vast sea that once covered what is now Idaho. As the sea receded, the current basin and range topography formed. Left behind was the skeleton of a spiral-tooth shark, *Helicoprion*, the only such animal discovered in Idaho. Castings of several dinosaurs are also on display.

Exhibits of southeastern Idaho fossil vertebrates from the Tertiary and Quaternary periods include a well-represented collection of fauna from the American Falls Reservoir

Helicoprion fossil, Idaho Museum of Natural History.
Courtesy of the Idaho Museum of Natural History.

area to the northwest of Pocatello. Other exhibits chronicle the evolution of the bison in North America, illustrated with fossilized remains. The paleontological exhibits have some rare and interesting specimens, including an extinct giant bison, *Bison latifrons*. This huge animal, which survived into the late Wisconsinan period some 20–30 thousand years ago, had horn-cores spanning 7 feet (2.1 m), more than three times as far apart as the modern bison. A mounted skeleton of a saber-toothed cat is also on display.

The anthropological collection has excellent displays on the Native Americans who have lived in this area for more than 10 000 years. Among artifacts visitors can see are stone tools, baskets, and other evidence of ancient lifestyles, found in an ice cave in a lava tube. The museum collections include more than 4000 pieces of basketry and other artifacts collected in this area. Displays feature Tlingit basketry, Nez Perce agriculture before the modern historic period, and the hunter-gatherer Great Basin Native Americans (including the Shoshone, Bannock, and North Paiute peoples). Pocatello is adjacent to the largest surviving Shoshone-Bannock community. The museum is well-known for its collection of baskets, beadwork, and buckskins.

In a display of regional animals that are part of the sagebrush desert food chain, several mounts are a century old and many of the species are endangered or threatened.

There are regularly changing exhibits as well. The museum is located in the university's Hutchinson Quadrangle.

Address: Idaho Museum of Natural History, 698 East Dillon Street, Pocatello, ID 83201.
Phone: (208) 282-3168.
Days and hours: Spring and summer: Tue.–Fri. 10 a.m.–6 p.m., Sat. 9 a.m.–5 p.m., Sun. noon–5 p.m.; fall and winter: Tue.–Fri. noon–6 p.m., Sat. 9 a.m.–5 p.m., Sun. noon–5 p.m. Closed Mon., Easter Sunday, 7/4, Thanksgiving and the following Fri., 12/24, 12/25, 1/1.

Fee: Yes.
Tours: Self-guided.
Public transportation: None.
Handicapped accessible: Yes.
Food served on premises: Snacks. There are fast-food restaurants in Pocatello near the Quadrangle.
Website: www.isu.edu/imnh/

Montana

BOZEMAN

Museum of the Rockies

In 1972, this facility opened at Montana State University in Bozeman to inform the public about the prehistory, history, and culture of the northern Rocky Mountains. It has more than 90 000 square feet (8400 m²) of displays, through which "Museum of the Rockies brings the world to Montana and shares Montana with the world," according to its website.

Bozeman, a small Rocky Mountain city, was named for John Bozeman, who brought the first wagon train to the region and settled in the Gallatin Valley. Native Americans called the area the Valley of the Flowers and knew it as a sacred hunting ground and neutral territory. The Europeans brought agriculture, making this land some of the area's most productive. Because the city is not far from some Rocky Mountain passes, it became a center for miners and westward-bound travelers.

The exhibits follow the theme "One Place Through All Time." They start with the age of the dinosaurs 80 million years ago and move on to cover the Plains Native Americans, the pioneers and homesteaders, and recent events in Montana. The staff members share their research with the public in each exhibit. The museum, famous for its exploration work, has impressive paleontology and archeology exhibits in the Siebel Dinosaur Complex. One display is a life-size restoration of a nesting colony of *Maiasaura peeblesorum* dinosaurs found in Choteau, Montana. The diorama shows the eggs, babies, and parents, presenting a rare view of dinosaur parenting skills, and includes proper foliage for the period and era. *Maiasaura*, which means "good mother lizard," is the official Montana state fossil. The hadrosaur on display is complete with a backbone and a scalloped fringe of skin running down the back. The skulls of a Triceratops and a *styracosaur*, as well as a variety of other fossils, are also on view.

This museum has a world-class collection of *Tyrannosaurus rex* fossils, among them the largest and smallest *T. rex* skulls found. Here visitors also can see "Montana's *T. rex*," one of the most complete fossils of this species, 12 feet (3.7 m) tall and about 40 feet (12 m) long. This gallery includes the Bowman Dinosaur Viewing Laboratory, where visitors can view volunteers at the museum preparing fossils for study and display by removing the rock encasing them.

Enduring Peoples is a gallery devoted to the Native Americans who have long inhabited the region. The life and culture of the Plains Natives Americans are illustrated in displays of tools, weapons, clothing, toys, and other artifacts.

The Paugh History Hall examines changes in culture and society in the northern Rockies, from Native Americans and fur traders, to gold miners and settlers. At first the northern Rockies was a fur-trading area. Settlement came later. Objects used by pioneers and homesteaders are on display, such as the first "camper," a 1915 sheepherder's wagon; a wagon shop full of tools; and a fully furnished 1930s house and gas station. Historic photographs and pieces from the museum's large clothing collection help illustrate life here in the second quarter of the 20th century. The museum also has many pieces of Western art, including works of Charles Marion Russell and John Paxton from around the turn of the 20th century.

The facility's planetarium, with its 40-foot (12-m) dome and Digistar projection system, gives vivid views of the heavens. This planetarium blends the science of astronomy with the art of theater. The museum also sponsors a variety of adult-oriented lectures on science and anthropology.

Address: Museum of the Rockies, 600 West Kagy Boulevard, Bozeman, MT 59717.
Phone: (406) 994-2251.
Days and hours: Daily 9 a.m.–5 p.m. Closed Thanksgiving, 12/25, 1/1.
Fee: Yes.
Tours: Self-guided.
Public transportation: None.
Handicapped accessible: Yes. Wheelchairs are available on a first-come, first-served basis.
Food served on premises: Snacks.
Website: www.museumoftherockies.org

GARDINER

Yellowstone National Park

See Yellowstone National Park, Wyoming.

Wyoming

YELLOWSTONE NATIONAL PARK

Yellowstone National Park

In 1874, 44 years before the creation of the National Park Service, Congress set aside 2.2 million acres (890 000 hectares) of unique lands with more geysers and hot springs than anywhere else on Earth. Yellowstone, which is larger than the states of Delaware, Rhode Island, and the District of Columbia combined, is, in effect, the first national park. Although Iceland, New Zealand, and Kamchatka in Siberia all have concentrations of the thermal features found in Yellowstone, the total of all the geysers in these three places, plus all the additional geysers around the world, is fewer than those in Yellowstone. (See Lassen Volcanic National Park, Mineral, California.)

The Yellowstone River runs through the park, creating spectacular scenery marked by more than 150 waterfalls. The river also produces Yellowstone Lake, the nation's largest

Lower Falls from Artist Point in Yellowstone National Park.
Courtesy of the National Park Service and Diane Renkin.

mountain lake. Golden cliffs and yellow banks give the river its name. The Minnetaree Native Americans called the river *Mi tse a-da-zi* or Yellow Stone River. French fur trappers translated the Native American name into *roche jaune* or "yellow stone." Today, both the river and park are called Yellowstone.

An unusually large number and variety of wild animals live on these lands in their natural habitats. The park has the greatest concentration of large and small mammals in the lower 48 states. Living on these lands are seven species of hoofed animals, two species of bears, about 50 species of other mammals, 11 native species of fish, and six species of reptiles.

The park is divided into five distinct regions, each quite different. A roadway 145 miles (233 km) long forms a near figure-8 as it passes through the five sections and crosses the Continental Divide three times. (See Rocky Mountain National Park, Estes Park, Colorado.) The 45th parallel, marking the halfway point between the North Pole and equator, also runs through the park.

When this national park was established, the original purpose was for people to relax and restore their spirits away from the "real world." Although the recreational aspects are still important, the park's mission has expanded to include saving the local ecology. By the 1890s the bison were nearly extinct. In 1916 they were recovering, and two herds, a total of 278 animals, were recorded in the park. Today about 4500 bison wander the park freely. Yellowstone is recognized as an international model for reclamation of animals and is an International Biosphere Reserve and World Heritage Site.

The park headquarters in Mammoth Hot Springs was the site of Fort Yellowstone between 1886 and 1918, where a detachment of 200 cavalry officers were billeted. The former bachelor officers' quarters is now Albright Visitor Center, which has exhibits about the natural and human history of the park.

Geological highlights of Mammoth Country include cascading terraces formed by the calcium carbonate-laden water of the hot springs. For more than 8000 years they have precipitated up to 2 tons (1.8 tonnes) of travertine mineral each day, forming layer upon layer of beautiful stone. The colors of the hot springs, ranging from red to black, are caused by minerals, algae, reflections, light absorbed by colloids suspended in the water, and light absorbed by the water itself.

Norris Geyser Basin, in the southern part of Mammoth Country, once held a soldier station. The first structure, built in 1886, was replaced in 1897 with the current building, which was modified about 20 years later and has since been restored. It holds a museum that explains the evolution of the park ranger from soldier to modern specialist.

The first formal exploration of this area was done by Philetus W. Norris, the second superintendent of Yellowstone (1877–1882). A self-taught historian, archaeologist, and scientist, he provided information on the geysers and hot springs, built the first roads to some of the scenic wonders, and encouraged development of game management—the first such program on federal lands. The rustic Norris Museum was opened in 1930 as one of the first trailside museums in the park.

To some, the geysers are the park's key feature. Exhibits at the Norris Museum explain the hot springs, geysers, mud pots, and fumaroles in the area, as well as the life forms. The boiling water and thin land crust cause new pots, springs, and geysers to form regularly. The tallest geyser in the world, Steamboat Geyser, is in this section of the park.

Mammoth Country's Norris Area is where two major underground faults meet. Earthquakes and tremors are common here. The surface constantly changes as water flows and then stops because of shifting subterranean features. Geysers are formed when surface water seeps deep into the ground and is heated by magma or molten lava near the surface. The heater water then rises and bubbles or gushes out of the surface of the rock as hot springs, boiling pots, and geysers. Although the hot, mineral-laden water can be harsh to plants and animals, some species have adapted, and flourish in this environment.

The most popular tourist stop in the Old Faithful Area is Old Faithful, a geyser that, as its name implies, erupts regularly—approximately every 94 ± 10 minutes. A downloadable NPS Yellowstone App gives eruption predictions. Other hydrothermal features, including fumaroles, mud pots, and hot pools, dot the area. Castle Geyser, perhaps the park's oldest geyser, has the largest cone of any geyser in the park. When it erupts—about every 9–10 hours—the water may reach a height of 90 feet (27 m). A noisy steam phase, lasting about 1 hour, follows the eruption.

The focus of Lake Country is Yellowstone Lake. One of the largest alpine freshwater lakes in the world, it is 430 feet (131 m) deep and has a shoreline 110 miles (177 km) long. The lake formed over the last 2 million years, when a series of violent volcanic eruptions caused a vast basin, or caldera, to form. Volcanic activity is still possible: the land in this part of the park is rising at just under 1 inch (2.5 cm) per year. The subterranean turmoil is belied by the beauty and majesty of the landscape and wildlife. The lake is framed at the eastern edge by the Absaroka Mountain Range. Among the animals living here are moose, waterfowl, trout, and whitefish. Fishing Bridge Visitor Center, at the northern end of the lake, has exhibits on birds, wildlife, and the lake's ecology.

During the glacial period, Yellowstone Lake was a much larger lake. The glaciers dammed the valley, causing a large lake. As they melted, water flowed out of the valley, leaving behind a smaller lake, as well as silt and impermeable clay that is now known as Hayden Valley. The valley's rich underbrush and shrubbery abounds with wildlife: moose, bison, grizzly bears, white pelicans, and trumpeter swans. The valley also has some thermal features, one of which is Mud Volcano.

Canyon Village is dominated by the Grand Canyon of Yellowstone, which was formed by the Yellowstone River as it cut through rhyolite, a glassy volcanic rock. Hot water acting on the volcanic rock has brought out shades of gray, pink, and orange in the predominantly golden-colored cliffs. It is worth stopping here to see the spectacular sights: the river 1000 feet (300 m) below the canyon rim, the 109-foot (33.2-m) Upper Falls, and the towering 308-foot (93.9-m) Lower Falls.

Roosevelt Country is named after President Theodore Roosevelt, who had a favorite campsite in the area. (See Theodore Roosevelt National Park, Medora, North Dakota.) The hills are covered with sagebrush and the forests are filled with fir, pine, and aspen. Bears, deer, elk, and bison are plentiful. This area has a petrified forest with well-preserved, ancient subtropical plants and standing trees of cold climates, such as spruce and fir. The weather variations, from subtropical to cold, were a result of volcanic activity of the past. From Tower Falls, visitors get a good view of the canyon, with its brown-to-tan columnated basalt walls formed by volcanic action.

Yellowstone National Park also maintains a large Heritage and Research Center at the northern edge of the park in Gardiner, Montana, to store and preserve artifacts and documents related to the park. Included in the facility are a museum holding over 720 000 items ranging from obsidian points to skulls of the first wolves reintroduced into the park; an archive of thousands of documents, maps, and photographs; a library of books, newspapers, and brochures; and a herbarium with over 17 000 specimens. Opened in 2005, one-hour public tours of the site are offered weekly, and researchers may visit as well with advance notice. There are rotating displays in the lobby.

Address: Yellowstone National Park, WY 82190. Heritage and Research Center: 20 Old Yellowstone Trail South, Gardiner, MT, 59030.
Phone: (307) 344-7381. Heritage and Research Center: (307) 344-2662.
Days and hours: Park is open daily. Many roads and facilities are closed during the winter. Check the website for details. Heritage and Research Center: Mon.–Fri. 8 a.m.–5 p.m. except holidays.
Fee: Yes.
Tours: Self-guided. A variety of ranger-led talks and tours are offered. Reservations are required for the Heritage and Research Center tours. Check the website for the schedule.
Public transportation: None. Commercial tour buses may be available.
Handicapped accessible: Partly. Many visitor centers are fully accessible; others are partly accessible; still others are not. Wheelchairs are available for rental on a first-come, first-served basis at many sites in the park.
Food served on premises: Restaurants, cafés, snack shops.
Special note: Many roads and facilities are closed during the winter. Reservations for summer camping and lodging inside the park should be made months in advance.
Website: www.nps.gov/yell/index.htm

9

Pacific

Alaska

Anchorage Museum

Founded in 1968, the Anchorage Museum, with 247 000 square feet (22 900 m²), embraces a variety of subject matter about Alaska and the circumpolar North. From art to ecology, from history to science, this facility—the largest in the state—studies the land, art, history, and peoples found within Alaska. The front façade is built from fritted glass (finely porous glass particles sintered together) 12 inches (30 cm) thick as insulation against the cold. Outside the frit is another sheet of glass with silvering to add ultraviolet protection for the artifacts inside, and inside is a non-silvered sheet of glass.

One of the major permanent exhibits found here is Living Our Cultures, Sharing Our Heritage: The First Peoples of Alaska in the Smithsonian Arctic Studies Center. Housing over 600 objects, the hall contains seven large display cases highlighting the various indigenous peoples of Alaska: Iñupiaq, Eastern Siberia, St. Lawrence Island Yupik, Yup'ik, Unangax̂, Sugpiaq, Athabascan and Eyak, Tlingit, Haida, and Tsimshian. Upon entering the exhibition, a map indicates where each group lives. Wall units showcase objects such as headgear, baskets, bowls, and masks, with touchscreens that allow visitors to see more information about each artifact. Each case is divided into three areas: Community & Family, Ceremony & Celebration, and Living from the Sea, Land, & Rivers, also with interactive screens providing more information. Items in these cases are on loan from the Smithsonian Institution. (See Smithsonian Institution, Washington, DC.) Large television screens display interviews with people who discuss how time has changed their lifestyles.

Among the important technological artifacts of note are the original Yupik "parka" overcoat, made from crested auklet bird skins found on St. Lawrence Island. There is a collection of snow goggles to protect the eyes against glare, in which a narrow slit is cut in the eye-coverings to block most light. A full-sized Bering Sea kayak is also on display (along with hand-crafted models). The kayak is the end-product of 6000 years of Alaskan knowhow, carefully engineered to be low-weight, fast, and rugged in rough seas. Each traditional kayak is a bespoke craft, tailored to the individual measurements of its user's body. It is made of sealskins sewn together and stretched over a wooden frame, tied together with sinew.

An interactive Discovery Center focuses on the natural history of Alaska, including the local geology, water, and biology. The hall is divided into several sectors: internal geology, external geology, and finally, biological adaptations to Alaska. In the underground geology area, a shake table demonstrates earthquakes, in which visitors can build small structures and see how they withstand vibrations. A seismograph, a small demonstration of how dirt liquefies during an earthquake and collapses buildings, and a map showing locations of historic temblors in Alaska are here as well. Volcanoes are explained

Smithsonian Arctic Studies Center, Anchorage Museum.
Courtesy of the Anchorage Museum.

using Nanoleaf® technology. A visitor touches the tiles to "build up pressure" inside a cutout model of Mount Redoubt, a stratovolcano visible from Anchorage. Eventually an "ash plume" escapes the volcano, accompanied by sound effects.

In the surface geology area, an augmented reality sandbox—a topography table—demonstrates how surface processes (wind and water) shape the land, via a video projector that superimposes colored elevations in real time as visitors adjust the "mountains" and "lakes" by hand (see Phillip and Patricia Frost Museum of Science, Miami, Florida). Alaska Aeolian landscapes (wind-blown deposits) such as sand dunes, dirt, and sediments, are discussed in one display. Several demonstrations of convection and air pressure help to discuss Alaskan weather formations. Northern latitudes offer the aurora borealis and summer days without end (and likewise, long winter nights)—a computer monitor displays the daily aurora forecast.

In the biology area, an aquarium showcases marine species from the Alaska coast, including sea stars, king crabs, and sea anemones.

A hydroponics garden shows how to grow vegetables indoors during the Arctic winter. On site is also a small planetarium in a 24-foot (7.3-m) dome, with regular star shows plus other presentations on auroras, northern geology, and ecology.

The museum offers occasional adult-oriented lectures and workshops on science and indigenous technology. A variety of changing exhibits are also on view. The library and archives are available for visits by scholars and the public.

Address: Anchorage Museum, 625 C Street, Anchorage, AK 99501.
Phone: (907) 929-9200.
Days and hours: May–Sep.: Daily 9 a.m.–6 p.m.; Oct.–Apr: Tue.–Sat. 10 a.m.–6 p.m. (with extended hours on Fri.), Sun. noon–6 p.m., closed Mon.
Fee: Yes. Extra fee for planetarium.
Tours: Self-guided and—with advance notice—guided group tours.
Public transportation: Public buses stop at the museum.

Handicapped accessible: Yes.

Food served on premises: Café and restaurant.

Special note: Check the website or call for current planetarium schedules, lectures, workshops, changing exhibits.

Website: www.anchoragemuseum.org

FAIRBANKS

University of Alaska Museum of the North

The sole research and teaching museum in Alaska, University of Alaska Museum of the North, traces its origins to naturalist Otto Geist and his travels through the Alaskan Territory in 1926, collecting archaeological and ethnographic specimens. The museum opened in 1929, and expanded dramatically after Alaska became a state in 1959. An eye-catching structure dating to 1980, evocative of glaciers and alpine ridges, houses the collection of over 2.2 million artifacts, with an additional wing added in 2005. Among its highlights is the largest collection of northern-latitude reptiles, dinosaurs, and early mammals in the world. Research under museum auspices is conducted in fields including archaeology, geology, entomology, ethnology, mammalogy, and ornithology.

On the main level, an introduction to the largest state in the United States is provided in the Gallery of Alaska, organized into its five ecological regions: Western Arctic Coast, Southwest, Interior, Southcentral, and Southeast. Visitors circulate past the nearly 9-foot-tall (2.7-m) stuffed brown bear named "Otto" at the gallery's entrance.

Among the artifacts in the Interior room is a display of woolly mammoth skull plus tusks, showcasing the 31 species of mammals that occupied the interior grasslands, along with the largest public display of samples of gold nuggets and *objets d'art*. Here also is a 36 000-year-old mummified *Bison priscus* (steppe bison) named "Blue Babe," discovered by miners in 1979 near Fairbanks. Blue Babe is one of only three permafrost taxidermy discoveries in the world that are on display. Athabaskan Native American clothing and tools are shown. Two 30-minute movies are shown during the year in the auditorium: in the summer, *Dynamic Aurora* explains the aurora borealis (northern lights); in the winter, the film *Extreme Weather* discusses how global warming is changing the Earth's weather.

In the Southeast zone, Tlingit, Haida, and Tshimshian Native American objects and clothing are displayed. Birds are highlighted in the Southcentral biome, and the geological history of Alaska over several hundred million years is explained here.

The Western Arctic Coast area boasts nine species of marine mammals (such as walrus, seals, bowhead whales, and polar bears) on display, and includes an exhibit of Eskimo clothing and ivory artwork. The end of the Age of Dinosaurs and the dawn of the Cretaceous Era is explained through scientific techniques of collection.

The story of the Aleut and Alutiiq native peoples is discussed in the Southwestern room, and includes many examples of Aleut basketry. The Southwest islands are home to seabirds, which are shown in this room, along with northern fur seals.

The upper floor of the museum serves as an art gallery, showing ancient to modern artworks over the course of two millennia created by Native Americans and other residents. One section includes clothing, tools, masks, and regalia made by Native Americans.

Highlights from the museum's own researchers are showcased in the Collections Gallery.

Outdoor exhibits include the Kolmakovsky Blockhouse, from a Russian-American Company trading post, built in 1841. After the United States purchased Alaska in 1867, the building was used by other traders, but by 50 years later it was abandoned. It was shipped to the University in 1931 (minus the lower three log courses and roof, which had deteriorated significantly); it was reconstructed using original techniques in the 1980s and underwent extensive conservation in 2010–2011.

There are also changing exhibits on geology, biology, anthropology, and art; an auditorium offers regular movies with natural-history themes.

Address: University of Alaska Museum of the North, 1962 Yukon Drive, Fairbanks, AK 99775.
Phone: (907) 474-7505.
Days and hours: Jun.–Aug.: 9 a.m.–7 p.m.; Sep.–May: 9 a.m.–5 p.m.
Fee: Yes. Extra fee for movies.
Tours: Self-guided and guided group tours. Downloadable audio tour is also available, including sounds of wildlife and samples of native languages.
Public transportation: MACS bus transit system via red, blue, and yellow lines to the museum; some hotels offer shuttle buses as well.
Handicapped accessible: Yes. Several wheelchairs are available on a first-come, first-served basis.
Food served on premises: Café.
Website: www.uaf.edu/museum/

HAINES

Hammer Museum

Founded in 2002 by Dave Pahl, who has worked as a blacksmith, shipwright, construction worker, and restorer of old tools, this facility is the only museum in the United States whose mission is "to preserve the history of the hammer," according to its website. Dave and his wife, Carol, bought a small building in 2001 to house the museum, originally the head carpenter's house for when Fort William H. Seward was being built during the first decade of the 20th century. To build his home, the head carpenter used spare parts from the fort's construction. When Mr. Pahl bought the structure, the foundation was rotting and the building was sliding down the hill. The house needed major renovations, so he excavated underneath the structure in order to install a foundation.

During the work, under the house Dave Pahl found a centuries-old Tlingit warrior's pick, or "slave killer," which is a particular type of ceremonial hammer used by this Native American tribe. The Tlingit captured slaves when they raided other tribes. The slaves were kept for a season or two and then were either allowed to leave, killed as a sacrifice, or adopted into the tribe. Usually the slaves were sacrificed (using the slave killer) after someone important died, such as an elder or a chief, to act as their servants in the afterlife. Mr. Paul viewed this discovery of a native hammer as a good portent for the future museum.

By 2007, the Pahls installed the iconic 19-foot 8-inch (6-m) high statue of a claw hammer in front of the eclectic building. The statue's "handle" was fabricated from a spruce log 26 inches (66 cm) in diameter, and the "head" was constructed from fiberglass and Styrofoam. In 2008 the museum changed from a private collection to a nonprofit foundation.

The museum owns about 9000 hammers, of which 2000 are on display on a rotating basis, mostly mounted on the walls. Some smaller hammers are inside glass display cases. One wall of the exhibit displays patented hammers with copies of their patents. Several manikins—gifts from the Smithsonian Institution from a former woodworking exhibit—are posed to show authentic techniques for certain types of hammers. (See Smithsonian Institution, Washington, DC.)

Hammers in the collection range from ancient to modern. The oldest hammer, a ball-shaped hammer of dolerite (a fine-grained dark igneous rock), was found at the third pyramid at Giza, in Egypt, dating to 2500 B.C.E. Among the newest hammers is a digital tap hammer used in the 1970s by the Boeing Corporation (see Future of Flight Aviation Center & Boeing Tour, Mukilteo, Washington) to detect internal faults and cracks acoustically during construction of an airplane; and a hammer used on the International Space Station, on loan from NASA. Other hammers on display include musical hammers for tuning pianos and playing dulcimers, a Meissen porcelain meat-tenderizing hammer, cobblers' hammers, and coopers' hammers.

Other fascinating implements include a bank check-canceling hammer from the 1880s used to void checks by cutting an "X" through them (possibly the source of the phrase "to cut a check"), and a tack hammer that automatically dispenses tacks magnetically, so no fingers get near the hammer's target. Visitors may see aluminum ice-breaking hammers that stewardesses used on airlines for mixed drinks from the late 1970s, and taffy hammers from the early 20th century, used to break up large sheets of taffy.

Docents are on hand to answer questions about the collection. The museum identifies hammers that owners bring in, and offers occasional workshops.

Address: Hammer Museum, 108 Main Street, Haines, AK 99827.
Phone: (907) 766-2374.
Days and hours: May–Sep.: Mon.–Fri. 10 a.m.–5 p.m., Sat. 10 a.m.–2 p.m. Closed Sun. Open at 10 a.m. on days when cruise ships arrive. Closed Oct.–Apr. except for research.
Fee: Yes.
Tours: Self-guided.
Public transportation: None.
Handicapped accessible: Partly. Call the museum for specific questions.
Food served on premises: None. There are several restaurants within a couple of blocks of the site.
Website: www.hammermuseum.org

California

ARCADIA

Los Angeles County Arboretum and Botanic Garden

The historic Rancho Santa Anita, part of the San Gabriel Mission, is where this 127-acre (51.4-hectare) arboretum is situated. It lies at the base of the San Gabriel mountains, directly over the Raymond Fault, one of many fault lines that honeycomb this part of the continent. The fault allows underground water to surface in a spring, which feeds a picturesque lagoon surrounded by trees and plants. The arboretum has a mild southern-California climate and fertile, well-watered land.

The Historical Section of the arboretum preserves some of California's history and reflects the state's heritage. Exhibits explore how Native Americans, Mexicans, and Europeans lived and used the land around them. About 3000 years ago, stone-age people speaking a Shoshonean language settled this area. When the Spanish missionaries arrived here in the 18th century, they found a village called Aleupkigna, whose 150 inhabitants relied on the rich land and plant life for their existence. The missionaries named these people the Gabrielino Indians. In the early 19th century, a local settler named Hugo Reid recorded much of what we know today about this culture, which eventually died out. Reid, who became a naturalized Mexican citizen and a Catholic, married a mission Native American and eventually received title to Rancho Santa Anita.

A typical Gabrielino Native American village, re-created in the Historical Section, illustrates how these people lived. It includes several *wickiups*: huts made of willow poles that are set into the ground in a circle, bent at the top, and tied together. The framework is covered with layers of tule reed mats woven with bay leaves, which repel insects. The 1840 Hugo Reid house is an authentic reconstruction of the first permanent structure on this land. The building is constructed of handmade, sun-dried adobe bricks, which have been white-washed inside and out with lime from crushed seashells. Its roof is smeared with *brea*, the local word for tar that bubbles up naturally from the ground. (See Natural History Museum of Los Angeles and La Brea Tar Pits, Los Angeles, California.) Today, in what was Reid's garden, there is an orchard of peach, plum, fig, pomegranate, and olive trees. One pomegranate tree, south of the courtyard wall, was planted during Reid's time. A hedge of prickly cactus surrounding the house and grounds is of the type that served as a living fence for the cattle on the ranch.

In 1875, Elias Jackson "Lucky" Baldwin purchased the property for investment and farming, and began introducing trees and shrubs from around the world to southern California. Visitors can admire the exterior of the Queen Anne Cottage that Baldwin built in 1885–1886. In 1893, the *Los Angeles Times* said of the property, "There are few more beautiful than the grounds about the home.... He took this vast estate when it was practically a desert and he made it into a land of flowers, trees, and fruit-bearing orchards." The Victorian Coach Barn that belonged to Baldwin was considered an extravagance at the time. In the Historical Section visitors also see the Santa Anita Depot, built by the Santa Fe Railroad in 1890. The depot was moved here in 1970 and refurbished.

The Historical Section and other parts of the arboretum have served as a set for many movies and television productions, including *On the Road to Singapore* (1939) with Bob Hope and Bing Crosby, *Tarzan and the Huntress* (1946), and the television show *Fantasy Island*, which used the Queen Anne Cottage as its centerpiece.

The local climate provides an ecosystem in which a wide range of plants flourish. Among the 30 000 permanent plants in the arboretum, representing more than 7000 species, are rare and exotic trees and shrubs from around the world. The plants are set in an informal landscape, organized into geographic regions and specialized gardens.

Through the arboretum's program of collecting and testing new plant material, more than 150 flowering plants have been introduced to the landscaping community. Visitors can learn about ways to use trees, shrubs, and ground cover, or just enjoy this peaceful retreat.

Herbs in the 1.5-acre (0.61-hectare) Herb Garden are organized according to use: medicinal, culinary, fragrance, and dyes. The serpentine Braille Terrace has a waist-high wall of fragrant and touchable plants. The Demonstration Garden for All Seasons contains 7000 square feet (650 m^2) of trees, shrubs, and perennials that provide a seasonal blend of color. One demonstration area illustrates how to arrange ornamental plants and vegetables in small pots. The Henry C. Soto Water Conservation Garden, with its attractively landscaped drought-tolerant plants, emphasizes balance between natural resources. The Celebration Garden and Perennial Garden offer landscaping ideas for family living, showcasing gardening in the southern California climate. A Rainbow Serpent Garden evokes the aboriginal creation myths of Australia, and is home to Australian plants. There is also an Aquatic Garden.

The arboretum has greenhouses full of begonias and shade plants. The Tropical Green House displays a vast number of flowering orchids, hybrids, ferns, bromeliads, and other colorful plants, with an emphasis on propagating endangered species.

Among other major collections are many species of acacia, a tree or shrub that tolerates smog and drought, 39 species of cassia, a drought-resistant shrub with golden yellow flowers, and 150 out of 500 known types of eucalyptus trees. This is one of the largest collections of eucalyptus trees outside Australia. Nearly 30 species of the official tree of the city of Los Angeles—erythrira or coral tree—grow at the arboretum as well.

The arboretum has a library, open to all visitors. Holdings in the library include over 30 000 books and over 100 journals, mainly on Mediterranean and other semi-tropical plants, but also on general gardening and garden layouts, medicine, ethnobotany, and plant lore.

Address: Los Angeles County Arboretum and Botanic Garden, 301 North Baldwin Avenue, Arcadia, CA 91007.
Phone: (626) 821-3222.
Days and hours: Arbortetum: Daily 9 a.m.–4:30 p.m.; grounds close at 5 p.m. Closed 12/25. Library: Tue.–Fri. 8:30 a.m.–5:30 p.m., Sat. 8:30 a.m.–5 p.m., Sun. noon–4 p.m. Closed Mon. and 12/25.
Fee: Yes.
Tours: Self-guided and docent-led guided. Call or see the website for the current schedule.
Public transportation: Metro Gold Line comes to Arcadia, and take the Arcadia Transit Green Line to the arboretum.
Handicapped accessible: Yes. Wheelchairs are available to rent on a first-come, first-served basis.
Food served on premises: None. There are restaurants within about a mile of the site.
Website: www.arboretum.org

ESCONDIDO

San Diego Zoo Safari Park

Both the San Diego Zoo and the San Diego Safari Park are operated by the nonprofit organization San Diego Zoo (SDZ) Global®. The organization, whose motto is "leading the fight against extinction," claims to be the largest zoological society in the world. (See San Diego Zoo, San Diego, California.)

A sister facility to the zoo, the massive San Diego Zoo Safari Park opened in 1972, about 35 miles (56 km) away from the San Diego Zoo. Originally it was to be used exclusively for breeding and conserving rare and endangered plants and animals, but financial considerations forced the zoological society to change its plans and admit visitors. The park's 1800 acres (730 hectares) of hills and valleys are similar to biogeographic regions of Africa and Asia. Approximately 1500 mammals of 325 species and 1500 birds of 125 species live here.

San Diego Zoo Safari Park has the world's largest veterinary hospital, and is also a center for quarantining imported wild animals into southern California. Next to the park is another organization run by SDZ Global, the Institute for Conservation Research, which runs a "Frozen Zoo" for the park. A frozen zoo is a facility that stores genetic material from animals, especially those that are endangered or even extinct, in order to preserve genetic diversity.

In this innovative design, birds intermix with the world's largest collection of hoof stock from Africa and Asia. The only way for visitors to see this hoof stock is to take guided tram tours through the park to see recreated savannas of eastern and southern Africa, Asian plains, a North African desert, and an Asian water hole. Park staff members point out that in this zoo, people are put in moving cages and the animals roam freely. Oryx, water buffalo, sitang, and gazelles form herd hierarchies and function as they would in the wild. The tram rides also pass Asian and African elephants, Sumatran tigers, Grevy's zebras, rhinoceroses, ostriches, South African cheetahs, and dozens of endangered species, all part of the breeding program at this park. The park has been instrumental in breeding the highly endangered California condor.

Visitors can walk along the 1.75-mile (2.82-km) Kilimanjaro Trail to see some of the larger animals and several gardens. Along the trail is the 3.6-acre (1.5-hectare) Walkabout Australia with its South Pacific plants, birds, and mammals, such as red-necked wallabies and western gray kangaroos. Among the gardens are the California Native Plants Trail, Herb Garden, Bonsai Pavilion, Epiphyllum House, and Baja Garden containing hundreds of endangered specimens.

Nairobi Village features shows and exhibits of small animals and birds. The village has one of the world's largest suspension flight aviaries and a waterfowl lagoon with hundreds of birds. Tropical Asia features animals, birds, and plants from Asia, Indonesia, and Papua New Guinea. Lowland gorillas live on a grassy knoll near the complex.

An optional balloon ride allows visitors to experience a bird's-eye view of the park.

Address: San Diego Zoo Safari Park, 15500 San Pasqual Valley Road, Escondido, CA 92027.
Phone: (760) 747-8702.
Days and hours: Daily. Hours vary seasonally: see the website.
Fee: Yes. Balloon rides cost extra.
Tours: Guided, except for localized trails. See website for details on various animal encounters.
Public transportation: MTS buses run from Escondido to near the site, then walk 0.4 mile (0.6 km).
Handicapped accessible: Yes. Wheelchairs and scooters are available to rent on a first-come, first-served basis.
Food served on premises: Restaurants, cafés, bar, snack shops.
Website: www.sdzsafaripark.org

La Jolla

Birch Aquarium at Scripps

For a magnificent view of the Pacific Ocean, there are few better spots than this aquarium, which promotes public understanding of the ocean, marine science, and sea life. The aquarium is maintained by the Scripps Institution of Oceanography, part of the University of California–San Diego. The institution is named for Ellen Browning Scripps and her brother, the noted newspaper publisher E. W. Scripps, both of whom were significant financial supporters of research and education efforts carried out here.

Scripps Oceanography has maintained a public aquarium since its inception in 1903. The Birch Aquarium complex is named for Stephen Birch, a mining-industry executive, and his wife, Mary, both formerly of Mahwah, New Jersey, who set up a foundation supporting programs in education, science, and medicine—this aquarium among them. Scripps Oceanography carries out a variety of scientific research around the world, including the Keeling Curve showing the rise of atmospheric carbon dioxide since 1958, a global seismic network, and ocean-monitoring floats. It maintains a research fleet of ships, one of the largest in the United States, that plies the world's oceans.

The aquarium, geared toward everyone from preschoolers to adults of all ages, is made up of two side-by-side buildings with a re-created tide pool between them. The complex has more than 35 naturalistic tanks and 60 habitats depicting the marine communities of California and the Gulf of California, once known as the Sea of Cortez. Many exhibits are interactive.

The museum covers oceanography with an eye toward helping the public understand the role of Scripps Oceanography in improving our changing world. Chemistry, currents, tides, waves, climates, weather, and relationships between marine organisms are among the subjects explored in permanent and temporary exhibits, all of which highlight current Scripps research.

Exhibits include displays of over a dozen seahorse species, plus a unique 18-foot (5.5-m) wide tank designed to breed leafy and weedy seadragons, cousins to seahorses. Over 5000 seahorses have been bred here to share with other institutions. There is a re-creation of a ship, R/V *Sally Ride*, from Scripps's research fleet. A variety of virtual-reality displays allow visitors to "tour" R/V *Sally Ride* (named for astronaut Sally Ride) and explore mapping the ocean floor.

In the Hall of Fishes, visitors follow the California Current from the cold waters of the Pacific Northwest to the temperate waters of California. Exhibits feature the tropical coral-rich waters of Mexico, and the Indo-Pacific Ocean. Among the exhibits here is an experimental coral reef to assist scientists to test equipment before using it on wild coral reefs.

The highlight of the hall is the two-story, 70 000-gal (260 000-L) Kelp Forest with giant kelp (*Macrocystis pyrifera*). A machine produces waves, and natural sunlight pours in through the tank's open top to give visitors the feeling of being underwater. California moray eels, kelpfish, garibaldi, and sheephead are among the many species in the Kelp Forest, which visitors can view from both the top of the tank and a specially designed viewing gallery below the water line. A nearby display explains the kelp-forest ecosystem. A special gallery features marine life acquired on research expeditions. Other large tanks display sea life from the Gulf of California. Three tanks represent Magdalena Bay, Cabo San Lucas, and Los Islotes Reef.

The outdoor displays at Tide Pool Plaza overlooking the Institution of Oceanography's campus and research pier describe the unique nature of this ecosystem and show such species as the soft-bodied sea hare, hermit crabs, sea stars, and sea anemones. From December through May, visitors may see gray whales as they migrate.

The outdoor Smargon Courtyard has oceanographic exhibits and special ocean-viewing area with signs explaining ocean phenomena. One exhibit examines leopard sharks (native to the La Jolla area) and other elasmobranchs (similar species of sharks and also rays). Another exhibit examines renewable energy sources, such as the wind, Sun, and ocean waves, to power civilization.

Address: Birch Aquarium at Scripps, 2300 Expedition Way, La Jolla, CA 92037.
Phone: (858) 534-3474.
Days and hours: Daily 9 a.m.–5 p.m.
Fee: Yes.
Tours: Self-guided. Call or see the website for the current schedule of animal feeding times.
Public transportation: Buses run to La Jolla Shores Drive, then walk several blocks uphill to the site.
Handicapped accessible: Yes. Wheelchairs are available on a first-come, first-served basis.
Food served on premises: Café.
Website: aquarium.ucsd.edu

LOS ANGELES

Natural History Museum of Los Angeles County and La Brea Tar Pits

More than 35 million specimens and artifacts from over 4.5 billion years of the Earth's history make the Natural History Museum of Los Angeles County the third-largest natural-history collection in the United States. The collection is housed in part in the original 1913 museum building, which is on the National Register of Historic Places. The museum at La Brea Tar Pits, about 6.5 miles (10 km) away, is a satellite facility adjacent to the La Brea Tar Pits, from which millions of fossils have been excavated over the years.

Natural History Museum of Los Angeles County
The Natural History Museum of Los Angeles County is engaged in research, education, acquisition, and care of its collection. It is also a significant cultural resource for the community and nation. The main foyer houses the only megamouth shark specimen on public view, one of only two ever found. The facility has one of the largest invertebrate paleontology collections, and the largest catalogued vertebrate collection, in North America. Dinosaur Hall has the only example of a *Tyrannosaurus rex* growth progression, a *Camptosaurus* or "duckbill" dinosaur, and an *Allosaurus*, among 300 different fossils on display.

Four habitat halls acclaimed for their realism and detail have mounts of exotic animals from Africa, North America, and other areas, placed in their natural environments. One especially impressive scene shows a growling grizzly bear along a stream bank, holding a freshly killed fish in its paws.

The museum, which does research on marine creatures, holds the second-largest collection of marine mammal fossils and mounts in the world. The ichthyology collection

Natural History Museum of Los Angeles County.
Courtesy of the Natural History Museum of Los Angeles County.

contains more than 3 million preserved adult fish and larvae. The Natural History Museum features the world's largest collection of mounted, southwestern moths and butterflies, and North America's largest collection of ants. A herpetology collection focuses on North and South America, Africa, and Australia. There is even a shell collection, among the largest in the world, showcasing over 200 specimens.

A comprehensive Gem and Mineral Hall has a walk-through gem vault displaying choice pieces of gold, gem crystals, colored gems, meteorites, and minerals from California. A gold exhibit in a stylized gold mine explores the natural history of this mineral and probes people's associations with it through chemistry, crystallography, and mining. In the exhibit are artifacts and specimens from around the world.

The impressive Hall of Birds features mounts from the museum's extraordinary ornithology collection of more than 100 000 birds. Three walk-through habitats—the Tropical Rainforest of Central Peru, the Canadian Prairie Marsh, and Condorminium, the California mountain condor exhibit—all feature animated animals. Visitors can learn about how birds fly; there are artifact drawers containing eggs, nests, wings, and feet.

At the Discovery Center, visitors can do fossil rubbings, use a microscope, and observe live animals.

In the anthropology and history collections are such interesting technological pieces as a bell made by Paul Revere and the world's oldest existing banjo. Becoming Los Angeles, an exhibit about the formation of modern Los Angeles, is complete with a late 1930s city model, including vehicles, textiles, scientific instruments, and artifacts from the railroad, movie, and oil industries.

Address: Natural History Museum of Los Angeles County, 900 Exposition Boulevard, Los Angeles, CA 90007.

Phone: (213) 763-3466.

Days and hours: Daily 9:30 a.m.–5 p.m. Closed 7/4, Thanksgiving, and 12/25.

Fee: Yes.

Tours: Self-guided. There are evening talks held at the museum; call or see the website for the current schedule.

Public transportation: Metro Expo Line stops at the museum; Metro buses stop near the museum, then walk.

Handicapped accessible: Yes.

Food served on premises: Café.

Website: www.nhm.org

La Brea Tar Pits

The La Brea Tar Pits and museum (established in 1977 as the George C. Page Museum) is composed of more than 100 tar pits, a building that covers tar pits that are undergoing excavation, a park, and several Pleistocene animal replicas. Native Americans knew of the tar pits well before they were first reported to Europeans in the 1760s. In 1828, Antonio Jose Rocha and Nemecio Dominguez requested a land grant from Spain allowing them to found Rancho La Brea on the site while still giving the public access to the tar pits. In 1916, Los Angeles County received more than 23 acres (9.3 hectares) of Rancho La Brea lands for a park, with the caveat that the scientific features were to be preserved. George C. Page came to Los Angeles from Nebraska as a teenager in 1917, and within a few years was a successful entrepreneur. In 1972 he persuaded the city to allow

La Brea Tar pits.
Courtesy of La Brea Tar Pits.

him to build a museum in the park to house fossils found in the tar pits, which are now a National Historic Landmark.

The tar pits are sinkholes filled with bituminous asphalt, a blackish, semisolid material. For centuries, animals that wandered too close and stepped in the tar pits became trapped. Excavations since the late 1800s have uncovered fossils of millions of vertebrates and invertebrates, 140 plant species, and 420 animal species. Paleontologists continue to uncover fossils of extinct species, the earliest of which are about 38 000 years old. They have also found fossils of animals no longer living in California.

Museum displays contain fossils of everything from microscopic plants to ground sloths. The museum has some of the best-preserved mammoth remains in the nation.

Most of the pits were first opened for excavations between 1913 and 1915. Visitors can view the pit site and Project 23 all year, but excavations continue at Pit 91 exclusively in summer.

In the museum, visitors can watch the Fossil Lab in operation; see newfound specimens being cleaned, prepared, catalogued, and studied; and view the storage area. Exhibits in the museum explain the southern California environment of the Pleistocene Epoch 2.5 million years ago, when oceans covered the region. Evidence recovered from the tar pits indicates that about 38 000 years ago, after the oceans retreated, the first terrestrial plants and animals appeared here. An Ice Age Encounters multimedia presentation includes an animated life-sized puppet of a saber-toothed cat.

Address: La Brea Tar Pits, 5801 Wilshire Boulevard, Los Angeles, CA 90036.
Phone: (213) 763-3499.
Days and hours: Daily 9:30 a.m.–5 p.m. Closed 7/4, Thanksgiving, and 12/25.
Fee: Yes.
Tours: Self-guided. Guided tours of the pits are available.
Public transportation: Metro Expo Line stops at the museum; Metro buses stop near the museum.
Handicapped accessible: Yes. Wheelchairs are available for rent on a first-come, first-served basis.
Food served on premises: Snacks. There are nearby eateries.
Website: www.tarpits.org

MARIPOSA

California State Mining and Mineral Museum

Mariposa, a town in the Sierra Nevadas at the southern end of the Mother Lode Gold Belt, is the western gateway to Yosemite National Park. The town began nearly two centuries ago when gold was first discovered in the mountains, sparking the California Gold Rush of 1848. Gold was not the only mineral found, however. Discoveries of silver, platinum, tungsten, copper, lead, zinc, and many other minerals brought prosperity to the area and changed the history of the region and of California.

Appropriately, Mariposa became the home of the California State Mining and Mineral Museum and its collections of rocks, gems, metals, and minerals. Among its holdings are 1327 specimens that once belonged to the California State Geological Society. Many of these samples were used for identification and assay, and come from the mining districts of the Mother Lode. These specimens were turned over to the San Francisco–based California State Mining Bureau when it was formed in 1880. From 1899–1983 the collection grew to more than 13 000 specimens, and was kept at the Ferry Building in San Francisco. In 1989 the collection was moved to this new facility in Mariposa. The museum is now a state park—the only California state park with no significant land.

Part of the building looks like a turn-of-the-twentieth-century stamp mill for crushing gold ore, with a headframe and an inclined track for hauling ore. Weathered metal roofing and large wooden beams add authenticity. The museum's two sections are connected by a re-created mine tunnel. Many of the pieces housed here are historically important and of high quality. Specimens on display come from each of the 53 counties in California, every state in the nation, and several parts of the world.

The entrance area has displays about the collection's history. A photographic exhibit illustrating the history of the old San Francisco Ferry Building and the early days of California has such artifacts as original sample bags and field notes. One display has some of the more dramatic specimens in the collection, such as gold and pegmatite from California; orange wulfenite from Tiger, Arizona; linarite from Cerro Gordo, California; sulfur crystals from Italy; fluorite from England; and stibnite from Japan.

There are samples of benitoite, the California state gem; gold, the state mineral, including the Fricot Nugget weighing 13.8 pounds (6.26 kg); and serpentine, the state rock; as well as replica of a saber-toothed cat, the state fossil. A special vault exhibits rotating displays of gold and other valuable items needing additional security. Many exhibits show minerals as they are found in nature, paired with products made from the extracted metal. For example, azurite, a blue mineral of basic copper carbonate, is surrounded by pennies; hematite, a brownish-red ore of iron oxide, is displayed with a pile of nails, paper clips, and razor blades. The gallery also explains the Dana systematic classification of minerals, and displays mineral samples to touch.

The 200-foot (61-m) mine tunnel has life-size dioramas of gold-mining techniques used over the past century and a quarter, including underground drilling, blasting, and mucking. Visitors learn about drillers' advances, from single jacking to modern power equipment, and see period gold-ore cars and trucks.

The final section shows mining and the development of mineral resources. Visitors can learn about mining technologies such as gold-panning, dredging, mine sluicing (flushing the broken mineral through a channel of water), and hydraulic mining—which is illegal. The different techniques needed for mining various minerals—borax, gold, and mercury—are illustrated. A working model of a 1904 five-stamp mill used for the design of this museum building is on display. The re-created assay office shows where miners learned whether they had hit pay dirt after months or years of dangerous and lonely work. It contains a hand-painted safe from the turn of the 20th century, a mortar and pestle for grinding mineral samples, scales to weigh gold samples, a furnace for melting gold, and some furniture.

Address: California State Mining and Mineral Museum, 5005 Fairgrounds Road, Mariposa, CA 95338.
Phone: (209) 742-7625.
Days and hours: Thu.–Sun., in winter 10 a.m.–4 p.m., in summer 10 a.m.–5 p.m. Closed Mon.–Wed.
Fee: Yes.
Tours: Self-guided.

Public transportation: None.
Handicapped accessible: Yes.
Food served on premises: Picnic area. Restaurants are available in Mariposa, about 2 miles (3.2 km) north.
Website: www.parks.ca.gov/?page_id=588

MINERAL

Lassen Volcanic National Park

Lassen Peak, at the juncture of the Cascade Mountains and the Sierra Nevadas, lies within a series of mountain ranges in the Pacific Northwest that form part of the Ring of Fire, a chain of volcanoes extending around the Pacific Ocean. Tectonic plates of the Pacific region are riding over each other, and, at their edges, volcanic action is causing violent upheavals. As the earth's crust is forced deep beneath the continent, some of it melts, and the resulting magma becomes part of the volcanic activity. (See Mount St. Helens National Volcanic Monument, Castle Rock, Washington; Mount Rainier National Park, Ashford, Washington; and Lava Beds National Monument, Tulelake, California.)

The volcanic activity of Lassen Peak is on the northern flank of Mount Tehama. During the last eruption, from 1914 to 1917, Lassen Peak blew a mushroom cloud seven miles (11 km) into the stratosphere. It is the world's largest plug-dome volcano, rising some 2000 feet (600 m) to a height of 10 457 feet (3187.3 m) above sea level. The dome is made of dacite, an igneous rock that is 67% silicon dioxide and 15% aluminum dioxide. After the 1980 eruption of Mount St. Helens, the significance of this park intensified as a place to study how the land recovers after eruptions. Self-guided walks, hiking trails, and roadside pullouts pass great lava pinnacles, large volcanic lava mountains, and jagged craters.

Remains of early human habitation can be found throughout the park, including stone points, knives, and pieces of metal. Native Americans of the Atsugewi, Yana, Yahi, and Maidu groups often summered here, hunting deer and other summer game, and then moved on before the rugged winter set in.

The first recorded European to pass through the area was Jedediah Smith as he traveled westward in 1828. Peter Lassen, for whom the park is named, received a land grant of five leagues, or 22 000 acres (8900 hectares), from the Mexican government in 1844 and established Rancho Bosquejo. He brought settlers to the region with the plan of establishing a city, which never materialized. The Gold Rush of 1848 brought more settlers. In 1851, William Nobles established a trail to California through what is now the park. Remnants of this trail can still be seen.

Because of this area's importance as an active volcanic landscape, it became a national park in 1916. (See Hawai'i Volcanoes National Park, Hawaii National Park, Hawai'i.) Here people can study volcanic phenomena and associated thermal features, including a great lava plateau with isolated volcanic peaks, cooled lava flows, jagged craters, steaming sulfur vents, fumaroles, and boiling mud pots. In the geothermal areas the waters are above 100°C, but there are no true geysers. The yellow color of the land is from sulfur left behind as hydrogen sulfide gas escapes from the ground. Orange-yellow deposits are caused by metal sulfates created by the bacteria living in the water pools. (See Yellowstone National Park, Yellowstone National Park, Wyoming.) Glaciated canyons, streams, and lakes are all part of this beautiful region.

A road through the park passes over mountains and through meadows as high as 7400 feet (2300 m), circles three sides of Lassen Peak, and offers spectacular views. It also provides access to more than 150 miles (240 km) of hiking trails that pass by lakes and volcanic and geothermal features. At the Kohm Yah-mah-nee Visitor Center (whose name originates in the Mountain Maidu language, and means "snow maiden"), visitors can see a 20-minute introductory film about the park's history and geology. A relief map illustrates the four types of volcanoes, and a seismometer is on display. Road guides are for sale here. The Loomis Museum near Manzanita Lake is a historic structure with photographs of the eruption of 1914–1915, the original photographic equipment used, artifacts, and more.

The park, at the junction of the Cascade and Sierra Nevada ranges, contains ecosystems of both ranges. The almost 800 plant species identified in the park provide food and shelter for more than 250 vertebrate species and many invertebrates. In the park's lower forests there are ponderosa pine, incense pine, sugar pine, white fir, and Douglas fir. Common shrubs are gooseberry, currant, and snowberry. Upper forests, those above 8000 feet (2400 m), have red firs and white pines inhabited by lodgepole chipmunks.

The part has subalpine and alpine zones, accessible only by hiking trail, where whitebark pine and mountain hemlock bend close to the ground from strong winter winds. (See Rocky Mountain National Park, Estes Park, Colorado.) Flying overhead may be Clark's nutcrackers, mountain chickadees, and other mountain birds.

Abundant wildlife populates the glaciated canyons, lakes, streams, and wildflower meadows of the lower forests. Among the species found here are flickers, flycatchers, nuthatches, western gray squirrels, great horned owls, mule deer, black bears, coyotes, and mountain lions. In the chaparral areas, where thick brush grows on steep mountainsides, are mountain quail.

Because of the high elevation, for travelers who have heart or other health problems, it is wise to check with a physician before visiting. Even in summer, jackets or sweaters are recommended because days and evenings are often cool. Road access can be limited in winter.

Address: Lassen Volcanic National Park, Mineral, CA 96063.
Phone: (530) 595-4480.
Days and hours: Daily 24 hours a day. Kohm Yah-mah-nee Visitor Center: May–Oct.: daily 9 a.m.–5 p.m.; Nov.–Apr.: Wed.–Sun. 9 a.m.–5 p.m., closed Mon.–Tue. Closed Thanksgiving and 12/25.
Fee: Yes.
Tours: Self-guided and ranger-led guided. Call or see the website for the current schedule.
Public transportation: None.
Handicapped accessible: Partly. Many facilities are, but some trails may not be.
Special note: High altitudes may be a health problem for some. Inclement weather is common in winter.
Food served on premises: Café at the Kohm Yah-mah-nee Visitor Center, snacks.
Website: www.nps.gov/lavo/index.htm

MONTEREY

Monterey Bay Aquarium

The town of Monterey sits on a peninsula formed by Carmel Bay, the Pacific Ocean, and Monterey Bay—a small waterway with a deep chasm that contains deep-sea life. This cold-water well attracted sardines at one time, making the port a major fishing and sardine-canning center in the United States. At its peak in 1945, Monterey had 18 canneries and 20 reduction plants. John Steinbeck made the city and its canneries famous in his book *Cannery Row*, but overfishing, changes in currents, and other environmental factors caused the sardines and Monterey's canning industry to disappear.

Today, as home of the Monterey Bay Aquarium, the city is famous for another sort of fish industry. This well-known fish museum was built on the site of the Hovden Cannery, the largest and last cannery on Cannery Row. When it closed in 1973, Stanford University purchased the property to protect the shoreline next to its research center, Hopkins Marine Station.

The aquarium was founded by Lucile and David Packard, one of the founders of the Hewlett-Packard Company. Their daughter and son-in-law—both marine biologists—and friends came up with the idea of converting the old cannery into an aquarium. Although it was not practical to use the original cannery building, some of its features—such as an old boiler, a pumphouse, and a warehouse—were incorporated into the aquarium's design. The aquarium shows a film about the history of the canneries and has historical displays illustrating the canning process.

Because Monterey Bay is one of the spectacular marine regions of the world, it is the major focus of the aquarium. Through scientific research, exhibits, and public education, the aquarium staff members hope to increase public awareness of this beautiful, unique ecosystem. The aquarium building, right at the water's edge, incorporates part of the bay and shoreline in its exhibits. It contains 35 000 creatures of 550 species in about 200 exhibits, from sandy sea floors to granite reefs, shale reefs, sloughs, kelp forests, and the open sea.

The centerpiece exhibit is the world's first accurate re-creation of a living kelp forest community. Surge machines and water jets invented partly by David Packard provide the constant water motion need to keep the giant kelp alive. Sardines, mackerel, and many other fish swim among the kelp fronds in the 28-foot (8.5-m) high tank, which visitors can view from three different levels. (See Birch Aquarium at Scripps, La Jolla, California.) Scuba divers hand-feed the fish as guides narrate the action. Communication between the divers and a guide above water allows spectators to ask questions while divers are under water.

This Aquarium is involved in a unique sea-otter rescue and rehabilitation program. After injured and stranded animals are rescued, they are treated and, when possible, later released back into the wild. After release, the otters are monitored so scientists can learn more about their nutrition, behavior, and physiology.

A 90-foot (27-m) long, hourglass-shaped exhibit called Monterey Bay Habitats includes deep reefs, the sandy sea floor, shale reefs, and a wharf. Bubble-shaped viewing windows showcase a variety of sea life, such as large sharks, bat rays, salmon, and exhibits to teach about the ecological role of marine animals and the various species living in the bay.

The Sandy Shore exhibit has an open-air, walk-through aviary with dunes, a beach, and shore vegetation. It is populated by snowy plovers, stilts, avocets, sandpipers, and killdeer. The Rocky Shore has interactive displays such as a touch pool, a wave-crash exhibit, and a surge channel with surf perches and other fish. This exhibit also has an underwater video camera that visitors can guide to get a close look at marine life in a tide pool. The images are projected on monitors on each side of the exhibit. A bat-ray petting pool gives visitors an opportunity to touch these creatures.

The Open Sea wing explores life in the open ocean. A series of exhibits describes ocean ecology and the food web that supports life in the sea. A 1.2-million-gallon (4.5 million-L) tank, 35 feet (11 m) tall, illustrates the waters where the Monterey Bay meets the open sea. It houses sharks, ocean sunfish, green sea turtles, and schools of yellowfin tuna. This tank was the largest window in the world for viewing life in an aquarium tank when built in 1996: 13 inches (33 cm) thick and 15 feet (4.6 m) high by 54 inches (137 cm) wide.

Other exhibits of the Open Sea showcase puffins; Ocean Travelers gallery focuses on the problem of marine plastic pollution, and some solutions. Video microscopes in the Plankton Lab give close-up views of the tiny plants and animals that live in the ocean. This aquarium was the first to raise moon jellyfish, which it displays in a large tank. At the Wildlife Viewing Station, telescopes set by panoramic viewing windows provide views of harbor seals and wild shorebirds on the nearby shoreline. Visitors can stand on observation decks to see the kelp beds and native wildlife. During migration season, whales often pass by the facility.

Address: Monterey Bay Aquarium, 886 Cannery Row, Monterey, CA 93940.
Phone: (831) 648-4800.
Days and hours: Daily 10 a.m.–5 p.m. Closed 12/25.
Fee: Yes. Guided tours are extra.
Tours: Self-guided. A variety of guided behind-the-scenes tours are available. Call or see the website for the current schedule.
Public transportation: MST trolley stops at the aquarium daily during the summer, and weekends only the rest of the year. Buses also stop near the site.
Handicapped accessible: Yes. Wheelchairs are available on a first-come, first-served basis.
Food served on premises: Restaurant, café, and snack bar.
Website: www.montereybayaquarium.org

MOUNT HAMILTON

Lick Observatory

The domes of the Lick Observatory are at the summit of Mount Hamilton, at over 4200 feet (1300 m) the highest peak in the Diablo Mountain Range. Each evening the nearby city of San Jose turns on low-pressure sodium lights, installed in cooperation with astronomers concerned about bright city lights interfering with their views of outer space.

The observatory was founded in 1888 after James Lick, a wealthy San Francisco businessman and real-estate baron, approached Professor George Davidson, president of the California Academy of Sciences, with a plan to build the world's largest and most powerful telescope. (See California Academy of Sciences, San Francisco, California.) Although at the time it was popular to place telescopes near major universities, Davidson convinced Lick that a mountaintop would be a far better place. Mount Hamilton was selected after an extensive search, and large telescopes and superior observing conditions were joined for the first time. From then on, observatories were modeled after this arrangement.

Although Lick donated the largest gift for scientific research up to that time, $700 000 for this project, he died before construction began. His remains were moved to the top of Mount Hamilton in 1887, where they lie marked with a bronze table at the base of the telescope.

The construction of the observatory took eight years and can at best be described as formidable. Workers had to blast the top of the mountain with black powder and move thousands of tons of rocks to level the site. Heavy gear had to be hauled up the mountain by wagons and animals, and a kiln needed to be built nearby for firing bricks.

Upon completion, it was the first mountaintop observatory occupied year-round. Because it took five hours by horse-drawn wagon to go from San Jose to the observatory, scientists and staff members had to live on the mountain. In 1900, a school was built for the staff members' children. By 1910 motor vehicles provided a faster link to civilization. In 1996, improved transportation allowed staff members and families to move to the University of California campus at Santa Cruz. The winding road visitors use today follows the same route first laid out before the turn of the century.

The original main building serves as a Visitor Center today. A guided tour of the Lick 36-inch (91-cm) Great Refractor telescope, at one end of the building, leaves from the Visitor Center regularly each day. At the other end is the Nickel 40-inch (102 cm) telescope. The Visitor Center has a small display of meteorites, photographs taken using the telescopes, and information about construction at the site. There is also a Visitor's Gallery at the Shane Telescope Building that contains an area from which to view the telescope, as well as exhibits about the telescope and the work achieved with it.

Lick's 36-inch (91-cm) Great Refractor under the large, iron dome is now a century and a quarter old, but it is still occasionally active, mostly as a teaching tool. In its early years it was the most powerful telescope in the world, as Lick had requested, and it has been updated since to remain useful for educational programs. One lens is made of very clear optical crown glass, and the other lens is of harder, denser, lead-containing flint glass. Shortly after the refractor was installed, a spectrograph was added so the hot gases of nebulae could be studied, and the Sun's movement could be accurately measured in relation to the nebulae. This telescope discovered the fifth moon of Jupiter in 1892, the first Jovian moon found since Galileo.

The 36-inch (91-cm) Crossley reflector, once used for historically important studies of stellar evolution and the expansion of the universe, has a camera so that today it can be used for direct photograph and measuring the brightness of stars. It was built by Andrew Common in Great Britain, who sold it to Edward Crossley of Halifax, England. In 1895, after Crossley realized the weather in Halifax was unsuitable for astronomical observations, he donated the instrument to the Lick Observatory. In 1934 this instrument became one of the first to have an aluminized reflector installed. The aluminum does not tarnish easily and reflects more ultraviolet light than silver mirrors. This reflector is not used for research purposes anymore.

The 40-inch (102-cm) Nickel reflector, completed in 1980, was built in the observatory's laboratories on the University of California campus in Santa Cruz. It resides in a dome that once held a 12-inch (30-cm) Clark refractor. It is named after Anna L. Nickel, who donated a large sum to the observatory. Attached to this telescope is a CCD (charge-coupled device) camera, and spectrographs.

The largest telescope, the 120-inch (305-cm) Shane reflector, observes distant quasars and the birth of stars. Named for astronomer C. Donald Shane, it began recording data in 1959. The lens, held in storage, was ground from a Pyrex® blank created in 1933 for the Palomar Observatory. It took eight years to grind and polish the lens for its current use. At the prime focus of the telescope sits a CCD camera, which digitally records the images.

Since 1718, when Edmond Halley observed that certain stars move in the heavens, astronomers have sought ways to measure this motion. In 1934, the 20-inch (51-cm) Carnegie Double Astrograph was installed for this purpose. It contains a camera larger than a person, which produces images on 17-inch (43-cm) glass plates, and two telescopes that simultaneously take photographs, one in blue light, and the other in yellow, to measure the subtle shifts of stars. Like the Crossley reflector, the astrograph is now retired.

The Katzman Automatic Imaging Telescope is a 30-inch (76 cm) robotic system designed to automatically search for supernovae, quasars, variable stars, and check for motion of asteroids. The Automated Planet Finder is another robotic system with a 2.4 m (94 in.) telescope searching automatically for exoplanets (planets orbiting other stars), especially Earth-like rocky ones, by tracking with high precision a star's wobble caused by the orbiting planet.

Among the innovations Lick Observatory has introduced are: the first adaptive-optics guide star, which uses a laser to create a bright point of ionized sodium atoms in the upper atmosphere for better continuous focusing of telescopes to cancel out atmospheric turbulence; and discoveries of exoplanets. The facility also developed the first astronomical digital imager, which has replaced photographic film; and first bounced a laser off the Moon to precisely measure its distance from the Earth.

The steep, winding, 18-mile (29-km) drive to the top of Mount Hamilton to visit the observatory offers a magnificent view of the Sierra Nevadas—weather permitting.

Address: Lick Observatory, 7281 Mount Hamilton Road, Mount Hamilton, CA 95140.
Phone: (408) 274-5061.
Days and hours: Thu.–Sun. noon–5 p.m. Closed Thanksgiving, day after Thanksgiving, 12/24, 12/25.
Fee: Free. Tickets for night-viewing cost extra.
Tours: Self-guided. Free talks are offered in the dome of the 36.-inch (91-cm) Great Refractor regularly in the afternoons. See the website for a schedule. Notify the observatory in advance if a large group is arriving.
Public transportation: Several shuttle bus companies offer rides from San Francisco and San Jose.
Handicapped accessible: In Visitor Center only.
Food served on premises: Snacks and vending machines.
Special note: Inclement weather occasionally causes road closures. Bring a jacket for the cooler weather at higher elevations. No fuel facilities are available.
Website: www.ucolick.org/main/visit/info.html

MOUNTAIN VIEW

Computer History Museum

In 1975, Gordon Bell, vice president of engineering Digital Equipment Corporation (DEC) and DEC cofounder Ken Olsen founded a small museum devoted to computers at its corporate home in Massachusetts in 1975. By 1978 the museum was known as The Digital Computer Museum (TDCM) in the corporate lobby of DEC. TDCM set up a storage facility in Mountain View, California, along with the Computer Museum History Center in 1996. Several years later, the Massachusetts organization closed and transferred most of its holdings to Mountain View. This California-based nonprofit renamed itself the Computer History Museum, and bought its current building in 2002, previously used by the Silicon Graphics firm. The museum covers the art and science of computation from ancient times to the modern era, primarily focusing on hardware—but with more than a nod to software—and has what may be the largest collection of computing machinery in the world.

The extensive main exhibit, Revolution: The First 2000 Years of Computing, includes over a dozen and a half galleries of artifacts. The exhibit begins with methods of calculation, from slide rules to abaci, from finger-counting techniques in Eastern Asia to adding machines and slide rules. A video explaining the unique Antikythera Mechanism shows how this ancient Mediterranean device appears to be the oldest known astronomical calculator. Among the unusual, lesser-known devices presented are sectors (pre-slide-rule instruments from the 1500s), a replica of Schickard's 1623 calculator, and Leibniz's 1673 Stepped Reckoner.

Computer History Museum.
© Doug Fairbairn Photography, courtesy of the Computer History Museum.

The era of programmed machinery began in the 19th century, starting with Jacquard's 1801 punched-card loom, which evolved into Hollerith's census-tabulation machinery for the 1890 Federal Census, to IBM's standardized cards for much of the 20th century. The exhibit continues with analog computers such as Arnold Nordsieck's 1956 Differential Analyzer. Analog computers rely on the physical motion of parts or varying amounts of electrical current for calculations. Their heyday lasted into the 1960s, but they are still employed for certain purposes, or as hybrids with digital computers.

In the 1930s, engineers and mathematicians such as George Stibnitz tinkered with electrical relays as the basis for digital computing, for example, his Model K Adder ("K" for being built on his kitchen table) in 1936. By the early 1940s, vacuum-tube technology was advanced enough for John Mauchly to build ENIAC, the first electronic digital computer, and by 1948, storage tubes provided memory for stored program computers.

From this point forward, the exhibit explores main-frame computers of the 1950s and 1960s, the technology behind computer memory and digital logic, minicomputers, artificial intelligence, gaming, art, music, personal computers, networks, and more. Along the way, visitors see examples such as the Heathkit H8 kit for hobbyists to build their own microcomputer, and an Apple 1 with 4 k of RAM inside a crude wooden case from 1976. There are examples of Texas Instruments® programmable slide rules from the 1970s, and an original "mouse" built by Doug Engelbart in the late 1960s. Other peripherals include early laser printers, tablets and pens, gaming joysticks, and even toggle switches in Altair's early 8800 computer from 1976. Important artifacts are the Cray-1 supercomputer from 1975 and a Neiman-Marcus® Kitchen Computer from 1969 (sold for $10 000 and weighing 100 lbs./45 kg, advertised as a recipe storage device), possibly the first computer for the general public.

The final gallery in the Revolution exhibit is devoted to the world wide web. First dimly imagined in the late 1950s, such modern innovations as hypertext only slowly came to be adopted in the 1980s. A variety of early on-line networking schemes are shown, including Teletext from the United Kingdom, LexiNexis® legal database from 1967, and Compuserve in the United States.

Another gallery in the museum is Make Software: Change the World!, describing a number of major software applications that are used globally, such as MP3 audio compression, Photoshop® image processing blurring the lines between reality and fantasy in imagery, magnetic resonance imaging software as a diagnostic tool for medicine, car crash simulation to show how automobile safety may be improved, Wikipedia® community-based encyclopedic database, texting for rapid communication and banking in developing nations, and World of Warcraft® as an illustration of the business of computer gaming. The museum collects important historical software, such as Apple MacPaint, the APL language, and Word® for Windows® 1.1.

A third, smaller gallery reviews the life and work of Ada, Countess of Lovelace, and her interaction with Charles Babbage and his Analytical Engine, especially her invention of programming as a sequence of steps to solve mathematical problems. Also in that gallery is a history of autonomous vehicles, along with an actual Google® self-driving car.

There are two demonstration labs for historical computers. One room houses a restored, working IBM 1401 mainframe business computer from 1959, including card readers, tape drives, and printers. The other room contains a restored, working 18-bit PDP-1 minicomputer (PDP stands for Personal Data Processor) built by DEC in 1963. One of the first computer games, SpaceWar!, was played on a PDP-1. These computers and the SpaceWar! game are run on a weekly or monthly basis for public view.

The Museum also contains a large research archives, which hold over two linear miles (3 km) of documents, plus nearly 100 000 artifacts, films, photographs, and ephemera. Scholars may visit with advance notice.

Address: Computer History Museum, 1401 North Shoreline Blvd. Mountain View, CA 94043.
Phone: (650) 810-1010
Days and hours: Wed.–Sun. 10 a.m.–5 p.m. Closed Sun.–Mon. (Open Tue. during the summer.)
Fee: Yes.
Tours: Self-guided and docent-led guided. Call or see the website for the current schedule. Remote docent-led virtual tours using a robot are possible with four-week advance reservation.
Public transportation: Shuttle bus is available from the Mountain View Transit Center.
Handicapped accessible: Yes.
Food served on premises: Café.
Website: www.computerhistory.org

OAKLAND

Oakland Museum of California

This beautiful, three-tiered modern structure punctuated by terraced gardens, terraces, and hidden patios is an excellent starting place for visiting California. Each tier is devoted to one of three aspects of California: natural science, history, and art.

Among the holdings in the bottom tier is the Gallery of California Natural Science, housing a collection that outgrew the former Snow Museum of Natural History. Henry A. Snow was a turn-of-the-twentieth-century explorer and big-game hunter who collected about 80 000 animal and plant specimens from around the world.

The hall is organized into the seven places of California that highlight the differences in ecology: Oakland urban environment, Sutter Buttes, Mount Shasta, Yosemite, Cordell Bank National Marine Sanctuary, Tehachapis, and Coachella Valley. (See Yosemite National Park, Yosemite National Park, California.) Inside the entrance, a relief map illustrates the ecosystem, which is further described in display cases and dioramas. Each zone is explored in a separate room. Plants, animals, and insects are set in very accurate, detailed displays depicting natural habitats. Some displays include bird calls and other sounds of nature. Visitors who bring a flashlight are better able to examine the subtleties of the underground world of animals and insects.

The Aquatic California Gallery has 24 permanent displays on the plants, insects, and fish in the underwater environments of California: oceans, marshes, rivers, lakes, snowbanks, and hot springs. A River Runs Down exhibit tracks a stream from its beginnings as a snowbank melting in the mountains. This changing stream flows into an alpine lake, joins a hot spring, and continues down to a spring-season vernal pool. Vernal pools during both the rainy and dry seasons are displayed side by side. The stream proceeds to a delta and out into San Francisco Bay, where visitors learn about the tides and their effects.

A Gallery of California History, the facility's middle tier, houses an anthropology and history collection that was first displayed in 1910 at the Oakland Public Museum in the Camron-Stanford House. This hall focuses on various periods of California culture, from prehistory to modern times. The prehistory exhibit, Before Other People Came, displays tools, weapons, basketry, medical equipment, and other devices. A scale model of a granary shows how Native Americans preserved acorns for later use. Exhibits related to the Spanish-Mexican period cover exploration of California, life in missions and rancheros, early navigation technology, and family life. Coming for Gold examines the Gold Rush; Creative Hollywood describes the earlier American periods, including interactive technology on how filmmaking works. The last section, California: To Be Continued … , contains a Silicon Valley garage describing high-tech industries. (See Computer History Museum, Mountain View, California.) The mining industry is represented with a complete 1870s assay office, as well as mining tools and patent models. Displayed is an 1896 John A. Meyer horseless carriage—the first built in California—and a model of the Caterpillar tractor, which was created in California.

The uppermost tier houses the Gallery of California Art, with works by California artists or containing California themes. Among the pieces are an early 19th-century globe, photographs of Yosemite Valley from the 1870s, a hand-colored etching of a California vulture by John James Audubon, and a collection of paintings depicting California native populations, wildlife, and scenery.

Because the museum emphasizes system and broad principles, the most effective way to see the galleries is by a variety of guided tours, which are led by well-trained, knowledgeable docents.

Address: Oakland Museum of California, 1000 Oak Street, Oakland, CA 94607.
Phone: (510) 318-8400.
Days and hours: Wed.–Sun. 11 a.m.–5 p.m., Fri. 11 a.m.–9 p.m. Closed Mon.–Tue., 7/4, Thanksgiving, 12/25, and 1/1.
Fee: Yes.
Tours: Self-guided and guided.
Public transportation: The site is one block from the Lake Merritt BART station. AC Transit buses stop at the site.
Handicapped accessible: Yes.
Food served on premises: Café.
Website: museumca.org

PASADENA

Mount Wilson Observatory

When astronomer George Ellery Hale discovered Mount Wilson's clear skies and steady air, excellent for viewing celestial bodies, he decided to build an observatory here. Hale had completed a 40-inch (102-cm) refractor telescope at Yerkes Observatory near Chicago in 1897, and now wanted a site for a bigger and better telescope. The Mount Wilson Observatory was begun in 1904. For the first half of the century, this observatory was the foremost astronomical center in the world.

During the first 10 years of the observatory's existence, men and equipment were hauled up the mountain on muleback. Then a toll road was cut for trucks to use to bring new telescopes, people, and equipment to the top. Today the observatory can be reached via a long, winding road. Tours of the facility allow visitors to see the telescopes with which many great discoveries were made.

In 1908, Hale installed a 60-inch (152-cm) reflecting telescope at the observatory. The creation of this giant telescope was not an easy task, with its 15-foot (4.6-m) shaft and 2-ton (1.8-tonne) worm gear, 10 feet (3 m) in diameter. True to his motto, "make no small plans," Hale also placed a 100-inch (254-cm) telescope on Mount Wilson that from 1917 to 1948 was the largest in the world. Other pieces of equipment were added over the years, such as a 150-foot (46-m) solar tower telescope installed in 1912, which made many major astronomical discoveries possible.

This research facility has a long list of significant accomplishments to its credit. It was the first observatory to compare the Sun with other stars in our galaxy. In 1908, Hale discovered the magnetic nature of sunspots here. Harlow Shapley determined the Earth's position in the Milky Way galaxy from this observatory in 1917. In the 1920s, Nobel Prize–winner Albert Michelson reflected a beam of light between Mount Wilson and Mount San Antonio to measure the speed of light, an experiment that provided important information to both astronomers and physicists. Michelson measured the diameter of stars from this mountain, as well. In the 1950s the site prepared a map of the Moon for the Apollo lunar landing. Other work in the 1950s included finding three moons of Jupiter and determining the nature of magnetic fields on stars.

During 1924–1925, the Mount Wilson facility was expanded to include the Hale Solar Laboratory in Pasadena, where Hale continued doing research after his retirement. The Mount Wilson Institute, a nonprofit organization that focuses on education and research for the Mount Wilson Observatory, is housed at both locations.

The current thrust of the observatory is to study nearby stars in the Milky Way. Four historic instruments at the observatory are used in these studies. The 60-inch (152-cm) reflecting telescope measures starspot cycles on nearby stars and compares them with 11-year sunspot cycles. A 60-foot (18-m) solar tower telescope installed in 1911, with a dome added in 1914, studies the wavelike motions of the Sun's visible surface (a scientific discipline called solar seismology, which was created at Mount Wilson). The 150-foot (46-m) solar tower telescope in use since 1912 shows magnetic field recordings with long-term changes in sunspot cycles. In the 1920s Edwin Hubble used the 100-inch (254-cm) Hooker telescope to show that galaxies are rushing away from each other at speeds proportional to their distance, a finding that led to our modern theory of the expanding universe. Work by these two men at this observatory eventually brought scientists to the conclusion that the universe began between 10 and 20 billion years ago.

Only the 150-foot (46-m) solar tower telescope, which has been used to record magnetic fields on the Sun every clear day since 1912, and the 60-inch (152-cm) reflecting telescope are open to the public.

Near the dome of the Hooker telescope is the CHARA (Center for High Angular Resolution Astronomy) Exhibit Hall. Here the CHARA stellar interferometer array is explained, along with Michelson's original 20-foot (6.1-m) beam interferometer from the 1920s. The modern CHARA system uses six telescopes placed all over Mount Wilson, whose light is combined, so the effective size is like a single telescope 0.2 miles (0.3 km) in diameter. There is also an Astronomical Museum built in 1937, which displays photographs taken with the telescopes, a scale model dating from the 1920s of the observatory, and other artifacts.

Address: Mount Wilson Observatory, Pasadena, CA 91102.
Phone: (424) 289-0325.
Days and hours: Apr.–Nov.: daily 10 a.m.–4 p.m. Closed Dec.–Mar. except for special tours.
Fee: Free. Guided tours and group night-time observations incur a cost.
Tours: Self-guided and guided. Guided tours Sat. and Sun. at 1 p.m. Group night-time observations are possible.
Public transportation: None.
Handicapped accessible: Yes.
Food served on premises: Café.
Special note: Inclement weather may cause road closures.
Website: www.mtwilson.edu

SACRAMENTO

California State Railroad Museum

Sacramento celebrates railroading's colorful history at these reconstructed Central Pacific Railroad buildings and 100 000 square-foot (9300-m²) museum. Besides learning about California railroading and its influence on western US history, visitors can see more than 20 carefully restored pieces of rolling stocks and hundreds of historic artifacts.

In 1854, civil engineer Theodore Judah arrived in Sacramento to build the first railroad west of the Mississippi. This 22-mile (35-km) system, called the Sacramento Valley Railroad, was completed in 1859. He then proposed another route that would cut through the Sierra Nevadas and link California to the rest of the country. At first he was unable to procure federal funding, so in 1861 he gathered local businessmen Leland Stanford, Charles Crocker, Collis P. Huntington, and Mark Hopkins at the Huntington & Hopkins Store in Sacramento to finance the venture. These men, who later became known as "the Big Four," launched the Central Pacific Railroad. (See The Huntington Library, Art Collections, and Botanical Gardens, San Marino, California; Mariners' Museum and Park, Newport News, Virginia; and Blenko Glass Company, Milton, West Virginia.)

In 1862, President Lincoln determined that the Union Pacific should build a rail line from the Nebraska Territory westward to the California border, and that the Central Pacific should build a rail line from Sacramento eastward across California. The first railroad to reach the California border would be permitted to build the rest of the rail line across the country. After many blasting accident and construction mishaps, the Central Pacific met the Union Pacific at Promontory Summit, Utah, on May 10, 1869. To mark the event, Governor Stanford of California and T. C. Durant, vice president of the Union Pacific, drove a gold spike into the ground. Sacramento became the western terminus for the railroad industry; those going on to San Francisco boarded a steamboat for a ride on the Sacramento River.

The museum has a re-creation of the hardware store where the Big Four met in 1861 and a roundhouse. We suggest that visitors, upon entering the museum, first see a 30-minute video presentation on the birth and development of the transcontinental railroad. Just as the presentation ends with a railroad scene in the Sierra Nevadas, the projection screen rolls up to reveal a life-size diorama of the same Sierra Nevada scene, complete with granite outcroppings, crates of spikes, piles of rails, a surveyor's tent, and Locomotive No. 1, the 150-year-old *Gov. Stanford*, about to enter a snowshed. Visitors leave their theater seats and step into this diorama.

The *Gov. Stanford*, a 4-4-0 locomotive, arrived in Sacramento in 1863 after being manufactured in Philadelphia by the Norris Locomotive Works, disassembled, and shipped around Cape Horn. When restoring this and other equipment, the museum follows three rules: the equipment must be operable, the restoration must be reversible or removable, and all work must be documented.

The Virginia and Truckee (V & T) Railroad Locomotive No. 12, the *Genoa*, built in Philadelphia in 1873, is one of the oldest operating locomotives in the United States. Between 1873 and 1903 it pulled passenger trains from the Comstock Lode, Carson City, and Reno, Nevada. This locomotive is exhibited pulling V & T Combination Car No. 16 across an original 1884 truss bridge, one of many bridges the railroad built so that it could be dismantled and moved elsewhere when necessary.

V & T Railroad Locomotive No. 13, the *Empire*, built in 1873, is a beautiful piece of highly decorative industrial art, typical of the period. Mirrors surround it so visitors can view and admire the entire locomotive.

The museum has gone to great pains to put the rolling stock in context. A reefer car at the loading platform, complete with agricultural products, provides a sense of railroading's role in making California a major agricultural state and in changing the eating habits of the country. The Great Northern Railroad Post Office Car No. 42 from the 1950s contains exhibits illustrating changes in postal delivery from the 1800s to the present. The *St. Hyacinthe*, a 1929 Canadian National Railway sleeping car, simulates the experience of streaking through dark towns at night.

Audiovisuals, computers, and life-size figures are incorporated into many of the exhibits to teach about railroading. Costumed docents are stationed throughout the museum to offer information and answer questions. Many of the museum artifacts are labeled, but because this information is sketchy, we advise that you take a guided tour. Demonstrations throughout the museum explain railroad history. On weekends in the summer the museum offers short excursions on historic, steam-powered trains. The roundhouse next to the museum building, an important part of railroad history where steam locomotives were housed and serviced, has six tracks radiating from a Union Pacific turntable 80 feet (24 m) in diameter. (See B&O Railroad Museum, Baltimore, Maryland.)

The museum is within Old Sacramento State Park, so nearby are a reconstruction of the Central Pacific Railroad Passenger Station, the western terminus of the transcontinental railroad; and the Central Pacific Railroad Freight Depot, among other historic buildings.

The museum also has a Library & Archives with a Reading Room. Holdings here include Californian and western railroading materials such as books, periodicals, drawings, manuscripts, maps, photographs, and timetables.

Address: California State Railroad Museum, 125 I Street, Sacramento, CA 95814.
Phone: Museum (916) 323-9280; Library & Archives (916) 323-8073.
Days and hours: Museum: Daily 10 a.m.–5 p.m. Closed Thanksgiving, 12/25, and 1/1. Library & Archives: Tue.–Sat. 1 p.m.–5 p.m. Closed state holidays.

Fee: Yes.

Tours: Self-guided and guided.

Public transportation: The Gold Line stops at Sacramento Valley Station, and walk 0.3 mile (0.5 km) to the site.

Handicapped accessible: Yes.

Food served on premises: None, but Sacramento has many restaurants.

Special Note: Some archival materials are held offsite. The Library & Archives need one week advance notice to procure them.

Website: www.californiarailroad.museum

SAN DIEGO

San Diego Air & Space Museum

One of more than a dozen museums in the 1200-acre (490-hectare) Balboa Park, this museum preserves, restores, and displays vintage aircraft and other vehicles, highlights San Diego's role in aviation, and examines the contributions of aerospace technology to society. The park, part of Alonzo Horton's master plan when he developed San Diego in 1867, contains many buildings constructed for the Panama-California International Exposition of 1915 and the California-Pacific International Exposition of 1935. Located within the 1935 Ford Building—an excellent example of Art Deco Streamline Moderne

Pavilion, San Diego Air & Space Museum.
Courtesy of San Diego Air & Space Museum.

architecture built by the Ford Motor Company (see The Henry Ford, Dearborn, Michigan), this museum displays a good collection of aerospace material.

The museum has 85 vintage aircraft, about 1400 scale models, a small collection of vintage automobiles, trucks, motorcycles, and bicycles, and more than 40 000 aviation-related pieces of memorabilia from which the artifacts on display are selected. Among the exhibits is a reproduction of the 1911 Montgomery Glider, *Evergreen*, which John J. Montgomery created during his decades of experimenting for the first controlled flight by a "heavier-than-air" craft. The flight took place in 1883 at Otay Mesa, not far from San Diego. Montgomery, sometimes known as the father of basic flying, continued to study and experiment with gliders for 28 years, and developed an advanced plane that could accommodate an engine for powered flight. He made several test runs in 1911. On the last test, before the engine was installed, the *Evergreen* crashed and Montgomery died.

A World War I SPAD S.VII, built in Norwich, England, by Mann, Egerton and Company, is on view. This model originated in 1916 when the SPA factory placed a 150-horsepower (112 kW) engine on a new fighter plane and added a single .303 Vickers gun in line with the pilot's right eye. Nineteen of the machines, including this one, went directly to the United States from England.

Also on view is the original Ryan X-13 Vertijet, built in 1957, which could take off vertically. This machine needed to be hauled to the takeoff site by a large truck and never became a success because it required a large landing strip, not always available in emergencies. This aircraft, however, contributed greatly to knowledge of Vertical Take-off and Landing Aircraft, such as the Hawker Harrier, which is also on display at the Gillespie Field Annex.

The Japanese World War II Mitsubishi A6M7 fighter, known as *Zero*, is the model that decimated the US Navy at Pearl Harbor. The plane on exhibit is one of four shipped to the United States after the war. More than 10 000 of these planes were made, but they had several weaknesses, including poor diving qualities and no armor plating. Once the US military learned about these problems, the plane was no longer very useful to the Japanese. (See National Museum of the United States Air Force, Wright-Patterson Air Force Base, Ohio.)

The museum has the only existing Consolidated PT-13 Husky, a plane that replaced the PT-1 as the standard model for flight training in the 1930s. PT is the acronym for "primary trainer."

The 1963 Lockheed A-12 Oxcart displayed outside is the tenth in a series of a dozen or so. These planes, first used in 1962, were built secretly for the Central Intelligence Agency—a Mach 3 replacement for the U-2 airplane. This plane was used for over a third of a century, holding speed and altitude records.

A French plane, the Le Rhône, is particularly interesting because it has a rotary engine that turns with the propeller. In most planes the engine is bolted to the body and only the propeller spins.

There are a bank of aircraft rides and flight-simulators for visitors to test drive. A 3D movie theater shows flight-related films.

The International Air & Space Hall of Fame honors those who created or flew the aircraft. Memorabilia on display include training suits, a piece of lunar rock, and the personal effects of a respected enemy, Baron Manfred von Richthofen, the "Red Baron" of World War I fame. One exhibit salutes the women of aviation. Many of the docents on staff are retired aviators or aerospace engineers who willingly share their expertise.

Gillespie Field Annex displays more aircraft, plus offers a second restoration facility (in addition to the basement of the Ford Building). In 2006 the museum acquired a low-speed wind tunnel stationed at Lindbergh Field (the former name of San Diego International Airport), where the facility conducts aircraft research (low-speed flutter tests) for commercial and general aviation purposes.

The museum also houses one of the largest aerospace-related libraries and archives in the United States, with over 20 000 books and periodicals, over 5000 films, two million images, and many special collections. The library and archives are open to the public by appointment.

Address: San Diego Air & Space Museum, 2001 Pan American Plaza, San Diego, CA 92101.
Phone: (619) 234-8291.
Days and hours: Daily 10 a.m.–4:30 p.m. Closed Thanksgiving and 12/25.
Fee: Yes. Flight simulators cost extra.
Tours: Self-guided and guided free docent tours. Guided group tours are available with advance registration.
Public transportation: MTS buses stop near the site.
Handicapped accessible: Yes. Wheelchairs are available on a first-come-first-served basis.
Food served on premises: Restaurant.
Website: sandiegoairandspace.org

San Diego Zoo

At the close of the Panama-California International Exhibition of 1915–1916, scattered animal collections in Balboa Park were gathered together to form the San Diego Zoo. Reminiscing many years later, Dr. Harry M. Wegeforth said, "I was returning to my office after performing an operation at the St. Joseph Hospital. I drove down Sixth Avenue and heard the roaring of the lions in the cages at the Exposition then being held in Balboa Park. I turned to my brother, Paul, who was riding with me, and half jokingly, half wishfully, said, 'Wouldn't it be splendid if San Diego had a zoo! You know, I think I'll start one.'" He formed the Zoological Society of San Diego, which oversees the zoo. In 1922, the animals began moving into new, consolidated facilities. Over the years the zoo acquired many more animals, some donated by ship captains returning from exotic ports around the world.

Among the many firsts at this zoo is the world's largest aviary, the Scripps Aviary, which opened in 1923. When Mbongo and Ngagi, featured in the film *Congorilla*, arrived in 1931 the zoo became the very first to house gorillas. Four koalas, which were added in 1952—the first koalas in a zoo outside Australia—were the stars of the film *Botany Bay*. The captive breeding program, an important endeavor at the zoo, had first successes with proboscis monkeys, soft-shelled tortoises, and thick-billed parrots. The program bred pygmy chimpanzees, the native vulture, and the condor for the first time in the United States.

These animal and plant collections are world-renowned for being among the rarest and most diverse ever gathered in one place. Most of the 4000 animals of 800 species live in barless, outdoor natural habitats separated from the public by moats. This is possibly the only place in North America where people can see Manchurian brown bears, western tufted deer, Mandarin sika deer, Siberian weasels, Collett's black snake, blue-tailed monitor lizards, Chinese monal pheasants, and herds of Somali wild asses and eastern kiangs. The zoo has many of the largest breeding colonies, as well.

The zoo is organized into bioclimatic zones, or groupings of animals and plants from similar climates, that make use of the local temperate climate. Among the plants in this tropical garden setting are some that augment the animals' diets. Many of the plants are rare and endangered. Most of the major walking paths featuring specific types of animals are termed Trails.

The internationally famed Australian exhibit has the largest colony of koalas outside Australia. They are part of a captive-breeding program that the San Diego facility pioneered, leading the way for other zoos. Other animals in this exhibit are Tasmanian devils, the great gray bowerbird, and the green tree python.

A three-acre (1.2-hectare) Tiger Trail complex is a Southeast Asian tropical rain forest containing more than 100 mammals, birds, and reptiles and thousands of plants. The Canyon Dry Riverbed is a mist-shrouded complex with tigers, tarsiers, fishing cats, and many interactive educational displays that describe the animals, present animal sounds, illustrate rain-forest destruction, and use fossils and skeletons to compare animals with people.

By the Sun Bear Trail, a 1.5-acre (0.61-hectare) tropical rain forest in the central canyon has naturalistic enclosures with trickling brooks, waterfalls, and thousands of trees, vines, and shrubs. The setting provides a home for sharp-clawed Malayan sun bears, endangered lion-tailed macaques of India, and exotic birds. Thousands of ginger plants infuse the forest with the pleasant aroma of their native lands.

Botanical specimens from Rwanda and Gabon fill the six section of the Gorilla Tropics, an African rain forest. An audio system re-creates the sounds of the rain forest. A spacious gorilla enclosure, designed to encourage natural animal behavior, contains lowland gorillas, pygmy chimpanzees, crowned eagles, and cherry-crowned mangabeys. Inside the glass enclosure are pools, waterfalls, and grassy hillsides. A path around the exhibit allows easy viewing. The Scripps Aviary, with its rare and unusual birds and plants, is part of this display. A walkway goes through several levels of the aviary, which is covered with a special blue mesh that is almost invisible against the sky. Three additional aviaries display a variety of tropical birds. Together, the four aviaries hold 75 species of birds rarely seen in zoos, including shoebill storks, African pygmy geese, touracos, and trumpeter hornbills.

Hippo Beach is probably the largest hippopotamus exhibit in the world. Egyptian papyrus lines the beach, which surrounds a 150 000-gallon (570 000-L) pool. Visitors view these huge animals from below the water line through a 105-foot (32-m) long observation window. The Northern Frontier exhibit has a re-created tundra for polar bears, Siberian reindeer, Arctic foxes, snowy owls, and a variety of northern ducks and other birds. This two-level indoor exhibit has a 130 000-gallon (490 000-L) pool cooled to 65°F (18°C) and a 27-foot (8.2-m) long viewing window that reveals the polar bears as they swim.

The zoo has a sister facility, the San Diego Zoo Safari Park, about 35 miles north. (See San Diego Zoo Safari Park, Escondido, California.)

Address: San Diego Zoo, 2920 Zoo Drive, San Diego, CA 92101.
Phone: (619) 231-1515.
Days and hours: Daily. Hours vary seasonally: see the website.
Fee: Yes.
Tours: Self-guided and guided. Special botanical tours run occasionally; see the website for details.
Public transportation: Amtrak and COASTER trains stop at the Santa Fe Depot; public buses and trolleys stop at the zoo.

Handicapped accessible: Yes. Wheelchairs and scooters are available on a first-come, first-served basis. Free shuttle bus service around the site.

Food served on premises: Restaurants, cafés, snack shops.

Website: zoo.sandiegozoo.org

SAN FRANCISCO

Cable Car Museum

Within months after gold was discovered at Sutter's Mill in 1848 and the California Gold Rush started, San Francisco grew from a sleepy town to a bustling port city and commercial center. The ground on which this hilly, rainy city was built quickly turned into a sea of mud, poorly suited for travel and transportation. To cope with this problem, 3.5 miles (5.6 km) of redwood-plank road were installed through the city in 1849, and a horse-drawn omnibus known as the Yellow Line was put into service as the city's first public transportation.

This system was far from perfect. The horses were expensive to buy and maintain, tired easily on the many hills, could barely pull the omnibus up the steepest streets, had a life expectancy of only four and a half years, and left droppings that posed health risks for the citizens. When metal rails and wheels were introduced in the 1850s, the horses' workload became much easier. Eventually steam engines replaced the Yellow Line horses, but their clatter upset horses pulling private carriages.

Englishman Andrew Hallidie, a wire-rope manufacturer, solved the city's transportation problem in 1873 by inventing and installing the first practical mechanized streetcar, a cable car, guided by wire cables running in a continuous loop through conduits in the streets. The cable car could move up and down a 16% grade with no difficulty. It was an immediate success.

The Cable Car Museum, located in the three-story Carbarn where cable cars are kept when not in use, has exhibits about the cable car's history. The Carbarn contains the power source for the cables and two areas from which to view cable car operations. A visitors' balcony overlooks the cable-winding apparatus; a window allows visitors to watch the cables as they move in and out of the building. A video provides a detailed explanation of the cable car system.

The process invented by Hallidie is the same one used today to move these vehicles, which remain a reliable source of transportation. The cars move slowly, 9 miles per hour (14 km/h) maximum, climbing the steep hills. Because San Francisco's cable car system was the first in the United States and is the last to survive in the world, it is designated a National Historic Landmark.

Today the city has four closed-loop cable car systems. From the Carbarn the cables run through a conduit beneath the city streets, pulling the cable cars along. When the cars reach the end of the line, they are hand-turned on a large wheel to position them for the trip back to the Carbarn. The cables are made today almost the same way they were in the beginning, with six strands of 19 wires each. They are powered by a 510-horsepower (380 kW) electric motor, a gear reducer, a driving sheave, and an idler sheave. Each system operates independently of the others. Underneath each car, a mechanism controlled by the conductor grips and releases the cable. Over time the cables stretch, so they require frequent tightening to keep the system operating smoothly.

Exhibits in the museum present memorabilia and photographs of the city and its cable cars. Three vintage cable cars are on display: The Sutter Street dummy or grip car No. 46 and trailer No. 54; the 1873 Clay Street Hill car No. 8; and Hallidie's car, the world's first cable car.

The museum, in a busy commercial and tourist section with limited parking, is reached easily by public transportation—including cable car.

Address: Cable Car Museum, 1201 Mason Street, San Francisco, CA 94108.
Phone: (415) 474-1887.
Days and hours: Apr.–Oct.: 10 a.m.–6 p.m; Nov.–Mar.: 10 a.m.–5 p.m. Closed Thanksgiving, 12/25, and 1/1.
Fee: Free.
Tours: Self-guided
Handicapped accessible: Yes.
Food served on premises: None. Many restaurants are within a few blocks of the museum.
Website: www.cablecarmuseum.org/index.html

California Academy of Sciences

This Academy, the oldest scientific institution in the West, came into being in 1853, just five years after the town of Yerba Buena changed its name to San Francisco. It was formed by a group of naturalists who regularly met to discuss their passion and who later put together a cabinet of specimens housed in a Congregational Church. When the collection outgrew the space in 1891, it was moved to an elegant building constructed with funds donated by James Lick. (See Lick Observatory, Mount Hamilton, California.)

The 1906 San Francisco earthquake and fire destroyed the building and the majority of its holdings. That same year a museum expedition to the Galápagos Islands brought back world-class specimens of flora and fauna to form the core of the Academy's new collection. It remains one of the most significant collections from that region. In 1916, the Academy reopened in Golden Gate Park as a complex with three separate parts: the Natural History Museum, the Steinhart Aquarium, and the Morrison Planetarium.

This institution has a long-standing program of studying, displaying, preserving, and interpreting nearly 46 million specimens. It maintains an Institute for Biodiversity Science and Sustainability collection that is, in essence, a reference library for studying natural diversity and evolutionary biology. This is a key program given today's rapidly changing ecosystems, in which specimens with potential commercial and medical value could be lost forever if not identified and catalogued.

In the Kimball Natural History Museum, interactive exhibits foster increased understanding and respect for nature. The Tusher African Hall, a classic gallery, has a series of interesting dioramas, the highlight of which is a barrier-free African safari with authentic sounds. Giraffes, elands, zebras, and other African wildlife are frozen in time at a water hole during the dry season. Other dioramas display mammals, birds, and additional creatures that inhabit a savanna. These settings explain the relationship between predator and prey. At one end of the hall is Human Odyssey, tracing the evolution of humanity. Here are found a touchscreen Human Migration Map showing the paths traveled by various bands of people around the world, plus a selection of humanoid and human skull casts (and reconstructed appearances).

California Academy of Sciences.
© 2015 California Academy of Sciences.

An exhibit called Giants of Land and Sea covers the ecology of northern California, including a virtual trip to the top of a redwood tree, and a fog room (which demonstrates how fog irrigates the redwoods). Megafauna such as humpback, minke, and sperm whale skulls are on display. Earthquakes, a significant concern in this part of the continent, are simulated in a display called Shake House. Visitors stand on a platform to feel the intensity of several famous San Francisco temblors.

Color of Life is an interactive exhibit explaining the role that color plays in the life of animals, whether to attract mates or confuse predators (or prey). An infrared display shows how snakes see in infrared to find body heat of their prey. Special imaging techniques allow visitors to examine beetles and butterfly wings on the microscopic scale.

In the Project Lab, visitors can see a laboratory fitted with typical equipment that museum staff members use for preparing specimens for display. Cameras mounted overhead transmit views of the painstaking work to visitors outside. The Naturalist Center is a hands-on cabinet of curiosities for all ages, along with a variety of books.

Osher Rainforest is contained within a large glass dome 90 feet (27 m) high with over 1600 plants and animals. The internal environment is kept at a temperature of $28 \pm 1°C$ ($82 \pm 2°F$) and a humidity of at least 75%. Free-flying birds and butterflies, beetles, and boas are housed here.

The Steinhart Aquarium, a classic European-style exhibition hall, contains almost 40 000 living specimens of over 900 species. Northern California Coast habitat, a doughnut-shaped 100 000-gallon (380 000-L) tank, has fish from the open sea and San Francisco Bay. In the Discovery Tidepool, visitors can get their hands wet and touch sea creatures.

Water Planet is an exhibit of seven clusters of tanks with changing fauna to show how they adapted to feed, move, reproduce, defend themselves, and sense the environment.

The aquarium exhibits a colony of African penguins—an animal approaching extinction in the wild—and maintains an extensive Species Survival Plan® along with several partner institutions.

Alligator snapping turtles and freshwater fish live amid the tropical foliage and water in the Swamp. A favorite attraction is the 17-foot (5.2-m) jaw of a great white shark that was caught in Monterey Bay.

A Foucault pendulum swings daily at the museum. Morrison Planetarium presents traditional sky shows, with sophisticated special effects, throughout the day on a 75-foot (23-m) dome. The roof of the museum is even accessible to visitors. Here are found weather-monitoring instruments, solar panels, and hills covered with native plant life to help make the building a LEED-certified structure.

Address: California Academy of Sciences, Golden Gate Park, 55 Music Concourse Drive, San Francisco, CA 94118.
Phone: (415) 379-8000.
Days and hours: Mon.–Sat. 9:30 a.m.–5 p.m., Sun. 11 a.m.–5 p.m.
Fee: Yes. Tours and planetarium cost extra.
Tours: Self-guided; guided Behind-the-Scenes tour.
Public transportation: From Amtrak, take the Muni Streetcar to Embarcadero Station, then take buses which stop near the Academy.
Handicapped accessible: Yes. Wheelchairs are available on a first-come, first served basis.
Food served on premises: Café and restaurant.
Website: www.calacademy.org

Exploratorium

This unique hands-on science center was founded in 1969 by the noted physicist and educator Frank Oppenheimer, brother of J. Robert Oppenheimer, the father of the atom bomb. Frank Oppenheimer believed in the "three I's" of science exhibition—innovation, interaction, and involvement; the result is the Exploratorium, a place of extraordinary interactive exhibits illustrating principles of science, technology, human perception, and art. The goal of the museum is to create inquiry-based experiences in order to transform learning.

The museum sits on Piers 15 and 17, built in 1931 and 1917, on San Francisco's Embarcadero. During the 1950s the water between the two was filled in, but when the museum moved in 2013 from its previous home at the Palace of Fine Arts to the Embarcadero, the infill was carefully removed to allow more open water and pedestrian areas. Inside the Exploratorium's warehouse-like structure in Pier 15 is a huge 218 000-square-foot (20 300-m^2) area for offices, classrooms, eating facilities, machine shop, and more, of which 75 000 square feet (7000 m^2) are exhibition areas. Oppenheimer's approach to museum exhibition might be called controlled chaos without preordained results for the experiments. It is as much laboratory as exhibit hall, a place to become familiar with the principles that govern nature.

Many areas of science are demonstrated in the Exploratorium, including physics, chemistry, probability, mechanics, and genetics. Phenomena exhibited here range from absorption to harmonics to wave forms. Often several approaches are used to illustrate

Exploratorium.
© Exploratorium, www.exploratorium.edu

the same principle, with staff members studying the public's response to exhibits to help them improve interactive exhibit pieces. New exhibits are regularly introduced and tested. In the center of the museum is the Exhibit Development Shop, an open industrial shop where visitors watch employees make and refurbish displays and activities. One function of the Exploratorium is to create exhibits for other institutions, as well as teach educators how to demonstrate science topics.

Exhibits fall into six general areas, but as in nature, the dividing lines are blurred. To encourage visitors to think about what they see, a popular display (especially for photographs) near the entrance to the museum galleries features a toilet and a water fountain, both connected to the same water supply. From which do visitors drink—and why?

Human Phenomena is the gallery where psychology and human behavior are explored, as visitors examine a variety of experiments either alone, or with a partner. Many exhibits involve "The Science of Sharing," i.e., collaboration, whether to work with or against another visitor. Others study reaction times or emotional responses.

The Tinkering Studio is a maker's center primarily for children. It also houses some unusual contraptions and artistic endeavors with some kind of mechanical workings, such as chains that create wave patterns, clocks, and so on. There are also circuit boards for people to create electrical circuits with bulbs, batteries, switches, and motors.

Seeing & Listening Gallery examines light, vision, sound, and hearing. Whether it is a monochromatic room where all color is washed out, or a giant mirror to view one's image, the curators believe that natural phenomena should be investigated by the visitors themselves.

Living Systems Gallery explores life from microscopic cells to entire ecosystems. Microscopes show fruit flies, stem cells, and zebrafish. Plankton are on display, along with a variety of tanks and terrariums showcasing living creatures.

Outside on the pier are Outdoor Exhibits where the weather and tidal patterns form part of the museum. A block of concrete with a chunk of iron wedged inside is slowly disintegrating as the rust expands. The wind plays the seven strings of a 27-foot (8.2-m) tall Aeolian harp. The clarity of the bay's water is examined by lowering a "secchi disk," a series of metal rings, into the water.

Inside on the upper level is the Fisher Bay Observatory, which explores the geography, history, and ecology of San Francisco Bay. An environmental station records weather and ocean conditions. Binoculars are available for viewing the bay's activity from the museum.

The Artist-in-Residence program helps the public appreciate the beauty of science and the interrelationship between art and science. A Sun Painting exhibit, developed by this program, consists of a polar axis light tube on the building's roof that sends light to a mirror and then to prisms, which create a spectrum that plays on a screen. Each visitor who walks through the display generates unique patterns of light and color.

Address: Exploratorium, Pier 15 (Embarcadero at Green Street), San Francisco, CA 94111.
Phone: (415) 528-4444.
Days and hours: Tue.–Sun. 10 a.m.–5 p.m. Closed Mon. except for certain holidays. Thursday evenings for adults 6 p.m.–10 p.m.
Fee: Yes.
Tours: Self-guided and guided. See the website for schedule of guided tours.
Public transportation: Streetcar stops at the site. The BART Embarcadero stop is about 0.7 mile (1.1 km) and the Ferry Building is 0.6 mile (1 km) by foot.
Handicapped accessible: Yes.
Food served on premises: Restaurant and café.
Special note: The Exploratorium is so large that a full day may be necessary to explore all the exhibits.
Website: www.exploratorium.edu

San Francisco Botanical Garden

The 1087 (440 hectares) of sandy land that San Francisco came to possess in the 1870s, full of rocky outcroppings and seepage areas, is now one of the best-known landmarks in the city: Golden Gate Park. Although the botanical gardens and park were planned at the same time, it took more than 70 years, and a bequest in the 1920s from the wealthy Helene Strybing, to make it a reality. This 55-acre (22-hectare) living museum of plants contains many botanical species from New Zealand that were displayed in 1915 at the Panama-Pacific International Exposition in the park. It also has plants from around the world, species typical of California, and many medicinal plants.

Because this area is characterized by rainy, mild winters, and foggy, dry summers, the garden focuses on five places in the world with similar climates: California, southern Africa, the western and southwestern coast of Australia, the central coast of Chile, and the Mediterranean. The Arthur Menzies Garden of California Native Plants has 3.5 acres (1.4 hectares) of botanicals from coastal and inland California, many of which have adapted to the local summers. The manzanita, for example, developed a tough, leathery leaf that retains moisture. The California buckeye evolved a different coping technique: dropping its leaves in the heat and leafing out in cooler, rainy months.

Overlooking the Wildfowl Pond, San Francisco Botanical Garden.
Courtesy of San Francisco Botanical Garden.

The South Africa Garden grows plants that are native to the southern top of Africa, such as the Proteaceae. Their leathery leaves sometimes develop fine, white hairs that reduce evaporation. *Xanthorrhoea*, part of the Southwestern Australian Plant Collection, grows flower spikes 2–10 feet (0.6–3 m) tall that sit atop a grassy tuft of leaves. Aboriginal Australians used the roasted spikes as food or to produce a fermented beverage, *mangaitj*. The New Zealand Garden features the *Metrosideros excelsa*, or New Zealand Christmas tree, with red-tipped aerial roots, and *Dicksonia squarrosa*, a tree fern with 8-foot (2.4-m) leathery fronds.

The Succulent Garden focuses on plants from arid climates such as New Zealand and Australia. It also has puyas from Chile and agaves from the southwestern United States and elsewhere, echeverias and the broad-crowned Mexican hand tree from Mexico, and aloes from southern Africa.

Quillaja saponaria, the soap-bark tree, is part of the Chilean Collection. The tree's bark can be peeled and mixed with water to make a lather for bathing and washing clothes. The mayten tree, *Maytenus boaria*, is a decorative tree now growing along the streets of San Francisco.

The Mesoamerican Cloud Forest is testing the adaptive abilities of plants from the high-altitude jungles (6000–10 000 ft.; 1800–3000 m) of tropical Mexico and Central America, regions with misty climates similar to that of San Francisco. Many of the 500 plant species here are endangered in their native habitats, including the summer-blooming salvia and *Cuphea* bloom. Epiphytes, begonias, and blooming *Fuchsia boliviana* grow among the logs.

The botanical garden is known for outstanding collections of deciduous Asiatic magnolias and magnolia relatives. *Magnolia campbellii*, one of the garden's two signature plants, was brought to the grounds from England in 1924. It was the first of this kind of magnolia to flower in the United States. The other signature plant, *Acer pentaphyllum*, a maple tree native to China, is a rare adult specimen outside of China.

The garden also features rhododendrons, azaleas, and camellias. It has an extensive outdoor display of Vireyas or Malaysian rhododendrons, plants from the Malay Archipelago to the northern tip of Australia. They are rarely found outdoors because they fare poorly in cool temperatures.

The Redwood Grove has more than 100 species of native California plants, including a stand of California coast redwoods. One of these trees, planted in 1898, is among the oldest specimens in the garden.

The botanical garden has reproduced a small section of the Katsura Imperial Garden of Kyoto, Japan. Visitors can follow the stone walk or the wooden platform of the Moon Viewing Garden. It is a fine place for quiet contemplation. Other specialty gardens include the Fragrance Garden, Ancient Plant Garden, Children's Garden, and Celebration Garden.

The Helen Crocker Russell Library of Horticulture has a broad reference collection and some rare books, such as *Gerard's Herbals* (1633), *The Queens' Closet Opened* (1671), and *Orchidacea of Mexico and Guatemala* (1843). Each year changing botanical art exhibits are mounted in the library.

Address: San Francisco Botanical Garden, Golden Gate Park, 1199 9th Avenue, San Francisco, CA 94122.
Phone: (415) 661-1316.
Days and hours: Garden: Daily. Hours vary seasonally; see the website for details. Library: Wed.–Mon. 10 a.m.–4 p.m. Closed Tue. and major holidays.
Fee: Yes. Free for local residents with proof of residency.
Tours: Self-guided and guided. Docent-led tours occur daily; see the website for details.
Public transportation: Streetcars and buses stop near the site.
Handicapped accessible: Partly. Main trails are accessible.
Food served on premises: None. A number of eateries are within several blocks south of the site.
Website: www.sfbg.org

San Francisco Maritime National Historical Park

San Francisco sits on a peninsula that juts out between the Pacific Ocean and San Francisco Bay, one of the great natural harbors of the world. The city's 24 miles (39 km) of waterfront makes it an important water-based commercial center.

Maritime National Historical Park is located along the Embarcadero, a road curving around San Francisco Bay that is lined with wharfs, fleets of crab boats and trawlers, seafood restaurants, and bay cruisers. The park, just across the street from Ghirardelli Square, includes a museum building and historic ships anchored along a wharf. It not only preserves and teaches about the ships that helped define the Pacific coast, but also preserves the technology of the people who sailed these waters and fished and traded here.

The oval-shaped museum building looks much like a luxury liner at anchor in the bay. It was originally constructed as a bathhouse in 1939 but was not used for this purpose very long because the bay water proved too cold for most swimmers. The museum houses hundreds of artifacts, photographs, documents, and charts covering West Coast

San Francisco Maritime National Historical Park.
Courtesy of National Park Service.

seafaring history from 1840 to the present. The main floor displays ship models and parts of original San Francisco Bay–based ships and boats. A comprehensive exhibit covers the technology and history of regional steamships. Other exhibits explore the history of the California Gold Rush and its effect on San Francisco, the importance of whaling to the region's economy, and Cape Horners, ships that sailed around the dangerous Cape Horn to reach San Francisco from the Atlantic Ocean. On the second level of the museum are exhibits about classic yachts, whaling, and Cape Horners.

Historic commercial ships at the Hyde Street Pier make up the largest collection of historic vessels on the West Coast. These vessels, from the last quarter of the 19th century and first quarter of the 20th, illustrate the transition from sail to steam power and from wood to steel construction. Visitors can board them to learn more about sailing technology and West Coast maritime history.

The 1886 square-rigged *Balcutha*, launched in Scotland, is typical of hundreds of ships that stopped in San Francisco while bringing in coal, wine, hardware, and other goods from Europe, and taking grain from California around Cape Horn. The schooner *C. A. Thayer*, built in 1895, sailed the seas until 1950 and is one of two surviving ships from a fleet of over 900 that carried lumber from the Pacific Northwest. When steam replaced sail in the lumber trade, *C. A. Thayer* sailed to the Alaskan salmon salteries and codfish banks. This was the last commercial sailing ship working from a West Coast port in the United States.

The sidewheel ferry *Eureka*, with its massive four-story team engine, was built in 1890 to convey freight trains across San Francisco Bay. In 1922 it became the world's largest passenger ferry, carrying as many as 2300 passengers per trip. The schooner *Alma*, launched in 1891, hauled bulk cargo to market, such as lumber, coal, manufactured goods, and fruit from Big Valley farms, until roads and bridges replaced her services. This flat-bottomed boat is the last San Francisco scow schooner still afloat.

Although built in England in 1914, the paddle tug *Eppleton Hall* is like the tugs that towed ships in San Francisco Bay during the Gold Rush. The tug's side-lever engines are descendants of engines used in the sidewheel steamers that once plied the West Coast waters from Panama to San Francisco. The 1907 *Hercules*, built in New Jersey, made her first voyage towing a ship through the Straits of Magellan. Her name reflects the power and strength of her triple-expansion steam engine and steel hull. One of the first of a new breed of ships, she spent her working days hauling lumber-laden schooners to Puget Sound, towing ships through the Golden Gate, and moving barges to and from Hawai'i.

Hyde Street Pier also has a working small-boat shop where visitors can watch and ask questions as boats are being constructed and repaired. Other museum-owed ships are docked along the waterfront, such as the *Pampanito*, berthed at Pier 45. On board this Balboa-class submarine, made for long-range trips during World War II, visitors see torpedo tubes, a 5-inch (13-cm) deck gun, and the crew quarters. At Pier 3 is the *Jeremiah O'Brien*, operated by the National Liberty Ship Memorial. This ship, which once carried troops and cargo around the world, is the last surviving unaltered member of a fleet of more than 2500 Liberty ships built during World War II.

Address: San Francisco Maritime Historical Park (Visitor Center), 499 Jefferson Street (at Hyde), San Francisco, CA 94109. Maritime Museum, 900 Beach Street (at Polk), San Francisco, CA 94109.
Phone: Visitor Center (415) 447-5000. Maritime Museum (415) 561-7100.
Days and hours: Visitor Center: Daily 9:30 a.m.–5 p.m. Maritime Museum: Daily 10 a.m.–4 p.m. Both are closed Thanksgiving, 12/25, and 1/1.
Fee: Free for museum, fee for pier.
Tours: Self-guided and guided. See website for schedule.
Public transportation: Buses run to within a block or so of the site.
Handicapped accessible: Partly. Limited accessibility on boats.
Food served on premises: None. The Fisherman's Wharf area is replete with restaurants.
Website: www.nps.gov/safr/index.htm

San Francisco Zoo & Gardens

Mark Twain once wrote, "The coldest winter I ever spent was a summer in San Francisco." This region's unique climate is at least as challenging for staff members of the San Francisco Zoo & Gardens as it was for Mark Twain. The zoo's weather is influenced greatly by its proximity to the Pacific Ocean and by frequent heavy onshore winds. Even with these difficult conditions, the staff has managed to create a habitable environment for more than 2000 animals and birds and 600 species of native and exotic plants from Africa, Asia, and Australia.

Frequent fog and 35-mile-per-hour (56 km/h) winds leave a salt residue on the plant life, posing a significant challenge to botanists and gardeners. Horticulturists landscape with plants that can tolerate these conditions, such as the Australian *Myoporum laetum* and *Eucalyptus globulus*. Indoor exhibits, such as the Lion House and the South American Tropical Aviary, contain tropical plants. Much of the vegetation, both indoors and outdoors, serves as dinner for the animals; for example, koalas enjoy eucalyptus leaves. Employees at the Insect Zoo grow specialized plants suitable for the propagation of arthropods and butterflies.

When the zoo began in 1929, the Works Progress Administration (WPA), a federal agency that hired the unemployed, built much of it. The zoo once housed animals in the

African Savanna at San Francisco Zoo & Gardens. Marianne Hale/San Francisco Zoo & Gardens.

classic way but has since evolved into a cageless facility with naturalistic habitat enclosures organized into five zoogeographic zones.

African Savanna is a three-acre (1.2-hectare) habitat for reticulated giraffe, greater kudu, Grant's zebra, ostriches, and African crowned crane. Visitors enter this area through a covered passageway and emerge into the savanna. The habitat allows the animals to interact and move around. An immense giraffe barn—accessible to visitors where they can see giraffes up close—looms over the habitat.

The exhibit Great Ape Passages affords visitors the chance to see chimpanzees and orangutans in various habitats, whether nearby in an indoor dayroom, above in two elevated, meshed passageways, or outdoors scampering over climbing structures and platforms.

Koala Crossing has an Australian outback open habitat containing eucalyptus trees, grasses, and other foliage. The koalas in this exhibit have been breeding successfully since their arrival from Australia in 1985. Videos and graphics teach about the natural history of marsupials, with particular emphasis on koalas.

A walkway connecting Koala Crossing and Australian Walkabout is lined with interactive education displays about marsupials and Australian bird species. Australian Walkabout is a 2.2-acre (0.89-hectare) grassy environment featuring red kangaroos, common wallaroos, wallabies, and emus.

The lushly landscaped Gorilla Preserve, built in 1980, was the first exhibit in the country displaying gorillas in family groups. A wall of viewing windows permits visitors to watch these animals close up without disturbing them, and graphics explain their habits and behavior. Penguin Island features one of the most successful breeding colonies of Magellanic penguins in the world. More than 120 Magellanic chicks have successfully hatched here over the years. The 200-foot (61-m) pool and island with nesting burrows resemble the Patagonian coast of Argentina.

The zoo is an accredited member of the Association of Zoos and Aquariums, and participates in Species Survival Plan® (SSP) programs. As part of SSP, many of the zoo's animals may be on loan to other institutions for breeding, and it has rare and endangered animals on display from other institutions. Over 50 of the zoo's breeding programs are part of an SSP program.

The Insect Zoo has everything from velvet ants to Madagascar hissing cockroaches. In the exhibit Nature Trail, visitors have close contact with small animals, turtles, birds, and toads.

Themed gardens for visitors to enjoy include the Native California Gardens, a garden of African plants, and a Prehistoric Garden showcasing ancient plants such as horsetails, ferns, and mosses. Because the Sunset District of San Francisco was a region of sand dunes, there is also a Dune Garden filled with coyote brush, beach strawberry, and more.

Visitors can ride on the zoo's antique Dentzel carousel, built in 1921, or the Little Puffer 22-inch (56-cm) gauge steam train, built around 1904 by Cagney Brothers' Miniature Railroad Company of New York City. It ran at the zoo from 1925 to 1978, was placed into storage for a gorilla exhibit, then restored and put back into service (running on natural gas) in 1998.

Address: San Francisco Zoo & Gardens, Sloat Boulevard at the Great Highway, San Francisco, CA 94132.
Phone: (415) 753-7080.
Days and hours: Daily, winter 10 a.m.–4 p.m., summer 10 a.m.–5 p.m. Last admission 1 hour before closing.
Fee: Yes. Tours, carousel, and train cost extra.
Tours: Self-guided and guided, including bicycle tours. See website for schedule.
Public transportation: L Taraval buses stop at the entrance to the zoo.
Handicapped accessible: Yes. Wheelchairs are available for rent on a first-come, first served basis.
Food served on premises: Cafés.
Website: www.sfzoo.org

San Marino

The Huntington Library, Art Collections, and Botanical Gardens

In the midst of the urban sprawl known as the Los Angeles area is The Huntington Library, Art Collections, and Botanical Gardens, a cultural and educational institution on 207 acres (84 hectares) of beautifully landscaped, rolling lawns. Founded by businessman Henry E. Huntington in 1919, it features one of the finest libraries in the world, a major art collection, and a significant botanical garden. The complex continues to reflect Huntington's interests.

Collis Huntington, one of the Big Four investors in the Central Pacific Railroad and also an investor in the Southern Pacific Railroad, left a fortune to his nephew, Henry. (See California State Railroad Museum, Sacramento, California.) Henry sold his share of the Southern Pacific Railroad stock and invested his money in Los Angeles's Pacific Electric Railway Company, which became the largest electric transit system in the world by the 1930s. Its red streetcars made possible the region's sprawling layout. Within 10 years, Henry Huntington had doubled his fortune. In 1903, when he moved his headquarters

Beautiful Science—Medicine, Huntington.
Courtesy of The Huntington Library Art Collections and Botanical Gardens.

to San Marino, he purchased a 600-acre (240-hectare) ranch and built an imposing house that is now the Huntington Art Gallery. After selling his shares in the streetcar business, he spent the rest of his days building a large collection of books and European art, a grand facility in which to put them, and an impressive botanical garden. The center continues to expand these facilities.

The Library Exhibition Hall, open to the public on a walk-in basis, displays many library treasures, including books concerning the history of science. Over 200 pieces in these displays rotate through the year, and the pages are frequently changed to avoid light damage.

Visiting the research library is by appointment only for qualified scholars. This facility emphasizes European and American history, literature, and art. Included among the millions of pieces are materials from the ancient world, Asia, and the Middle East. The science and technology works cover the 13th–21st centuries. A medieval collection has particularly extensive medical writings. The library has the first printed edition of Euclid's *Elementa geometriae* (1482) and the Jenson editions of Pliny the Younger's *Historia naturalis* (1472). A world-class collection of books printed in Great Britain during the 16th and 17th centuries contains the *Philosophical Transactions* of the Royal Society (1665–1857). Among the 18th-century holdings is the French Academy's *Machines et inventions* (1735–1777). Other works fall in the categories of chemistry, geology, botany, zoology, American medicine through the Civil War period, astronomy, mathematics, transportation, science fiction, and physics—these include Benjamin Franklin's *Experiments and Observations on Electricity* (1751–1754).

Many of the 15 specialized gardens on the 207-acre (84-hectare) estate were designed by William Hertrich, a landscape gardener and superintendent of the Huntington from 1904–1948. Hertrich imported exotic plants from around the world to test their survival potential in southern California's climate. Today the great diversity and quality of the collection make these gardens among the best in the world. They contain more than 14 000 kinds of plants in specialty gardens, theme gardens, and gardens with national motifs, such as Chinese and Japanese. The Camellia Collections contain some 1200 cultivars—one of the world's largest collections. The Rose Garden traces the history of these flowers over 2000 years. A Palm Garden has over 200 species that adapt easily to the area's cool, wet winters and hot, dry simmers.

An extraordinary assortment in the 10-acre (4-hectare) Desert Garden began soon after Henry Huntington bought the estate. Hertrich went to Arizona, New Mexico, and Texas for a three-month collecting expedition and brought back several railroad cars filled with desert plants, helping to provide the Desert Garden with 2500 species arranged by geographic region. It has the world's largest outdoor group of mature cacti and other succulents and an unparalleled variety of shapes, colors, and textures. Visitors will see 100 specimens of puyas plants here, which are native to the Andes of South America and are among the most primitive of the bromeliads.

The Shakespeare Garden has flowering plants typically found in Elizabethan gardens, including many mentioned in Shakespeare's plays. It contains old damask and musk rose varieties dating back to Roman and Renaissance times that would have been familiar to Shakespeare.

The Herb Garden has plantings of medicinals, teas, flavorings for wines and liqueurs, and herbs used in cooking, cosmetics, dyes, and insect repellents.

Gardens with ecological themes include the Jungle Garden, Subtropical Garden, and Australian Garden. Because the local climate brings little rain and some winter frost, the Jungle Garden is not a true jungle, but with its ginger, ferns, palms, bamboos, and tree trunks entwined by orchids and bromeliads, it captures the essence. The Australian Garden has 100 of the 600 varieties of eucalyptus found in Australia. Plants from around the world, many not typically found in Southern California, grow in the Subtropical Garden.

Although the British, American, and European works of art in the Huntington Gallery are not scientific in content, they are well worth viewing. Among them are American paintings from 1730–1930, Renaissance paintings, and 18th-century French sculpture, tapestries, porcelain, and furniture.

Address: The Huntington Library, Art Collections, and Botanical Gardens, 1151 Oxford Road, San Marino, CA 91108.
Phone: (626) 405-2100.
Days and hours: Wed.–Mon. 10 a.m.–5 p.m. Closed Tue., and Thanksgiving, 12/24, 12/25, and 7/4.
Fee: Yes.
Tours: Self-guided and guided docent-led tours and downloadable audio tours.
Public transportation: Buses stop about 1 mile (1.6 km) away; then walk or take a taxi.
Handicapped accessible: Partly. Some areas of the gardens are steep. Wheelchairs are available on a first-come, first-served basis.
Food served on premises: Café, snack bar, restaurants.
Website: www.huntington.org

SANTA ROSA

Luther Burbank Home & Gardens

Charles Darwin's *The Variation of Animals and Plants under Domestication* and other Darwin works greatly impressed a frugal, self-reliant, and persistent 19th-century farmer from Lancaster, Massachusetts, named Luther Burbank. While harvesting little red potatoes, Burbank found a rare Early Rose potato seedball and applied insights he had gained from Darwin to develop it into a new blight-resistant potato, firm and dependable. This variety, introduced to the public as the Burbank Potato, came to be known as the Idaho Baker, the most popular potato grown in the United States.

For $150 Burbank sold the rights for this plant to a seed grower and distributor, and set off for California in 1875. He settled in Santa Rosa and in 1884 purchased four acres (1.6 hectares) on which he located his home, gardens, and carriage house. Not far from Santa Rosa, near Sebastopol, he bought 18 additional acres (7.3 hectares) called the Gold Ridge Farm, where he did much of his research. Burbank said he considered Santa Rosa "the chosen spot of all the earth as far as nature is concerned." The house where he lived from 1884 to 1906, as well as his carriage house, greenhouse, and gardens, is now open to the public. (See The Henry Ford, Dearborn, Michigan.)

After arriving in California, Burbank was soon experimenting on a variety of plants to improve the quality of their fruit or flowers. He would find novel characteristics, cross-breed, and then re-cross-breed. After developing the variation he wanted, Burbank re-combined the varieties until the desired trait was stabilized. From his work on the two

Luther Burbank's Home: Carriage House and Greenhouse.
Courtesy of Luther Burbank Home & Gardens.

properties he introduced over 800 new plant varieties, including hundreds of ornamental flowers, 200 fruits, and many vegetables, nuts, and grains. Some plants were brought over from other countries, and others were developed, hybridized, or improved from types already known in the United States.

The exact number of plants Burbank found or developed is not known. He kept limited records for his personal use, not for scientific history, and did not publish papers on his findings. In his time patent law did not cover horticultural developments, so no relevant legal documents exist.

Burbank's modified Greek Revival house is filled with artifacts and memorabilia from his lifetime. The china cabinet is made with Paradox walnut, *Juglans hindsii* × *Juglans regia*, a hybrid he developed for its rapid growth and beauty. A Paradox walnut tree still grows in the garden. In the Music Room is the desk Burbank used. A house near this one, where he lived with his second wife, Elizabeth, was torn down in 1967. That year Elizabeth moved into this original house, where she remained until her death in 1977.

The renovated carriage house, which originally stored carriages and horse tack, is now a museum with exhibits on the life and work of Burbank. The greenhouse Burbank designed and built in 1889 is where many of his experiments took place. It now contains a re-creation of his office, many of his tools, and changing exhibits.

Only 1.6 acres (0.65 hectares) remain of his outdoor laboratory, the garden. It displays some of Burbank's work and, as requested by his widow, is a memorial garden with a pictorial exhibit on the man and the history of the property. A tour of the garden reveals roses that he introduced, fruit varieties that he created, and plants that he refined. The garden has mature specimen trees, a sensory area, nut trees, an orchard, and demonstration beds with Burbank-influenced varieties of lilies, dahlias, zinnias, asters, gladioli, Watsonia, vegetables, and herbs. Also on view are the Burbank Russet potato and the Shasta daisy.

For more than 20 years, Burbank tried to create a spineless cactus that could be used as cattle forage in the desert. He finally succeeded, producing several such varieties, some with beautiful fruit. Visitors can see them in the Spineless Cactus section. The Drought Tolerant Garden, adjacent to the California live oak tree, displays plants that require little water in summer and are compatible with oaks. Many of these, including evening primrose and California fuchsia, are similar to plants on which Burbank worked.

The Victorian Garden has botanicals evident in photographs from Burbank's days, such as lilacs, spirea, and saucer magnolias, all popular during the 19th century.

According to his wishes, Burbank was buried on the Santa Rosa property in an unmarked grave in front of the house. A cedar of Lebanon tree he planted in 1893 shaded the grave site from his death in 1926 until 1989, when it became diseased and was removed.

Address: Luther Burbank Home & Gardens, 204 Santa Rosa Avenue, Santa Rosa, CA 95404.
Phone: (707) 524-5445.
Days and hours: House: Apr.–Oct.: Tue.–Sat. 10 a.m.–4 p.m., Sun. 11 a.m.–3 p.m. Closed Mon. Gardens: Daily 8 a.m.–dusk.
Fee: Free for gardens; fee for museum.
Tours: Self-guided and guided in the gardens. Audio tours are also available for the gardens. Guided tours of the house.
Public transportation: Buses from San Francisco stop at the Santa Rosa Transit Mall. From there, walk 0.2 mile (0.3 km) to the site.

Handicapped accessible: Yes.

Food served on premises: None. There are some restaurants mostly north and west of the site, a few blocks away.

Website: www.lutherburbank.org

TULELAKE

Lava Beds National Monument

Lava Beds National Monument is part of Medicine Lake volcano, which is in the southern Cascade Mountain Range and the Pacific Ring of Fire. (See Lassen Volcanic National Park, Mineral, California; and Mount Rainier National Park, Ashford, Washington.) This volcano has been active for more than a million years. Features of the land show evidence of past volcanic activity: cinder cones, small cones of pyroclastic cinder material; shield volcanoes, wide, low lava domes formed by quiet lava flows; pahoehoe, smooth and ropy lava; and aa, rough, clinker-like lava. Also found here are stratovolcanoes of lava and ash in conical layers, and an extraordinary number of lava tubes, caves formed when the outer portion of a lava stream hardens while interior lava keeps flowing. (See Hawai'i Volcanoes National Park, Hawaii National Park, Hawai'i.)

There are more than 200 lava tubes at this site, formed over 30 000 years ago when Mammoth Crater exploded. Ladders have been installed and trails cleared in many of the caves to make exploring safer.

Some caves collapsed over the centuries, allowing water and light to enter and plant and animal life to emerge. Others are constantly below freezing and covered in ice; visitors must explore them carefully.

The dramatic, rugged landscape, formed by molten basaltic lava, now has communities of sagebrush-grasslands, juniper-chaparral, and coniferous forest dominated by ponderosa pine. These plants and dozens of others provide homes for predatory birds like great horned owls, bald eagles, hawks, and falcons. More bald eagles winter here than in any other place in the lower 48 states. The region has smaller birds such as doves, hummingbirds, and chickadees, as well.

Two amphibians, the Boreal toad and the Pacific tree frog, have been identified in the park. Reptiles living here include lizards, skinks, boas, and rattlesnakes. The grasslands contain small mammals such as yellow-bellied marmots, squirrels, mice, moles, and kangaroo rats. The rare sage grouse is also found here. Some caves contain one or more of eight species of bats known to inhabit the park.

Tule Lake National Wildlife Refuge, adjoining the monument, is a stopover for tens of thousands of birds migrating along the Pacific Flyway in late spring and fall. The name *tule* (pronounced *TOO-lee*) is derived from the Spanish for bulrush or reed. Modoc Native Americans who lived on this land used large bulrushes growing along the lake for homes, boats, and other items. Native Americans left petroglyphs and pictographs at many spots in the park.

European settlers first arrived here in the 1850s. In 1872–1873 a major Indian war took place—the only one fought in California. Several of the park's visitor areas relate to the war. Captain Jack's Stronghold is named after a leader of the Modoc people, Kintpuash, whom the settlers called Captain Jack. After the US Army destroyed their camp, the Modocs spent the winter in this spot, protected by natural lava barriers that were difficult for the soldiers to penetrate. At Hospital Rock, a high point in the surrounding flat land, two army officers were wounded while negotiating with Modoc warriors. Canby's Cross is where General Canby, sent to negotiate a peace treaty, was killed in April 1873. He is the only US general to die during the Indian Wars. The Modocs believed that if they killed the generals the soldiers would simply go away, but at the end of this standoff the Modocs were sent to a reservation in Oklahoma. Flint arrowheads from the Native Americans are scattered throughout the park, and there are remnants of settlers' cabins, shepherds cabins, illegal Prohibition-era stills, and Civilian Conservation Corps camps from the 1930s.

The Visitor Center presents geology exhibits, overviews of natural and regional history, and displays of Native American artifacts and soldiers' memorabilia. One intriguing display has bullets with teeth marks left by wounded soldiers who were asked to "bite the bullet" during surgery. The center has an original watercolor by William S. Simpson, a journalist and illustrator sent to the United States by *The Illustrated London News*. He went to the site of the Modoc Wars in 1873 and produced one of the few detailed pictures of the region and events at the time they occurred.

Address: Visitor Center: 1 Indian Well Campground Trail, Tulelake, CA 96134.
Phone: (530) 667-8113.
Days and hours: Park: Daily. Visitor Center: Daily summer 9 a.m.–5 p.m., spring and fall 9 a.m.–4:30 p.m., winter 10 a.m.–4 p.m. Closed 12/25.
Fee: Yes.
Tours: Self-guided and guided. Reservations are required for cave tours.
Public transportation: None.
Handicapped accessible: Partly. Visitor Center and some caves are accessible.
Food served on premises: Snacks. Restaurants are in Tulelake and Merrill, Oregon, about 25 miles away.
Special note: Visitors should wear long trousers, long sleeves and jackets, and closed-toed shoes for cave-exploring. Bring hardhats, gloves, flashlights, and kneepads. (Hardhats and flashlights are available at the Visitor Center.) Cave temperatures are about 55°F (13°C). White-nose syndrome is a problem for bats, so check with the park about bringing your gear. Stop at the Visitor Center before exploring caves, and do not go caving alone. The Mushpot Cave next to the Visitor Center is the only lava tube in the park with installed lighting.
Website: www.nps.gov/labe/index.htm

YOSEMITE NATIONAL PARK

Yosemite National Park

Five hundred million years ago, a huge sea similar to the Gulf of Mexico covered what is now the great valley of California and the Sierra Nevada. Over the eons the spectacular landscape that includes Yosemite National Park was shaped by erosion from the sea, movement of the Earth's crust, volcanoes, glaciers, and weathering.

When Yosemite Valley was formed, it was filled with a glacier that carved out the weak granite and left behind monoliths such as El Capitan and Cathedral Rock. The terminal moraine left behind by the receding glacier dammed the Merced River and formed ancient Lake Yosemite. Over time, sediments filled the lake and left the current valley floor. Mirror Lake, a contemporary lake, is filling with silt and evolving into a meadow.

The land ranges from 2000 to 13 000 feet (600–4000 m) above sea level and has five exposure-determined life zones: foothills, low montane, upper montane, subalpine, and alpine. Each has distinctive plants and animals, although some species migrate between zones.

In 1864, President Abraham Lincoln deeded Yosemite Valley and Mariposa Grove to the state of California. Naturalist and photographer John Muir and magazine editor Robert Underwood Johnson realized that the area surrounding the protected land was being destroyed by settlers and persuaded the US government to authorize a national park in 1890. John Muir, President Theodore Roosevelt, and railroad magnate Edward Harriman convinced California to return ownership of Yosemite Valley and Mariposa Grove to the federal government and fold them into the new national park. These areas are best known for their alpine meadows, giant sequoia groves, and wide range of biomes.

Today, visitors can enjoy waterfalls, towering cliffs, rounded domes, enormous monoliths surrounded by sheer canyon walls, and a flat valley floor that has meadows of wildflowers, flowering shrubs, oak woodlands, and mixed-conifer forests. There are thick foothills, chaparral, conifer forests, as well as alpine rock, plus lots more. Over 400 species of vertebrates are found here—possibly because of the large diversity of available habitats—including 17 species of bats and 262 species of birds. The American Bird Conservancy has labeled Yosemite National Park as a Globally Important Bird Area.

The area near the south entrance, known as Wawona from the Native American word meaning "big tree," originally was a Native American encampment. In 1857 Galen Clark built a hostel called Clark's Station here. It is gone now, but the Wawona Hotel, established in this area in 1875, is still in operation. Today Wawona houses the Pioneer Yosemite History Center and has a collection of relocated historic buildings and horse-drawn coaches from the 1800s–1900s.

A small building next to the Valley Visitor Center houses a museum with exhibits about the park's natural history. Behind the center is the Indian Cultural Museum and a reconstructed Ahwahnee Indian Village. A self-guided trail with signs explains the life of these people, called Miwok, the first to inhabit Yosemite. The nearby Museum Gallery features art and historic exhibits. Other exhibit facilities in the valley area are the Visitor Center Arts Exhibit and Happy Isles Nature Center.

Glacier Point, an overlook of the area, is a superb place to start a visit to Yosemite Valley. Perhaps the best example of a glacier-carved canyon in the world, it has a sheer cliff more than 3000 feet (910 m) tall that opens over the valley. Across the valley is Yosemite Falls.

The road through Tioga Pass, constructed in the early 1880s as a mining road, has been rebuilt to accommodate modern vehicles. At 9945 feet (3031 m), it is the highest automobile pass in California. Visitors with health problems should consult with their doctors before ascending to this altitude. The road goes to Tuolumne Meadows, 8575 feet (2614 m) above sea level, the largest subalpine meadow in the Sierra Nevada. The rugged scenery is filled with lakes, fragile meadows, domes, and peaks. The area is full of wildlife, bighorn sheep graze the meadows, and wildflowers abound in early summer.

Mariposa Grove, the largest of the three sequoia groves in the park, features the 1800-year-old (give or take a few centuries) Grizzly Giant. Sequoias are the third-longest-lived species, after bristlecone pines of Nevada, Utah, and eastern California; and Alerce trees from Chile. In the past, these trees were not sufficiently protected against human abuse: Two sequoias with tunnels cut through their trunks so tourists could drive through died as a result. For years great effort went into protecting these giants from fire. Only later did park rangers realize that fire is part of the reproductive cycle of these trees.

In the more than a century and a half of this park's existence, pure stands of trees in the black oak woodlands of the valley have decreased significantly. The National Park Service has initiated a major effort to restore these black oak woodlands, has reintroduced bighorn sheep, and is working to save peregrine falcons as well. The National Park Service moves black bears away from populated parts of the park by trapping them and turning them loose in less-populated sections.

Over 200 miles (320 km) of road are in the park, providing access to nature's wonders. The park provides free shuttle buses to some of the popular tourist destinations. Hiking trails, walking paths, and roadside pullouts offer an opportunity to discover the beauty of the park on foot. Many conservation problems have plagued Yosemite Valley over the years, some caused by its high concentration of visitor services. Studies are being conducted to balance the needs of the land and demands of the public.

Address: Yosemite National Park, Yosemite National Park, CA 95389.
Phone: (209) 372-0200.
Days and hours: Daily. Hetch Hetchy Entrance Station is open only during daylight.
Fee: Yes.
Tours: Self-guided and guided. Bus tours and ranger-led walks.
Public transportation: Amtrak runs a combination bus-train service to Yosemite Valley; Greyhound buses run to Merced, then transfer to YARTS buses to Yosemite Valley. A free shuttle system runs through eastern Yosemite Valley.
Handicapped accessible: Partly. Wheelchairs and scooters are available for rent at Yosemite Lodge and Curry Village on a first-come, first-served basis.
Food served on premises: Restaurants.
Special note: Elevation can cause health problems with some visitors. Inclement weather during winter can close roads.
Website: www.nps.gov/yose/index.htm

Hawai'i

HAWAII NATIONAL PARK

Hawai'i Volcanoes National Park

This national park is unique for letting visitors see (relatively) up close the most awesome and dangerous places on Earth: active volcanoes. The geology is unique, but also the variety of ecosystems within a small area is mind-boggling: from quasi-moonscapes where almost nothing grows on recent lava formations; to lush, humid tropical jungle, all at a relatively cool temperature compared with the hot coastline because of the average altitude (4000 ft., or 1200 m). Vulcanism of the hot spot in the ocean created the Hawaiian islands over the past 70 million years—and this process continues.

The park encompasses 505 square miles (1310 km^2) of land, including Mauna Loa (a shield volcano) and Kīlauea, a very active volcano. By the 1840s, Kīlauea became a tourist venue, and hotels were built overlooking the rim. Lorrin Thurston invested in one of the hotels, a descendant of which is now Volcano House, and became a proponent of creating a national park. By 1916, Congress approved and this land became the first US National Park in a territory.

In the Kīlauea Visitor Center is an introduction to the area. Exhibits present various ecosystems, formation of the islands, and invasive species. Rangers conduct tours of the geology, flora, and fauna, starting at the visitor center. There are seven biomes in the park, ranging from the seacoast and lowlands up to sub-alpine and alpine habitats in Mauna Loa, whose summit reaches 13 677 feet (4169 m) and comprises nearly 20 000 cubic miles (83 000 km^3). The islands are the most isolated in the world, so that many species have evolved in isolation into a spectacular array of organisms—many of which lack the ability to fend off invasive species from elsewhere. Among the best known are the endangered nēnē (Hawaiian goose) and the ʻōhā wai (*Clermontia lindseyana*) tree.

From Volcano House, which is a hotel and restaurant built in 1941, a view of the Halemaumau Crater of Kīlauea is possible: steam by day and a faint red-orange glow by night. A better view of the crater is had from the overlook at the Thomas A. Jaggar Museum, which also has more detailed discussions and displays about vulcanism and the local geology. Immediately next to the Jaggar Museum is the Hawaiian Volcano Observatory with its notable observation tower, run by the US Geological Survey, but not open to the public.

Among the fascinating geological features in the park are the lava tubes, formed by underground lava flow. Visitors can tour Nāhuku (Thurston Lava Tube), while other lava tubes are closed because they are fragile. There are ranger-led tours of the steam vents at Haʻakulamanu (Sulphur Banks), which emit sulfurous water vapor with a characteristic odor—surface water seeps down to the hot rocks, and vaporizes, carrying sulfur-laden minerals in the gases.

A popular activity is to rent bicycles in Kalapana and ride the gravel path to the edge of the National Park to view the lava flowing into the ocean on the southeastern shore of the Big Island. The vista and its orange glow is particularly spectacular at dusk and early evening, but bicycling can be strenuous over the low hills and unpaved parts of the path for those not accustomed to bicycling.

Another day hike is across a dormant crater. The path winds along the lush rim of the crater and descends until plant life abruptly vanishes at the edge of the basin. A marked path takes hikers across a moderately flat desolate landscape. We recommend hiking boots for these walks.

Address: Hawaiʻi Volcanoes National Park, 1 Crater Rim Drive, Hawaii National Park, HI 96718.
Phone: (808) 985-6000.
Days and hours: The park is open continuously. Kīlauea Visitor Center: Daily 9:00 a.m.–5:00 p.m.
Fee: Yes.
Tours: Self-guided and guided. Call or see the visitor center for daily ranger-led tours.
Public transportation: Hele-On Bus runs from Hilo to the park entrance twice a day.
Handicapped accessible: Partly. Some trails are accessible, including the Sulphur Banks; many dipping down into the craters have steep or unstable inclines and are not. Wheelchairs are available at the Kīlauea Visitor Center.
Food served on premises: Restaurant in Volcano House.

Special note: The park (or sections thereof) is subject to closure based upon volcanic activity and noxious fumes. Though the temperature is moderate, the sun is vertical; hats, sunscreen, and water bottles are recommended at all times.
Website: www.nps.gov/havo/index.htm

HILO

'Imiloa Astronomy Center of Hawai'i

Located in the University of Hawai'i–Hilo's Science and Technology Park above the main campus, the 'Imiloa Astronomy Center's mission is "to honor Maunakea by sharing Hawaiian culture and science to inspire exploration." The center attempts to show the relationships between traditional Hawaiian culture and modern astronomy, especially that science being done on Mauna Kea, the volcano held sacred by Hawaiians as an elder, linking sea to sky. Opened in 2006 in a unique structure symbolizing various aspects of the local geology and life, the site is fully bilingual: all information is posted in both the English and Hawaiian languages. The word *'imiloa* itself comes from two Hawaiian words: *'imi* (to seek) and *loa* (long or deep). The combination means "seeking knowledge in depth." Three connected titanium-covered conical buildings represent three active volcanoes, Mauna Loa, Mauna Kea, and Hualālai, on the Big Island, and house 40000 square feet (3700 m²) of display area. In the main circular lobby of the central cone, "Voyage of the Navigator," a colorful 14-foot (4.3-m) tile mosaic in the floor, designed by Hilo artist Clayton Young, depicts Polynesians discovering the Hawaiian islands. It was made with 140 000 tesserae from the Italian firm Bisazza. Above, the transparent glass

'Imiloa Astronomy Center.
Courtesy of the 'Imiloa Astronomy Center.

tip of this conical building allows the nearly vertical tropical sun to shine on the floor mosaic inside.

The main hall's Origins and Voyages exhibits presents two views of astronomy and creation of the universe, that of modern international science, and that of the traditional local civilization, via many interactive displays. Topics include cosmology, black holes, spectroscopy and the electromagnetic spectrum, and the various tools that both modern astronomers and traditional Hawaiians use for celestial navigation and observation. Traditional tools, including adzes of stone and abraders from shells, plus plant fibers used to bind materials together for canoe construction, are shown with a model of a *wa'a* (canoe). There is a description of how the Polynesian navigator used the *ka lani pa'a* (the fixed celestial sphere) to guide a sea journey via the Four Star Families that appear in sequence from sundown to sunrise. A six-foot (1.8-m) Science on a Sphere® globe presents realistic images of various planets in our solar system, plus a variety of phenomena, including the Earth's weather and global warming. There are video feeds from the Mauna Kea Observatories at the summit of the world's highest mountain (about 33 500 ft. or 10 200 m, measured from its base under the ocean to summit).

A planetarium 52 feet (16 m) in diameter, with full-dome video projection and the first Sky-Skan 3D system, accommodates 120 visitors, with the exclusive show "Maunakea: Between Earth and Sky," which describes how Polynesians navigated by the stars, as well as other shows on a variety of astronomical topics.

A Cyber-CANOE (collaborative, analytics, navigation, and observation environment) room is a virtual-reality environment with three large ultra-high-resolution liquid-crystal displays that allow visitors to experience regular interactive sessions on a variety of scientific topics with a presenter.

Behind the center is a Native Garden, populated with a variety of local flora plus "canoe plants," so called because they were brought over by the Polynesians many centuries ago on their canoes. As visitors walk up the hill, the various island ecosystems are portrayed using more than 50 species, going from seaside up a volcanic cone.

Address: 'Imiloa Astronomy Center of Hawai'i, 600 'Imiloa Place, Hilo, HI 96720.
Phone: (808) 932-8901.
Days and hours: Tue.–Sun. 9 a.m.–5 p.m. Closed Mon. Closed on Thanksgiving, 12/25, and 1/1.
Fee: Yes. Extra fee for guided tours.
Tours: Self-guided and guided (with advance booking).
Public transportation: Taxi from Hilo.
Handicapped accessible: Yes.
Food served on premises: Restaurant.
Special Note: The center offers occasional science-related lectures. Call for the schedule.
Website: imiloahawaii.org

Pacific Tsunami Museum

Located in a solid-looking stone structure dating to 1930, formerly owned by First Hawaiian Bank on the main seaside street in Hilo, the Pacific Tsunami Museum opened in 1998. Jointly founded by University of Hawaii professor Walter Dudley and tsunami survivor Jeanne Branch Johnston, the mission of this small but unique museum is that "through education and awareness no one should ever again die in Hawai'i due to a

Pacific Tsunami Museum.
Courtesy of Pacific Tsunami Museum.

tsunami." The word *tsunami* is from Japanese, a combination of *tsu* (harbor) and *nami* (wave).

Unlike earthquakes, storms, and forest fires, tsunamis are relatively unknown by the general public, but are the leading cause of death by natural disaster in Hawai'i. Hilo itself is situated at the apex of a funnel-shaped bay ideal for focusing "tidal waves," and has experienced a number of disastrous tsunami events. The waterfront of Hilo has been left mostly open grass, because of the recurrent destruction from tsunamis.

A tsunami is abruptly moving seawater in a series of waves usually caused by an earthquake or landslide. Such a "wall of water" can travel up to 600 miles per hour (970 kph). When the volume of seawater reaches a suitably shaped coastline, the wave bunches up and grows in height, sometimes 50–100 feet (15–30 m). The wavelength of a tsunami can be 100 miles (160 km). This wall of ocean can suddenly destroy an entire town, and can kill hundreds of people.

Among the exhibits at the museum are a scale model of Hilo in 1946, before the April Fools' Day tsunami; descriptions of various tsunamis of the 19th and 20th centuries; detailed discussions of Pacific tsunamis of the 1950s and 1960s (including a parking meter bent at more than 90-degree angle from the 1960 Hilo tsunami); and other tsunamis of the Indian Ocean and Japan of the 21st century. Besides being an exhibit about tsunamis, the museum also gives advice on how to protect oneself from an approaching tsunami whether local (from the islands, with an arrival time in minutes), or distant (from Japan, Alaska, or Chile, with an arrival time in hours). Posters ask people to prepare with an emergency kit to last for seven days, warn people to flee uphill and inland immediately if they observe warning signs, and monitor via television and radio the situation after a tsunami hits.

One room is dedicated to understanding the physics of tsunamis, with interactive displays, historical instruments and buoys, and The Energy of Moving Water wave machine that shows how tsunamis form. In the former Philadelphia-built bank vault turned into the small Donna Saiki Theater, a 23-minute introductory video plays. Note the water mark on the front door of the museum indicating the height of the water in the 1946 disaster.

The museum also involves survivors of tsunamis themselves. One exhibit includes recordings and videos of survivors describing how they were rescued. Some of the docents in the museum are survivors, who will answer questions about their experiences. The museum also runs regular public community outreach and education to schools about safety and disaster-preparedness when dealing with a tsunami. Tsunami safety brochures are available at the site.

Address: Pacific Tsunami Museum, 130 Kamehameha Avenue, Hilo, Hawaii 96720.
Phone: (808) 935-0926.
Days and hours: Tue.–Sat., 10 a.m.–4 p.m. Closed Mon.
Fee: Yes.
Tours: Self-guided. Docents are available.
Public transportation: Taxi.
Handicapped accessible: Yes.
Food served on premises: None. A variety of restaurants are in Hilo, within a few blocks of the museum.
Website: www.tsunami.org

HONOLULU

Bernice Pauahi Bishop Museum

Established in 1889 by Charles Reed Bishop (1822–1915), a philanthropist, the Bishop Museum now is the largest museum in Hawai'i and boasts the world's largest collection of artifacts from all the Polynesian islands across the Pacific. This facility holds over 25 million objects portraying the natural and anthropological worlds, ranging from botany and entomology to zoology and malacology (study of mollusks). The museum claims one of the largest collections of insects in the United States (over 14 million specimens). Bishop set up the institution as a memorial to his late wife, Princess Bernice Pauahi Bishop, who died in 1884 as the last legal heir of the Hawaiian royal Kamehameha family. Designated the Hawai'i State Museum of Natural and Cultural History, it has a campus composed of a number of buildings plus a small native garden, all surrounding a circular Great Lawn.

The historical three-story Hawaiian Hall was constructed in 1898 in a Romanesque style, portraying life on the islands. First floor exhibits examine life before contact with Europeans, particularly *Kai Ākea*, the beliefs, legends, and divine beings. Second-floor displays discuss *Wao Kanaka*, the people's everyday life, that is, land and nature. The top floor showcases *Wao Lani*, the gods and important events in Hawaiian history, such as various rulers and their overthrow, through artifacts including the feather cape worn by King Kamehameha I, created from 450 000 feathers from roughly 80 000 Hawai'i mamos (now extinct birds). All of the historical Victorian-era cabinetry of richly dark koa wood is restored, and displays a variety of items such as fish-hooks, weapons, clothing, and stone tools. Hanging in the huge atrium is a life-size papier-mâché model of a sperm whale, and a real sperm whale skeleton 55 feet (17 m) long.

Connected to Hawaiian Hall is another Victorian building, Pacific Hall (also built in 1898), which examines the culture throughout Oceania, and how Hawai'i fits into

Hawaiian Hall, Bishop Museum.
Courtesy of the Bishop Museum.

that civilization. On the first floor are displayed a variety of Polynesian technological items including woven mats, model canoes, and artworks. Videos of scholars describing oceanic life are shown continuously. The second floor examines the interrelationships and migrations among the various Pacific peoples, including comparative linguistics, oral history, and archaeological discoveries.

Other galleries in this complex have changing exhibitions, 19th-century Hawaiian art and rare books, and *Kāhili* (feathered standards for local chiefs).

The Jabulka Pavilion includes the first planetarium built in Polynesia (1961), with shows depicting celestial navigation by Polynesians, and the night sky. There is a Science on a Sphere® globe with four projectors presenting weather, tsunamis, and solar activity. The modern Science Adventure Center explores geology and biodiversity on the Hawaiian Islands, such as a model walk-through volcano and a variety of aquariums with local aquatic life.

Outside is a small Native Hawaiian Garden divided into three zones: coastal, dry-forest, and plants—such as a breadfruit tree—brought by Polynesian settlers.

The Library and Archives, open to the public, hold photographs, rare manuscripts, sound recordings, maps, and more.

Address: Bernice Pauahi Bishop Museum, 1525 Bernice Street, Honolulu, HI 96817.
Phone: Museum (808) 847-3511; Library and Archives (808) 848-4148.
Days and hours: Museum: Daily 9 a.m.–5 p.m. Closed on Thanksgiving and 12/25. Library and Archives: Tue. and Thu. 1 p.m.–4 p.m., Sat. 9 a.m.–noon.
Fee: Yes. Planetarium shows are extra.
Tours: Self-guided and guided. Audio tours are available to visitors with smart-phones.

Public transportation: Buses run to within a few blocks of the site. The Waikīkī Trolley Purple Line also stops at the museum.
Handicapped accessible: Yes.
Food served on premises: Café.
Special Note: A variety of lectures and events on Hawaiian arts and technology occur regularly; call or see the website for details.
Website: www.bishopmuseum.org

Oregon

ASTORIA

Fort Clatsop & Visitor Center

After acquiring the Louisiana Territory, President Thomas Jefferson wanted to assert the young nation's power along the Pacific Coast by challenging the British in the northwest and the Spanish to the south. He sent an expedition to explore the new lands, travel the Missouri River to its source, and find the most direct route to the west coast. This trip, the famous Lewis and Clark Expedition, was the first major exploration the United States undertook and financed. (See Monticello, Charlottesville, Virginia.)

Fort Clatsop, a living-history museum, is a reconstruction of the site where the expedition party spent one difficult winter. Lewis and Clark named the fort after the friendly local Native American tribe, the Clatsops.

On the expedition, also known as the Corps of Discovery, 45 people spent 1804–1806 making scientific observations and collecting plants, minerals, and weather data. The leaders, Meriwether Lewis and William Clark, were members of the US Regular Army. Lewis also was the president's personal secretary. The Corps of Discovery studied and conducted diplomatic talks with Native American tribes, mapped geographic features, and recorded important observations and events. They brought back journals and artifacts providing a detailed accounting of the area surveyed and the people they came across.

In late November 1805 the expedition reached the mouth of the Columbia River. Here they established their winter headquarters This well-forested spot provided wood for fires, lumber for building the fort, and tall trees to protect against ocean gales. Animals for food were plentiful in the area. Because the fort was only 15 miles (24 km) from the Pacific Ocean, it also provided access to salt, which was needed for the winter and the trip home. To collect salt, the men boiled sea water and let the steam evaporate until only salt residue remained.

The group took about three weeks to build a 50-square-foot (4.6-m²) structure according to floor plans that Clark drew. It consisted of three connecting huts facing four connecting huts, joined at each end with a palisade. The central section of the fort had a small parade ground. The Corps of Discovery spent the next three and a half months there. The cold, raw rain continued almost every day. As the men moved about, the fringe on their buckskin leather clothing shed the ever-present rainwater. Their diet was monotonous and they were plagued by fleas.

Captain Lewis spent much of his time recording precise observations about local plants, animals, and the environment. These descriptions allowed future scientists to identify his discoveries and recognize the new species he observed, such as the ring-necked duck, the whistling swan, and the white sturgeon. His botanical characterizations include about three dozen plant species previously unknown to Western science. His writing delves into the economic, geographic, and medical potential of these plants. The Academy of Natural Sciences in Philadelphia has many specimens Lewis collected on the Pacific Coast. When the next generation of Chinook and Clatsop peoples were virtually wiped out, his careful observations of their society became even more valuable.

The maps and charts drawn by Captain Clark were the first accurate guides to the lands between present-day North Dakota and the Oregon coast. Clark was also the principal illustrator of plants and wildlife that the party found.

The site is part of the Lewis and Clark National and State Historical Parks. The current fort reconstruction from 2006 was based on notes and floor plans etched into the elk-skin cover of Clark's field notes. This reconstruction, built in the Virginia/Kentucky log-cabin style, is believed to be on or near the original site of the fort. Every detail has been authentically reproduced to present a clear idea of life in these cramped quarters during a most dreary winter. During the summer the site has costumed interpreters demonstrating the almost forgotten tools and skills the explorers used to survive. A previous 1955 reconstruction was destroyed in a blaze in 2005. Programs at the reconstructed fort include information on flintlock muzzle loaders (the guns used by the original expedition) and other technology of the period.

Discovery walks introduce visitors to plant species Lewis observed. At the expedition's canoe landing, visitors learn how low dugout canoes were made and used. The Visitor Center has exhibits and artifacts relating to the Lewis and Clark Expedition. One highlight of the Visitor Center is a high-prow cedar-log canoe built about 1900 by the Makah Nation, typical of the Pacific Northwest canoes used in the time of the expedition. A similar canoe at the front of the center was crafted by the Chinook Nation in 2000. One display case shows hand tools contemporaneous to the explorers, including gimlets, adzes, axes, and braces. Another display has weaponry similar to that carried by the expedition, such as muskets, rifles, and pistols. Mapping tools of the era, such as chronometers and octants, are also on display.

A short drive from the fort is the Salt Works, the original site where members of the expedition made salt. There is a reproduction of a salt cairn similar to the one they used. The original cairn was built of boulders cemented together with native clay. Five kettles placed on the cairn produced about three quarts (3 L) of salt a day. Salt was used not only to season the food for the Corps, but to preserve meat from spoilage.

Address: Fort Clatsop and Visitor Center, Lewis and Clark National Historical Park, 92343 Fort Clatsop Road, Astoria, OR 97103.
Phone: (503) 861-2471.
Days and hours: Summer 9 a.m.–6 p.m.; rest of the year 9 a.m.–5 p.m. Closed 12/25.
Fee: Yes.
Tours: Self-guided [including a cell-phone audio tour, (503) 207-2240] and guided. Ranger-led walks and fort-related demonstrations: see the website for a schedule.
Handicapped accessible: Yes. Wheelchairs and scooters are available on a first-come, first-served basis.
Public transportation: None.

Food served on premises: None. Several restaurants are found in Astoria.
Website: www.nps.gov/lewi/planyourvisit/fortclatsop.htm

CAVE JUNCTION

Oregon Caves National Monument & Preserve

The existence of this cave came to light in 1874 when Elijah Davidson and his dog Bruno were on a hunting expedition and followed a bear through the Siskiyou Mountains of southern Oregon. Davidson saw the bear slip into the cave opening. Probably Native Americans knew of the cave, which is referred to in the plural—Oregon Caves—even though only one large cave has ever been found.

The cave was opened by developers in the 1890s but proved to be of little commercial value because it was too remote for tourists. In 1909, poet Joaquin Miller dubbed the cave the "Marble Halls of Oregon," and the publicity that followed alerted federal officials and President Taft of its importance. Taft proclaimed it a national monument in 1909 because of "unusual scientific interest," putting it under the control of the US Forest Service. What makes this cave special is that not only is the cave within the most geologically complex area north of central Mexico, but also that visitors can view the geological complexity from the inside.

To tour the cave, visitors take a 90-minute ranger-guided walk that covers a little over a mile above and below the surface. Although the path only ascends 200 feet (61 m), the tour is strenuous because the cave entrance is 4000 feet (1200 m) above sea level and inside the cave there are more than 500 steps to climb. The tour is not recommended for those with claustrophobia or health problems.

This area began its evolution about 250 million years ago when volcanic islands attracted bacterial reefs, which built up calcium-rich sediment. The accumulated strata turned into sedimentary rock. Via motion of tectonic plates, the rock drifted northeastward and then was forced under the North American plate. Pressure and folding caused the reef to recrystallize into marble, which eventually cracked and fissured, forming the Siskiyou Mountains. As the land folded and uplifted about 17 million years ago, erosion carried away layers of rock, permitting water to seep into the marble and down the fractures. The region's naturally acidic water dissolved some of the marble and created large fissures, corridors, and rooms starting about 500 000 years ago. Over time the water level dropped, leaving many passages. Groundwater continued to seep into and dissolve the marble, then evaporate as it reached the large, air-filled openings below. Over the millennia, the resulting calcite deposits left behind beautiful decorations on the ceilings, walls, and floors, called dripstones and flowstones.

Oregon Caves is the only cave known to have rocks from all major rock groups, in six types: igneous (both volcanic and plutonic), sedimentary (clastic and chemical), plus metamorphic (both regional and contact). Contact metamorphic and regional marble are found here, both coarse-grained metamorphic rocks derived from limestone.

Indigenous species of insect and arachnid life inhabit the cave: pseudoscorpions, springtails, grylloblattids, millipedes, and harvestmen. Ice bugs, or grylloblattids, *Grylloblatta oregonensis* and *Grylloblatta siskiyouensis*, were first identified in the late 20th century. Collembola, wingless insects known as springtails, were originally found in fossils from the Devonian period. Townsend big-eared bats are among the eight species of bats that inhabit the cave, and rabbits and bushy-tailed woodrats have been seen near the entrance. Pacific giant salamanders live in a moist environment; they reside in the cave to escape from the dry summer heat. Grizzly bears once lived in the area, but do not any longer, though black bears—and cougars—do.

Visitors who tour the cave must dress warmly because the temperature inside ranges from 41–50°F (5–10°C). Some of the paths are wet and slippery, so sturdy shoes are recommended. The surface of the park consists of two mountainside biomes. Below 4000 feet (1200 m) is mixed forest with oaks and other broad-leafed trees, and such conifers as pines and firs. Above 4000 feet (1200 m) are conifers only, such as Port Orford-cedars and the stately Douglas fir. A system of marked trails crosses the forest floor. Hikers are likely to see black-tailed deer, flying squirrels, golden mantled squirrels, Townsend chipmunks, woodpeckers, Steller's jays, and blue grouse. The high biodiversity is caused by the convergence of three climatic zones: Pacific Northwest from the north, Mediterranean from the south, and Semiarid from the west.

Address: Oregon Caves National Monument & Preserve, 19000 Caves Highway, Cave Junction, OR 97523.
Phone: (503) 861-2471.
Days and hours: Surface areas daily.
Fee: Free for surface areas of park; fee for underground.
Tours: Daily guided tours, late Mar.–early Nov. See website for details.
Handicapped accessible: Limited. Watson's Grotto is accessible; more distant rooms are not.
Public transportation: None.
Food served on premises: Café and restaurant.
Special note: High elevation and many steps make this a strenuous site. Check with your doctor if you have health issues, or knee, back, or foot problems. Reservations for guided tours recommended (though not required) during the summer season.
Website: www.nps.gov/orca

FLORENCE

Sea Lion Caves

The only known home on the North American mainland for wild Steller sea lions is midway along the Oregon coast, at Sea Lion Caves. The two-acre (0.8 hectare) site consists of a system of caves constantly pounded by the Pacific Ocean, and rocky ledges overlooking the sea. Most often, Steller sea lions live on offshore rocky islands.

Walkways around the site's headquarter building have sweeping vistas of the coast. From the headquarters building, visitors take stairs, ramps, and an elevator to descend more than 200 feet (61 m) through the cliffs to sea level to enter the cave. To explore the caves, visitors follow paved walkways. One walkway leads outside to the North Entrance Viewpoint, which offers a good view of the rocky ledge rookeries and shoreline.

The eroding action of the ocean and faulting of the Earth's crust helped form this vast cavern. During the great volcanic activity that occurred in the Pacific Northwest 25 million years ago, in the Miocene Period, lava eruptions buried the region in igneous and basalt rocks. These lava flows formed headlands and steep coastal cliffs. Under the

Sea Lion Caves.
Courtesy of Steve Saubert.

lava the Earth's crust rose, pushing lighter sedimentary sandstone and igneous gran-
ites upward, while at the same time, more lava flowed. The result was a series of layers
of rock, ash, and sediment. Visitors walking through the cave can see the stratification
formed over the millennia and hear the sea pound the coastline, continually carving the
rocky shore.

The sea lions here are known by a variety of names: Northern sea lions, Steller sea
lions, or by the Latin, *Eumetopias jubatus*. German naturalist Georg Wilhelm Steller, the
first scientist to identify and study these aquatic animals, was a team member on the 1741
Russian expedition to Alaska led by Danish explorer Vitus Bering. On this important
trip, Bering discovered the 51-mile (82-km) strait that separates Alaska and Siberia.
Afterward, the strait was named for Bering, and the sea lion was named for Steller.

Steller sea lions, part of the family Otariidae, have external ears that open and close
in or out of the water, and hind flippers that point forward. These distinguishing features separate them from the Phocidae family of seals, which have no external ears and
rear-pointing flippers. Sea lions are pinnipeds: they have web-like fins used for propul-
sion. The Latin roots of the word *pinniped* mean "feather-footed." The four web-like flippers have a skeletal structure similar to that of the legs of land animals. Because the rear
flippers work independently of each other, sea lions can walk on land. Their swimming
speeds have been clocked at up to 17 miles per hour (27 km/h).

The sea lions of these caves feed chiefly on bottom fish from the local waters, such as
skate, small sharks, squid, and various species of rock fish. Sea lions can dive to a depth
ranging from 80 to 100 fathoms (480–600 ft.; 150–180 m) and can remain underwater
4–5 minutes. These animals are known to swim as much as several hundred miles (nearly
500 km) in search of food, either singly or in small groups, but they generally do not migrate or have a migration season.

During spring and summer, these warm-blooded mammals breed and birth their young on the rocky ledges. Masses of sea lions are often seen diving into the water and lounging on rocks to warm in the sun. During fall and winter, the sea lions are inside the caves. The greatest threats to young pups' survival are not disease or predators, but rather being trampled by massive adult sea lions or drowning when they are pushed into the Pacific Ocean before they learn to swim. (See Miami Seaquarium, Miami, Florida.)

Visitors enter the caves two-acre (0.8-hectare) main room, which is about 125 feet (38 m) high and has multicolored walls from mineral stains, algae, and lichens. Spectators watch the sea lions from walkways a safe distance from the animals. Display cases show the sea lion's skeletal structure and compare it with that of other animals. Care has been taken to keep the caves and surrounding area as natural as possible while allowing access to visitors.

The area also serves as a rookery for birds, including gulls, cormorants, and guillemots. The pigeon guillemot, a rare variety, is a migratory bird that stops here for mating season beginning in early April. Brandt's cormorants are found throughout the cave area. Their habitat ranges from Baja California to southern British Columbia. Sometimes double-crested cormorants and pelagic cormorants appear on the rocky ledges. California gulls, herring gulls, and western gulls visit this area of the Pacific Coast. Visitors who bring binoculars may see whales, the only natural predators of the sea lion. Gray whales spend the summer feeding close to shore, and occasionally a killer whale appears near the caves.

An opening in the cave provides a picturesque view of Heceta Head Light Station, just north of Sea Lion Caves. The station was built in 1894 to aid seamen navigating the rugged coastline.

Address: Sea Lion Caves, 91560 Highway 101, Florence, OR 97439.
Phone: (541) 547-3111.
Days and hours: Daily, hours vary by season; call for exact hours. Closed Thanksgiving and 12/25.
Fee: Yes.
Tours: Self-guided.
Handicapped accessible: No.
Public transportation: None.
Food served on premises: Snacks only.
Special note: Caves are cool and breezy; jackets suggested.
Website: www.sealioncaves.com

PORTLAND

Oregon Zoo

In the late 19th century, a Portland pharmacist collected animals as a hobby and kept them in the back of his shop. In 1887 they were presented to the city as a gift and exhibited in Washington Park. Over the years this zoo grew and the philosophy behind it evolved to encompass conservation, protection, and propagation of endangered species. The zoo conducts research, participates in Species Survival Plans® (SSP), and hoses 65 threatened and endangered species, 20 of which are part of the SSP program. Currently the zoo has five major zones: Fragile Forests, Asia, Africa, Pacific Shores, and the Great

Northwest. The Oregon Zoo has won national awards for its program design and naturalistic animal settings.

The Penguinarium has an award-winning natural habitat of endangered Humboldt penguins, animals that normally live along the rocky shores of Peru. The Penguinarium is heated to 70°F (21°C), the warmest indoor animal habitat in the zoo, and is designed to encourage breeding. Real and gunnite rocks are used to create rocky shores. A wave machine and water jets simulate the surf and currents these penguins would encounter in their natural home. Visitors can watch the penguins above and below the water line through glass walls, and can hear their braying from speakers positioned around the exhibit. A series of displays interprets the penguins' behavior and ecosystem. Life-size models illustrate the characteristics of the 18 species of penguins in the world, and a map indicates where they are found. All 18 species live in the Southern Hemisphere. The exhibit is also home to a flock of Inca terns.

Built of basalt, Cascade Crest re-creates the alpine ecosystem found in the Cascade Mountains. The habitat includes bent alpine trees and a snow cave, and houses mountain goats. Nearby is the Cascade Stream and Pond, in which beavers and otters play and swim; visitors can see them from both above and below the water surface. Birds that live here include herons and egrets.

Elephant Lands includes the only elephant museum in the world, featuring displays about the history of people and the Asian elephant. This is one of the largest indoor facilities to house elephants. In the six-acre (2.4-hectare) facility, the elephants enjoy a 160 000-gallon (610 000-L) pool plus mud wallows. Because of its renowned elephant-breeding program, the zoo holds the SSP studbook for these animals. Dozens of elephants have been born here, more than at any zoo in the world.

Africa Rainforest combines West African animals with art, culture, and plants. Living among the rich vegetation are fruit bats, banded mongoose, a rock python, colobus monkeys, Cape clawless otters, and several types of birds and waterfowl. In the Bamba du Jon Swamp building, where thunder, lightning, and torrential downpours are regular occurrences, visitors find slender-snouted crocodiles, monitor lizards, lungfish, and frogs. The building has a ranger station where visitors learn about the people of the rain forest and the threats to animals living there.

The four-acre (1.6-hectare) Africa Savanna recreates the dry, open bush country of East Africa. Animals exhibited include the impala, DeBrazza's monkeys, marabou stork, reticulated giraffes, and, of course, lions and cheetahs. The savanna also has an indoor aviary.

There is an insect zoo inside the Nature Exploration Station; an exhibit of Amur tigers; a primate exhibit with chimpanzees, orangutans, and white-cheeked gibbons; and a Family Farm where visitors can have encounters with domesticated animals.

The Washington Park and Zoo Railway, which opened in 1958, offers a six-minute trip past Elephant Lands and Family Farm on its Zoo Loop line. The rails are narrow-gauge, with the cars and locomotives built at 5/8 scale. One of the locomotives is the *Zooliner*, a replica of the diesel Aerotrain of the 1950s. (See National Railroad Museum, Ashwaubenon, Wisconsin.) The only remaining US Railway Postal Station is found here.

Address: Oregon Zoo, 4001 Southwest Canyon Road, Portland, OR 97221.
Phone: (503) 226-1561.
Days and hours: Daily. Fall, winter, and spring 9:30 a.m.–4 p.m., summer 9:30 a.m.–6 p.m.
Fee: Yes.

Tours: Self-guided. For various animal encounters and programs see the website.
Handicapped accessible: Yes. Zoomer shuttle available on request around the zoo. Wheelchairs and scooters are available for rent on a first-come, first-served basis.
Public transportation: TriMet trains and MAX buses stop near the zoo.
Food served on premises: Cafés and snack shops.
Website: www.oregonzoo.org

WINSTON

Wildlife Safari

In a beautiful, secluded valley in southern Oregon, more than 600 animals and birds from over 100 species roam in large, uncaged areas similar to their natural habitats. Besides being a drive-through site for the public to see exotic wildlife close up, the Wildlife Safari breeds North American endangered species and has a long-standing program of animal research in veterinary medicine and related sciences. This facility is one of a small number of parks that successfully maintain wildlife herds.

The drive-through portion of the park has four loosely delineated sections: Asia, Africa, The Americas, and Wetlands. In these areas, some species that originate in one part of the world live with species from other parts.

At one time, nine tiger subspecies populated the swamps, rain forests, and grasslands of Asia, ranging as far north as Siberia: Sumatran, Amur, Bengal, Indochinese, South China, Javan, Bali, Caspian, and Malayan. The South China subspecies is considered functionally extinct in the wild, because no wild individuals have been seen for decades. Javan, Bali, and Caspian subspieces are completely extinct. Only about 8000 Bengal

Wildlife Safari.
Courtesy of Jacob Schlueter.

tigers are left in the world. This facility is doing cutting-edge AI research with critically endangered Sumatran tigers.

Cheetahs, the fastest of all land animals, once lived in many parts of the world but are now threatened and live mainly in small sections of eastern and southern Africa. The park's special compound for breeding cheetahs to revive their dwindling numbers has been very successful. Keeping cheetahs in captivity is not new: records indicate that this practice has existed since about 2000 B.C.E. The park has the most successful cheetah-breeding program in the United States.

The Asia Compound houses several common animals and some rare and endangered species. The mouflon sheep, smallest of the wild sheep, was once abundant throughout central Europe. Now its range is limited to the Mediterranean islands of Sardinia and Corsica. These sheep feed heavily on plants—some of which are deadly for human beings—including some members of the nightshade family. Another animal on display is the white fallow deer, from Europe, Asia Minor, and Iran. An interesting characteristic of these deer is webbing between the antler points. Sika deer once traveled freely over much of Asia, but today are found in parks, farms, and reserves in Japan, Taiwan, Europe, and America. Two varieties of Asian antelope live on the grounds: blackbuck and nilgali, both originating in India. Nilgali translates as "blue cow" and refers to the animal's dark blue-gray color. Among the birds in the Asia compound are African crowned, demoiselle, and sarus cranes, and flightless emus and rheas.

A relative of the American bison, the Tibetan yak, resides here in one of the largest herds in North America. Several varieties of camels live on the grounds. The rare and endangered two-humped Bactrian camels originate in the Gobi Desert and mountainous regions of Asia. These hardy animals survive very cold to very hot conditions. The guanaco, a South American wild camel, was bred to produce the modern domestic llama. Two large islands contain monkey populations: white-handed gibbons on one, and siamang gibbons on the other.

The Americas exhibit has herds of American bison, pronghorn antelope, musk oxen, desert bighorn sheep, Roosevelt elk, mule deer, white-tailed deer, black-tailed deer, moose, and caribou. Birds on display include snow geese, Canada geese, wild turkeys, and sandhill cranes. A separate compound contains Alaskan brown bears and black bears.

The African Exhibit, landscaped with wooded hills and grassy pastures to look much like the Serengeti and Masai Mara, has Cape eland, the world's largest antelope, Thomson gazelles, white-bearded gnu (or as they are commonly known, wildebeest), Damara zebra, and ostriches. The Barbary sheep, or aoudad, which look very much like goats because of their curved horns and beards, are the only wild sheep in Africa. The pharaohs of ancient Egypt domesticated Egyptian geese, which are abundant in the safari park. Also found here are several animals with unique markings and unusual horns, such as sable antelope, gemsbok, and springbok. At the beginning of the 20th century, the southern white rhinoceros was nearly extinct, but with the help of this and other parks it is slowly returning.

The African elephant is quickly becoming extinct in the wild because of habitat destruction and poachers looking for the animal's ivory tusks. This park contributes to research on elephants.

The park has some river horses, or hippopotamuses, which normally live in African rivers and lakes, and a lion compound for the "king of beasts." Once common throughout Europe, lions started disappearing during classical Greek times and are now found only in parts of Africa and the Gir Forest of India.

A walk-through village has services for the public and additional animal exhibits. In the Petting Zoo, visitors can touch Cameroon pygmy goats, four-horned sheep, black-buck, and other animals. There are exhibits of ring-tailed and black-and-white ruffed lemurs, golden lion tamarins, bobcats, and timber wolves, and a Viewing Pond where white-handed gibbons, Canada geese, black swans, and mute swans live. The animal clinic, nursery, and eggery have viewing windows for visitors. Live animal programs and videos are presented in an indoor theater, a discovery room, and an outdoor area.

Because these animals are wild and unpredictable, visitors must stay in their cars to tour the park and cannot enter in convertibles. Vans are available for rental if needed. Some of the animals are shy and do not approach the roadways, so visitors should bring a pair of binoculars. During the winter, the staff members conduct guided van tours through the park.

Address: Wildlife Safari, 1790 Safari Road, Winston, OR 97496.
Phone: (541) 679-6761.
Days and hours: Mid-Nov.–mid-Mar.: drive-through daily 10 a.m.–4 p.m., village daily 9 a.m.–4 p.m.; mid-Mar.–mid-Nov.: drive-through daily 9 a.m.–5 p.m., village 9 a.m.–6 p.m. Closed Thanksgiving and 12/25.
Fee: Yes.
Tours: Self-guided. Guided tours for large groups with prior arrangements. Animal encounters: see the website for schedule.
Handicapped accessible: Yes.
Public transportation: None.
Food served on premises: Café.
Website: wildlifesafari.net

Washington

ASHFORD

Mount Rainier National Park

The Cascade Mountain Range, stretching from Lassen Peak in northern California to Mount Garibaldi in British Columbia, Canada, is part of the "Ring of Fire," a series of volcanoes that encircle the Pacific Ocean. The Cascades contain several volcanoes—extinct, active, and dormant—including the dormant Mount Rainier, located in southwestern Washington State. (See Lassen Volcanic National Park, Mineral, California; and Lava Beds National Monument, Tulelake, California.) Native Americans living in lowlands around the mountain, who hunted and harvested berries, herbs, bulbs, and beargrass on its slopes, called it *Takhoma* or *Tahoma*, meaning "highest mountain," "great snowy peak," or "the mountain." In 1792, when Captain George Vancouver discovered Puget Sound, he spotted "a remarkably high mountain covered with snow" and named it after his friend, Rear Admiral Peter Rainier, a man who never saw the mountain.

The elevation of Mount Rainier ranges from 1800 to 14 410 feet (270–4392 m). The mountain was considerably taller about a half-million years ago after eruptions added to its height, but when the volcanic activity diminished, explosions, collapses, and erosion reduced the altitude. Mount Rainier's massive size prevents moist Pacific maritime air from flowing inland, causing moisture to be dumped on the mountain. Large amounts of rain and snow fall each year, making this one of the wettest places in the world. The mountain is often engulfed in fog.

Nutrient-rich pumice and ash laid down by the ancient eruptions contribute to the region's rich soil. The abundant plant life that grows here supports a plentiful animal population.

Mount Rainier has the largest single-peak glacial system in the contiguous United States, made up of 25 large, named glaciers and many smaller, unnamed ones. Between 1750 and 1850 these rivers of ice reached far into the valleys, but by the 1920s they began a retreat that lasted 30 years. Beginning in the 1950s, some of the glaciers advanced again. Advances and retreats can be measured by the position of rock deposits called moraines, by scratches in the bedrock, and by tree lines separating older and newer forests.

The park's 300 miles (480 km) of hiking trails provide an intimate view of the region. Hiking is best between mid-July and the end of September, when the trails are usually free of snow. For the hearty, one of the best ways to see all aspects of the peak is to hike the 93-mile (150 km) Wonderland Trail that circles the mountain. As the trail rises from 2300 and 6900 feet (700 and 2100 m), it passes through major life zones, among which are moist rain forest, giant old-growth forest, and sub-alpine meadow. Hikers usually take 10–14 days to complete the trail.

The park has many shorter, self-guided trails as well. Carbon River Rain Forest Loop Trail, named for the area's coal deposits, is approximately one-third mile (0.5 km) long and explores a rare inland rain forest. Temperate rain-forest climate and heavy rainfall help create the most luxuriant forest in the park. The 17-mile (27-km) Carbon Glacier Trail permits a close look at a glacier.

The Sunrise area, 6400 feet (2000 m) high, is in the mountain's rain shadow and therefore receives less rain than other parts of the park. From here are spectacular views of nearby Mount Blair, Glacier Peak, and Mount Adams. The three-mile (5-km) Sourdough Ridge Trail crosses a subalpine meadow, with its fertile soil and fragile ecosystems. A Visitor Center has exhibits on subalpine and alpine environments.

The Ohanapecosh River surrounds lowland forest of Douglas fir, western red cedar, and hemlock trees, which hikers see as they follow the Grove of the Patriarchs Trail, 1.1 miles (1.8 km) long. A second, 0.4-mile (0.6-km) trail called Hot Springs Nature Trail explores the forests and hot mineral springs, once the site of a health spa. The Visitor Center in this part of the park focuses on the forests of the Northwest.

Longmire, the first part of the park developed, is where James Longmire opened the Mineral Spring Resort in 1884. When the park was established in 1899, the resort became part of it. A museum that recounts the park's history is housed in the original administration building. Nearby, the 0.7-mile (1.1-km) Trail of the Shadows gives an overview of the area's human and natural history.

The Paradise area, 5400 feet (1600 m) high, gets on average almost 52 feet (16 m) of snow annually and holds a world record of 93 feet (28 m) of snow in 1971–1972. From Nisqually Vista Trail there is a good view of Nisqually Glacier if it is not shrouded in fog or clouds. Hikers who follow this trail through high-country alpine meadows learn how the weather affects the landscape, plants, and animals. Several other trails start here, each focusing on a different aspect of the region. Naturalist walks and talks are given along the trails and at the Visitor Center. The Visitor Center also has audiovisual presentations and exhibits.

A major road system goes through the park. From Nisqually Entrance on the western side of the park, the road starts at an altitude of 2000 feet (610 m) and reaches 5400 feet (1600 m) as it heads east to Henry M. Jackson Memorial Visitor Center at Paradise. Along the way it traverses several life zones, including the Humid Transitional, with its high Douglas fir, cedar, and hemlock; the Canadian, an open, less-shaded zone; and the Hudsonian, with thin, porous, volcanic soil supporting mountain hemlock and subalpine firs.

Common wildlife here are elk, mountain goats, black-tailed deer, black bears, snow-shoe hares, hoary marmots, and pika. Elk, introduced to the Cascade Mountain Range in the 1920s, spend the summer in the area. However, they overbrowse the forest and subalpine meadows, which is threatening to do irreversible damage to these ecosystems.

Address: Mount Rainier National Park, Ashford, WA 98304.

Phone: (360) 569-2211.

Days and hours: Daily. Jackson Visitor Center, Dec. and Apr.: 10 a.m.–4:15 p.m. weekends and holidays; Jan.–Mar.: Fri.–Sun. and holidays 10 a.m.–4:15 p.m.; May–Nov.: daily. Ohanapecosh Visitor Center: daily late Jun.–mid-Sept. Sunrise Visitor Center: daily Jul.–mid-Sept.

Fee: Yes.

Tours: Self-guided and guided. For ranger-led hikes and programs, see the website for schedule.

Handicapped accessible: Partly. Most of the visitor centers are accessible, but trails may not be.

Public transportation: None.

Food served on premises: Restaurants, café, snack bar.

Special note: Check road conditions fall, winter, and spring. Weather is often foggy or rainy, and temperature is often cool. Selected sections are closed in winter. The only park road open throughout the year is State Route 706 from the Nisqually entrance to Longmire.

Website: www.nps.gov/mora/index.htm

CASTLE ROCK

Mount St. Helens National Volcanic Monument

The world was reminded of the Cascade Mountains' volcanic nature when Mount St. Helens erupted in March 1980 after more than a hundred years of tranquility. Over the past 4000 years, this volcano in the southwestern part of the state, the youngest volcano in the Cascade Range, has been more active and explosive than any other in the 48 contiguous states. (See Mount Rainier National Park, Ashford, Washington; Lava Beds National Monument, Tulelake, California; and Lassen Volcanic National Park, Mineral, California.)

After the volcano reawakened, earthquakes and steam explosions occurred for several months thereafter. The entire north side of the mountain slid away on May 18, 1980, in an enormous landslide, the largest witnessed in human history. Before the landslide settled, a blast of hot rock shot northward. The eruption began, with steam and ash rising 15 miles (24 km). The landslide and eruption opened a crater more than 2000 feet (610

m) deep and between 1.2 and 1.8 miles (1.9–2.9 km) across. Exploded debris mixed with water from the mountain's melting snow and ice, forming lahars (volcanic mudflows) that covered 25 square miles (65 km²). The North Toutle River Valley was buried in mud 50–600 feet (15–180 m) deep. These geological processes reduced the mountain's height from 9677 feet (2950 m) to 8363 feet (2549 m). Erosion has since reduced the summit further.

From 1980 to 1986 and 2004 to 2008, lava eruptions formed a muffin-like "lava dome." The dome is now over 1000 feet (300 m) high and 3000 feet (910 m) in diameter.

In 1982, the site became a national monument for research, education, and recreation. Visitors drive toward the mountain high on valley walls with spectacular views of the volcano, lahar deposits, and timber sheared off from the geologic forces. Passersby see nature healing itself at a rapid rate. The US Forest Service planted grasses to stabilize slopes, and many trees for later use. Lake ecology is evolving back into a typical subalpine environment, bare landscape is becoming green, subalpine meadows are blooming with wildflowers, and animals are returning.

As lahars slipped down the mountain, they took trees from their roots and engulfed roads, bridges, and houses. Fifty-seven people lost their lives in this calamity, and about 2300 square miles (600 km²) of prime timberland became a wasteland. (Much wood was salvaged by timber companies.)

The eruption significantly altered the shape, size, biology, and chemistry of the mountain lakes. As huge amounts of sediment and nutrients from avalanche debris and pumice particles settled in the lakes, bacteria flourished. In Spirit Lake, one of the best-known lakes here, oxygen was quickly used up and new bacteria that required metal, other gases, and new nutrients came to dominate. As time went on, however, the oxygen supply returned and the biology and chemistry of the lake reverted to normal alpine and subalpine standards.

About 1800 years ago, Mount St. Helens changed its eruptive style, emitting pahoehoe basalt, a rock more typical of Hawaiian volcanoes than of those on the mainland. (See Hawai'i Volcanoes National Park, Hawaii National Park, Hawai'i.) The Cave Basalt lava flow cooled and shrank last within its interior, leaving behind the 12 810-foot (3902-m) long Ape Cave, the longest intact lava tube in the continental United States. The low slope of the tube is far easier to walk; the upper end is covered with debris falling from the ceiling.

Run by Washington State Parks, a Visitor Center at Silver Lake, near Castle Rock, has exhibits, some interactive, that explain volcanoes and the events that happened at Mount St. Helens. Two films that include television-news footage and eyewitness accounts introduce this eruption. Other exhibits and dioramas cover the prehistory and history of the region, the people who once lived in the area, and the ways Native Americans, trappers, miners, and people in search of recreation have used the land. A trail 0.6 mile (1 km) long runs near Silver Lake, and includes a boardwalk from which to view migratory birds.

The US Forest Service runs Johnston Ridge Observatory, only five miles (8 km) from the volcano, which includes exhibits, eyewitness accounts, and a movie about the volcano. During the summer, interpreters hold talks. The Boundary Trail offers stunning views into the crater.

The three-hour drive to Windy Ridge Viewpoint offers many opportunities to admire the power and majesty of nature. At the top of the ridge visitors have a grand view to the peak of the mountain, only 4 miles (6.4 km) away.

Address: Mount St. Helens National Volcanic Monument, 3029 Spirit Lake Highway, Castle Rock, WA 98611.

Phone: (360) 449-7800.

Days and hours: Daily 6 a.m.–10 p.m. Silver Lake Visitor Center: Mar. 1–May 15: daily 9 a.m.–4 p.m.; May 16–Sep. 15: daily 9 a.m.–5 p.m.; Sep. 16–Oct. 31: daily 9 a.m.–4 p.m.; Nov. 1–Feb. 28: Thu.–Mon 9 a.m.–4 p.m. Closed Martin Luther King, Jr., Day, Presidents' Day, Veterans Day, Thanksgiving and day after Thanksgiving, 12/25, and 1/1. Johnson Ridge Observatory: Daily mid-May–Oct. Coldwater: Nov.–May: Sat.–Sun. 10 a.m.–4 p.m.; Jun.–Oct.: special programs only. Closed 12/24, 12/25.

Fee: Yes.

Tours: Self-guided at the Mount St. Helens Visitor Center. Ranger-led talks at the Johnson Ridge Observatory.

Handicapped accessible: Partly. Visitor centers are accessible, but trails may not be.

Public transportation: None.

Food served on premises: None.

Special note: Climbing permits may be required. Winter brings inclement weather and road-closures.

Websites: National Monument: www.fs.usda.gov/recarea/giffordpinchot/recreation/recarea/?recid=34143

Silver Lake: parks.state.wa.us/245/Mount-St-Helens

Johnson Ridge: www.fs.fed.us/visit/destination/johnston-ridge-observatory

GRAND COULEE

Grand Coulee Dam

The Columbia Basin Project was started in the 1930s to improve the water supply, which was unpredictable because of erratic rainfall in the Columbia River Basin. The principal component of this project is Grand Coulee Dam, completed in 1941 by the Public Works Administration under President Franklin D. Roosevelt. The dam is one of a string of 11 built along the Columbia River, the fourth-largest river (by volume discharge) in the United States. Grand Coulee Dam delivers water for irrigation, provides flood-control, produces electric power, and forms a lake for recreation. A guided tour of the dam facility lays out its history and how it functions.

About 50 million years ago, the land that was to become the Columbia Basin was covered with an inland sea. Fissures in the rocks opened between 16.5 and 6 million years ago, spewing very fluid, fast-flowing lava that left layers as thick as 5000 feet (1500 m). This resulted in a flat, smooth plain. During the Pleistocene epoch, starting about 2 million years ago, glaciers advanced and retreated across the region, reaching as far south as what is today's Coulee City. An ice dam, at the end of a large glacier, blocked the Clark Fork River, and formed a glacial lake in western Montana. Beginning about 18000 years ago, the ice dam broke repeatedly. This had been holding back vast amounts of meltwater. The rushing water flooded eastern Washington State many times, eroding topsoil, and carried boulders over the landscape, depositing them in open prairies. These rock formations are now called "haystacks" by local farmers. After the last Ice Age ended, glaciers retreated and the river receded to its original channel.

Grand Coulee Dam.
Courtesy of the Bureau of Reclamation, Grand Coulee Dam.

The deeply cut river channels high on the plateau, left from the Ice Age, are now dry. They lie some 500–1600 feet (150–490 m) higher than the current riverbed and are a constant reminder of the region's dramatic geologic history. These dry old riverbeds are called coulees (from French *coulée*, meaning "flow"), and the Grand Coulee is the largest of these Channeled Scablands.

The source of the Columbia River is the glacial lakes of British Columbia. A quarter of million square miles (650 000 km²), an area as big as Texas, drain into this river, which flows for 1200 miles (1900 km) before emptying into the Pacific Ocean along the Washington-Oregon coast.

Early farmers and ranchers in the Columbia River Basin had two major difficulties: poor transportation for delivering their goods to market, and an unreliable water supply. Three railroads built across the basin in the late 19th and early 20th centuries solved the transportation problem, but the water supply remained inconsistent until the Columbia Basin Project was completed.

Grand Coulee Dam supplies water to over half a million acres (200 000 hectares) of farmland where over 90 different crops, including potatoes, sugar beets, alfalfa, wheat, and vegetables, are grown. While driving through the Columbia River Basin to the dam, visitors see thriving apple orchards on well-irrigated land juxtaposed with desert-like landscape.

Although power production was a not a major function of the dam in the original plan, it is now the largest producer of electricity in the United States. Four powerhouses have a total capacity of more than 6.8 megawatts.

The Visitor Center provides an introductory overview of the Columbia Basin Project and Grand Coulee Dam. Exhibits describe the dam's construction—including interviews with construction workers; the hydroelectric power system; the effects the dam had on Native American life; and how management of the competing needs of the locale are dealt with, from flood-control to irrigation, from recreation to fish migration. Some of the exhibits are interactive.

The guided tour starts begins at the Tour Building on the east side of the river, and takes visitors by bus to the upper level of the Pump-Generator Plant. From the plant's lower level, visitors see huge units that lift water from Roosevelt Lake to Banks Lake and the connecting irrigation system. This part of the tour explains how the plant generates electricity. From the top of the spillway, 350 feet (110 m) above the river, visitors have a panoramic view of the river, the power plants, and the town of Coulee Dam.

The dam creates the huge Franklin D. Roosevelt Lake, which extends 151 miles (243 km) to the Canadian border. It is used for boating and fishing, and has recreational facilities along its banks. Many other dams along the Columbia River also have informative displays for the public.

Sun Lakes–Dry Falls State Park (which is not part of the Grand Coulee Dam facilities), 7 miles (11 km) southwest of Coulee City, contains an interesting geologic formation from the Ice Age: Dry Falls. This dry waterfall, 3.5 miles (5.6 km) wide and 400 feet (120 m) tall, was one of the greatest waterfalls in all of history. By comparison, Niagara Falls is smaller—a mile (1.6 km) wide and only 165 feet (50.3 m) high. The park has a visitor center with exhibits explaining the geologic significance of this unique landscape, and a glass wall that provides an excellent view of Dry Falls.

Address: Grand Coulee Dam Visitor Center, State Highway 155, Grand Coulee, WA 99133.
Phone: (509) 633-9265.
Days and hours: Visitor Center: daily 9 a.m.–5 p.m.; extended hours from Memorial Day through Sep. Tours: Daily from Apr. 1–Oct. 31. Closed Thanksgiving Day, 12/25, and 1/1.
Fee: Free.
Tours: Self-guided in the Visitor Center; guided only for the tour.
Handicapped accessible: Yes.
Public transportation: None.
Food served on premises: None.
Special note: Times of tours vary; tours may be canceled because of work along the tour route or inclement weather.
Website: www.usbr.gov/pn/grandcoulee/visit/index.html

MUKILTEO

Future of Flight Aviation Center & Boeing Tour

Boeing Corporation, manufacturer of airplanes for a century, began in 1916 when William Boeing opened the Boeing Airplane Company with a staff of six employees. The first plane he produced was the B & W (Boeing and Westervelt), a float-equipped biplane. Soon Boeing purchased Heath Shipyard, a failing business, and turned it into an airplane production space. The old tar-paper building in the shipyard received a coat of red paint and the nickname the Red Barn. (See Museum of Flight, Seattle and Everett, Washington.) Within 14 years this was the largest complex devoted exclusively to producing aircraft. In 1933, Congress forced the Boeing group to split into three separate corporations: Boeing Corporation, which manufactures airplanes; Pratt and

Whitney, which manufactures airplane engines; and United Airlines, which carries passengers and freight. Guided bus and walking tours at Boeing's airplane manufacturing plant allow visitors to see different stages of airplane assembly and learn about the company.

In the last half of the 20th century, Boeing Corporation created a fleet of military and commercial airliners that are a standard around the world. The B-17 (the Flying Fortress) and B-52 had a significant effect on our military's war and peace efforts. (See National Museum of the United States Air Force, Wright-Patterson Air Force Base, Ohio.) Over 25 000 jetliners were put into service since the first 707 rolled off the production line.

In the 1960s, Boeing Corporation purchased about 780 acres (320 hectares) of land north of Seattle for facilities to produce its 747 series of aircraft. Later Boeing started making the 767 series of airplanes at the site as well. Thousands of airplanes from the 747 and 767 series were manufactured here. The nearby Renton Division of Boeing produces the 737 and 757 series. Some of the 787 Dreamliner series are built here, as well as at a plant in South Carolina. Airlines do not send their planes back to this plant for refurbishing and refitting; that work is done at neighboring Paine Field.

The 747 is one of the largest and longest-range jetliners in service. Its tail is six stories high, and it can fly up to one-third of the earth's circumference without refueling. Boeing produces a variety of different models used for either freight or passenger service. The 747-200 series can carry 366 passengers. The 747-300 series, with a stretched upper deck, can handle 400 passengers. A special high-performance model of the 747 circled the globe in 37 hours.

The 767 series, which has twin jets, was introduced in 1982. This model is a medium-range plane with a capacity of 215–300 passengers. It carries freight or passengers up to 7000 miles (11 000 km) nonstop. The 787 series is also midsize with twin engines, with a range of up to 9800 miles (16 000 km), and is significantly lighter than the 767, being made with composite materials.

The main assembly building is 11 stories high and covers 98 acres (40 hectares) of land. This is the largest building in the world by volume: 472 million cubic feet (13.3 million m^3). The tour focuses primarily on the assembly building, which is kept at a constant 65–70°F (18–21°C) so it is comfortable for workers and visitors. This property also has offices, warehouses, paint hangars, and production buildings.

Tours begin at the Future of Flight Aviation Center, where visitors can see working models of jet engines, a variety of aircraft components, and videos on the history of jet aviation. A rooftop observation deck provides visitors with views of the local aircraft, Paine Field, and the Boeing factory nearby.

The 90-minute tour starts with a 25-minute show describing the early history of the company and showing aircraft production. A long walk through tunnels and an elevator ride take visitors to a platform high above the production floor. From this vantage point, visitors get a sense of the production pace and can better appreciate the facility's impressive size and scale. Sounds of rivet guns can fill the air as airplane bodies are assembled, wings are attached, and interiors are fitted for passengers or freight. During a brief stop at the paint hangar, guides provide information about the planes' destinations.

The afternoon tour may be more interesting because the morning tour often passes through the assembly building during employees' morning coffee break. Summer tours fill early, so advance reservations are highly recommended. Rest rooms are available only at the Aviation Center before or after the tour.

Address: Future of Flight Aviation Center and Boeing Tour, 8415 Paine Field Blvd, Mukilteo, WA 98275.

Phone: (800) 464-1476.

Days and hours: Aviation Center: daily 8:30 a.m.–5:30 p.m.; Boeing Tour: 9 a.m.–3 p.m.

Fee: Yes.

Tours: Self-guided at the Aviation Center; guided only for the Boeing Tour.

Handicapped accessible: Yes. Provide advance notice for wheelchairs on the Boeing Tour. Wheelchairs are available on a first-come, first-served basis.

Public transportation: Buses from Seattle stop near the Aviation Center, then walk 0.6 mile (1 km) to the site.

Food served on premises: Café in Aviation Center.

Special note: No children under 4 feet (1.2 m) tall are permitted, and those under 16 must be supervised. No photography is permitted on the tour.

Websites: www.futureofflight.org

SEATTLE

Washington Park Arboretum

In 1934, the University of Washington and the Seattle Parks Department created this 230-acre (93-hectare), urban, living museum of woody plants that can grow in the mild Puget Sound climate. The arboretum began in 1895 when Professor Edmond S. Meany planted some trees on the campus of the University of Washington. Today the plant collections and educational programs are part of the larger University of Washington Botanic Gardens, including the nearby Center for Urban Horticulture, managed by the

Azalea Way, Washington Park Arboretum.
Courtesy of University of Washington Botanic Garden.

University. The grounds at Washington Park are managed cooperatively by the city and the university.

The Olmsted Brothers landscaping firm, famous for creating urban parks throughout the country, developed a master plan for the arboretum in 1936. (See Frederick Law Olmsted National Historic Site, Brookline, Massachusetts.) Azalea Way, Rhododendron Glen, and the Lookout are elements of the Olmsted design that are still in place. Much of the original garden, trails, roads, and buildings were built by federal Works Progress Administration workers.

Among topics under study at the University of Washington Botanic Gardens Center for Urban Horticulture is the effect of urban microclimates on the growth of plants in urban and urbanizing conditions. How to improve the growth of plants in the harsh urban environment is important because plants cleanse the air, temper the climate, and create a buffer against noise, wind, and glare. The center is investigating stresses on root systems and shoot growth in crowded environments, the technology of micropropagating landscape plants (a technique that allows propagation of a plant from a few cells of another plant), invasive plants, and the restoration of degraded environments. The center's research suggests that several Chilean varieties of plants may do well in the Seattle area.

This arboretum has over 40 000 specimens of about 4500 different plants, including hundreds of varieties of conifers, cherries, and magnolias, as well as maples, mountain ash, and camellias. The mild climate allows species from Europe, New Zealand, China, and South America to grow here alongside native varieties. Few other arboreta prominently display in an outdoor setting so many plants from such a variety of places.

The Witt Winter Garden features witch hazels, rhododendrons, camellias, winter hazels, birches, and late-winter blooms such as *Cornus mas* and Lenten roses (Helleborus spp). Although the peak blooming season typically begins in late March and continues into June, something is always flowering along the park's many trails.

Rhododendron Glen has many rhododendron species and hybrids, ranging from dwarfs to tree forms. The best time to see these plants in flower is from March through July. In Loderi Valley are large-leaved rhododendrons that bloom in March and April, as well as Loderi hybrids growing amidst magnolias.

Foster Island, a waterfront area of Lake Washington's Union Bay, is a wildlife sanctuary that contains oaks, alders, pines, and birches. Azalea Way, best seen from March to June, is a three-quarter-mile (1.2-km) path bordered by flowering cherries, azaleas, and dogwoods. Woodland Garden has a vast collection of Japanese maples growing in a shady valley where a meandering stream connects two small ponds. Visitors who stop by this section in fall will see fine foliage colors.

A display of five ecogeographic gardens from Pacific Rim climates similar to Seattle, the Pacific Connections Garden, dates from 2008. Gardens include Cascadia (the Pacific Northwest), New Zealand, southeastern Australia, Chile, and China. The New Zealand Forest, a 2.5-acre (1-ha) display of plants native to the inland South Island of New Zealand, is the largest display of New Zealand plants outside of that country.

The Japanese Garden in Washington Park was once part of the University's collections, but is now an independent entity with an admission fee. It was designed by Juki Iida, who created over a thousand other such gardens worldwide. The 3.5-acre (1.4-hectare) site is a condensed representation of Japan, with mountains, lakes, rivers, tablelands, and a village. Some 500 huge boulders were moved here from sites near Snoqualmie Pass in the Cascade Mountains. Hundreds of azaleas and rhododendrons grace the garden, along with camellias, evergreens, flowering fruit trees, mosses, and ferns. Tea ceremonies are performed here. This serene spot is closed in winter.

Addresses: Washington Park Arboretum, 2300 Arboretum Drive East, Seattle, WA 98112
Phone: Arboretum: (206) 543-8800. Japanese Garden: (206) 684-4725.
Days and hours: Arboretum: daily dawn–dusk. Visitors Center: daily 9 a.m.–5 p.m. Japanese Garden: Apr. and Sep.: Mon. noon–6 p.m., Tue.–Sun. 10 a.m.–6 p.m.; May–Aug.: Mon. noon–7 p.m., Tue.–Sun. 10 a.m.–7 p.m.; Oct.: Sun. noon–5 p.m., Tue.–Sat. 10 a.m.–5 p.m.; Nov.: Mon. noon–4 p.m., Tue.–Sun. 10 a.m.–4 p.m. Closed Dec.–Mar.
Fee: Arboretum is free; Japanese garden has a fee.
Tours: Self-guided and guided. See the website for schedule of guided tours. Audio tours are available at the website.
Handicapped accessible: Yes. Wheelchairs are available.
Public transportation: METRO Buses stop at the sites.
Food served on premises: None.
Websites: Arboretum: botanicgardens.uw.edu/washington-park-arboretum/ or uwbotanicgardens.org
Japanese Garden: www.seattlejapanesegarden.org

Woodland Park Zoo

The Woodland Park Zoo, which covers 92 acres (37 hectares), is an award-winning, state-of-the-art facility with naturalistic habitats for 300 species of animals. Habitats are divided by climate and vegetation in bioclimatic zones where animals from different species are grouped together. Mixed-species environments, typical of life in the wild, stimulate natural animal behaviors like foraging and breeding, and make it possible for scientists to learn more about ways to conserve these animals and promote their survival.

The zoo supports more than 35 wildlife conservation projects in the Pacific Northwest and across six continents to save threatened and endangered species, and takes part in over a hundred Species Survival Plans® (SSPs) and other species-recovery projects. This zoo breeds dozens of endangered species and some threatened species as part of the SSPs.

An award-winning western lowlands gorilla exhibit houses these very social, endangered animals. The exhibit provides a realistic environment similar to the gorillas' native habitat, the tropical forests of western equatorial Africa, where they can live and breed as they do in the wild. It contains trees and vegetation from the Pacific Northwest that look similar to those in their natural habitat. Because these sensitive animals sometimes need to escape the public and each other, the exhibit contains a big, grassy mound, plus other features where they can get away for privacy. Fiberglass trees and vines supplement the vegetation in the sheltered area.

The African Savanna, a replica of the Serengeti Plain, has a water hole where several species of animals mingle: giraffes, zebras, gazelles, Egyptian geese, and others. Sandy-colored patas monkeys spend most of their time hiding in the grass. This is one of the largest displays of these monkeys in the nation.

What looks like huge rock outcroppings in the lion habitat disguise indoor sleeping dens and a food-preparation area for the lions. The rocks, of sprayed concrete dyed a realistic color, are heated so these big cats can stay outdoors for a greater portion of the year.

Tropical Asia, an extensive, award-winning exhibit on four acres (1.6 hectares), is divided into three major zones. One zone is the Assam Rhino Reserve, which houses greater one-horned rhinoceroses, as well as Asian forest tortoises, Visayan warty pigs, and demoiselle cranes. A second zone is the 2.7-acre (1.1-hectare) Trail of Vines that recreates the forests in Malaysia, and a forest canopy in northern Borneo. Here visitors see orangutans, pythons, Malayan tapirs, and siamangs, an endangered species of gibbon. Orangutans are severely endangered animals. The third zone is the Banyan Wilds, in which tigers, small-clawed otters, and sloth bears live, along with birds.

Australasia examines how geography and generic isolation cause animals to adapt to the bioclimate in which they live. The exhibit features such species as wallaroos, emus, and kookaburras.

The Tropical Rain Forest is a 2.5-acre (1-hectare), multilevel, indoor and outdoor exhibit showcasing the rain forests of South America and Africa. Here live nocturnal mammals, reptiles, insects, colobus monkeys, and two species of lemurs. This plant-shrouded habitat also houses spotted ocelots, poison dart frogs, free-flying birds, and western lowland gorillas.

South-central Alaska is illustrated in the award-winning Northern Trail exhibit. The *taiga*—a Russian word meaning "land of little sticks"—displays wolves in a setting with stunted trees and charred stumps of wood. A tundra area has snowy owls, and brown bears. The interpretive center in the tundra area resembles an Athabaskan house, a shelter the Athabaskans typically built along rivers. The montane area, a rugged, mountainous region, houses mountain goats and elk. A display of Steller's sea eagles is situated here.

Temperate Forest is an exhibit of the world's cooler forests, such as are found in the Pacific Northwest, exhibiting a variety of animals such as Asian cranes, maned wolves, red pandas, and Chilean flamingos. A re-created wetlands hosts a variety of swamp-dwelling creatures. Here is also BugWorld, which is the zoo's insect and spider collection.

Addresses: Woodland Park Zoo, 5500 Phinney Avenue North, Seattle, WA 98103.
Phone: (206) 548-2500.
Days and hours: Daily. Oct.–Apr.: 9:30 a.m.–4 p.m.; May–Labor Day: 9:30 a.m.–6 p.m.; Sep.: Mon.–Fri. 9:30 a.m.–4 p.m., Sat.–Sun. 9:30 a.m.–6 p.m. Closed 12/25.
Fee: Yes.
Tours: For Animal Encounters and other programs, see the website for schedules and details.
Handicapped accessible: Yes. Wheelchairs and scooters are available for rent.
Public transportation: METRO Buses stop at the zoo entrance.
Food served on premises: Café and snack shops.
Websites: www.zoo.org

SEATTLE AND EVERETT

Museum of Flight

The Museum of Flight started in 1966 when a small group of aviation enthusiasts formed the Pacific Northwest Aviation Historical Foundation to preserve aircraft and related artifacts and educate the public. Finding people with an interest in aviation was not difficult in this part of the country, where aviation has long provided an important link with more remote areas in Canada and the northwestern United States, including Alaska. The Boeing Company had its headquarters in Seattle since its beginning in 1916 until 2001, when the headquarters relocated to Chicago.

When William Boeing started this company, he bought out a tar-paper building—the E. W. Heath Shipyard—alongside the Duwamish River and turned the entire building into a production facility for designing and building airplanes. Many of Heath's craftsmen stayed employed at the new firm, transferring their skills from ships to aircraft. The original barn, called the Red Barn after its 50th anniversary, moved to its present location, was faithfully restored to its original condition in the early 1980s, and is now part of the museum. (See Future of Flight Aviation Center & Boeing Tour, Mukilteo, Washington.)

The museum occupies 25 acres (10 hectares) of the Boeing Field/King County International Airport, right next to the Boeing Company Military Flight Center. From the museum's viewing balcony overlooking the airport, visitors see much of the airport's daily operations.

The museum's East Campus is composed of interconnected buildings: the historic William E. Boeing Red Barn and the modern glass-and-steel T. A. Wilson Great Gallery, plus a two-story wing devoted to World War I and II fighter aircraft. The Red Barn, now on the National Register of Historic Places, contains exhibits chronicling the early history of aviation and airplane manufacture. Both floors feature exhibits describing Boeing's early design and assembly efforts up to 1958. Full-size dioramas show wood carvers and seamstresses making materials for the first B & W (Boeing and Westervelt) plane. Visitors also see exhibits explaining early propeller designs and memorabilia of the early days of flight.

The Great Gallery houses more than 50 airplanes and space vehicles, about 20 of which are suspended from the ceiling. Among them is a reproduction of Boeing's first aircraft, the B & W; and an 18-seat restored Boeing 80-A trimotor that was the first commercial passenger aircraft to carry female flight attendants. Aircraft on display that were not built by Boeing include a reproduction of the 1902 Wright glider; a Curtiss Jenny, a favorite of pilots during and after World War I; a three-passenger Fairchild F-24 originally owned by ventriloquist Edgar Bergen; and a workhorse of the 1920s that carried mail across the nation, a Stearman C-3B biplane. Exhibits cover the history and technology of flight.

The facility also has planes of the 1930s, such as the Stinson SR flown by daring bush pilots. This aircraft helped bring provisions and mail to those in the Northwest Territories and Alaska—areas with rough mountain terrain and no proper roads.

A variety of other unusual and historic aircraft are on view. The Bowers Fly Baby 1A is an experimental aircraft from 1962 that is easy to assemble. The 1928 Boeing P-12 used innovative design features like steel and aluminum tubing in the fuselage structure. The Boeing 247D of 1934 has an all-metal body and was capable of going 200 miles per hour (320 km/h). Aerocar III, designed in 1968, can convert from a car into an airplane in 10 minutes.

Among the more modern aircraft on display are a Lockheed M-21 Blackbird and its D-21B drone, a rare set of spying aircraft of the 1960s. The Space Gallery of the museum includes exhibits on the Apollo command module, Mercury space capsule, and the Russian Resurs 500 space capsule from 1992. The museum's theater daily shows movies on the history of aviation and space.

The West Campus of the facility includes an Aviation Pavilion, which shows off airplanes including the Boeing 247D of 1934, with an all-metal body and able to fly at 200 miles per hour (320 km/h); the first Boeing 727, 737, and 747; the first Air Force One jet, a Concorde supersonic airplane, a B-17 and B-29, B-47, and a Boeing 787 (number 3). Next to the Aviation Pavilion is the Charles Simonyi Space Gallery, devoted to the full-scale space shuttle trainer from NASA Johnson Space Center, shuttle programs, and current spaceflight programs. The space shuttle trainer offers 30-minute detailed tours. Beyond the Pavilion is Vietnam Veterans Memorial Park, in which is a B-52 bomber used in that war.

The Restoration Center & Reserve Collection at Paine Field in Everett is part of the museum complex. Here many aircraft are preserved and restored, and some of the rare aircraft are stored. Self-guided visits to the Restoration Center are possible.

A Library and Archives with 5000 cubic feet (140 m³) of material include nearly 70 000 books plus many periodicals, photographs, manuscripts, drawings, technical manuals, log books, and blueprints. The archives also contain over four million images.

Addresses: The Museum of Flight, 9404 East Marginal Way South, Seattle, WA 98108. The Museum of Flight Restoration Center & Reserve Collection, 2909 100th Street SW, Everett, WA 98204.

Phone: Museum: (206) 764-5700. Restoration Center: (425) 745-5150. Archives: (206) 764-5874.

Days and hours: Museum: Daily 10 a.m.–5 p.m. Closed Thanksgiving and 12/25. Restoration Center: Wed.–Sun. 9 a.m.–4 p.m.

Fee: Yes.

Tours: Self-guided and guided. Shuttle Trainer, movie theater, and Restoration Center cost extra.

Handicapped accessible: Yes. Wheelchairs are available.

Public transportation: METRO Buses stop at the museum.

Food served on premises: Café.

Website: www.museumofflight.org

VANTAGE

Gingko Petrified Forest

A petrified forest is a rare phenomenon. Ginkgo Petrified Forest, on the shore of the Wanapum Reservoir, is exceptional because of the number and diversity of its petrified logs. Perhaps 60 different species of trees make up this forest, more than at any other such site. Petrified wood is Washington's official state gem. (See Petrified Forest National Park, Grand Canyon, Arizona.)

The petrified samples include trunks, limbs, and leaves of maples, Douglas firs, spruces, sequoias, pines, walnuts, and elms. Of particular note are the eight petrified gingko logs found within the region—three within the park—which give the park its name. Once common on this continent, ginkgo trees became extinct in North America but survived in eastern Asia. Later, ginkgoes were introduced as an ornamental tree and are now found across the North American landscape. From information gathered at this site, scientists learn what kinds of trees once grew in the region and whether they vary from current species. This information helps researchers deduce many characteristics of the soil, weather, and other environmental conditions in ancient times.

About 40 million years ago, during the Eocene epoch, the Cascade Mountains began to rise and cut off rain and moisture to regions east of the range. Some 15–20 million years ago, during the Miocene epoch, the region was lake and swampland filled with many species of trees. Dying trees and broken limbs washed down streams and were carried by mudflows to be deposited on the lake bottom, where they might have rotted—were it not for clay silt sealing out the air. The lakes and rivers dried up and the forest died out. Lava from rising Cascade volcanoes again from 5–7 million years ago poured out over the landscape, trapping logs in an anaerobic environment. Because the trees were waterlogged, they did not burn. Over time the buried forests and logs became preserved as their woody, organic matter was replaced by silica-based minerals.

During the last Ice Age, glaciers collected vast amounts of water as they moved south from what are now Canada, Idaho, northern Washington, and Montana. As the glaciers met warmer southern weather, they began to melt, and water became trapped behind them. These ice dams eventually broke, in some cases releasing torrents of water that carried away surface material and exposed the petrified logs.

The park has three major areas: the Heritage Area, the Natural Area, and the Wanapum Recreation Area. An Interpretive Center at the Heritage Area has a geology exhibit, an overview of park history, and displays of palagonite, or pillow lava, formed when the lava that covered the region met lake water. Visitors can see a collection of unpetrified wood samples, more than 30 varieties of cut and polished petrified pieces, and other minerals. A pamphlet for self-guided trail walks is available here.

Ancestors of the region's native Wanapum Native Americans made tools and other objects out of petrified wood. The name Wanapum means "river people." Adjacent to the Interpretive Center is a collection of about 60 petroglyphs, or rock carvings, of hunter and gatherer symbols. Studies suggest that the carvings are more than 200 years old. The ancient petroglyphs were moved to this site when water from the Wanapum Reservoir began flooding the area north of the park. Originally the rock carvings were along cliffs overlooking the Columbia River about 1.5 miles (2.4 km) north of the park.

The Natural Area, a short distance from the Interpretive Center, has a one-and-a-quarter-mile (2-km) interpretive trail that follows part of the prehistoric lakebed in which the petrified logs formed. Along the trail, hikers see many examples of petrified wood, each identified by a trail marker. Among the highlights of this trail is a petrified ginkgo log.

Three miles (5 km) of hiking trails go through the central Washington sagebrush environment, a biome representative of the dry regions east of the Cascade mountains. A small Trailside Museum tells the story of petrification.

The Wanapum Recreation Area has a host of wildlife, including elk, deer, and coyote. During the spring, summer, and fall, visitors may spot such reptiles as the blotched lizard, the gopher snake, and the Northern Pacific rattlesnake, and hear sage thrashers, sage sparrows, magpies, and other birds in the sage. Visitors often see the great bald eagle, hawks, ravens, and various species of swallows flying overhead.

Addresses: Gingko Petrified Forest State Park, 4511 Huntzinger Road, Vantage, WA 98950.
Phone: (509) 856-2700.
Days and hours: Daily, but vary throughout the year. See website for details. The Trailside Museum is closed in winter.
Fee: Yes (Discover Pass required).
Tours: Self-guided.
Handicapped accessible: Partly: the Interpretive Center is accessible; the hiking trails are not.
Public transportation: None.
Food served on premises: None. Snack shop in Vantage.
Websites: parks.state.wa.us/288/Ginkgo-Petrified-Forest

Registered Trademarks

Index

For the benefit of digital users, indexed terms that span two pages (e.g., 52–53) may, on occasion, appear on only one of those pages.

Adler Planetarium, 253
agriculture, 11, 14, 40–41, 46–47, 48, 60–62,
　　100–1, 103, 159–60, 186–87, 198–202,
　　207–8, 217, 219, 241–42, 261, 271–72,
　　305–6, 314, 316, 318, 353, 408–9, 420,
　　452, 489, 497
Akeley
　　Carl, 103, 255, 256, 301
　　Delia, 255
Albany, New York, 95–96
Albuquerque, New Mexico, 374–79
Alexandria, Virginia, 202–4
Allen, Dudley Peter, 282
America's Stonehenge, 50–52
American Clock and Watch Museum, 1–3
American Museum of Natural History, 101–4
American Museum of Science and
　　Energy, 244–47
American Precision Museum, 58–60
American Saddlebred Museum, 227–28
amphibians, 3–4, 5, 8–9, 24–25, 29, 43, 76–
　　77, 103, 112, 141, 149, 161–62, 169, 172,
　　176, 178, 180–81, 192, 197, 215–16, 230,
　　252, 255, 258, 310, 313–14, 324, 363,
　　387, 400, 411, 412, 443, 460, 473, 485–
　　86, 489, 503
anatomy, 23, 79, 80, 135, 137, 139, 174–75,
　　192, 228–29, 257, 280
Anchorage, Alaska, 426–28
Anchorage Museum, 426–28
anemones, 5, 76, 172, 185, 259, 435
Annapolis Junction, Maryland, 70–72
anthropology, 8, 103, 104, 159, 193, 255, 268,
　　269, 280, 301–2, 323–24, 327, 331, 349–
　　50, 355, 364, 391, 400, 410, 420, 421, 426,
　　436, 449, 458, 481–82, 489
Appleton, Wisconsin, 292–93
aquariums, 3–5, 24–26, 33, 35, 76–77, 115,
　　149, 172, 174, 176, 185, 192, 198, 215–
　　16, 230, 238, 244, 258–59, 300, 322,
　　327, 363–64, 389, 434–35, 442–43,
　　459–60, 482
arboretums, 30–31, 49, 116–17, 361–62, 373–
　　74, 430–32, 500–1
Arcadia, California, 430–32

archaeology, 1, 8, 29, 50–52, 112, 145, 199, 217,
　　220, 347–48, 349–50, 400, 410, 481–82
Arizona-Sonora Desert Museum, 363–64
arms and armor, 11, 44–46, 48, 58–59, 60,
　　96, 108–9, 125, 129, 214, 217, 220, 287,
　　290–91, 302, 326, 327, 331–32, 359–61,
　　367, 375–76, 380, 454, 466, 481, 484
Aschauer, John, 59
Asheboro, North Carolina, 189–91
Ashford, Washington, 492–94
Ashwaubenon, Wisconsin, 294–95
Astoria, Oregon, 483–85
astronomy, 24, 33, 50–51, 88, 91–92, 93, 102,
　　120, 138, 160–61, 180, 211, 229, 269, 308,
　　342–44, 364–65, 379, 384–85, 400, 403,
　　444, 450–51, 478–79
Atlanta, Georgia, 181–86
Atlanta Botanical Garden, 181–83
Audubon, John James, 199–200, 239, 256,
　　325, 449
Audubon Aquarium of the Americas, 238
Audubon Butterfly Garden and
　　Insectarium, 240–41
Audubon Louisiana Nature Center, 240
Audubon Zoo, 239–40
Auenbrugger, Leopold, 282–83
Augusta, Maine, 10–12
automobiles, 14, 118–19, 168, 209, 241, 260–
　　61, 271, 274, 278, 284–85, 305, 325, 344,
　　386, 413, 436, 439, 449, 454
aviation, 99, 106–8, 118–19, 137, 160–61,
　　174, 194–95, 204–6, 209, 210, 241–42,
　　260–61, 264–66, 283, 290–91, 307–8,
　　322, 330–31, 333–34, 366–67, 403, 430,
　　453–55, 498–99, 503–5

Baltimore, Maryland, 72–77
Baltimore Museum of Industry, 72–74
Bangor, Maine, 12–15
Bartlesville, Oklahoma, 330–32
Bartram
　　John, 133–34, 207–8
　　William, 133–34
Bartram's Garden, 133–35
Batavia, Illinois, 248–50

Bath, Maine, 15–16
Beaumont
   Texas, 381–85
   William, 303
bees, 23, 191, 310
Bell, Alexander Graham, 99, 159–60
Bernice Pauahi Bishop Museum, 481–83
Bicycle Museum of America, The, 286–88
bicycles, 14, 59, 60, 195, 209, 283–84, 286–88,
   325, 454, 477
Billings Farm and Museum, 60–63
biology, 8, 23, 111, 143, 185, 193, 197–98,
   229, 240, 261, 280, 322, 349–50, 400,
   427, 460–61
Birch Aquarium at Scripps, 434–35
Birmingham, Alabama, 164–65
Black Hills Mining Museum, 338–39
Blanchard, Thomas, 44
Blaschka
   Leopold, 29, 97–98
   Rudolf, 29, 97–98
Bloomfield Hills, Michigan, 268–69
Boeing, William, 498–99, 504
Bonner Springs, Kansas, 305–7
Booth
   Ellen Scripps, 268, 269
   George, 268, 269
Boston, Massachusetts, 22–31
botany, 8, 27, 29, 54, 56, 104–6, 110–11, 112–
   13, 117, 133–34, 139, 146, 166, 168, 169,
   170, 181–83, 223, 269, 310, 313, 316,
   319–21, 326–27, 346–48, 354–55, 361–
   62, 396–97, 400, 407–9, 425, 471–72,
   475–76, 481, 484, 500–1, 505
Boulder, Colorado, 402–3
Boyce Thompson, William, 361–62
Boyce Thompson Arboretum, 361–62
Bozeman, Montana, 421–22
Bradbury Science Museum, 379–81
Brailsford, William, 186, 187
Bristol, Connecticut, 1–3
Bronx Zoo, 113–14
Brookfield, Illinois, 251–52
Brookfield Zoo, 251–52
Brooking, Albert, 325
Brookline, Massachusetts, 26–27
Brooklyn Botanic Garden, 104–6
Bruce, Archibald, 89
Bruetsch, Walter, 263
Brunswick
   Georgia, 186–88
   Maine, 17–18
Bryce, Utah, 348–52, 399
Bryce Canyon National Park, 351–52, 399

Burbank, Luther, 272, 471–72
Bureau of Engraving and Printing, 157–59
Burroughs, John, 168, 382
Busch Gardens Tampa, 177–78
butterflies, 19, 23, 56, 66, 112, 114, 159, 169,
   193, 197, 238, 239, 240, 302, 321, 324,
   325, 333, 391, 408–9, 435–36, 459
B&O Railroad Museum, 74–76

Cable Car Museum, 457–58
California Academy of Sciences, 458–60
California State Mining and Mineral
   Museum, 438–40
California State Railroad Museum, 451–53
Cambridge, Massachusetts, 28–31
canals, 31–32, 121–22
Canaveral National Seashore, 178–81
Canterbury, New Hampshire, 48–50
Canterbury Shaker Village, 48–50
Carillon Historical Park, 283–86
Carnegie Museum of Natural History, 146–47
Carnegie Science Center, 147–48
Carver
   George Washington, 165, 166, 272,
     305, 314–16
   Moses, 314–16
Castle Rock, Washington, 494–96
Cave Junction, Oregon, 485–86
Central Park Zoo, 114
Chantilly, Virginia, 204–6
Charleston, South Carolina, 198–200
Charlotte, North Carolina, 192–94
Charlottesville, Virginia, 206–8
chemistry, 89, 94, 109, 130, 131, 138, 142–44,
   166, 168, 192, 193, 214, 237, 285, 340,
   384–85, 391, 434, 436, 460–61
Chesapeake Bay Maritime Museum, 81–83
Cheyenne Mountain Zoo, 403–5
Chicago, Illinois, 253–61
Chisholm, Minnesota, 276–80
Clark, William, 207, 483, 484
Cleveland
   Grover, 135, 161
   Ohio, 281–83
clocks, 1–2, 30, 37, 109, 122–23, 139, 207, 217,
   220, 334
Cole Land Transportation Museum, 12–15
Collection of Historical Scientific
   Instruments, 30
College of Physicians of Philadelphia, 135–36
Colonial Williamsburg, 216
Colorado Springs, Colorado, 403–7
Colorado Springs Pioneers Museum, 405–7
Columbia, Pennsylvania, 122–24

Computer History Museum, 446–48

computers, 23–24, 30, 59, 72, 84, 92, 107, 160, 174, 232, 245–46, 312, 380, 446–48

coral, 5, 24–25, 56, 76, 77, 115, 149, 159, 171–72, 174, 176, 185, 215, 259, 322, 389, 412, 434

Corning, New York, 96–99

Corning Museum of Glass, 96–99

Cornwall, Pennsylvania, 124–26

Cornwall Iron Furnace, 124–26

Cosmosphere International Science Education Center & Space Museum, 307–9

Cranbrook Institute of Science, 268–69

Crawford W. Long Museum, 188–89

cryptology, 70–72

Curtiss, Glenn Hammond, 99

Custer, South Dakota, 334–36

da Vinci, Leonardo, 192–93, 290

Darwin, Charles, 29, 66, 110, 111, 131, 283, 319, 471

Dayton, Ohio, 283–86

Dearborn, Michigan, 270–72

Deere, John, 266–67

Delaware Museum of Natural History, 65–67

Denver, Colorado, 407–15

Denver Botanic Gardens, 407–9

Denver Museum of Nature & Science, 409–10

Denver Zoo, 410–12

Desert Botanical Garden, 354–55

Desert of Maine, The, 18–19

Diamond, Missouri, 314–16

dinosaurs, 8–9, 23, 46, 66, 102, 145, 146, 184, 185, 255, 268–69, 280, 297, 312, 327, 332, 347, 377–78, 390, 400, 409–10, 421, 428, 435

Discovery Place, 192–94

Discovery World, 299–301

Dittrick Medical History Center, 281–83

Dover, Delaware, 64–65

Doylestown, Pennsylvania, 126–28

Drake, Edwin Laurentin, 153, 154–55, 383

du Pont, Eleuthère Irénée, 68, 133

Dyche, Lewis Lindsay, 310, 325

Eastman, George, 119–20

ecosystems and ecology, 9, 10, 23, 24, 25, 30, 38, 56, 60, 66, 76, 95–96, 103, 106, 115, 141, 146, 147, 149, 162, 169–70, 171–72, 173–74, 176, 180–81, 196–97, 212, 214, 215, 238, 239, 240, 245–46, 249, 280–81, 301, 312, 314, 322, 349, 350–51, 352–53, 355, 357, 363–64, 380, 387, 400, 404, 408,

410, 423, 424, 428, 431, 434, 441, 442, 443, 448, 459, 462, 476, 477, 479, 489, 501

Edgar Mine, 418–19

Edison, Thomas Alva, 56–57, 64, 93–94, 96–97, 159–60, 167–69, 271, 293, 381–82

Edison and Ford Winter Estates, 167–69

Edison Museum, 381–82

Egypt, 9, 23, 29, 109, 147, 214, 255, 280, 410, 430

Einstein, Albert, 80, 135, 376

Eisenhower, Dwight D., 294, 305

electricity and electronics, 23, 32–33, 64, 70, 71–72, 74, 75, 77–78, 83–85, 86, 91–93, 123, 130–31, 137, 138, 159–60, 168, 246, 271, 272, 288, 299–300, 333, 369, 381, 447, 497–98

Elverson, Pennsylvania, 128–30

engineering, 26, 32, 43, 58, 59, 88, 90, 91, 109, 124, 142, 145, 164, 174, 179, 192, 207, 231–32, 233, 234–35, 244, 265, 267, 270–71, 285, 297–98, 299–300, 322, 333, 346, 351, 352–53, 359–61, 368–69, 373, 382–83, 391–93, 496–98

Enigma, 71

entomology, 8, 35, 56, 103, 112, 145, 149, 159, 161, 162, 193, 197, 240, 310, 324, 325, 326–27, 398, 400, 402, 411, 435–36, 448, 459, 468, 481, 485–86, 489, 503

Erie Canal Museum, 121–22

Escondito, California, 432–33

Estes Park, Colorado, 415–17

ethnology, 8, 29

Everett, Washington, 503–5

Everglades National Park, 169–71

evolution, 9, 23, 29, 102, 103, 105, 106, 144, 159, 185, 226, 268–69, 326, 327, 364, 458, 477

Ewing, New Jersey, 83–85

Exploratorium, 460–62

Fairbanks
　　Alaska, 428–29
　　Franklin, 56, 57

Fairbanks Museum and Planetarium, The, 56–58

Fermi, Enrico, 246, 248, 376, 379–80

Fermi National Accelerator Laboratory, 248–50

Fernbank
　　Museum of Natural History, 183–86
　　Science Center, 183–86

Field Museum of Natural History, 254–56

fire trucks, 13–14, 101, 127, 139, 217

Firestone, Harvey, 168, 271–72, 382

fish, 3–5, 24–25, 43, 76–77, 112, 115, 159, 169, 172, 174, 176, 185, 192, 196–97, 198, 212, 215, 230, 238, 244, 258–59, 300, 310, 325, 327, 347, 387, 389, 390, 400, 403–12, 419, 423, 434, 435–36, 442–43, 448, 459–60, 489

fishing, 10, 11, 15, 81–82, 103, 112, 214, 238

Flagstaff, Arizona, 342–44

Fleischmann Planetarium and Science Center, 372–73

Florence, Oregon, 486–88

Fonthill, 126–28

Ford, Henry, 93, 155–56, 167, 168–69, 270, 271, 272

Ford Rouge Factory Tour, 272

Forest History Center, 275–76

Fort Clatsop & Visitor Center, 483–85

Fort Crawford Museum, 302–4

Fort Eustis, Virginia, 208–10

Fort Meade, Maryland, 70

Fort Myers, Florida, 167–69

Fort Sam Houston, Texas, 385–87

Fort Worth, Texas, 387–89

Fort Worth Zoo, 387–89

fossils, 8–9, 23, 29, 46, 57, 66, 89, 95–96, 102, 103, 112–13, 144, 145, 146, 159, 174, 183–85, 197, 198, 207, 255, 280, 296–97, 300, 301, 302, 304, 310, 312, 322, 326–27, 336–37, 347, 370–71, 377–78, 379, 390–91, 395, 400, 409–10, 418, 419–20, 421, 435–36, 438, 456, 505–6

Francis, James, 32

Franklin
  Benjamin, 2, 30, 33, 97–98, 130–31, 133, 134, 137, 138, 139, 393–94, 469
  New Jersey, 87–89

Franklin Institute, The, 136–38

Franklin Mineral Museum, 87–89

Frederick Law Olmstead National Historic Site, 26–27

Freeport, Maine, 18–19

Fresnel lens, 37, 214, 289

Fuller, R. Buckminster, 175, 271, 300, 321

Future of Flight Aviation Center & Boeing Tour, 498–500

Galveston, Texas, 389–90

gardens, 1, 6, 23, 26–27, 41, 46–47, 49, 51, 53–54, 62, 66, 68, 69, 104–6, 110–11, 117, 119–20, 133–34, 136, 140, 141, 168, 175, 179, 181–83, 185, 199, 217, 263, 269, 319–21, 333, 354–55, 363, 364, 369, 373–74, 396–97, 407–9, 427, 430–32, 433, 448, 462–64, 468, 470, 472, 479, 482, 501–2

Gardiner, Montana, 422

gems, 10, 29, 56, 102–3, 159, 193, 196–97, 255, 371, 390, 400, 410, 417, 436, 438–39

geology, 18–19, 29, 56, 66, 96, 102–3, 112, 146, 159, 170, 184, 193, 196–97, 212, 223, 269, 275, 277, 278, 295–97, 302, 311–12, 328–29, 334–35, 337, 340, 345–47, 348–50, 351–52, 353, 363, 364, 369, 377, 378–79, 380, 384–85, 394–95, 400, 417–18, 422–23, 424, 425, 426–27, 428, 440–41, 459, 473, 474, 475, 476–77, 482, 485, 492–93, 494–95, 498, 506

George Eastman Museum, 119–21

George Washington Carver National Monument, 314–16

Georgetown
  Colorado, 412–15
  South Carolina, 200–2

Georgetown Loop Railroad & Mining Park, 413–15

Gingko Petrified Forest, 505–6

glass, 11–12, 48, 96–98, 117, 123, 132, 143, 170, 174, 203, 221–22, 272, 413, 426
  biological models, 29, 97–98
  See also *Fresnel lens*

Glenn H. Curtiss Museum, 99–100

Glore, George, 317–18

Glore Psychiatric Museum, 317–19

Goddard, Robert, 307–8, 334

Golden, Colorado, 417–19

Gramophones and Phonographs, 64–65, 84, 94, 168, 381

Grand Canyon, Arizona, 346–52

Grand Canyon National Park, 348–50

Grand Coulee, Washington, 496–98

Grand Coulee Dam, 496–98

Grand Rapids, Minnesota, 275–76

Gray, Asa, 319, 321

Greece, 9

Greenfield Village, 271–72

Groton, Connecticut, 5

Hagley Museum and Library, 67–70

Haines, Alaska, 429–30

Hale, George Ellery, 449, 450

Hallidie, Andrew, 457

Hammer Museum, 429–30

Hammondsport, New York, 99–100

Hampton, Virginia, 210

Harley-Davidson, Inc., 155–56

Harry C. Miller Lock Collection, 234–35

Harvard Museum of Natural History, The, 28–29

Harvard Semitic Museum, 29

Hastings
  Museum, 325–26
  Nebraska, 325–26

Hawaii National Park, Hawaiʻi, 476–78
Hawaiʻi Volcanoes National Park, 476–78
Henry, William, 150–51
Henry Ford Museum of American
    Innovation, 270–71
Henry Ford, The, 270–72
Hermany, Charles, 233
Hibbing, Minnesota, 276–80
Higgins, Pattillo, 382–83, 385
Hilo, Hawaiʻi, 478–81
Historic Jamestowne, 220–21
Historic Speedwell, 85–87
History Colorado Center, 412–13
Hofwyl-Broadfield Plantation State Historic
    Site, 186–88
Hollerith, Herman, 80, 447
Homestead, Florida, 169–71
Honolulu, Hawaiʻi, 481–83
Hoover, Herbert, 211–12, 305
Hoover Dam, 352, 368–70
    Arizona, 352
    Nevada, 368–70
Hopemont, 229
Hopewell Furnace National Historic
    Site, 128–30
Hornaday, William T., 161, 310
Hot Springs
    Arkansas, 223–25
    South Dakota, 336–38
Hot Springs National Park, 223–25
Houston, Texas, 390–94
Houston Museum of Natural Science, 390–91
Hubble, Edwin, 342, 450
Hull-Rust Mahoning Mine, 276–80
Huntington
    Collis P., 213, 221, 451, 468–69
    Henry E., 468–69, 470
Huntington Library, Art Collections, and
    Botanical Gardens, The, 468–70
Hutchinson, Kansas, 307–9

Idaho Museum of Natural History, 419–21
Idaho Springs, Colorado, 417–19
ʻImiloa Astronomy Center of
    Hawaiʻi, 478–79
Indiana Medical History Museum, 261–64
Indianapolis, Indiana, 261–64
industry, 7–49, 52–54, 58–60, 67–69, 72–74,
    81–82, 90–91, 93, 97, 98, 99, 100–1,
    112, 120, 121–22, 124–30, 150–51,
    153–56, 159–60, 164–65, 202, 221–22,
    238, 241, 260–61, 266–67, 270–72, 274,
    292–93, 332, 356–57, 384–85, 389, 391,
    393–94, 406, 413, 426, 429–30, 435, 436,
    481–82, 498–99

InfoAge Science History Learning
    Center, 91–93
International Museum of Surgical
    Science, 256–58
Intrepid Sea, Air & Space Museum, 106–8

Jackson, Mississippi, 241–43
Jamestown Settlement, 219–20
Jefferson
    Georgia, 188–89
    Thomas, 32, 70, 131, 133, 134, 206–8,
        223, 483
jellies, 4, 25, 174, 192, 238, 296, 389, 443
Jewel Cave National Monument, 334–36
John G. Shedd Aquarium, 258–60
John Pennekamp Coral Reef State
    Park, 171–73
Johnson Victrola Museum, 64–65
Johnston, Jeanne Branch, 479–80
Joseph Priestley House, 130–32

Kendall, Nicanor, 58, 59
Kennebunkport, Maine, 19–22
Kennedy, John F., 178, 360
Kennedy Space Center Visitor
    Complex, 178–81
Kentucky Horse Park, 225–27
Kettering, Charles, 285
Key Largo, Florida, 171–73
Kill Devil Hills, North Carolina, 194–96
Kitt Peak National Observatory, 364–66
KU Natural History Museum, 309–11

La Brea Tar Pits, 437–38
La Jolla, California, 434–35
laboratory, 69, 89, 93–94, 132, 168, 192, 197,
    246–47, 255, 263, 271, 322, 357, 382, 400,
    409, 421, 459
Lackawanna Coal Mine, 149–51
Lake Vermilion–Soudan Underground Mine
    State Park, 276–80
Lassen Volcanic National Park, 440–41
Lava Beds National Monument, 473–74
Lawrence, Kansas, 309–11
Lead, South Dakota, 338–41
Lewis
    Meriwether, 207, 483, 484
    Miles, 1, 2
Lexington, Kentucky, 225–29
Lick, James, 443–44
Lick Observatory, 443–45
Lincoln
    Abraham, 80, 235, 271, 451, 475
    Nebraska, 326–28
Lindbergh, Charles, 118–19, 160, 265

Linnaeus, Carl, 111, 133, 310, 321
Linthicum, Maryland, 77–79
living-history museum, 6, 31, 40, 42, 46–49,
       60–61, 128–30, 217, 219, 241–42,
       271–72, 275
Long, Crawford W., 188–89
Los Alamos, New Mexico, 379–81
Los Angeles, 435–38
Los Angeles County Arboretum and Botanic
       Garden, 430–32
Louisville, Kentucky, 230–34
Louisville Zoological Garden, 230–31
Lowell
    Massachusettes, 31–33
    Percival, 342, 343, 344
Lowell Heritage State Park, 31–33
Lowell National Historical Park, 31–33
Lowell Observatory, 342–44
Lucas, Anthony, 382–83, 385
lumbering, 10–11, 15–16, 241, 242, 275–76, 278
Luray, Virginia, 211–13
Luther Burbank Home & Gardens, 471–73

MacMillan, Donald B., 17
Madison, Wisconsin, 295–97
Maine Maritime Museum, 15–16
Maine State Museum, 10–12
Mammoth Site, 336–38
Manitowoc, Wisconsin, 297–98
Maria Mitchell Association, 33–35
Mariners' Museum and Park, 213–15
Mariposa, California, 438–40
maritime history, 5–7, 15–16, 81–82, 87, 159–
       60, 213–14, 236–37, 288–89, 297–98,
       464–66, 479, 482
Marsh-Billings-Rockefeller National
       Historical Park, 60–63
May, Wilbur D., 373, 374
Mayflower II, 40, 41–42
medicine, 30, 49, 79–80, 135–36, 137, 139–40,
       160, 188, 202–4, 217, 223–24, 228–29,
       232–34, 236–37, 242, 256–58, 261–63,
       272, 280, 281–83, 303–4, 317–18, 325,
       376, 385–86, 405–6
Medora, North Dakota, 328–30
Memphis, Tennessee, 243–44
Mercer, Henry Chapman, 126–28
Mercer Mile, The, 126–28
Mercer Museum, 126–28
Merritt Island National Wildlife
       Refuge, 178–81
meteorology and climatology, 58, 91–92, 102–
       3, 137, 198, 400, 402–3, 428, 434, 460,
       462, 479
Metropolitan Museum of Art, 108–10

Miami, Florida, 173–77
Miami Seaquarium, 175–77
Michaux, André, 133, 199
Michelson, Albert, 450–51
Michigan Iron Industry Museum, 273–74
Middleton, Henry, 198–99
Middleton Place, 198–200
Midgley, Charles, 285
Miller
    J. Clayton, 234
    Harry C., 234, 235
Mills, Enos A. 415, 416–17
Milton, West Virginia, 221–22
Milwaukee
    Public Museum, 301–2
    Wisconsin, 299–302
Mineral, California, 440–41
minerals, 9, 10, 19, 23, 29, 56, 57, 66, 87–88,
       89, 96, 102–3, 112–13, 145, 146, 159, 166,
       193, 196–97, 269, 277, 278, 288, 296, 310,
       325, 327, 332, 364, 370–71, 390, 400, 410,
       417–18, 436, 438–39
Mines Museum of Earth Science, 417–18
mining, 87–88, 89, 125, 126, 129, 146, 149–50,
       151–52, 260, 273–74, 276–79, 338–39,
       340, 370–71, 400, 413–14, 418–19, 436,
       439, 449, 495, 506
Minnesota Discovery Center, 276–80
Mississippi Agriculture and Forestry
       Museum, 241–43
Mississippi River Museum, 243–44
Missouri Botanical Garden, 319–21
Mitchell
    Maria, 33–34
    William, 33–34
Monroe Moosnick Medical and Science
       Museum, 228–29
Monterey, California, 442–43
Monterey Bay Aquarium, 442–43
Monticello, 206–8
Moody Gardens, 389–90
Moravian Pottery and Tile Works, 126–28
Mordy, Wendell, 372–73
Morristown, New Jersey, 85–87
Morse, Samuel F. B., 75, 78, 85–86, 159–60
Mount Hamilton, California, 443–45
Mount Lebanon, New York, 100–1
Mount Rainier National Park, 492–94
Mount St. Helens National Volcanic
       Monument, 494–96
Mount Wilson Observatory, 449–51
Mountain View, California, 446–48
Mukilteo, Washington, 498–500
Muncie, Indiana, 264–66
Museum of Flight, 503–5

Museum of Science and Industry, 260–61
Museum of the American Printing House for the Blind, 231–32
Museum at Prairiefire, 311–13
Museum of Science, 22–24
Museum of the Rockies, 421–22
Musical Instrument Museum, 356–57
Mütter Museum, 135–36
Mystic, Connecticut, 3–8
Mystic Aquarium, 3–5

Nantucket, Massachusetts, 33–38
National Agricultural Center and Hall of Fame, 64–65
National Agriculture Aviation Museum, 241–43
National Aquarium, 76–77
National Center for Atmospheric Research Mesa Lab Visitor Center, 402–3
National Cryptologic Museum, 70–72
National Electronics Museum, 77–79
National Model Aviation Museum, 264–66
National Museum of Health and Medicine, 79–81
National Museum of Nuclear Science & History, The, 374–77
National Museum of the Great Lakes, 288–90
National Museum of the United States Air Force, 290–92
National Railroad Museum, 294–95
National Watch & Clock Museum, 122–24
Native Americans, 9, 11, 17, 29, 40, 41, 51–52, 55, 57, 81–82, 89, 95–96, 103, 116, 146, 153, 183–84, 188, 218, 219, 223, 238, 242, 243–44, 276, 280, 289, 301–2, 303, 304, 305, 325, 327, 328, 331, 332, 336, 338, 345–46, 347–50, 352, 355, 364, 366, 380, 391, 394, 395, 400, 410, 412–13, 415, 420, 421, 422–23, 426, 428, 429, 431, 437–38, 440, 449, 473–74, 475, 483, 484, 492, 495, 497–98, 506
Natural Bridge Caverns, 394–96
natural history, 8–9, 22–24, 35, 56–57, 65–66, 95–96, 101–3, 112–13, 144–45, 146–47, 159, 183–85, 196–98, 224–25, 254–56, 268–69, 275, 280–81, 301, 309–10, 311–12, 316, 325, 326–27, 363–64, 377–79, 399–400, 409–10, 416, 419–20, 426–27, 435–36, 448, 458–60, 475, 493
Natural History Museum of Los Angeles, 435–37
Natural History Museum of Utah, 399–401
nautical instruments, 6, 213–14, 219, 288–89, 297–98, 484
Negaunee, Michigan, 273–74

New Bedford, Massachusetts, 38–40
New Bedford Whaling Museum, 38–40
New Bremen, Ohio, 286–88
New England Aquarium, 24–26
New England Maple Museum, 54–56
New Haven, Connecticut, 8–10
New Mexico Museum of Natural Science and History, 377–79
New Orleans, Louisiana, 236–41
New Orleans Pharmacy Museum, 236–37
New York, New York, 101–16
New York
    Aquarium, 115–16
    Botanical Garden, 110–12
    State Museum, 95–96
Newport News, Virginia, 213–15
Nicholasville, Kentucky, 234–35
North Carolina
    Museum of Natural Sciences, 196–98
    Zoological Park, 189–91
North Salem, New Hampshire, 50–52
Northumberland, 130–32

Oak Ridge, Tennessee, 244–47
Oak Ridge Site–Manhattan Project National Historical Park, 244–47
Oakland, California, 448–49
Oakland Museum of California, 448–49
observatories, 24, 33–35, 88, 98, 138, 148, 186, 254, 269, 342–65, 391, 443–45, 477, 479, 495
octopus, 25, 76, 149, 258
Ogdensburg, New Jersey, 87–89
Oklahoma City, Oklahoma, 332–34
Old Rhinebeck Aerodrome, 118–19
Old Slater Mill National Historic Landmark, 31, 52–54
Old Sturbridge Village, 46–48
Olmsted
    Frederick Law, 26, 30, 101, 104, 110, 114, 117, 161, 239
    John Charles, 27, 104, 110, 117
Oppenheimer, J. Robert, 380, 460
Oracle, Arizona, 352–54
Oregon Caves National Monument & Preserve, 485–86
Oregon Zoo, 488–90
ornithology, 8, 9, 23, 35, 56, 66, 96, 103, 112, 116, 141, 145, 161, 162, 166, 169, 170, 172, 174, 176, 177, 180–81, 187, 190, 192, 195–96, 197, 212, 215, 230, 238, 239, 240, 244, 252, 255, 310, 313–14, 323, 324, 325, 327, 332, 348, 359, 362, 363, 387–88, 397, 398, 400, 411, 424–25, 428, 433, 436, 441, 442–43, 455, 456, 458, 459, 467, 473, 475, 488, 489, 491, 492, 502–3, 506

Overland Park, Kansas, 311–13
Oyster Bay, New York, 116–18

Pacific Tsunami Museum, 479–81
Packard, David, 442
Pahl, Dave, 429
paleontology, 8, 9, 112, 145, 146, 297, 310, 311,
    326–27, 346–47, 349–50, 384–85, 390,
    400, 435, 505–6
Paper Discovery Center, 292–93
Pasadena, California, 449–51
patents, 2, 43, 68, 69, 93, 100, 101, 121–22, 168
Pawtucket, Rhode Island, 52–54
Peabody Museum of Archaeology &
    Ethnology, 29
Peabody Museum of Natural History, 8–10
Peary, Robert E., 17
Peary-MacMillan Arctic Museum, 17–18
penguins, 3–4, 25, 114, 115, 149, 176, 230, 239,
    259, 314, 324, 389, 460, 467, 489
Penn, William, 139, 141
Pennsylvania Anthracite Heritage
    Museum, 149–51
Pennsylvania Hospital, 139–40
Penrose, Spencer, 403–4
Peterson, Roger Tory, 4
Petrified Forest National Park, 346–48
Philadelphia, Pennsylvania, 133–45
Philadelphia Zoo, 140–42
Phillip and Patricia Frost Museum of
    Science, 173–75
Phillips, Frank, 330–31
Phoenix, Arizona, 354–59
Phoenix Zoo, 357–59
photography, 27, 94, 99, 119–20, 143, 168,
    263, 381–82
Physick, Phillip Syng, 139, 303
physics, 24, 98, 120, 137–38, 147, 192, 193,
    228–29, 244–47, 248–50, 261, 277, 299,
    311–12, 333, 334, 340, 357, 374–76, 379–
    80, 386, 460–61, 479–81
Pima Air & Space Museum, 366–67
Pittsburgh, Pennsylvania, 146–48
Pittsburgh Zoo & PPG Aquarium, 148–49
planetarium, 6, 24, 58, 101, 102, 137, 138, 147,
    148, 175, 186, 193, 229, 240, 253–54, 269,
    308, 322, 326, 327, 334, 365, 372, 373,
    379, 391, 422, 427, 460, 479, 482
Planting Fields Arboretum State Historic
    Park, 116–18
Plimoth Plantation, 40–42
Plymouth, Massachusetts, 40–42
Pocatello, Idaho, 419–21
polymers, 69, 143

Polynesia, 9, 255, 410, 478–79, 481–82
Portland, Oregon, 488–90
Prairie du Chien, Wisconsin, 302–4
Priestley
    Joseph, 130–32
    Mary, 132
printing, 47–48, 73, 74, 75, 157–58, 159–60,
    217, 231–32, 261, 272, 283–84, 393–94
Printing Museum, The, 393–94
Prospect Park Zoo, 115

Queens Zoo, 116

radio, 64, 71, 74, 78, 83–84, 91–92, 99, 211,
    244, 265, 287, 288–89, 298, 326
railroads, 10–11, 12–14, 69, 74–75, 123, 138,
    149–50, 151–52, 164–65, 210, 241, 260–
    61, 271, 274, 277, 285, 294–95, 306, 324,
    334, 413, 414, 436, 451–52, 468, 489
Raleigh, North Carolina, 196–98
Rancho San Rafael Regional Park, 370–74
Reid, Hugo, 431
Reno, Nevada, 370–74
reptiles, 5, 29, 35, 43, 76–77, 103, 112, 114,
    141, 149, 161–62, 170, 176, 178, 192,
    196–97, 212, 215, 230, 239, 240, 252, 255,
    258, 310, 313, 324, 348, 363, 387, 400,
    411–12, 423, 428, 435–36, 455, 456, 459,
    473, 489, 503
Revere, Paul, 157, 393–94, 436
Rhinebeck, New York, 118–19
Rice Museum, The, 200–2
robotics, 147, 270, 377
Rochester, New York, 119–21
Rocky Mountain National Park, 415–17
Roebling
    Carl, 89–90
    Charles, 90
    John A., 89–90
    New Jersey, 89–91
Roebling Museum, 89–91
Roentgen, Wilhelm Conrad, 248, 282,
    376, 379–80
Rome, 9, 214
Roosevelt
    Franklin D., 211–12, 314–15, 496
    Theodore, 159, 328, 329–30, 331, 334,
    425, 475
Rush, Benjamin, 136, 139
Rutland, Vermont, 54–56

Sacramento, California, 451–53
Sahuarita, Arizona, 359–61
Saint Louis Zoo, 323–24

Salt Lake City, Utah, 399–401
San Antonio, Texas, 394–99
San Antonio Botanical Garden, 396–97
San Antonio Zoo, 397–99
San Diego, 453–57
San Diego Air & Space Museum, 453–55
San Diego Zoo, 455–57
San Diego Zoo Safari Park, 432–33
San Francisco, California, 457–68
San Francisco Botanical Garden, 462–64
San Francisco Maritime National Historical
    Park, 464–66
San Francisco Zoo & Gardens, 466–68
San Marino, California, 468–70
Sanford Lab Homestake Visitor
    Center, 340–41
Santa Rose, California, 471–73
Sarnoff, David, 83–84, 91
Sarnoff Collection, The, 83–85
Saugus, Massachusetts, 42–43
Saugus Iron Works National Historic
    Site, 42–43
Science Center of Minnesota, 280–81
Science History Institute, 142–44
Science Museum Oklahoma, 332–34
scientific instruments, 8, 17, 27, 30, 33–35, 86,
    88, 89, 132, 135, 143, 146, 166, 174–75,
    203, 207, 217, 228–29, 253, 258, 343–44,
    402, 417–18, 436, 450–51, 479, 484
Scowden, Theodore, R., 233
Scranton
    George, 150–51
    Pennsylvania, 149–53
    Seldon, 150–51
Scranton Iron Furnaces, 149–51
Sea Lion Caves, 486–88
sea lions, 3–4, 114, 175, 230, 486, 487–88
seals, 4, 25, 114, 115, 216, 259, 389, 443
Seashore Trolley Museum, 19–22, 33
Seattle, Washington, 500–5
Sedgwick County Zoo, 313–14
Shaker Museum and Library, 100–1
Shakespeare, William, 105, 219, 470
Shaw, Henry, 319, 321
shells, 24, 66, 112, 145, 159, 185, 193, 302, 325,
    390, 395, 400, 435–36, 481
Shenandoah National Park, 211–13
ships and boats, 5–7, 10, 15–16, 17, 38–39,
    40, 41–42, 73, 81–82, 85, 106–7, 115,
    121–22, 127, 147, 159–61, 172, 209,
    213–14, 219, 244, 260–61, 288–89,
    297–98, 300, 334, 376, 426, 434,
    464–66, 479, 481–82, 484
Shotoku, 393

Silver Spring, Maryland, 79–81
Slater, Samuel, 52, 266
Slipher, Vesto M., 342, 343, 344
Sloss, James Withers, 164
Sloss Furnaces National Historic
    Landmark, 164–65
Smithson, James, 159
Smithsonian Institution, 159–61
Smithsonian's National Zoo and Conservation
    Biology Institute, 161–63
snails, 25, 259
Snow, Huntington, 309–10
snowplows, 12, 22, 152
Soudan, Minnesota, 276–80
space, 6, 9, 34, 60, 102–3, 106, 107–8, 138, 147,
    160–61, 174, 179, 180, 186, 204–6, 211,
    253, 254, 261, 291, 307–8, 322, 334, 353,
    360–61, 367, 373, 380, 391–93, 410, 418,
    430, 504–5
Space Center Houston, 391–93
Species Survival Plan®, 4, 116, 141, 148, 230,
    239, 251, 259, 313, 323, 358–59, 387, 404,
    411, 460, 467–68, 488–89, 502
spiders, 35, 103
Spindletop Gladys City Boomtown, 383–84
Springdale, Utah, 348–52, 401
Springfield, Massachusetts, 44–46
Springfield Armory National Historic
    Site, 44–46
squids, 9, 66
St. Johnsbury, Vermont, 56–58
St. Joseph, Missouri, 317–19
St. Louis, 319–24
St. Louis Science Center, 321–22
St. Martin, Alexis, 303
St. Michaels, Maryland, 81–83
St. Paul, Minnesota, 280–81
Stabler, Edward, 202–3
Stabler-Leadbeater Apothecary Museum, 202–4
Stanford, Leland, 451
Staten Island Museum, 112–13
Steamtown National Historic Site, 151–53
Sterling Hill Mining Museum, The, 87–88
Steven F. Udvar-Hazy Center, 204–6
Stonington, Connecticut, 5
Sturbridge, Massachusetts, 46–48
Sunset Crater Volcano National
    Monument, 345
Superior, Arizona, 361–62
Syracuse, New York. 121–22

Tampa, Florida, 177–78
Terry, Eli, 1–2, 123
Texas Energy Museum, 384–85

textiles, 11, 31–33, 38, 41, 47, 48, 52–54, 59–60, 68, 108, 109, 143, 150, 217, 391, 400, 413, 436
Theodore Roosevelt National Park, 328–30
Thomas Edison National Historical Park, 93–94
Titan Missile Museum, 359–61
Titusville
    Florida, 178–81
    Pennsylvania, 153–55
Toledo, Ohio, 288–90
trolleys, 19–22, 293, 457–58
Truman, Harry S, 169, 305
Tucson, Arizona, 363–67
Tulelake, California, 473–74
Tuskegee, Alabama, 165–67
Tuskegee Institute National Historic Site, 165–67

University of Alaska Museum of the North, 428–29
University of Arizona Biosphere 2, 352–54
University of Nebraska State Museum, 326–28
University of Nevada, Reno, 370–74
University of Wisconsin–Madison Geology Museum, 295–97
U.S. Army Medical Department Museum, 385–87
U.S. Army Transportation Museum, 208–10

Vail
    Alfred, 85, 86, 159–60
    Stephen, 85
Vaux, Calvert, 26–27, 101, 104, 110, 114
Virginia Air and Space Center, 210
Virginia Aquarium & Marine Science Center, 215–16
Virginia Beach, Virginia, 215–16

Wagner, William, 144, 145
Wagner Free Institute of Science, 144–45
Wall, New Jersey, 91–93
Warren Anatomical Museum, 30–31
Washington
    Booker T., 165–66, 315–16
    District of Columbia, 157–63
    George, 29, 44, 70, 133, 134, 202–3, 305, 386
Washington Park Arboretum, 500–2
Waterloo, Iowa, 266–68
WaterWorks Museum, 232–34

Wedgwood, Josiah, 131, 132
West Orange, New Jersey, 93–94
whale oil, 35–36, 37
whales, 3–4, 15, 25, 35–37, 38–39, 77, 96, 103, 109, 153, 175–76, 196–97, 198, 215, 259, 435, 459, 481, 488
Whaling Museum, 35–38
Wichita, Kansas, 313–14
Wilbur D. May Botanical Garden and Arboretum, 373–74
Wildlife Conservation Society, 113–16
Wildlife Safari, 490–92
Williamsburg, Virginia, 216–21
Wilmington, Delaware, 65–70
Wilson, James Perry, 9, 103
Windsor, Vermont, 58–60
Winston, Oregon, 490–92
Wisconsin Maritime Museum, 297–98
W. M. Keck Earth Science and Mineral Engineering Museum, 370–72
Woodland Park Zoo, 502–3
Woodstock, Vermont, 60–63
Woolaroc Museum & Wildlife Preserve, 330–32
Wright
    Brothers, 118, 194–95, 211, 260–61, 264, 283–84, 286, 290
    Orville, 194–95, 272, 283, 290
    Wilbur, 194–95, 272, 290
Wright Brothers National Memorial, 194–96
Wright-Patterson Air Force Base, Ohio, 290–92
Wupatki National Monument, 345–46

Yellowstone National Park, 422–25
Yellowstone National Park, Wyoming, 422–25
York, Pennsylvania, 155–56
Yosemite National Park, 474–76
Yosemite National Park, California, 474–76

Zion National Park, 350–51, 401
zoology, 8, 24, 28, 66, 103, 112, 113–15, 140–41, 146, 159, 175–76, 192, 193, 225–28, 277, 332, 389, 391, 481, 486, 487–88, 490–92
zoos, 113–15, 116, 148–49, 161–62, 177–78, 189–91, 230–31, 239, 251–52, 313–14, 323–24, 357–59, 363, 387–88, 397–98, 403–5, 410–12, 432–503